Biome Ecology

Volume 2

Biome is a large-scale biotic community and its environment, defined using functional rather than biodiversity-focused criteria of delimitation. Biotic communities are structured by drivers at various scales and levels of complexity. Most of the research in (biotic) community ecology in the past has been concentrating on fine- and mid-scales involving habitats and their complexes at landscape level. Less attention has been paid to large scales—those involving (sub)continents, hemispheres, and indeed—the entire planet. Knowing the biome structure and its origins, are becoming vital to understanding complex patterns in relation to processes operating at large scales. Such processes involve climate dynamics, land-use changes, and effects of continental- and global-level nature conservation efforts.

Biomes are functional units of ecology and challenging to define and delimit because of poorly tangible nature of processes underpinning the functioning of the large-scale communities. These processes involve biomass production, nutrient cycling, disturbance, and recovery dynamics and the like. Several biome schemes are used to tackle the functionality of these large-scale units, yet there is still a lot of latitude left to formulate more scientifically robust and better formalized tools of delimitation of biomes. Not all parts of the world have a functional classification of ecosystems available. Undoubtedly, this dire situation is calling for action. More information, and more profound syntheses of the biome patterns are dearly needed.

The new book series will focus on synthesizing the knowledge of distribution and origins of biomes in various parts of the world and at the planetary scale. Further, the series will offer space for development of new approaches of delimitation of biomes and new tools of using this knowledge in predictive, as well as reconstructive, modelling for the purposes of the most urgent tasks facing the mankind, such as climate change, land-use dynamics, survey of natural resources, food production, and health.

Ladislav Mucina · Michael C. Rutherford
Editors

Zonal Biomes of Southern Africa

Springer

Editors
Ladislav Mucina
Harry Butler Institute
Murdoch University
Perth, WA, Australia

Department of Geography
and Environmental Studies
Stellenbosch University
Stellenbosch, South Africa

Michael C. Rutherford
Department of Botany and Zoology
Stellenbosch University
Stellenbosch, South Africa

High Peak, Derbyshire, UK

ISSN 2948-1465 ISSN 2948-1473 (electronic)
Biome Ecology
ISBN 978-3-031-63730-8 ISBN 978-3-031-63731-5 (eBook)
https://doi.org/10.1007/978-3-031-63731-5

This Springer imprint is published by the registered company Springer Nature Switzerland AG
The registered company address is: Gewerbestrasse 11, 6330 Cham, Switzerland

Dedicated to those who with passion, zeal, dedication, and knowledge contributed to how we know, use, and protect the Southern African vegetation and biomes today.

Acknowledgements

Ladislav Mucina is an incumbent chair in Vegetation Science and Biogeography, a joint venture between the Murdoch University and the Iluka Resources Ltd, Perth, Western Australia. Both institution share his sincere thanks for logistic and other form of support. Johan Wasserman acknowledges the international Ph.D. bursary received from the Murdoch University and bursary from CRCTiME. Mervyn Lötter thanks Wildlife Conservation Society—Mozambique for funding the revision of terrestrial ecosystems for Mozambique which informed the regional contribution towards the development of the zonal biomes of Southern Africa. Les Powrie thanks the SANBI and Stellenbosch University (Centre for Geographic Analysis) for long-term support. We appreciate receiving many less accessible literature sources from and the exchange of views on many aspects of the biomes of Southern Africa with Warren McCleland, Marc Stalmans and Jonathan Timberlake. Ralph Clark and David Gwynne-Evans kindly contributed photos.

Funding

Johan Wassserman
Murdoch University: International Tuition Fee Scholarship
Murdoch University: Murdoch Strategic Scholarship
CRCTiME: Ph.D. bursary top-up

Ladislav Mucina
Murdoch University and Iluka Resources Ltd: Iluka Chair in Vegetation Science and Biogeography

Contents

Contributors

John E. Burrows Calitzdorp, Western Cape, South Africa

Caley Higgs Department of Geography and Environmental Studies, Stellenbosch University, Matieland, Stellenbosch, South Africa

Mervyn C. Lötter Mpumalanga Tourism and Park Agency, Lydenburg, South Africa;
School of Animal, Plants and Environmental Sciences, University of the Witwatersrand, Johannesburg, South Africa

Ladislav Mucina Harry Butler Institute, Murdoch University, Perth, Australia; Department of Geography and Environmental Studies, Stellenbosch University, Matieland, Stellenbosch, South Africa

Leslie W. Powrie Department of Geography and Environmental Studies, Stellenbosch University, Matieland, Stellenbosch, South Africa;
Kirstenbosch Research Centre, South African National Biodiversity Institute, Claremont, Cape Town, South Africa

Michael C. Rutherford High Peak, Derbyshire, United Kingdom; Department of Botany and Zoology, Stellenbosch University, Matieland, Stellenbosch, South Africa

James L. Tsakalos Harry Butler Institute, Murdoch University, Perth, Australia; Plant Diversity and Ecosystems Management Unit, School of Biosciences and Veterinary Medicine, University of Camerino, Camerino, Italy

Adriaan van Niekerk Department of Geography and Environmental Studies, Stellenbosch University, Matieland, Stellenbosch, South Africa

Johan Wasserman Harry Butler Institute, Murdoch University, Perth, Australia

Abbreviations

A2	Subtropical Alpine Zone (a zonobiome)
A3	Tropical Alpine Zone (a zonobiome)
AT	Albany Thicket Biome
C3	C3-assimilation strategy of plants
C4	C4-assimilation strategy of plants
CB	Continental Biome
CEB	Continental Ecotonal Biome
E1	Equatorial Zone (a zonobiome)
E1(sub)	Subtropical Tropical Rainforest (a global biome)
E2	Tropical Seasonal Zone (a zonobiome)
E2(tdf)	Dry Tropical Forest (a global biome)
FYN	Fynbos Biome
FZM	Flora Zambeziaca Map
GB	Global Biome
ITCZ	Intertropical Convergence Zone
KZN	KwaZulu-Natal Province (South Africa)
MAP	Mean Annual Precipitation
MAT	Mean Annual Temperature
MBSA	Map of Biomes of Southern Africa (our study region)
NDVI	Normalized Difference Vegetation Index
NK	Nama-Karoo Biome
RB	Regional Biome
REB	Regional Ecotonal Biome
S1	Ethesial Zone (a zonobiome)
S2	Hot Desert Zone (a zonobiome)
S2(scd)	Southern Coastal Desert (a global biome)
S2(scs)	Southern Coastal Semidesert (a global biome)
S2(sid)	Southern Inland Desert (a global biome)
S2(sis)	Southern Inland Semidesert (a global biome)
SHPB	Sub-Tropical High-Pressure Belt
SK	Succulent Karoo Biome
SVA (sva)	Arid Savanna Biome (a global biome)
SVM (savm)	Mesic Savanna Biome (a global biome)
T3	Warm-Temperate Zone (a zonobiome)
T5	Tropical Montane Zone (a zonobiome)
TDF	Tropical Dry Forest

VegMap2006	National Vegetation Map (Mucina and Rutherford 2006)
VegMap2018	National Vegetation Map (version as presented by Dayaram et al. 2019)
VegMapMoz	Vegetation Map of Mozambique (Lötter et al. 2023)
VT	Vegetation Type (basic mapping unit in Mucina and Rutherford 2006)
VU	Vegetation Unit' (as defined by Mucina and Rutherford 2006)
ZB	Zonobiome
ZE	Zonoecotone

Mapping Zonal Biomes of Southern Africa: Data Sources and Methods

1

Ladislav Mucina, Michael C. Rutherford, James L. Tsakalos, Mervyn C. Lötter, Adriaan van Niekerk, Caley Higgs, and Leslie W. Powrie

Abstract

The biomes of the southern portion of Africa, defined as the area south of an axis joining the Kunene and Zambezi Rivers, are mapped for the first time. The area covers nearly 3.4 million km^2 and incorporates South Africa, Lesotho, eSwatini, Namibia, Botswana, Zimbabwe, and southern Mozambique (south of the Zambezi River). This chapter sets out this mapping project's aims, basics of the conceptual biome-classification framework, and detailed sources of the climatic, vegetation, and related spatial proxy data—all forming the basis of the biome mapping. The biome mapping procedures combine the existing vegetation maps (detailed for South Africa, Lesotho, eSwatini and southern Mozambique) and bioclimatic information summarised in the form of ecologically relevant bioclimatic indices that relate to vegetation units as mapped (and modified) by previous sources.

Keywords

Azonality/zonality framework · Biome hierarchy · eSwatini · Lesotho · Mozambique · Namibia · South Africa · Spatial climatic diagram · Zimbabwe · Zonobiome

Supplementary Information The online version contains supplementary material available at https://doi.org/10.1007/978-3-031-63731-5_1.

L. Mucina (✉) · J. L. Tsakalos
Harry Butler Institute, Murdoch University, 90 South Street, Murdoch, Perth 6150, Australia
e-mail: Ladislav.Mucina@murdoch.edu.au

L. Mucina · A. van Niekerk · C. Higgs · L. W. Powrie
Department of Geography and Environmental Studies, Stellenbosch University, Private Bag X1, Matieland, Stellenbosch 7602, South Africa

M. C. Rutherford
High Peak, Derbyshire, United Kingdom

M. C. Rutherford
Department of Botany and Zoology, Stellenbosch University, Private Bag X1, Matieland, Stellenbosch 7602, South Africa

J. L. Tsakalos
Plant Diversity and Ecosystems Management Unit, School of Biosciences and Veterinary Medicine, University of Camerino, Via Pontoni 5, Camerino 62032, Italy

M. C. Lötter
Mpumalanga Tourism and Park Agency, P/Bag X1088, Lydenburg 1120, South Africa

M. C. Lötter
School of Animal, Plants and Environmental Sciences, University of the Witwatersrand, Private Bag 3, Wits, 2050 Johannesburg, South Africa

L. W. Powrie
Kirstenbosch Research Centre, South African National Biodiversity Institute, Claremont, Private Bag X7, Cape Town, South Africa

L. Mucina and M. C. Rutherford (eds.), *Zonal Biomes of Southern Africa*, Biome Ecology 2,
https://doi.org/10.1007/978-3-031-63731-5_1

1.1 Southern Africa: Home to a High Diversity of Biotic Communities

Southern Africa, here pragmatically defined as the area south of an axis formed by the area south of the Kunene and Zambezi Rivers, is one of the most bio-diverse regions of the World—home to thousands of plant taxa forming a rich kaleidoscope of biotic communities. It is home to natural treasures and wonders, and home to millions of people. To use the resources of this region sustainably, preserving the ways of life of both the human population and a knowledge of the extent of natural diversity and its underpinnings are vital.

On a large scale, the biotic communities of Southern Africa have been captured in many biogeographical and ecological surveys. Indeed, at least one of the countries of this region—the largest and economically leading one—South Africa, has had a long tradition of natural surveys, enjoying the support of governmental bodies such as the South African National Biodiversity Institute (SANBI, formerly also known as National Botanical Institute and even earlier as Botanical Research Institute). Today, South Africa is not only a leader in surveying natural resources but also in their conservation. In addition, South Africa, through SABONET (Southern African Botanical Network) and transfer of knowledge, has drastically influenced the surveying of natural resources of its northern neighbours.

'Biome' is considered by many to be the concept addressing the largest spatial scale of biotic/natural complexity. Biomes have been delimited using a plethora of criteria (Mucina 2019). In Southern Africa, these criteria and their influence on the development of a biome system have been exhaustively reviewed by Rutherford and Westfall (1986, 1994), Rutherford (1997) and Rutherford et al. (2006a).

Since its publication in December 2006 (effectively January 2007), the biome account Rutherford et al. (2006a) has been used across South Africa, Lesotho, and Swaziland (now eSwatini). The biome system for Namibia and Botswana followed Rutherford (1997) but was strongly adapted and extended, as described under the Tropical Seasonal Zone and Hot Arid Zone in Chapters 3 and 7. Unfortunately, no universally followed biome system has been in place for Zimbabwe and Mozambique. The current work expands the area covered by Rutherford et al. (2006a, b) by over two and a half times.

1.2 Aims

Leaving aside the forest biomes, published by Mucina et al. (2022) and azonal biomes (work in progress), this study aims to revisit and expand the zonal biome scheme of Rutherford et al. (2006a) by applying the novel biome-delimitation climatic and pedological criteria (Mucina 2023). The zonal biomes of this study make up much of the land area of the subcontinent.

In particular, we shall:

1. revisit the identity of biomes of South Africa, Lesotho, and eSwatini in the light of the globally applicable hierarchical scheme of biomes involving zonobiomes, global biomes and continental biomes, and recognise ecotonal biome units;
2. expand the application of this scheme in countries north and northeast of South Africa, and produce the biome maps featuring this expanded region;
3. introduce a system of regional (and regional ecotonal) biomes—a subunit of continental (and continental ecotonal) biome for the entire study area;
4. define the bioclimatic envelopes in the form of crucial climatic statistics/profiles for all biomes;
5. present structured descriptions of all zonal regional biomes of Southern Africa; and
6. produce a shape file capturing the biome structure of Southern Africa.

1.3 The Study Area

Africa is the second largest continent on our planet (30.3 million km^2), showing the land distribution along two axes—one running west–east (the broader northern part of the continent), and the other running north–south (the narrower, southern part of the continent, reminiscent of a broad peninsula). Our study region covers the southern half of the 'peninsula'; we shall refer to it as 'Southern Africa' (Fig. 1.1). This region is naturally defined by the northern border formed by the east–west flowing Kunene River (straddling the border between Namibia, Angola, and Zambia (approx. 17° S), and west–east (and then northwest–southeast) flowing Zambezi River,

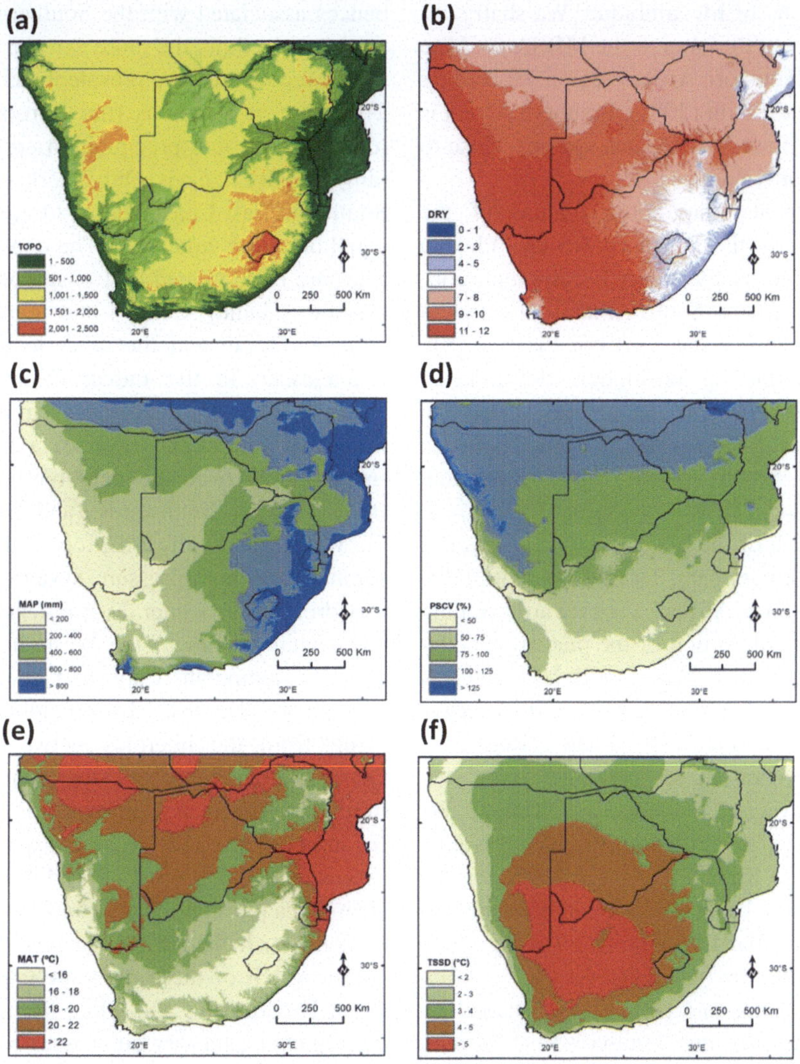

Fig. 1.1 Topography and selected climatic variables of the MBSA area. **a** Topography (m of elevation), **b** Dry months, **c** Mean annual precipitation, **d** Precipitation seasonality, **e** Mean annual temperature, **f** Temperature seasonality of the MBSA area. The topography data were extracted from the Shuttle Radar Topography Mission (SRTM) DEM (US Geological Survey 2018). The data serving creation of the maps featuring the selected climate variables were extracted from the WorldClim repository (https://www.worldclim.org/data/bioclim.html)

which first forms the border between Zimbabwe and Zambia and then bisects Mozambique into the southern and northern parts. This region also matches the concept of the 'southern African subcontinent' as often used in scientific and semi-popular literature.

On 3,390,413 km^2 (approx. 10% of the surface of the African continent), this region encompasses South Africa, Lesotho, eSwatini (formerly Swaziland), Namibia, Botswana, Zimbabwe, and southern regions of Mozambique. We shall subsequently refer to this area as the MBSA (= Map of Biomes of Southern Africa) region. According to van Wyk and Smith (2001), Southern Africa is home to more than 30,000 plant species, close to 10% of the global flora.

The dominant topographic feature of the MBSA region is the Southern African Plateau (SAP), bordered by the Great Escarpment, copying the coast of the South Atlantic and Indian Oceans at varying distances from the sea (50–200 km) and attaining the highest elevations in the Drakensberg/Maloti Mountains (Thabana Ntlenyana, elev. 3482 m) on the border between Lesotho and South Africa. Some prominent mountains associated with the Great Escarpment also occur north of the Limpopo River—Manica Highlands, Chimanimani. The SAP is, within the confines of South Africa, a westwards-sloping mesa. The central portion of the plateau is occupied by the endorheic Kalahari Basin, which is filled mainly by sandy sediments and extensive marginal calcrete flats. The MBSA Region is drained by several major perennial river systems, including Gariep/Orange and Kunene rivers, flowing to the Atlantic Ocean, Kavango discharging into Kalahari Basin and forming the inland Okavango Delta, and the mighty Zambezi and less spectacular Limpopo and Tugela which all discharge into the Indian Ocean.

The region's climate (Fig. 1.1) is shaped by the warm Mozambique/Agulhas Sea Current, modifying the thermic zonation of the eastern seaboards of the subcontinent and allowing penetration of the subtropical climatic conditions deep towards high latitudes. The other source of climatic influence is the Benguela Upwelling, which brings subantarctic cold waters along a long stretch of the Southern African Atlantic coast, thus exacerbating the aridity of the adjacent seaboards as far north as the city of Benguela in Angola. The disparate sea surface temperatures drive a steep climatic gradient, underpinning the west–east increase in total precipitation and the seasonality contrasts experienced along that transect. The southwestern portion of the MBSA area (the Cape) experiences winter rainfall (because of frontal disturbances associated with the Southern Westerlies). At the same time, the more extensive subtropical inland regions receive prevalent summer rainfall. There are vast regions (the Karoo and Albany regions) with low precipitation spread into two rainy and dry seasons. Only short stretches of the South and East Coast (Indian Ocean) experience equal or nearly equal rain. The non-seasonal patterns are underpinned by two major anticyclonic systems (Hadley Cells)—one above the South Atlantic Ocean and the other located south of Madagascar, in the Indian Ocean. While the South Atlantic Anticyclone is migrating along the north–south (and reverse) axis, it expands (in summer) over the neighbouring landmass. By expanding southwards (and hence pushing the Westerlies southwards, too) and eastwards in summer, the high-pressure system causes summer drought. In winter, as it retreats northwards, the frontal system of the Westerlies is restored over the temperate Cape and causes abundant rainfall. In the east, the precipitation patterns result from the interplay between the Indian Ocean Anticyclone and the yearly dynamics of the Intertropical Convergence Zone (ITCZ). In summer, the Anticyclone expands over the shores of southern Africa and brings rain-rich air masses (around the edges of the Anticyclone, in counter-clockwise movement) onto the continent. Much of these rains are captured by the Great Escarpment, and while the rest falls over the Highveld, the dry air masses reach Kalahari Basin. Two depressions form over the continent during this period. As the ITCZ moves south in austral summer, it also contributes (in the form of troughs) to tropical, rain-rich air (Congo Air) to increase the summer rainfall amounts in the eastern regions of southern Africa.

From a thermo-bioclimatic perspective, most of the MBSA region lies within the subtropical zone (characterised by the MAT spanning 18–21 °C) and only the southernmost reaches of the southern African 'peninsula' experience latitudinal warm-temperate conditions (spanning 14–18 °C). However, due to the high elevation attained by the Southern African Plateau (and some mountains along the Great Escarpment straddling the borders of Zimbabwe and Mozambique), warm-temperate conditions re-appear further north due to the temperature lapse rate with elevation.

The climatically seasonal regions supporting high biomass are exposed to large-scale natural disturbances such as fire and local mega-herbivory (sections of moist subtropical savanna).

These climatic and other drivers underpin the formation of one of the most intriguing biome-diverse regions of the World, home to several floras showing contrasting evolutionary origins and supporting high levels of endemism, especially in global biodiversity hotspots embedded within or reaching the study area, including Cape Floristic Region, Maputaland–Pondoland–Albany, Succulent Karoo, and Coastal Forests of Eastern Africa (Myers et al. 2000; Mittermeier et al. 2004).

The Southern African semideserts and deserts are a striking geographic feature of the region. These include the arid parts of Kalahari (for the controversial use of the term 'desert' in this context, see below), the Karoo and its extensions as far north as Kaokoland (also known as Kaokoveld), and the famous Namib Desert and less studied Gariep and Tankwa Deserts.

This chapter does not address the biome identity of the subantarctic ecosystems occurring on some islands under South African jurisdiction. These are treated in the light of biome classification by Mucina (2023).

1.4 Conceptual Framework: Biomes and Hierarchy

We use the zonality/azonality (Z/A) conceptual framework of Schimper (1898) and Walter (1942, 1976; see also Walter and Box 1976; Breckle 2002; Box 2016; Breckle and Rafiqpoor 2019; Mucina 2019), further called 'Walterian system' as the theoretical basis of our biome classification and mapping. This principle was also adopted by Rutherford et al. (2006a) in mapping the biomes of the Republic of South Africa, Lesotho, and Swaziland. This framework recognises zonal biomes driven by climate underpinning the bioclimatic zones (Box 2016) and azonal biomes. The Walterian framework distinguishes nine zonobiomes (I through IX; for a recent summary, see Box 2016). Motivated by inconsistencies of the latter framework in the Southern Hemisphere, Mucina (2023) has revised the zonobiome structure of the Southern Hemisphere. His system (Table 1.1) is also followed in this book.

Patches of some zonal biomes may occur embedded within other zonal biomes. However, these are linked to unique habitats that would create (micro- or meso-) climatic conditions like those in the native zonal range, which are examples of extrazonal vegetation (e.g., patches of Southern African Afrotemperate Forests (zonobiome T3) within the CB Fynbos (the zonobiome S1). The relevant concept of extrazonality is explained by Mucina (2023: Fig. 1).

Many zonal biomes (putatively matching the current climatic fabric) contain patches of vegetation, usually in special habitats, yet they do not show any extrazonal link to existing zonal vegetation. These are 'legacy biomes' (see Mucina 2019 for a brief review)—relics of the biomes that either do not exist anymore in a zonal fashion or are indicative of the previous extent of the existing biomes (the latter can also be considered extrazonal). For example, we mention so-called 'Fynbos Thickets' (Rebelo et al. 2006: 76–77) and patches of afrotemperate forests embedded within the Fynbos and Highveld grasslands (see Mucina et al. 2022).

Our system recognises four levels of zonal biomes: *Zonobiomes* (ZB), *Global Biomes* (GB), *Continental Biomes* (CB), and *Regional Biomes* (RB). The definition of these biome levels in relation to the climatic drivers, bioclimatic characteristics, and soil characteristics are found in Mucina et al. (2022) and Mucina (2023). Further, our system also recognises three levels

Table 1.1 The new zonobiome system (modified after Mucina 2023)

ZB	Abbr	Name	Macroclimatic drivers	Climatic diagnosis	Synonyms and corresponding concepts
		Tropical Zonobiomes			
E1	TRO	Equatorial Zone	Constant high temperature; dynamics of the ITCZ	Equally warm all-year-round; frost is rare and occurs only in marginal subtropical regions; all-year-round	Walter (1976: Zonobiome I)
E2	SAV	Tropical Seasonal Zone	Constant high temperature; ITCZ disturbances; rain-bearing monsoons	Warm all-year-round but cooler at low; precipitations occurring throughout the year but distinctly wetter in summer	Walter (1976: Zonobiome II)
		Subtropical Zonobiomes			
S1	MED	Ethesial Zone	Yearly tug-of-war (N–S/S–N) between the semi-stationary ridging high-pressure systems (STHB) and the Westerlies	Distinctly regular alternation of dry hot summers and wet cool winters	Walter (1976: Zonobiome IV)
S2	DES	Hot Arid Zone	Domination of the STHP for the most part of the year	Warm to very hot summers and cooler winters; poorly predictable rainfall events, dry all-year-round, experiencing winter rainfall on edges towards S1	Walter (1976: Zonobiome III)
		Cold Temperate Forest Zonobiomes			
B1	BOR	Boreal Zone*	Shifting dynamics of cold polar air masses; combined with maritime influence on edges	Cool summers; severe cold winters; summer rainfall	Walter (1976: Zonobiome VIII)
B2	OBZ	Oceanic Boreal Zone	Shifting dynamics of cold polar air masses; combined with maritime influence; strong Westerlies	Cool summers and cold winters; high rainfall especially in winter	Mucina et al. (2022: Zonobiome XII)
		Temperate Forest Zonobiomes			
T1	NEM	Nemoral Zone*	Weakened Westerlies due to continental effect and weakened temperate monsoon	Warm summers; frosty winters; no true dry season, precipitation maximum in summer	Walter (1976: Zonobiome VI)

(continued)

Table 1.1 (continued)

ZB	Abbr	Name	Macroclimatic drivers	Climatic diagnosis	Synonyms and corresponding concepts
T2	ANE	Austro-Nemoral Zone[#]	Weakened Westerlies due to rainshadow effect	Mild summers and winters; aseasonal precipitation, lower than in ZB T4	
T3	WFZ	Warm-Temperate Zone	Summer dynamics of the STHB underpinning the rain-bearing summer monsoons and extratropical cyclones	Warm to hot summers, mild winters; weak dry season; precipitation maximum in both boreal and austral summer	Walter (1976: Zonobiome V); Box (2016: zonal biome Ve); Breckle and Rafiqpoor (2019: Sub-Zonobiom an den Ostseiten der Kontinente)
T4	OTZ	Oceanic Temperate Zone	Strong influence of rain-bearing Westerlies on both hemispheres	Mild summers and cool winter; all-year-round rainfall, with maximum in both boreal and austral winter	Box (2016: zonal biome Vm); Breckle and Rafiqpoor (2019: Sub-Zonobiom an den Westseiten der Kontinente); Mucina et al. (2022: Zonobiome X)
T5	TMZ	Tropical Montane Zone	Temperature lag along ascending elevation and wind-bearing trade-winds	Warm-temperate thermic climate; All-year-round moist air; precipitation maxima at high sun	Earlier incorporated in ZB I
		Temperate Grassland Zonobiomes			
G1	STE	Northern Steppe Zone*	Weakened rain-bearing systems/large yearly temperature amplitudes	Warm summers; very cold and dry winter	Walter (1976: Zonobiome VII p.p.)
G2	CON	Cold Continental Zone*	High thermal continentality; Strong regional rain shadows	Dry year-round; sporadic rain; cold winters	Walter (1976: Zonobiome VII p.p.); Breckle and Rafiqpoor 2019: 'Sub-Zonobiom der Mittelasiatischen Wüsten + Sub-Zonobiom der Zentralasiatischen Wüsten')
G3	AST	Austro-Steppe Zone[#]	Rained-out Westerlies (rain shadow effect); adiabatic temperature increase	Cool summers and cold winters; low winter rainfall	Walter (1976: Zonobiome VII p.p.)

(continued)

Table 1.1 (continued)

ZB	Abbr	Name	Macroclimatic drivers	Climatic diagnosis	Synonyms and corresponding concepts
		Alpine Zonobiomes			
A1	ARC	Arctic Zone*	Polar night (boreal winter) and formation of polar cold masses; elevation-related temperature decrease	Cold and short summers; severe cold winters	Walter (1976: Zonobiome IX p.p.)
A2	PUN	Subtropical Alpine Zone	High-elevation temperature lapse; influence of STHB	Warm during the day; cold (often freezing) in the night; unimodal precipitation monsoonal and trade-wind systems	Walter and Box (1976: tentative 'zonobiome X' on a map); Breckle and Rafiqpoor (2019: 'Sub-Zonobiom der Kalten Hochplateauwüsten von Tibet und Pamir')
A3	PAR	Tropical Alpine Zone	High-elevation temperature lapse; influence of ITCZ	Warm during the day; cold (often freezing) in the night; no yearly temperature or precipitation seasonality; bimodal precipitation pattern peak due to ITCZ crossing the Equator twice	Breckle and Rafiqpoor (2019: Orobiom I)
A4	ANT	Antarctic Zone#	Polar night (austral winter) and associated formation of polar cold and dry masses	Cold dry summers and very cold dry winters	Walter (1976: Zonobiome IX p.p.)

Abbr: Informal abbreviation code of a zonobiome (ZB). The zonobiomes occurring only in the NH are indicated by an asterisk, while whole exclusive to the SH are indicated by #. The ZB codes including the letter E refer to the zones around the Equator; those including S refer to the Subtropics or rather the regions around the Tropics of Cancer and Tropics of Capricorn; those containing B refer to cold temperate regions know as boreal in the NH and anti-boreal in the SH; the letter T is involved in zones typically supporting temperate forests; the letter G introduces zones dominated by zonal grasslands. The letter A refers to the alpine zones, including those occurring above the timberline (tree line) as well as those in the Arctics and in Antarctica. ITCZ Intertropical Convergence Zone; STHB Sub-Tropical High-Pressure Belt

of ecotonal biomes: *Zonoecotones* (ZE; matching the rank of Global Biomes), *Continental Ecotonal Biomes* (CEB; analogous to the zonal Continental Biomes), and *Regional Ecotonal Biomes* (REB; analogous to the zonal Regional Biomes).

1.5 Mapping Sources and Compilation Procedures

The Maps are a result of a compilation including mapping sources covering regions of various extents and various levels of precision and detail within the mapping (study) area. The basic compilation procedure included:

1. recognition and implementation of baseline mapping sources for the countries (or groups of countries of the study area),
2. re-definition of the initially proposed biome categories to create a uniform and scientifically sound biome system, and
3. improvements (precision, accuracy, detail) of the mapping coverage of some units in previously poorly mapped areas using expert opinion and satellite imagery, including Google Earth.

1.5.1 South Africa, Lesotho, and eSwatini

The spatial extent of the regional biomes (RB) and regional ecotonal biomes (REB) map of South Africa follows the Vegetation Map of South Africa, Lesotho and Swaziland (hereafter referred to as 'VegMap2006'; Mucina et al. 2005; Mucina and Rutherford 2006), for the most part its 'VegMap2018' version (Dayaram et al. 2019 and the shape file available at www.sanbi.org) as the baseline. The construction of the map of the original 'biomes' of these three countries is described in detail in Rutherford et al. (2006a: Sect. 6.3 on pages 23–24).

Because we do not recognise some of the VegMap2018 edits of the original VegMap2006 coverage, we have opted to use the *vegetation units* (VUs) of the latter as the basis for the delimitation of some regional biomes and regional ecotonal biomes. The Supplementary Table S1.1 captures the crosswalk between the VegMap2018 (or VegMap2006) VUs and the units of the new biome hierarchy. The assignment to the bioregions was also consulted, although it was not (in most of the original VegMap2006 biomes) decisive for assigning a VU into a biome. In this context, it is essential to note that the VegMap2006 and VegMap2018 bioregions are conceptually different from the biomes featured in our paper.

Although we, in principle, followed the original extent of the VUs, it was necessary to recognise two new VUs and modified boundaries leading to some others (see Table 1.2).

The most significant changes to the original (VegMap2006) biome system:

1. The Savanna Biome (Rutherford et al. 2006b) was subdivided into CB Southern African Mesic Savanna and CB Southern African Arid Savanna.
2. The IOCB 'biome' (Mucina et al. 2006c) has been dissolved and incorporated into CB Southern African Mesic Savanna. Naturally, this does not apply to extant forests classified into one zonal and several azonal biomes found within the IOCB (see Mucina et al. 2022).
3. The Grassland Biome (Mucina et al. 2006b) has been subdivided into three biome units: zonal CB Drakensberg Alpine Grassland and two ecotonal biomes (CEB African Montane Grassland, CEB Dry Highveld Grassland).
4. The Desert Biome (Jürgens et al. 2006) has been subdivided into CB Namib Desert and CB Gariep Desert.
5. A new CB Tankwa Desert was recognised based on one new regional biome, Tankwa Desert, extracted from the original extent of the Succulent Karoo Biome (Mucina et al. 2006a). This high-level biome unit, which was hitherto unrecognised even at a vegetation-type level, was delimited using (in part) the extent of the Association 8 in van der Merwe et al. (2008). The Succulent Karoo was retained on Leeuberg

Table 1.2 New vegetation units (VU) and modifications of the boundary leading of theVM2018 VUs on the territory of South Africa

Vegetation unit	Notes
Dn 6 Holgat Coastal Duneveld	This is a new VU (the code is tentative) in proposal stage; this unit was coined to include the areas of the SKs 1 Richtersveld Coastal Duneveld north of the Holgat River
SKs 1 Richtervsled Coastal Duneveld	Through the extraction of the Dn 6 Holgat Coastal Duneveld, this VU becomes limited only to areas S of the Holgat River
DT1a Tanqua Desert	This is a new VU (the code is tentative); see the delimitation as described for the CB Tankwa Desert below
SKv 5 Tanqua Karoo	Through the recognition of the Tankwa Desert (as associated Tanqua Desert VU), the original extent of this unit has changed
Southern Cape Valley Thicket	The extent of this unit (originally coded as AT 1; now CVT1a Southern Cape Valley Thicket) has been modified in the valleys of Gouritz, Goekoe, Duivenhoks, and Breede Rivers
FS 7 Overberg Dune Strandveld	The extent unit has been modified to follow closely the coast
FS 8 Blombos Strandveld	The extent unit has been modified to follow closely the coast. Some boundaries were changed because of adjustments in the Southern Cape Valley Thicket
FS 9 Groot Brak Dune Strandveld	The extent unit has been modified to follow closely the coast. Some boundaries were changed because of adjustments in the Southern Cape Valley Thicket

and neighbouring Sterretjiesberg above the 500 m contour. The southern border of the Tankwa Desert extends to the 450 m elevational contour, and the division is not abrupt on a long and gently sloping plain. The eastern Tankwa Desert boundary extends between the elevations of 450 and 500 m, and it terminates at the AZi 7 Tankwa Wash Riviere unit in the south, adding a relatively narrow tapering belt to the plain below the steeper pediment area of the Skoorsteenberg area. It also includes the most westerly and lowest altitude (and driest) polygon of van der Merwe's unit 7.1 based on position relative to rain shadow, climate, topography, and elevation. The portion in the northwest was extended beyond the mapping domain of van der Merwe's Association 8 to replace the SKv 5 area above the elevation of 280 m.

6. The Albany Thicket (Hoare et al. 2006) has been recognised at the CEB level and the original VegMap2006. Several new regional biomes (see Chap. 9) have been recognised either by lumping several other original ATs units, re-assigning several other VUs into other continental biomes, or by recognising the VU AT 1 into an independent CEB called Southern Cape Valley Thicket. The crosswalk between the VegMap2006 units and the new RB system is captured in Table 1.3.
7. The Fynbos Biome (Rebelo et al. 2006) has been subdivided into three high-rank biome units: CB Fynbos, CEB West Coast Arid Strandveld, and CEB Cape Altimontane Fynbos.

1.5.2 Namibia

The Atlas of Namibia biomes map (Mendelsohn et al. 2002) and Giess (1971) served as an initial conceptual framework for Namibia, but not all units of these sources were recognised, and almost all boundary lines were modified. The SAFARI 2000 Project map (Rutherford et al. 2000) was partly used as a digital source. Our map has been developed using mainly the following sources and steps:

Table 1.3 A crosswalk between the VegMap2006 (and VegMap2018 in one case) between the VUs and the RB system as presented in this book)

VM06	VM18	VU name	RB or REB	CB or CEB
AT 10		Great Fish Noorsveld	AT1 Albany Thicket	CEB Subtropical Albany Thicket
AT 11		Great Fish Thicket	AT1 Albany Thicket	CEB Subtropical Albany Thicket
AT 12		Buffels Thicket	AT1 Albany Thicket	CEB Subtropical Albany Thicket
AT 13		Eastern Cape Escarpment Thicket	NK5 Escarpment Karroid Thicket	CB Nama-Karoo
AT 14		Camdebo Escarpment Thicket	NK5 Escarpment Karroid Thicket	CB Nama-Karoo
AT 2		Gamka Thicket	SK8 Little Karoo	CB Succulent Karoo
AT 3		Groot Thicket	NK4 Lower Karoo	CB Nama-Karoo
AT 4		Gamtoos Thicket	AT4 Sundays Dry Thicket	CEB Subtropical Albany Thicket
AT 5		Sundays Noorsveld	AT4 Sundays Dry Thicket	CEB Subtropical Albany Thicket
AT 6		Sundays Thicket	AT4 Sundays Dry Thicket	CEB Subtropical Albany Thicket
AT 7		Coega Bontveld	AT2 Albany Calcrete Bontveld	CEB Subtropical Albany Thicket
AT 8		Kowie Thicket	AT1 Albany Thicket	CEB Subtropical Albany Thicket
AT 9		Albany Coastal Belt	AT1 Albany Thicket	CEB Subtropical Albany Thicket
Azs 1		Algoa Dune Strandveld	AT3 Albany Strandveld	CEB Subtropical Albany Thicket
Azs 2		Albany Dune Strandveld	AT3 Albany Strandveld	CEB Subtropical Albany Thicket
FFd 11		Southern Cape Dune Fynbos	AT3 Albany Strandveld	CEB Subtropical Albany Thicket
	NK 14	Albany Broken Veld	AT1 Albany Thicket	CEB Subtropical Albany Thicket

VM06 (VegMap2006), VM18 (VegMap2018)

1. The delimitation of the CB Succulent Karoo Biome (RB Richtersveld-Aus Upland Karoo) in the Sperrgebiet and surrounding region from the Southern Namib Desert was based on higher precipitation and vegetation cover. Larger islands of Succulent Karoo vegetation within the Southern Namib Desert were identified and mapped (Williamson 1997; M.C. Rutherford, personal observation), with several other occurrences too small to map at our mapping scale.
2. The extent of the CB and RB Gariep Desert north of the Gariep River follows an approximate aridity limit used by Irish (1994) and Rutherford (1997) (east of the Richtersveld) and was also extended at appropriate low altitudes up the valleys of the Gariep, Fish, and Boom Rivers. In this way, it parallels the climatic conditions of this unit on the South African side (Mucina and Rutherford 2006).

3. The RB Southern Namib Desert was taken as extending along the coast as far north as the tiny enclave at Spencer Bay (Giess 1971). Further inland, the area is more transitional with decreasing influence of winter rain and the northern boundary was taken as the Koichab River, immediately S of the main sand sea (RB Namib Sand Sea Desert).
4. The eastern boundary of the RB Namib Sand Sea Desert was determined using mainly the Google Earth facility. The line of the Kuiseb River terminates the RB Namib Sand Sea Desert in the north (Giess 1971; Mendelsohn et al. 2002).
5. The adjustments to the borders between RB Northern Namib Desert and CB Nama-Karoo biome (RB Kaokoland Pediment Karoo) in the southern section of this RB follows detailed mapping borders of Jürgens et al. (2013) where the units 11 (species-poor Stipagrostis ciliata–Stipagrostis gonatostachys Central Desert Plains Grasslands), 12 (Stipagrostis obtusa Central Calcrete Desert Plains Grasslands) and 14 (Stipagrostis hirtigluma Desert Plains Grasslands) and all units west of these latter units were reclassified as belonging to RB Northern Namib Desert. Unit 16 Stipagrostis hirtigluma–Commiphora Stony and Rocky Desert Grasslands is a hyper-arid unit (MAP of about 65 mm indicated) mapped based on geology rather than vegetation. Most of this unit in the arid west was assigned to RB Northern Namib Desert and the less arid eastern parts to RB Kaokoland Pediment Karoo. Unit 17 Stipagrostis uniplumis Eastern Grasslands) had a similar aridity gradient (MAP of about 60 mm indicated) and was divided into western desert and eastern Nama-Karoo portions. All units east of the Units mentioned above were reclassified as belonging to the Nama-Karoo Biome.
6. The extension of Nama-Karoo northwards, mainly along the base of the western escarpment of Kaokoland, followed Irish (1994) conceptually, with the boundary between RB Northern Namib Desert and RB Kaokoland Pediment Karoo following the rainfall limits indicated for the southern section of this desert regional biome as described above. In this way, the more stylistic version of this belt of Nama-Karoo in Mendelsohn et al. (2002: Vegetation Types 6 and 12) was replaced.
7. The boundary between RB Kaokoland Pediment Karoo and RB Namaland-Bushmanland Karoo to the south was on topographical and biogeographical grounds near the watershed between the Gaub and Tsondab Rivers. In Namibia, the RB Namaland-Bushmanland Karoo was bounded west by the CB Namib Desert, as described above. It was bounded east by the CB Southern African Arid Savanna, as described below.
8. The RB Kalahari Duneveld corresponded to the Southern Kalahari vegetation type of Giess (1971) and Mendelsohn et al. (2000). However, the borders of this extensive dune field (originally digitised from Leistner 1967) were also recaptured in Botswana using the Google Earth facility where the dune cordons were readily recognised. The South African section of this unit uses the original (Rutherford et al. 2006b) extent of the RB Kalahari Duneveld.
9. The RB Namibian Central Upland Bushveld followed the high altitude Highland Savanna or Highland Shrubland units of Giess (1971) and Mendelsohn et al. (2002) on the Khomas Hochland but also extends northwards along the high-lying watershed while remaining west of the ferralic arenosols unit (Mendelsohn et al. 2002: soils map) which is strongly associated with the neighbouring RB Northern Kalahari Bushveld and with the RB Arid Kalahari Sandy Bushveld and RB Kalahari Duneveld to the southeast.
10. The RB Kaokoland Mopane is conceptually similar to the 'Mopane Savanna' unit mapped by Giess (1971). However, as described above, it borders on RB Kaokoland Pediment Karoo (not desert)

in the west. We also excluded from the mopane unit two sandy patches north-west of the Etosha Pan based on mapping by Mendelsohn et al. (2000). The mapped areas dominated by mopane in the eastern Caprivi followed the relatively detailed mapping of Mendelsohn and Roberts (1997) and formed part of the RB Zambezian Mopane.

11. Because the Kalahari sands of Namibia and Botswana correspond to an area with a relatively gentle north–south climatic gradient on a remarkably uniform substrate type (Scholes et al. 2002), species distributions were significant discriminant markers. At this gradient's relatively high rainfall end, the RB Baikiaea Woodland corresponded approximately to the northeastern Kalahari Woodland of Mendelsohn et al. (2002) but with borders redrawn as follows. Its southern border was approximated from species distributions, vegetation surveys at key sites (BIOTA Observatories; Jürgens et al. 2010, 2012), specific information for the areas north of Tsumkwe (Wanke 2006) and personal observations (M.C. Rutherford). The southern boundary of Baikiaea Woodland followed the approximate southern limits of species such as *Baikiaea plurijuga, Guibourtia coleosperma, Baphia massaiensis, Ozoroa schinzii, Psydrax livida, Schinziophyton rautanenii,* and *Pterocarpus angolensis* (see Curtis and Mannheimer 2005). In the Caprivi Strip, RB Baikiaea Woodland was distinguished from RB Zambezian Mopane and azonal wetlands using the data from Mendelsohn and Roberts (1997). In western Caprivi and further west in northern Namibia, *Baikiaea plurijuga*-dominated vegetation is discontinuous, and our RB Baikiaea Woodland mapping unit included the intervening areas of *Burkea africana*-dominated woodland in which the key associated species, such as *Guibourtia coleosperma, Baphia massaiensis,* and *Schinziophyton rautanenii* were important (Mendelsohn and Roberts 1997). In eastern Caprivi, a patch mapped as Burkea–Terminalia Woodland (Mendelsohn and Roberts 1997) was included in the surrounding Mopane Veld unit on account of the matrix of heavy clay with *Colophospermum mopane*, as well as of the lack of some tree species characteristic of Burkea Woodland (Hines 1997).
12. The RB Northern Kalahari Bushveld corresponds closest to the Northern Kalahari of Mendelsohn et al. (2002) but also includes northern parts of their Central Kalahari and eastern parts of their Thornbush Shrubland. The southern border of Northern Kalahari Bushveld with RB Arid Kalahari Sandy Bushveld also occurs in a zone with very gradual rainfall gradients on generally similar sandy substrates (Leser 1972; Cole and Brown 1976; Strohbach and Kutuahuripa 2014; Strohbach 2014). The southern boundary was pragmatically taken as the western extension of the Ghanzi Ridge in the Gobabis area. This also represented the approximate southern limits of species such as *Burkea africana, Combretum collinum, C. psidioides, Grewia avellana, Ochna pulchra, Philenoptera nelsii, Securidaca longepedunculata,* and *Senegalia ataxacantha.* An isolated sandy patch south of Ruacana, as mapped by Mendelsohn et al. (2000), was assigned appropriately to RB Northern Kalahari Bushveld.
13. The RB Arid Kalahari Sandy Bushveld was circumscribed in Namibia by the abovementioned borders. In addition, a large sandy area surrounded by RB Kaokoland Mopane northwest of the Etosha Pan, and mapped by Mendelsohn et al. (2000), was assigned to RB Arid Kalahari Sandy Bushveld.
14. The RB Namibian Dolomite Bushveld represents a relatively mesic dolomitic unit recognised by Giess (1971) as Mountain Savanna and Karstveld. We followed the borders of the higher rainfall part of the Otavi Group as provided by the geological map in Mendelsohn et al. (2002). This separates this regional biome from RB Northern Kalahari Bushveld to the east and the more arid RB Kaokoland Mopane to the west.

1.5.3 Botswana

The Vegetation Map of the Republic of Botswana (Bekker and De Wit 1991; Bekker et al. 1991) was used as a base map in preference to the Flora Zambesiaca Map (Wild and Fernandes 1968; further abbreviated as FZM), except for some areas near the eastern border where some extensions into Botswana from Zimbabwe were accepted, and some minor collations of the mapping line-work units were consequently required. Bekker and De Wit (1991) also carefully considered soils in their vegetation mapping.

1. It was necessary to separate Bekker and De Wit's (1991) A3a 'Terminalia sericea. Lonchocarpus nelsii/Acacia erioloba' unit into the higher rainfall northern part (RB Northern Kalahari Bushveld and RB Baikiaea Woodland; the latter missing from NW Botswana in Bekker and De Wit 1991) from the more extensive and arid central and southern part (RB Arid Kalahari Sandy Bushveld).
2. Other purely northern sandveld units of Bekker and De Wit (1991) corresponded to RB Northern Kalahari Bushveld. The southern boundary of RB Northern Kalahari Bushveld is formed by the Ghanzi Ridge eastwards to the south of the Makgadikgadi Pans. At the same time, the border of RB Baikiaea Woodland in the northwest was approximated from personal observations and distributional records of mainly *Baikiaea plurijuga* and *Guibourtia coleosperma* (see, for instance, Setshogo and Venter 2003). Both these borders find parallels where they extend into Namibia, and many of the border considerations described under Namibia above also apply in Botswana.
3. The eastern boundary of Arid Kalahari stretched to the edge of the deep Kalahari sands and abutted on part of RB Northwestern Limpopo Sandy Bushveld found in southeastern Botswana.
4. Units south and southwest of the A3a unit were assigned to RB Arid Kalahari Sandy Bushveld and RB Kalahari Duneveld, following similar considerations given under Namibia above.

1.5.4 Zimbabwe

A combination of approaches was used to map the zonal biomes for Zimbabwe. First, the FZM mapping (Wild and Fernandes 1968) coverage and species importance rankings (Rutherford et al. 2000) were used to develop a draft map that was refined and edge-matched to Mozambique using a variety of GIS layers.

1. The general rule followed was to extract units where *Baikiaea plurijuga* and its allied species were important as RB Baikiaea Woodland, where *Brachystegia* and *Julbernardia* species were important as miombo woodland, and where *Colophospermum mopane* was important as mopane veld.
2. Various published vegetation surveys were used for border confirmation, such as Hwange National Park (Rogers 1993) and communal lands in north and west Zimbabwe (Timberlake et al. 1993). The remaining savanna areas of Bushveld and Lowveld each had a much more comprehensive range of characteristic species. However, most of the Bushveld was found above the escarpment, and most of the Lowveld was below the escarpment. Small polygons were merged with their surrounding neighbouring type in a few minor instances.
3. Some units or parts of units of savanna character were reassigned to forest, grassland or azonal wetlands. Borders of several FZM forest types and grasslands were modified using satellite imagery and other information to fit the forest or grassland units. Some of the most important splits of FZM units relating to savanna were dividing two of the tree savanna units (Types 47 and 48) on floristic grounds into the RB Zimbabwe Sandy Bushveld and RB Zimbabwe Clayey Bushveld at higher altitudes and mainly

Lowveld at lower altitudes. The tree savanna Type 39, associated mainly with the lower upland areas on the western Mozambique border, was assigned mostly to the RB Zambezian Wet Miombo and to a lesser extent to the RB Zambezian Dry Miombo. The remaining scattered patches on the Zimbabwean Highveld were dissolved into the surrounding RB Zambezian Dry Miombo, pending further assessment. Just over half of Submontane and Montane Grassland (Type 68) was assigned to the RB Eastern Highlands Grassland with most of the remainder to the RB Zambezian Wet Miombo and the RB Zambezian Montane Miombo.

4. GIS datasets, particularly the climate data from WorldClim 2 (Fick and Hijmans 2017) and soil texture and nutrient variables from the 250 m resolution SoilGrid dataset (Hengl et al. 2015), were used to refine and improve on the course resolution FZM.
5. The high-elevation grassland on the Chimanimani Mountains was re-mapped using World View Esri imagery and elevation (SRTM 90 m).

The source of the regional biome concepts in the mapping units of the Flora Zambeziaca Map is shown in Table 1.4.

1.5.5 Mozambique

Although several local or regional vegetation maps or classifications were produced for Mozambique over the years, only those by Pedro and Barbosa (1955) and Wild and Barbosa (1967) provided national coverage. The availability of new GIS datasets, as well as spatial information on the distribution of plant taxa, such as the national forest inventory and the recently published 'Trees and Shrubs of Mozambique' (Burrows et al. 2018), supported the review and revision of the historical Wild and Barbosa vegetation map published 57 years ago.

The recently completed Vegetation Map of Mozambique (Lötter et al. 2023) identified 162 terrestrial vegetation units across Mozambique. This vegetation map was translated into regional biomes (for the region south of the Zambezi River) as captured in Table 1.5.

1.6 Cartographic Presentation

Cartographically, biomes are considered qualitative (nominal) information, and the communication objectives are thematic. In contrast to general-purpose maps (e.g., topographical maps), thematic maps focus on the geographic

Table 1.4 A crosswalk between our regional biome scheme and the Flora Zambeziaca mapping units

Regional biome	Flora Zambeziaca code and notes
Baikiaea Woodland	16, 38, 55b
Zambezian Dry Miombo	18, 19b, 23a, 23b, 29a, 29b, 30a, 39a, (40d, 67 in Mozambique)
Zambezian Wet Miombo	23, 29, 39, 67, 68
Zambezian Montane Miombo	68
Zimbabwe Sandy Bushveld	30, 35, 47, 48, 66
Zimbabwe Clayey Bushveld	18, 23, 39, 47, 48, 65, 66
Zambezian Sandy Savanna	29, 40, 47, 48, 50
Zambezian Clayey Savanna	47, 49
Limpopo Mopane	36, 47, 50, 61
Zambezian Mopane	12, 35, 50, 51, 70
Great Dyke Grassland	23, 66, 67
Lowveld Sandy Bushveld	47, 50
Eastern Highlands Grassland	7, 68, 69

Table 1.5 A crosswalk between the vegetation units of the Vegetation Map of Mozambique (VegMapMoz) and our regional biome scheme

Vegetation map of Mozambique	Regional biome
Chimanimani Montane Grassland	Eastern Highlands Grassland
Gorongosa Montane Grassland	Eastern Highlands Grassland
Manica Montane Grassland	Eastern Highlands Grassland
Limpopo Ridge Mopane Woodland	Limpopo Mopane
Southern Mopane Woodland	Limpopo Mopane
Limpopo Lowland Woodland	Lowveld Clayey Bushveld
Maputo Alluvial Vegetation	Lowveld Clayey Bushveld
Western Maputaland Clay Bushveld	Lowveld Clayey Bushveld
Lebombo Summit Sourveld	Lowveld Mountain Bushveld
Northern Lebombo Bushveld	Lowveld Mountain Bushveld
Southern Lebombo Bushveld	Lowveld Mountain Bushveld
Gaza Sandy Guibourtia Woodland	Lowveld Sandy Bushveld
Nwambiya-Pumbe Sandy Bushveld	Lowveld Sandy Bushveld
Tembe Sandy Bushveld	Lowveld Sandy Bushveld
Urronga Lowland Dry Woodland	Lowveld Sandy Bushveld
Maputaland Coastal Wooded Grassland	Subtropical East Coast Woodland
Cheringoma Coastal Palm Savanna	Tropical East Coast Woodland
Inharrime Coastal Palm Savanna	Tropical East Coast Woodland
Save Coastal Miombo	Tropical East Coast Woodland
Save Coastal Palm Savanna	Tropical East Coast Woodland
Bangomatete Rhyolite Dry Woodland	Zambezian Clayey Savanna
Buzi-Pungwe Alluvial Vegetation	Zambezian Clayey Savanna
Canxixe Lowland Dry Woodland	Zambezian Clayey Savanna
Dombe Basalt Dry Woodland	Zambezian Clayey Savanna
Rift Valley Lowland Woodland	Zambezian Clayey Savanna
Save Alluvial Vegetation	Zambezian Clayey Savanna
Save Lowland Dry Woodland	Zambezian Clayey Savanna
Stormberg Dry Woodland	Zambezian Clayey Savanna
Zambezi Alluvial Vegetation	Zambezian Clayey Savanna
Barue Escarpment Miombo	Zambezian Dry Miombo
Inhambane Coastal Miombo	Zambezian Dry Miombo
Mossurize Escarpment Miombo	Zambezian Dry Miombo
Pangue Dry Miombo	Zambezian Dry Miombo
Vilanculos Coastal Miombo	Zambezian Dry Miombo
Southern Inselberg Woodland	Zambezian Inselberg Woodland
Chimanimani Montane Miombo	Zambezian Montane Miombo
Luia Basalt Mopane Woodland	Zambezian Mopane
Mágoè Sandstone Mopane Woodland	Zambezian Mopane

(continued)

Table 1.5 (continued)

Vegetation map of Mozambique	Regional biome
Zambezi Valley Mopane Woodland	Zambezian Mopane
Guro Dry Woodland	Zambezian Sandy Savanna
Madanda Sandstone Dry Woodland	Zambezian Sandy Savanna
Maringue Sandstone Dry Woodland	Zambezian Sandy Savanna
Barue Plateau Moist Miombo	Zambezian Wet Miombo
Cheringoma Coastal Moist Miombo	Zambezian Wet Miombo
Cheringoma Escarpment Moist Miombo	Zambezian Wet Miombo
Cheringoma Plateau Moist Miombo	Zambezian Wet Miombo
Chimoio Moist Miombo	Zambezian Wet Miombo
Choa Escarpment Moist Miombo	Zambezian Wet Miombo
Dombe Escarpment Miombo	Zambezian Wet Miombo
Gorongosa Escarpment Moist Miombo	Zambezian Wet Miombo
Gorongosa Foothills Moist Miombo	Zambezian Wet Miombo

distribution and spatial variation of a feature or theme (Robinson et al. 1995), in this case, the biomes. The primary objective of the biome maps was to highlight the spatial relationships between the mapping units while ignoring any quantitative (e.g., climatic) differences, as these were considered during the delineation of the mapping units.

Given that map reading is a synoptic process where objects are visually related to one another, readers assign meaning to features and interpret them according to visual structure. A mental image is created with dark colours perceived as essential and bright colours as dominant (Murad-al-shaikh 2004). However, the perception of hierarchy must be avoided in a biome map, as no such unit is more important than another. Visual distinctiveness and the established colour scheme were considered in symbolising the biome data.

The colour scheme used in the VegMap2006 (Mucina and Rutherford 2006) was adapted to produce overview biome maps. The primary motivation for using this colour scheme as the basis was to facilitate direct comparison with the individual VUs in VegMap2006. Most of the colours used for the FSA biomes map were identical to those used for the VegMap2006 biome map. Developing a colour scheme for the biomes was particularly challenging as the ability to discriminate between colours reduces rapidly as the number of colours in a scene increases (MacEachren 2004). Also, due to the gamut of constraints of printers, only a limited number of unique colours could be employed. An additional constraint was that the colours selected for the continental biomes had to visually relate to the respective biome colours, allowing biomes to be coherently recognisable at the continental biome level. This constraint further reduced the pallet of available colours. Hatching schemes were not a cartographic option due to the complex shapes and small sizes of many of the polygons. Consequently, the choice of colour for each sub-biome was vital to ensure that adjacent units, or units within proximity, can be discriminated "at a glance".

The process of selecting appropriate colours for the regional biomes involved the biome colours as 'anchor colours', which were used as central hue values to mathematically generate the required number of contrasting colours for the regional biomes related to each biome. In total, 250 distinct colours were generated in the hue saturation lightness (HSL) colour space (van Niekerk and van der Merwe 2008).

Distinct colours (cool blues) were used to highlight azonal biome units. Minor hue and saturation modifications were made to the original Succulent Karoo colour definition to increase separability from some of the colours used for the Savanna bioregions.

National boundaries and country names were added to the maps for orientation purposes. Many patches of the biomes were annotated to improve identification. Smaller units with notable importance were also labelled. Map elements (incl. scale, north arrow, and legend) were added so they do not detract from the mapped information, with simplicity being the primary map design objective (Buckley and Vandegraf 2008).

We map the zonal biomes in their theoretical (modelled) extent. Here, we have 'dissolved' the patches of the forest patches and patches of azonal (other than pedobiomes, and, for convenient referencing, the prominent azonal features of the Etosha Pan in Namibia, the Okavango Delta, and the Makgadikgadi Pans system in Botswana) in the zonal biomes. This decision was motivated by creating a cartographic framework allowing for an effective superimposing of the azonal and forest biome polygons over the non-forest zonal biome map, avoiding boundary mismatches and creating 'slivers' at the final stage of the compilation.

1.7 Descriptions of the Regional Biomes

The rank of the Regional Biome is the basic biome unit of this study. Each regional biome has been, therefore, paid particular attention by detailed descriptions involving:

(a) The code and the name of the regional biome;
(b) Vegetation-mapping spatial synonymy (sourced from VegMap2006, VegMap2018, and VegMapMoz), where relevant;
(c) Schematic map (see Sect. 1.7.1 below);
(d) Spatial climatic diagram (see Sect. 1.7.2 below for more detail);
(e) Distribution (spatial extent of regional biome);
(f) Vegetation physiognomy and composition (focusing on vegetation patterns); the names of taxa follow in principle the Plants of South Africa online checklist (https://posa.sanbi.org/sanbi) and in cases where a taxon is not listed here (usually because it is found outside South Africa, Lesotho and eSwatini) then we follow Plant of the World Online (Plants of the World Online | Kew Science);
(g) Climate (featuring informational addition to the climate diagram);
(h) Topography, geology and soils; and
(i) Sources (listing published or otherwise publicly available works featuring the ecology and vegetation of the regional biome).

1.7.1 Construction of Schematic Maps of the Biomes

The biome maps were created in ArcGIS Pro 2.6.0 by extracting each biome from the zonal biome shapefile. Each biome was set to overlay water bodies and country boundaries within the MBSA area. To enhance the appearance of each biome, the colour of the polygons for that biome was set as red and the polygon outline width was set to ensure visibility of very small polygons.

1.7.2 Construction of the Climate Diagrams

A spatial climate diagram (Fig. 1.2) represents the climatic profile of each zonobiome, regional biome, and bioregion (Fig. 2.2 in Mucina and Rutherford 2006; Mucina 2023).

We created climate diagrams for each vegetation unit using multiple climate variables derived from modelled climatology patterns at high resolution for the Earth's land surface areas. These diagrams were created for all biomes at all levels of complexity in the MBSA area.

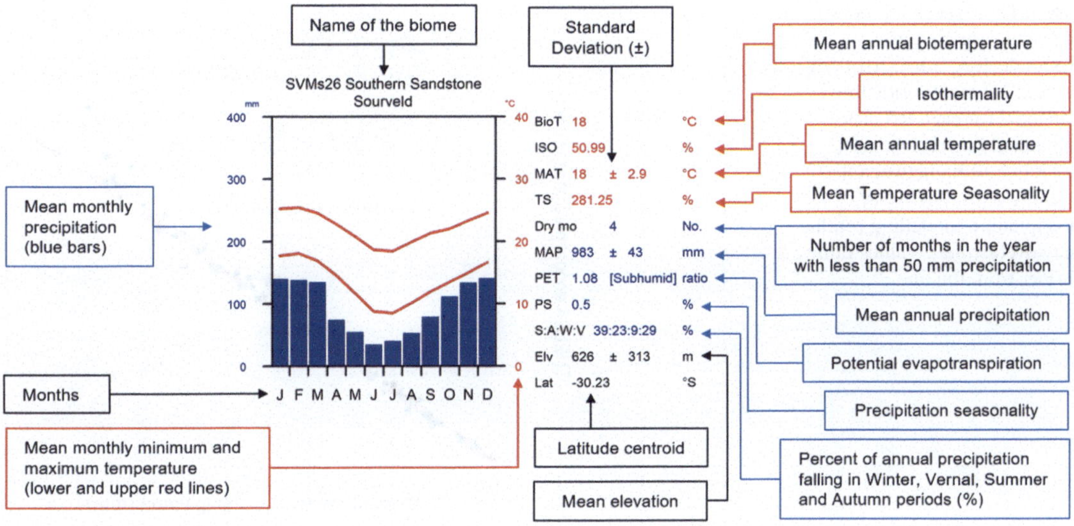

Fig. 1.2 Graphical explanation of the variables involved in the construction of a climate diagram

Below, we describe all variables used in constructing the climate diagrams and climate and elevation summary data. We used the R 4.3.0 statistical platform (R Core Team 2023) and a combination of contributed R packages specified below to source and process the climate data and diagrams. The detailed procedure of the climate diagram construction is described in Tsakalos et al. (2023).

1.7.2.1 Sources and Processing of the Climatic Data

We constructed the climatic profiles of the biomes using modelled climatology patterns at high resolution for the earth's land surface areas (CHELSA version 1.2; see Karger et al. 2017a, b; Beck et al. 2020) and WorldClim version 2 (Fick and Hijmans 2017) both at a 30 arc-second scale (~1 km, 0.00833°) and Mapzen's digital elevation product (~0.6 km, Hollister 2021). We obtained Mapzen data through Amazon Web Services using the elevatr R package (Hollister 2021), while we accessed and downloaded CHELSA and WorldClim data through their respective internet portals. Mapzen's product is unique in that it combines several sources of digital elevation models, including SRTM, the ArcticDEM (covering all areas north of 60°; Morin et al. 2016), EUDEM (digital elevation model over Europe; for review, see Mouratidis and Ampatzidis 2019), and others into a single product.

We used the CHELSA data (mean monthly mean, minimum and maximum temperature monthly temperature) because of its high accuracy in its temperature prediction over South Africa compared to other products (e.g., ERA-Interim: Dee et al. 2011; CRU: Harris et al. 2014; CHIRPS: Funk et al. 2015; see: Karger et al. 2017a, b). We opted to use WorldClim's precipitation because a preliminary inspection of CHELSA's values for Zambezian Inselberg Woodland rainfall and other vegetation units revealed values higher than our expectations (Fig. 1.3).

We used R (R Core Team 2023), and several contributed packages to link the new mapping product with the modelled climate surfaces. Specifically, we used the following steps:

Step 1. Import and stack the monthly raster datasets for the mean monthly mean, minimum and maximum temperature and precipitation using the 'rast' function from the *terra* R package (Hijmans 2022). In the case of the elevation data, no stacking was required.

Step 2. Reproject the mapping product onto the coordinate reference system using the 'procject' function of *terra*.

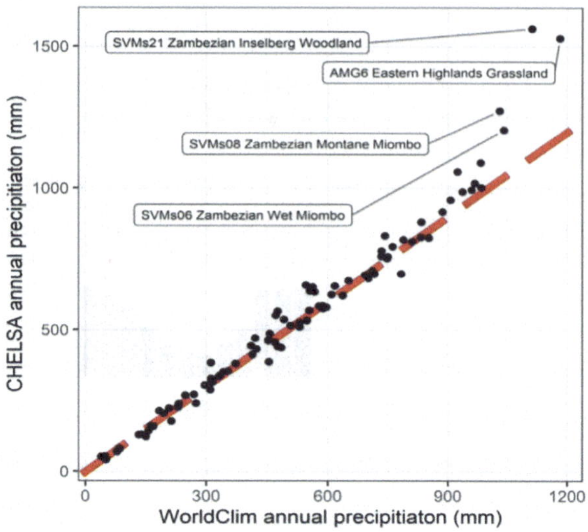

Fig. 1.3 Difference between the annual precipitation values derived from CHELSA version 1.2 and WorldClim version 2 for each RB. The red dotted line represents a 1:1 relationship between the X and Y axes; CHELSA's precipitation modelling routine may overestimate values falling further from the line

Step 3. Convert the mapping product from a vector shapefile into a raster product with zones that match the desired biome using the 'rasterise' function *terra*.

Step 4. Calculate the zonal statistics and summarised values (mean and standard deviation) of the rasters for each zone (i.e. biome) using the 'zonal' function from *terra*.

Several variables (including mean annual temperature, isothermality, temperature seasonality, mean annual precipitation, and precipitation seasonality) presented in the climate diagrams correspond to Nix's (1986) commonly used bioclimatic variables (see also Busby 1991). These variables, and others (totalling 19), have become colloquially referred to as 'bioclim' variables/Critically, these variables are supposed to represent features of climate that are physiologically relevant to plant growth and species distribution (Title and Bemmels 2018). To promote transparency in their calculation, we opted to calculate these variables using the extracted climatic data (see Steps 1–4 described above) by following Bede-Fazekas and Somodi's (2020) R code. While several automated approaches exist in R, their calculations are less transparent and may vary (e.g., Hijmans et al.'s 2021 *dismo* package and the associated 'biovars' function). In addition to these bioclimatic variables, we used base R coding to calculate Holdridge's (1966) bio-temperature and potential evapotranspiration (BioT and PET, resp.) and the number of dry months (Dry mo).

1.7.2.2 Climate Variables

We used the following climatic variables in the construction of the diagrams:

The red lines in a diagram indicate the patterns of the **Mean Monthly Minimum** and **Mean Monthly Maximum** temperatures. The blue bars depict **the Mean Monthly Precipitation**. All data come from the CHELSA version 1.2 and WorldClim version 2 and represent the variables tmin_1 to tmin_12, tmax_1 to tmax_12 and prec_1 to prec_12.

BioT (Biotemperature): Biotemperature is the temperature range in which the effective growth of plants occurs. We calculated the biotemperature as the sum of all monthly temperatures greater than 0 °C and less than 30 °C divided by 12. The values are small towards the poles, with low mean annual temperatures, whereas BioT is large in the tropics with the highest mean annual temperatures. Across the new biome scheme, BioT ranges from 24.2 to 24.47. For perspective, the Nival region (at latitudes of 9–67°22.5′) have BioTs close to zero, and the Tropics (i.e., occurring at latitudinal ranges spanning 13° north and south of the Equator) have values greater than 24 (see Holdridge 1966).

ISO (Isothermality): Quantifies how large the day-to-night temperatures oscillate relative to the summer-to-winter oscillations, given in percent. The variable in the climate diagrams represents 'BIO3', which is calculated using Bede-Fazekas and Somodi's (2020) R code. An ISO value of 100 indicates a minor temperature variability within an average month relative to the year. Across the new biome scheme, ISO ranges from 41.88 to 71.03 compared to the global range of 9.13 to 100, based on WorldClim V2.

MAT (Mean Annual Temperature): The mean temperature in °C uses the average monthly temperature averaged over twelve months. The variable in the climate diagrams represents 'BIO1', calculated using Bede-Fazekas and Somodi's (2020) R code. Across the new biome scheme, MAT ranges from 7.39 °C to 24.47 °C (the global range: –54.7 °C to 30.1 °C, based on WorldClim V2).

TS (Temperature Seasonality): This is the amount of temperature variation over a given year (or averaged years) based on the standard deviation (variation) of monthly temperature averages. The variable in the climate diagrams represents 'BIO4' Bede-Fazekas and Somodi's (2020) R code. Across the new biome scheme, TS ranges from 189.27 to 555.11 compared to the global range of 0 to 2364, based on WorldClim V2.

Dry Months (Dry mo): We defined the number of 'dry months' as the number of months in the year (between 0 and 12) for each biome with less than 50 mm of rainfall during the month.

MAP (Mean Annual Precipitation): Annual Precipitation, given in mm. The variable in the climate diagrams represents 'BIO12', calculated using Bede-Fazekas and Somodi's (2020) R code. Across the new biome scheme, MAP ranges from 38.96 to 1181.6 compared to the global range of 0 to 11,401 mm, based on WorldClim V2.

PET (Potential Evapotranspiration Ratio): The quantity of water given to the atmosphere within a zonal climate and upon a zonal soil by the area's natural vegetation. The PET variable, taken from Holdridge (1966), was calculated as the BioT/MAT × 58.93. When the ratio is < 1, rainfall exceeds potential evapotranspiration and vice versa. In the Climate Diagrams, PET is followed by the terminology (classes), corresponding to humidity provinces, defined by Holdridge (1966), specifically: Saturated (0.015625–0.03125), Subsaturated (0.03125–0.0625), Semi-saturated (0.0625–0.125), Super-humid (0.125–0.25), Perhumid (0.25–0.5), Humid (0.5–1), Subhumid (1–2), Semiarid (2–4), Arid (4–8), Perarid (8–16), Superarid (16–32), Semiparched (32–64).

PS (Precipitation Seasonality): A measure of the variation in monthly precipitation totals for the year. This index is the ratio of the standard deviation of the monthly total precipitation to the mean monthly total precipitation (also known as the coefficient of variation), expressed as a percentage. The variable in the climate diagrams represents 'BIO15' calculated from Bede-Fazekas and Somodi's (2020) R code. The values are higher in areas that experience extreme wet and dry periods and lower in areas with even rainfall throughout the year. Across the new biome scheme, PS ranges from 0.13 to 1.37 compared to the global range of 0 to 229, based on WorldClim V2.

S:A:W:V: These symbols represent the average seasonal rainfall in four seasons, defined as (for the Southern Hemisphere): S: summer (December to February), A: autumn (March to May), W: winter (June to August); V: spring or vernal (September to November). We calculated the percentage of rainfall within each of the seasons using base R coding.

Elevation (Elv): A measure of the elevation (m above sea level) from Mapzen's terrain tile product (described above). Across the new biome scheme, Elv ranges from 28 to 2793. At the same time, the global elevation ranges from 0 to 8849 m on Mount Everest/Sagarmatha.

Lat (Latitude): We derived the latitude (decimal degrees) as the biome *y*-coordinate centroid. We used terra's (Hijmans 2022) 'centroids' function to determine the y-coordinate centroid for each biome using the WGS 1984 (EPSG: 4326) coordinate system.

1.7.3 Sources of Geological Data for the Mozambique Regional Biomes

To gain a deeper understanding of the key geological features of the different biomes in Mozambique, we used a geological resource by the Mozambique Instituto Nacional de Geologia (1987). Using a geospatial intersection analysis, we compared this geological data set with our newly derived biome classification system. This intersection analysis resulted in a detailed summary, including the spatial extent and description of the geological composition (lithology) within each biome. The obtained results were integrated into the descriptions of the regional biomes.

References

Beck HE, Wood EF, McVicar TR et al (2020) Bias correction of global high-resolution precipitation climatologies using streamflow observations from 9372 catchments. J Clim 33:1299–1315

Bede-Fazekas S, Somodi I (2020) The way bioclimatic variables are calculate has impact on potential distribution models. Meth Ecol Evol 11:1559–1570

Bekker RP, De Wit PV (1991) Contribution to the vegetation classification of Botswana. FAO/UNDP/ Government of Botswana, Soil Mapping and Advisory Services, Gaborone

Bekker RP, De Wit PV, Fernando KS et al (1991) Vegetation map of the Republic of Botswana. Soil Mapping and Advisory Services Project. Government Printer, Gaborone

Box EO (2016) World bioclimatic classification. In: Box EO (ed) Vegetation structure and function at multiple spatial, temporal and conceptual scales. Springer International Publishing Switzerland, Cham, pp 1–52

Breckle S-W (2002) Walter's vegetation of the Earth, 4th edn. Springer-Verlag, Berlin

Breckle S-W, Rafiqpoor MD (2019) Vegetation und Klima. Springer Spektrum, Berlin

Buckley A, Vandegraf D (2008) Making better maps: cartographic design. Principles and practices for any map. ESRI, Redlands

Burrows J, Burrows S, Lötter M, Schmidt E (2018) Trees and shrubs of Mozambique. Print Matters Heritage, Noordhoek

Busby JR (1991) BIOCLIM—a bioclimate analysis and prediction system. In: Margules CR, Austin MP (eds) Nature conservation: Cost effective biological surveys and data analysis. CSIRO Publishing, Canberra

Cole MM, Brown RC (1976) The vegetation of the Ghanzi area of western Botswana. J Biogeogr 3:169–196

Curtis B, Mannheimer C (2005) Tree atlas of Namibia. National Botanical Research Institute, Windhoek

Dayaram A, Harris LR, Grobler BA et al (2019) Vegetation map of South Africa, Lesotho and Swaziland 2018: a description of changes since 2006. Bothalia 49(1):a2452. https://doi.org/10.4102/abc.v49i1.2452

Dee DP, Uppala SM, Simmons AJ et al (2011) The ERA-Interim reanalysis: configuration and performance of the data assimilation system. Quater J Roy Meteorol Soc 137:553–597

Fick SE, Hijmans RJ (2017) WorldClim 2: new 1-km spatial resolution climate surfaces for global land areas. Int J Climatol 37:4302–4315

Funk C, Peterson P, Landsfeld M et al (2015) The climate hazards infrared precipitation with stations—a new environmental record for monitoring extremes. Sci Data 2:1–21. https://doi.org/10.1038/sdata.2015.66

Giess W (1971) A preliminary vegetation map of South West Africa. Dinteria 4:1–114

Harris IPDJ, Jones PD, Osborn TJ et al (2014) Updated high-resolution grids of monthly climatic observations—the CRU TS3.10 Dataset. Int J Climatol 34:623–642

Hengl T, Heuvelink GBM, Kempen B et al (2015) Mapping soil properties of Africa at 250 m resolution: random forests significantly improve current predictions. PLoS ONE 10(6):e0125814. https://doi.org/10.1371/journal.pone.0125814

Hijmans RJ, Phillips S, Leathwick J (2021) dismo: Species distribution modelling. R package version 1.3–5. https://CRAN.R-project.org/package=dismo

Hijmans RJ (2022) terra: spatial data analysis. R package version 1.5–21. https://CRAN.R-project.org/package=terra

Hines C (1997) Vegetation. In: Mendelsohn JM, Roberts CS (eds) An environmental profile and atlas of Caprivi. Directorate of Environmental Affairs, Windhoek

Hoare DB, Mucina L, Rutherford MC et al (2006) Albany Thicket Biome. In: Mucina L, Rutherford MC (eds) The vegetation of South Africa, Lesotho and Swaziland. SANBI, Pretoria, pp 540–567

Holdridge LR (1966) Life zone ecology. Revised edition. Tropical Science Center, San José

Hollister JW (2021) elevatr: Access Elevation Data from Various APIs. R package version 0.4.1. https://CRAN.R-project.org/package=elevatr

Instituto Nacional de Geologia (1987) Carta Geológica. Escala 1:1,000,000. República Popular de Moçambique, Instituto Nacional de Geologia, Maputo https://maps.google.com/maps?q=-19.0,36.0&hl=es&z=14

Irish J (1994) The biomes of Namibia, as determined by objective categorisation. Navors Nasion Mus Bloemfontein 10:549–592

Jürgens N, Schmiedel U, Haarmeyer DH et al (2012) The BIOTA Biodiversity Observatories in Africa—a standardised framework for large-scale environmental monitoring. Environ Monitor Assess 184:655–678

Jürgens N, Desmet PG, Rutherford MC et al (2006) Desert Biome. In: Mucina L, Rutherford MC (eds) The vegetation of South Africa, Lesotho and Swaziland. SANBI, Pretoria, pp 300–323

Jürgens N., Haarmeyer DH, Luther-Mosebach J (eds) (2010) Biodiversity in southern Africa, vol 1. In: Patterns at local scale—the BIOTA Observatories. Klaus Hess Publishers, Göttingen

Jürgens N, Oldeland J, Hachfeld B et al (2013) Ecology and spatial patterns of large-scale vegetation units within the central Namib Desert. J Arid Environ 93:59–79

Karger DN, Conrad O, Böhner J et al (2017a) Climatologies at high resolution for the Earth land surface areas. Sci Data 4:170122. https://doi.org/10.1038/sdata.2017.122

Karger DN, Conrad O, Böhner J et al (2017b) Data from: climatologies at high resolution for the earth's land surface areas. Dryad Digit Reposit. https://doi.org/10.5061/dryad.kd1d4

Karger DN, Schmatz D, Dettling D et al (2020) High resolution monthly precipitation and temperature timeseries for the period 2006–2100. Sci Data 7:248. https://doi.org/10.1038/s41597-020-00587-y

Leistner OA (1967) The plant ecology of the southern Kalahari. Mem Bot Surv S Afr 38:1–172

Leser H (1972) Geoökologische Verhältnisse der Pflanzengesellschaften in den Savannen des Sandveldes um den Schwarzen Nossob und um Epukiro. Dinteria 6:1–41

Lötter M, Burrows J, Jones K et al (2023) Historical vegetation map and red list of ecosystems assessment for Mozambique. Version 2.0—Final report. USAID/SPEED+, AFD/FFEM, Maputo

MacEachren AM (2004) How maps work: Representation, visualisation, and design. Guilford Press, New York

Mendelsohn J, Roberts C (1997) An environmental profile and atlas of Caprivi. Directorate of Environmental Affairs, Windhoek

Mendelsohn J, el Obeid S, Roberts C (2000) A profile of North-Central Namibia. Environmental Profiles Project. Directorate of Environmental Affairs, Windhoek

Mendelsohn J, Jarvis A, Roberts C et al (2002) Atlas of Namibia: a portrait of the land and its people. David Philip, Cape Town

Mittermeier RA, Robles Gil P, Hoffmann M et al (2004) Hotspots revisited: Earth's biologically richest and most endangered ecoregions. CEMEX, Mexico City

Morin P, Porter C, Cloutier M et al (2016) ArcticDEM; a publicly available, high resolution elevation model of the Arctic. EGU Gen Assem Conf Abstr 18:8396. ArcticDEM. Available online: https://www.pgc.umn.edu/data/arcticdem/

Mouratidis A, Ampatzidis D (2019) European digital elevation model validation against Extensive Global Navigation Satellite Systems Data and comparison with SRTM DEM and ASTER GDEM in Central Macedonia (Greece). ISPRS Int J Geo-Inform 8:108. https://doi.org/10.3390/ijgi8030108

Mucina L (2019) Biome: Evolution of a crucial ecological and biogeographical concept. New Phytol 222:97–114

Mucina L, Lötter MC, Rutherford MC et al (2022) Forest biomes of Southern Africa. New Zeal J Bot 60:377–428

Mucina L, Jürgens N, le Roux A et al (2006a) Succulent Karoo biome. In: Mucina L, Rutherford MC (eds) The vegetation of South Africa, Lesotho and Swaziland. SANBI, Pretoria, pp 220–299

Mucina L, Hoare DB, Lötter MC et al (2006b) Grassland biome. In: Mucina L, Rutherford MC (eds) The vegetation of South Africa, Lesotho and Swaziland. SANBI, Pretoria, pp 348–437

Mucina L, Scott-Shaw CR, Rutherford MC et al. (2006c) Indian Ocean Coastal Belt. In: Mucina L, Rutherford MC (eds) The vegetation of South Africa, Lesotho and Swaziland. SANBI, Pretoria, pp 568–583

Mucina L, Rutherford MC, Powrie LW (eds) (2005) Vegetation map of South Africa, Lesotho and Swaziland. 1:1000000 scale sheet maps. SANBI, Pretoria

Mucina L, Rutherford MC (eds) (2006) The vegetation of South Africa, Lesotho and Swaziland. SANBI, Pretoria

Mucina L (2023) Biomes of the Southern Hemisphere. Springer Nature Switzerland, Cham

Murad-al-shaikh M (2004) Cartography with ArcGIS course manuals. Course version 1.2. ESRI, Redlands

Myers N, Mittermeier RA, Mittermeier CG et al (2000) Biodiversity hotspots for conservation priorities. Nature 403:853–858

Nix HA (1986) A biogeographic analysis of Australian elapid snakes. In: Longmore R (ed) Atlas of elapid snakes of Australia. Australian Flora and Fauna Series 7. Australian Government Publishing Service, Canberra

Pedro JG, Barbosa LAG (1955) A vegetação. In: Esboço do reconhecimento ecológico-agrícola de Moçambique, vol 2. Memorias e Trabalhos No. 23. Centro de Investigaação Cientifica Algodeira, Lourenço Marques [Maputo], pp 67–224

R Core Team (2023) R: a language and environment for statistical computing. R Foundation for Statistical Computing, Wien. https://www.r-project.org

Rebelo AG, Boucher C, Helme N et al (2006) Fynbos biome. In: Mucina L, Rutherford MC (eds) The vegetation of South Africa, Lesotho and Swaziland. SANBI, Pretoria, pp 52–219

Robinson AH, Morrison JL, Muehrcke PC et al (1995) Elements of cartography, 6th edn. John Wiley & Sons, New York

Rogers CML (1993) A woody vegetation survey of Hwange National Park. A report prepared for the Department of National Parks and Wild Life Management. Causeway, Harare

Rutherford MC (1997) Categorisation of biomes. In: Cowling RM, Richardson DM, Pierce SM (eds) Vegetation of southern Africa. Cambridge University Press, Cambridge, pp 91–98

Rutherford MC, Westfall RH (1986) Biomes of southern Africa—an objective categorisation. Mem Bot Surv S Afr 54:1–98

Rutherford MC, Westfall RH (1994) Biomes of southern Africa—an objective categorisation, 2nd edn. Mem Bot Surv S Afr 63:1–94

Rutherford MC, O'Farrel PO, Goldberg K et al (2000) SAFARI 2000: Sub-project 6: Species composition of southern African vegetation. CD-ROM, National Botanical Institute, Cape Town

Rutherford MC, Mucina L, Powrie LW (2006a) Biomes and bioregions of Southern Africa. In: Mucina L, Rutherford MC (eds) The vegetation of South Africa, Lesotho and Swaziland. SANBI, Pretoria, pp 30–51

Rutherford MC, Mucina L, Lötter MC et al (2006b) Savanna Biome. In: Mucina L, Rutherford MC (eds) The vegetation of South Africa, Lesotho and Swaziland. SANBI, Pretoria, pp 438–539

Schimper AFW (1898) Pflanzengeographie auf physiologischer Grundlage. Gustav Fischer Verlag, Jena

Scholes RJ, Dowty PR, Caylor K et al (2002) Trends in savanna structure and composition along an aridity gradient in the Kalahari. J Veg Sci 13:419–428

Setshogo MP, Venter F (2003) Trees of Botswana: names and distribution. SANBI, Pretoria

Strohbach BJ, Kutuahuripa JT (2014) Vegetation of the eastern communal conservancies in Namibia: II. Environmental drivers. Koedoe 56(1):Art.#1117. https://doi.org/10.4102/koedoe.v56i1.1117

Strohbach BJ (2014) Vegetation of the eastern communal conservancies in Namibia: I. Phytosociological descriptions. Koedoe 56(1):1116. https://doi.org/10.4102/koedoe.v56i1.1116

Timberlake J, Nobanda N, Mapaure I (1993) Vegetation survey of the communal lands: north and west Zimbabwe. Kirkia 14:171–270

Title PO, Bemmels JB (2018) ENVIREM: an expanded set of bioclimatic and topographic variables increases flexibility and improves performance of ecological niche modeling. Ecography 41:291–307

Tsakalos JL, Smith MR, Luebert F et al (2023) Climenv: download, extract and visualise climatic and elevation data. J Veg Sci 34:e13215. https://doi.org/10.1111/jvs.13215

US Geological Survey (2018) USGS EROS Archive – Digital Elevation – Shuttle Radar Topography Mission (SRTM). Earth resources observation and science center, reston. Available from: https://www.usgs.gov/centers/eros/science/usgs-eros-archive-digital-elevation-shuttle-radar-topography-mission-srtm

van der Merwe H, van Rooyen MW, van Rooyen N (2008) Vegetation of the Hantam–Tanqua–Roggeveld subregion, South Africa, Part 2: Succulent Karoo Biome related vegetation. Koedoe 50:160–183. https://doi.org/10.4102/koedoe

van Niekerk A, van der Merwe JH (2008) Vegetation map of South Africa, Lesotho and Swaziland: cartographic challenges and lessons learnt. Position IT 2008, pp 37–40

van Wyk AE, Smith GF (2001) Regions of floristic endemism in southern Africa. A review with emphasis on succulents. Umdaus Press, Hatfield

von Klima (1943) Boden und wirtschaftlicher Nutzung. Verlag von Paul Parey, Berlin

Walter H (1976) Globale Gliederung der natürlichen terrestrischen Ökosysteme. Flora 165:315–323

Walter H, Box EO (1976) Global classification of natural terrestrial ecosystems. Vegetatio 32:72–81

Walter H (1942) Die Vegetation des Europäischen Rußlands unter Berücksichtigung von Klima, Boden und wirtschaftlicher Nutzung. Verlag von Paul Parey, Berlin

Wanke A (2006) A Profile of Khaudum National Park (Namibia) with emphasis on the physical environment, woody vegetation, and water. Thesis, University of Pretoria, Pretoria

Wild H, Barbosa LAG (1967) Vegetation map of the Flora Zambesiaca area. M.O. Collins, Salisbury [Harare]

Wild H, Fernandes A (1968) Vegetation map of the Flora Zambesiaca area. Flora Zambesiaca supplement. M.O. Collins, Salisbury [Harare]

Williamson G (1997) Preliminary account of the floristic zones of the Sperrgebiet (Protected Diamond Area) in southwest Namibia. Dinteria 25:1–68

Major Zonobiome and Zonoecotone Patterns in Southern Africa

2

Ladislav Mucina, Michael C. Rutherford, Mervyn C. Lötter, James L. Tsakalos, Adriaan van Niekerk, Caley Higgs and Leslie W. Powrie

Abstract

This chapter is devoted to the description of the basic biome fabric of the study region (MBSA) including six zonobiomes, eleven global biomes, twelve continental biomes, and 87 regional biomes are recognised. Furthermore, six zonoecotones accommodate six continental ecotonal biomes and 25 regional ecotonal biomes. Maps and climatic diagrams of the high-rank (zonobiome, global, and continental) biomes and zonoecotones, are presented, setting the scene for the descriptions of regional biomes (zonobiomes and zonoecotones) in the chapters that follow. This chapter is accompanied by downloadable map files and relevant GIS shape files.

Keywords

Biome maps · Fynbos · Savanna · Forest · Temperate biomes · Tropical biomes · Zonoecotones

The complete scheme of the zonobiome and zonoecotone biome system for Southern Africa is presented in Table 2.1. The naming and coding convention for the zonoecotones follows

Supplementary Information The online version contains supplementary material available at https://doi.org/10.1007/978-3-031-63731-5_2.

L. Mucina (✉) · J. L. Tsakalos
Harry Butler Institute, Murdoch University, 90 South Street, Murdoch, Perth 6150, Australia
e-mail: Ladislav.Mucina@murdoch.edu.au

L. Mucina
Department of Geography and Environmental Studies, Stellenbosch University, Private Bag X1, Matieland 7602, Stellenbosch, South Africa

M. C. Rutherford
High Peak, Derbyshire, United Kingdom

Department of Botany and Zoology, Stellenbosch University, Private Bag X1, Matieland 7602, Stellenbosch, South Africa

M. C. Lötter
Mpumalanga Tourism and Park Agency, P/Bag X1088, Lydenburg 1120, South Africa

School of Animal, Plants and Environmental Sciences, University of the Witwatersrand, Private Bag 3, Wits, Johannesburg 2050, South Africa

J. L. Tsakalos
Plant Diversity and Ecosystems Management Unit, School of Biosciences and Veterinary Medicine, University of Camerino, Via Pontoni 5, Camerino I-62032, Italy

A. van Niekerk · C. Higgs · L. W. Powrie
Department of Geography and Environmental Studies, Stellenbosch University, Private Bag X1, Matieland 7602, Stellenbosch, South Africa

L. W. Powrie
Kirstenbosch Research Center, South African National Biodiversity Institute, Private Bag X7, Claremont, Cape Town, South Africa

L. Mucina and M. C. Rutherford (eds.), *Zonal Biomes of Southern Africa*, Biome Ecology 2,
https://doi.org/10.1007/978-3-031-63731-5_2

Table 2.1 A system of zonobiomes ZB, zonoecotones ZE, global biomes GB, continental biomes CB, and continental ecotonal biomes CEB for the zonal biome units occurring in Southern Africa

Rank	Code	Biome name	Zonality status
Zonobiome E1 Equatorial Zone			
ZB	**E1**	**Equatorial Zone**	**Zonal**
GB	***STF***	***Subtropical Rainforest***	***Zonal***
CB	STFa	*African Subtropical Forest*	Zonal
Zonobiome E2 Tropical Seasonal Zone			
ZB	**E2**	**Tropical Seasonal Zone**	Zonal
GB	***SVM***	***Mesic Savanna***	Zonal
CB	SVMs	Southern African Mesic Savanna	Zonal
GB	***SVA***	***Arid Savanna***	Zonal
CB	SVAs	Southern African Arid Savanna	Zonal
GB	***TDF***	***Tropical Dry Forest***	Zonal
CB	TDFa	East African Dry Coastal Forest	Zonal
Zonobiome S1 Ethesial Zone			
ZB	**S1**	**Ethesial Zone**	Zonal
GB	***MEDs***	***Southern Ethesial Woodland***	Zonal
CB	FYN	Fynbos	Zonal
Zonobiome S2 Hot Arid Zone			
ZB	**S2**	**Hot Arid Zone**	Zonal
GB	***DESc***	***Subtropical Coastal Hot Deserts***	Zonal
CB	DN	Namib Desert	Zonal
GB	***DESi***	***Subtropical Inland Hot Deserts***	Zonal
CB	DG	Gariep Desert	Zonal
CB	DT	Tankwa Desert	Zonal
GB	***SDESi***	***Subtropical Inland Hot Semi-Deserts***	Zonal
CB	NK	Nama-Karoo	Zonal
GB	***SDESc***	***Subtropical Coastal Hot Semi-Deserts***	Zonal
CB	SK	Succulent Karoo	Zonal
Zonobiome T3 Warm-Temperate Zone			
ZB	**T3**	**Warm-Temperate Zone**	**Zonal**
GB	***WTZs***	***Southern Warm-Temperate Forest***	***Zonal***
CB	ATFs	Southern African Afrotemperate Forest	Zonal
Zonobiome A2 Subtropical Alpine Zone			
ZB	**A2**	**Subtropical Alpine Zone**	**Zonal**
GB	***SAZs***	***Southern Subtropical Alpine Biome***	***Zonal***
CB	DAG	Drakensberg Alpine Grassland	Zonal
Zonoecotone E2(savm)–T3 Subtropical Warm-Temperate Wet Ecotone			
ZE		**Subtropical Warm-Temperate Wet Ecotone**	Zono-ecotonal
CEB	AMG	African Montane Grassland	Ecotonal

(continued)

Table 2.1 (continued)

Rank	Code	Biome name	Zonality status
Zonoecotone E2(sava)–[E2(savm)–T3]–S2 Subtropical Semiarid Seasonal–Semidesert Ecotone			
ZE		**Subtropical Semiarid Seasonal–Semidesert Ecotone**	Zono-ecotonal
CEB	DHG	Dry Highveld Grassland	Ecotonal
Zonoecotone S2–E1(sub) Subtropical Semiarid–Semidesert Ecotone			
ZE		**Subtropical Semiarid–Semidesert Ecotone**	Zono-ecotonal
CEB	AT	Subtropical Albany Thicket	Ecotonal
Zonoecotone S2(sdesc)–S1 Semidesert–Ethesial Coastal Ecotone			
ZE		**Semidesert–Ethesial Coastal Ecotone**	Zono-ecotonal
CEB	WCS	West Coast Arid Strandveld	Ecotonal
Zonoecotone S1–A4 Ethesial–Austro-Alpine Ecotone			
ZE		**Ethesial–Austro-Alpine Ecotone**	Zono-ecotonal
CEB	CAP	Cape Altimontane Fynbos	Ecotonal
Zonoecotone T3–(S2–E1(sub)) Semiarid Warm-Temperate Ecotone			
ZE		**Semiarid Warm-Temperate Ecotone**	Zono-ecotonal
CEB	CVT	Southern Cape Valley Thicket	Ecotonal

The codes of the zonobiomes follow Mucina (2023)

Mucina (2023). The spatial extent of each zonobiome is captured in Table 2.2 (in its detailed form available as a supplementary file Ch2 Mucina&Rutherford SUPPL Table S2.1). The information on the GIS data and metadata availability is given in the section on Data availability below.

We present two synoptic maps, one featuring the zonobiomes and zonoecotones biomes (Fig. 2.1; supplementary file Ch2 Mucina&Rutherford SUPPL Figure 2.1) and the other showing the extent of the global biomes (Fig. 2.2; supplementary file Ch2 Mucina&Rutherford SUPPL Figure 2.2). The map of the continental (and continental ecotonal) biomes of Southern Africa is presented in Fig. 2.3 (also supplementary file Ch2 Mucina&Rutherford SUPPL Figure 2.3) and the regional biomes are featured in Fig. 2.4 (also supplementary file Ch2 Mucina&Rutherford SUPPL Figure 2.4). The schematic maps accompanied by climate diagrams for biomes at the levels, including ZB, ZE, GB, CB, and CEB, are summarised in Fig. 2.5.

Although the study region spans southern latitudes only between 15°55' S and 34°83' S, the biomes of the mapping area constitute as many as six zonobiomes, eleven global biomes, twelve continental biomes, and 87 regional biomes. Six zonoecotones accommodate six continental ecotonal biomes (CEBs) and 25 regional ecotonal biomes (REBs; see Tables 2.1 and 2.2). The zonobiome structure includes zonobiomes E1, E2, S2, S1, T3, and A4 according to the coding system by Mucina (2023; zonobiomes I, II, III, IV, V, and IX, in the original Walter's terminology) and a new zonobiome—the Subtropical Alpine Zone (Mucina 2023).

2.1 Zonobiomes

Zonobiome E1 (original Walter's Zonobiome I) Equatorial Zone reaches the mapping region as GB Subtropical Rainforest, mainly found in Mozambique and along the South African coast (embedded within the subtropical GB Mesic Savanna and the ecotonal Albany Thicket) as far

Table 2.2 Statistical overview of the mapped number of polygons and the total area (in km^2) of each high-rank zonal biome of the Southern Africa

	Polygon count								Area							
	ZA	LT	SZ	MZ	NA	BW	ZW	Total	ZA	LT	SZ	MZ	NA	BW	ZW	Total
E1 Equatorial Zone	4659		268	408			209	5544	1430.17		147.62	8964.22			96.94	10,638.96
STF Subtropical Forest	4659		268	408			209	5544	1430.17		147.62	8964.22			96.94	10,638.96
STFa African Subtropical Forest	4659		268	408			209	5544	1430.17		147.62	8964.22			96.94	10,638.96
E2 Tropical Seasonal Zone	11,232		170	17,433	64	59	1683	30,641	415,131.53		13,500.69	314,764.90	495,335.90	578,248.57	387,639.84	2,204,621.43
SVA Arid Savanna	523			1301	43	23	274	2164	229,986.35			35,531.93	419,179.57	512,870.51	65,584.30	1,263,152.66
SVAs Southern African Arid Savanna	523			1301	43	23	274	2164	229,986.35			35,531.93	419,179.57	512,870.51	65,584.30	1,263,152.66
SVM Mesic Savanna	7266		170	3482	21	36	796	11,771	184,733.99		13,500.69	255,032.3	76,156.33	65,378.06	321,573.87	916,375.20
SVMs Southern African Mesic Savanna	7266		170	3482	21	36	796	11,771	184,733.99		13,500.69	255,032.3	76,156.33	65,378.06	321,573.87	916,375.20
TDF Tropical Dry Forest	3443			12,650			613	16,706	411.19			24,200.70			481.68	25,093.57
TDFa East African Dry Coastal Forest	3443			12,650			613	16,706	411.19			24,200.70			481.68	25,093.57
S2 Hot Arid Zone	149				30			179	9122.31				117,965.21			127,087.512
DESi Subtropical Inland Hot Desert	132				5			137	5626.81				8282.60			13,909.42
DG Gariep Desert	132				5			137	5626.81				8282.60			13,909.42

(continued)

Table 2.2 (continued)

	Polygon count								Area							
	ZA	LT	SZ	MZ	NA	BW	ZW	Total	ZA	LT	SZ	MZ	NA	BW	ZW	Total
DESc Subtropical Coastal Hot Desert	16				25			41	1120.41				109,682.60			110,803.01
DN Namib Desert	16				25			41	1120.41				109,682.60			110,803.01
DESi Subtropical Inland Hot Desert	1							1	2375.08							2375.08
DT Tankwa Desert	1							1	2375.08							2375.08
SDESi Subtropical Inland Hot Semi-Desert	692				45			737	270,643.25				208,338.21			478,981.45
NK Nama-Karoo	692				45			737	270,643.25				208,338.21			478,981.45
SDESc Subtropical Coastal Hot Semi-Desert	1339				6			1345	80,950.02				2010.01			82,960.03
SK Succulent Karoo	1339				6			1345	80,950.02				2010.01			82,960.03
S1 Ethesial Zone	7724							7724	81,379.75							81,379.75
MEDs Southern Ethesial Woodland	7724							7724	81,379.75							81,379.75
FYN Fynbos	7724							7724	81,379.75							81,379.75
T3 Warm-Temperate Zone	62,021		1	73			153	16,870	2531.62		0.51	186.41			125.11	2843.63
ATF Afrotemperate Forest	31,765		1	73			153	16,870	2531.62		0.51	186.41			125.11	2843.63

(continued)

Table 2.2 (continued)

	Polygon count								Area							
	ZA	LT	SZ	MZ	NA	BW	ZW	Total	ZA	LT	SZ	MZ	NA	BW	ZW	Total
ATFs Southern African Afrotemperate Forest	16,643		1	73			153	16,870	2531.62		0.51	186.41			125.11	2843.63
A2 Subtropical Alpine Zone	162	13						175	1327.93	2796.65						4124.58
SAZs Southern Subtropical Alpine Biome	162	13						175	1327.93	2796.65						4124.58
DAG Drakensberg Alpine Grassland	162	13						175	1327.93	2796.65						4124.58
E2(savm)–T3 Subtropical Warm-Temperate Wet Ecotone	12,197	6820	75	63			346	19,501	219,412.70	27,651.84	3714.22	834.73			2627.77	254,241.26
AMG African Montane Grassland	12,197	6820	75	63			346	19,501	219,412.70	27,651.84	3714.22	834.73			2627.77	254,241.26
E2(sava)–[E2(savm)–T3]–S2 Subtropical Semiarid Seasonal–Semidesert Ecotone	1112	1						1113	109,128.24	4.92						109,133.15
DHG Dry Highveld Grassland	1112	1						1113	109,128.24	4.92						109,133.15

(continued)

Table 2.2 (continued)

	Polygon count								Area							
	ZA	LT	SZ	MZ	NA	BW	ZW	Total	ZA	LT	SZ	MZ	NA	BW	ZW	Total
S2–E1(sub) Subtropical Semiarid–Semidesert Ecotone	4582							4582	25,320.56							25,320.56
AT Subtropical Albany Thicket	4582							4582	25,320.56							25,320.56
S2(sdesc)–S1 Semidesert–Ethesial Coastal Ecotone	147							147	2892.36							2892.35
WCAS West Coast Arid Strandveld	147							147	2892.36							2892.35
S1–A4 Ethesial–Austro-Alpine Ecotone	63							63	87.58							87.58
CAP Cape Altimontane Fynbos	63							63	87.58							87.58
T3–[S2–E1(sub)] Semiarid Warm-Temperate Ecotone	26							26	335.52							335.51
CVT Southern Cape Valley Thicket	26							26	335.52							335.51

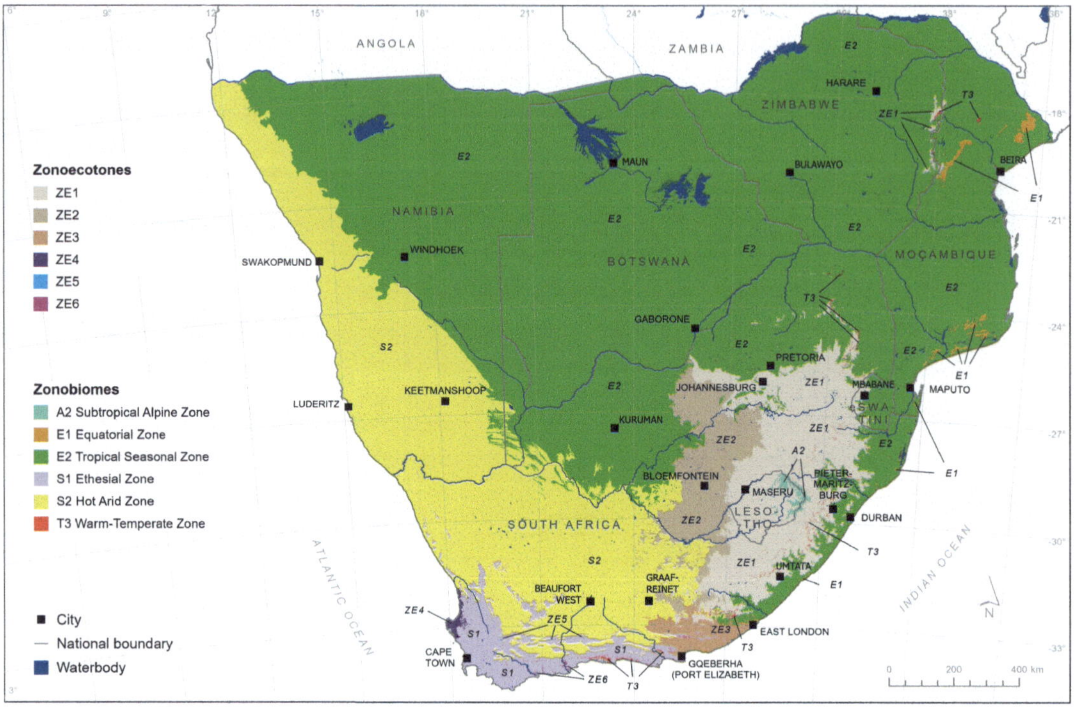

Fig. 2.1 The map of the zonobiomes (ZB) and zonoecotones (ZE) represented in Southern Africa. **ZE1**: E2(savm–T3) Subtropical–Warm-Temperate Wet Ecotone, **ZE2**: S1–A4 Ethesial-Austro-Alpine Ecotone, **ZE3**: S2(sdesc)–S1 Semidesert–Ethesial Coastal Ecotone, **ZE4**: S2–E1(sub) Subtropical Semiarid–Semidesert Ecotone, **ZE5**: T3–[S2-E1(sub)] Semiarid Warm-Temperate Ecotone, **ZE6**: E2(sava)—[E2(savm)–T3]–S2 Subtropical Semiarid Seasonal–Semidesert Ecotone

south as the Eastern Cape. The subtropical forests are discussed in Chapter 4.

Zonobiome E2 (original Walter's Zonobiome II) Tropical Seasonal Zone is by far the largest zonobiome in the MBSA area (hence Southern Africa). In Southern Africa, it is dominated by two global biomes—GB Mesic Savanna and GB Arid Savanna. In Rutherford et al. (2006a, b), these two biomes were classified as ‘Savanna Biome’. The third global biome (GB Tropical Dry Forest) of this zonobiome is dealt with in detail in Mucina et al. (2022) and Mucina (2023). The biomes of the tropical seasonal zone are described in Chapter 3.

Zonobiome S1 (original Walter's Zonobiome IV) Ethesial Zone is represented in the mapping area by the Fynbos Biome as defined by Rebelo et al. (2006). It encompasses several peinobiomes and psammobiomes, which are considered regional biomes in our present scheme. The biomes of this zonobiome are described in Chapter 5.

Zonobiome S2 (original Walter's Zonobiome III) Hot Arid Zone is also very diverse and represented in the mapping region by five continental biomes. For the first time, the Tankwa Desert has been singled out as a continental biome and hence excluded from the Succulent Karoo biome. The biomes of this zonobiome are described in Chapter 7.

Zonobiome T3 (original Walter's zonobiome V) Warm-Temperate Zone is characterised in Southern Africa by the afrotemperate forest. These were classified in Mucina et al. (2022) as putatively belonging to Walter’s zonobiome X, i.e. Oceanic Temperate Zone (Zonobiome T4 in Mucina 2023). However, it has subsequently been established (Mucina 2023) that these forests belong to the Zonobiome T3 (formerly called zonobiome V in Walter's terminology).

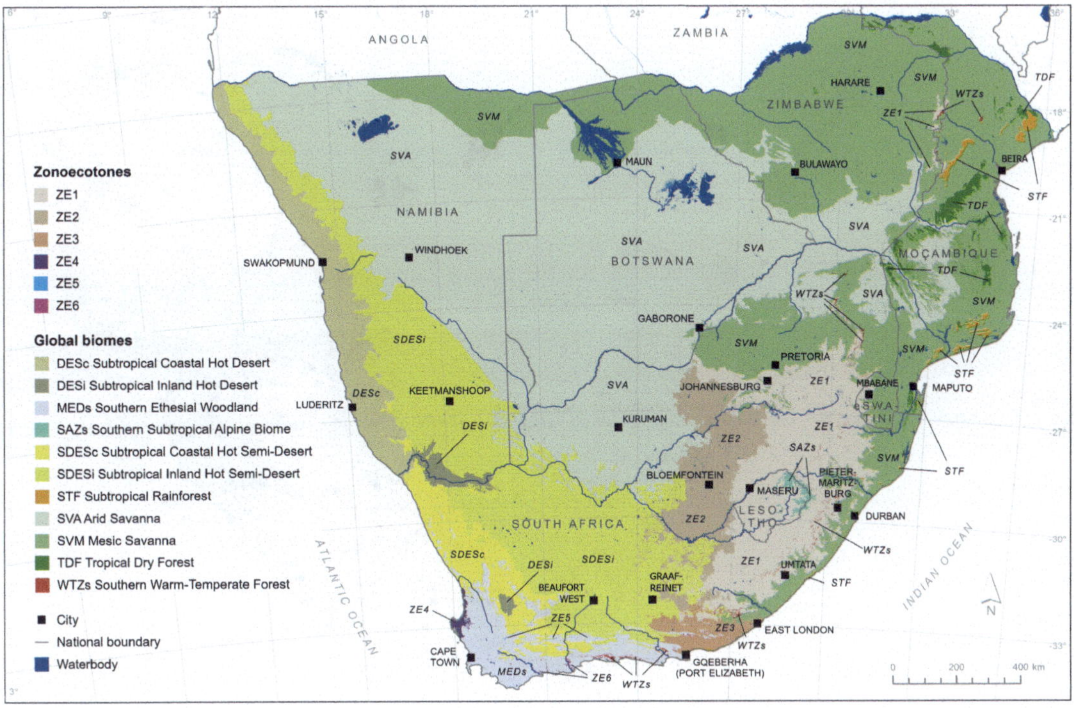

Fig. 2.2 The map of the global biomes (GB) of Southern Africa. For the full codes and names of the zonoecotones see the caption to Fig. 2.1

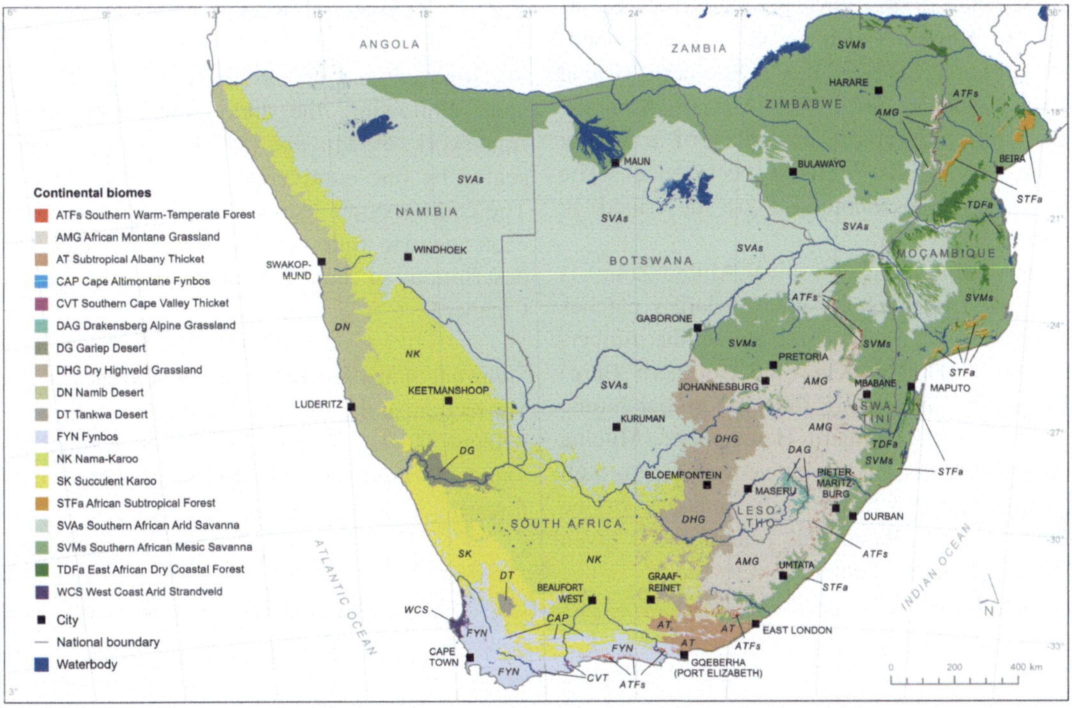

Fig. 2.3 The map of the continental biomes (CB) of Southern Africa

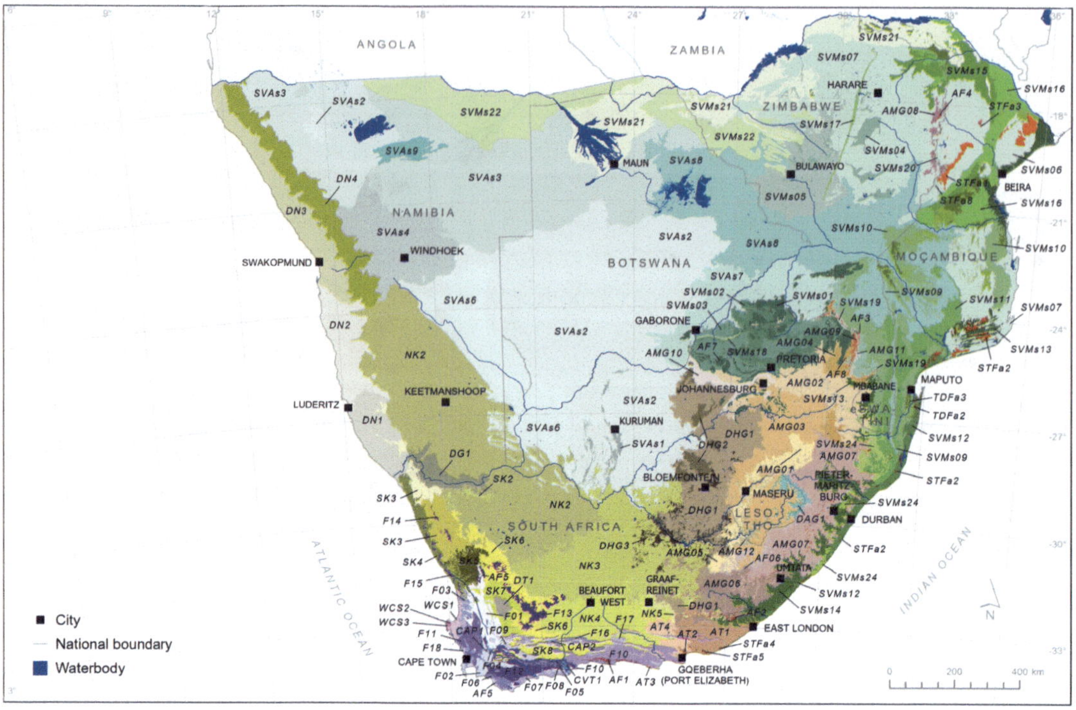

Fig. 2.4 The map of the regional biomes (RB) of Southern Africa

The patches of the afrotemperate forests outside the Knysna-Tsitsikamma region (embedded within the CB Fynbos, CB Southern African Mesic Savanna and CEB African Montane Grassland) are typical examples of a legacy biome (= a relict refugial occurrence). The biomes of this zonobiome are described in Chapter 6.

Zonobiome A2 (Subtropical Alpine Zone) is represented in the study area by larger patches of high-elevation grassland above the timberline in the Drakensberg (and Southern Berg) Mountains. Considering the new global biome scheme for the Southern Hemisphere (Mucina 2023), these grasslands are a geographic (as well as ecological and bioclimatic) analogue of the South American Puna. This zonobiome is given more attention in Chapter 8.

2.2 Zonoecotones

Six zonoecotones are recognised in the MBSA area. Following the hierarchical conceptual framework of Mucina (2023), each is represented by one continental zonoecotone biome. The coding symbolism designating these zonoecotones also follows Mucina (2023). Detailed descriptions of these zonoecotones are found in Chapter 9.

Subtropical Semiarid Ecotone S2–E1(sub)
We recognise the transitional nature of the CEB Albany Thicket as an example of a zonoecotone between the ZB S2 and ZB E1 (subtropical forest).

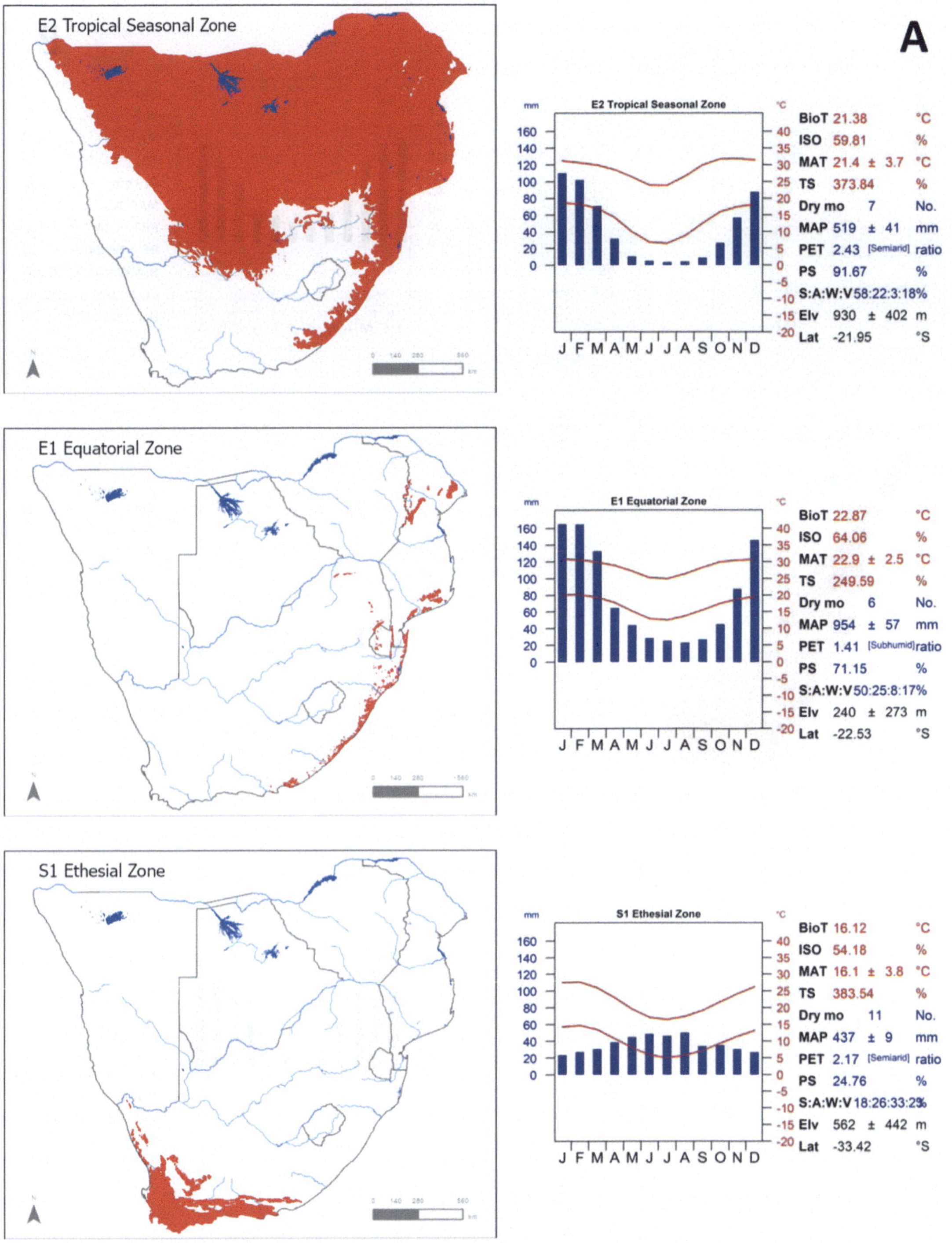

Fig. 2.5 The schematic maps and climate diagrams of all biomes at all hierarchical levels. **A** and **B** zonobiomes; **C** and **D** zonoecotones; **E** continental biomes within zonobiome E2; **F** and **G** continental biomes within zonobiome S2; **H** continental biomes within the zonobiome E2; **I** continental biome within the zonobiome E1; **J** continental biome within the zonobiome S1; **K** continental biome within the zonobiome T3; **L** and **M** continental biomes within the zonobiome S2; **N** continental biome within the zonobiome A2

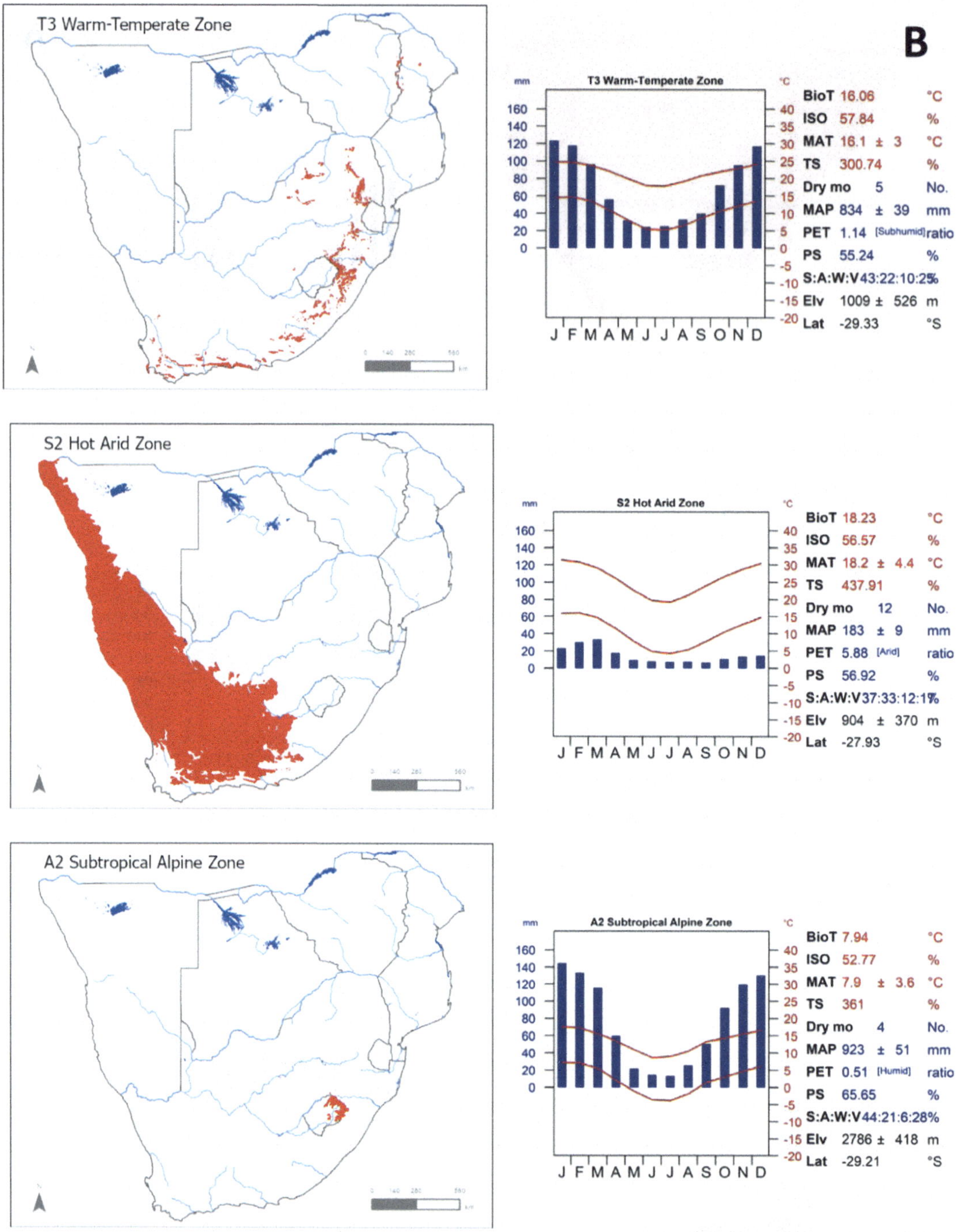

Fig. 2.5 (continued)

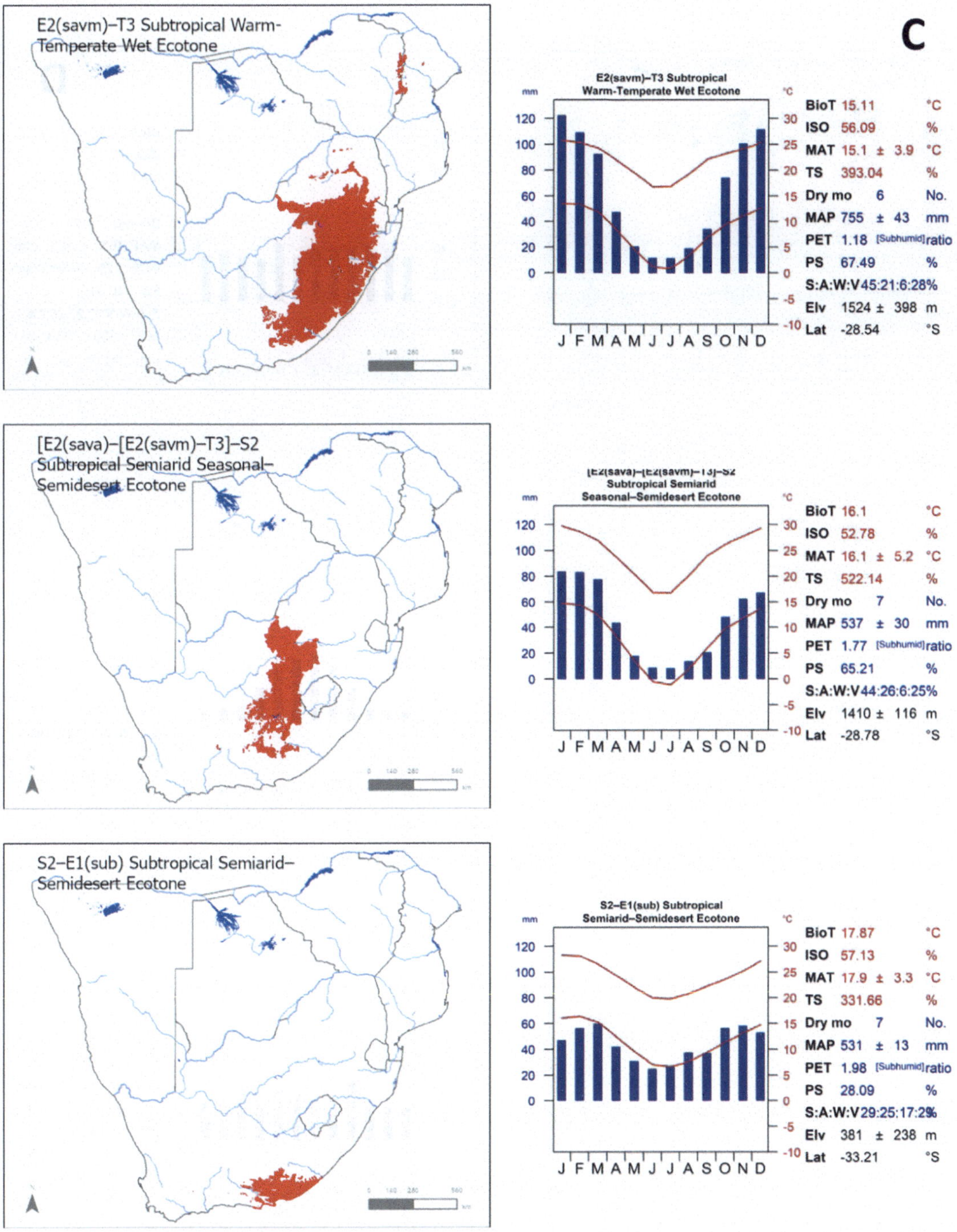

Fig. 2.5 (continued)

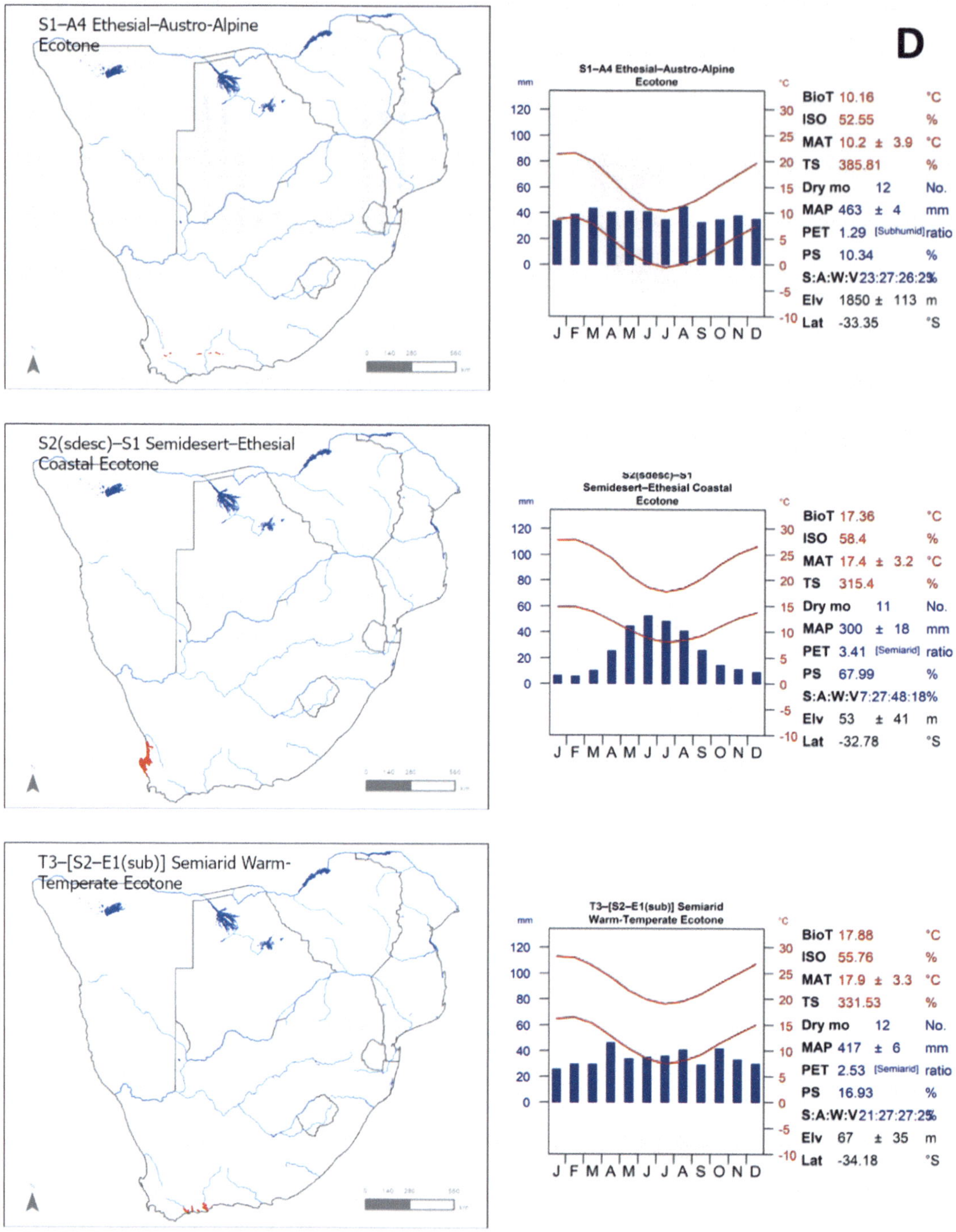

Fig. 2.5 (continued)

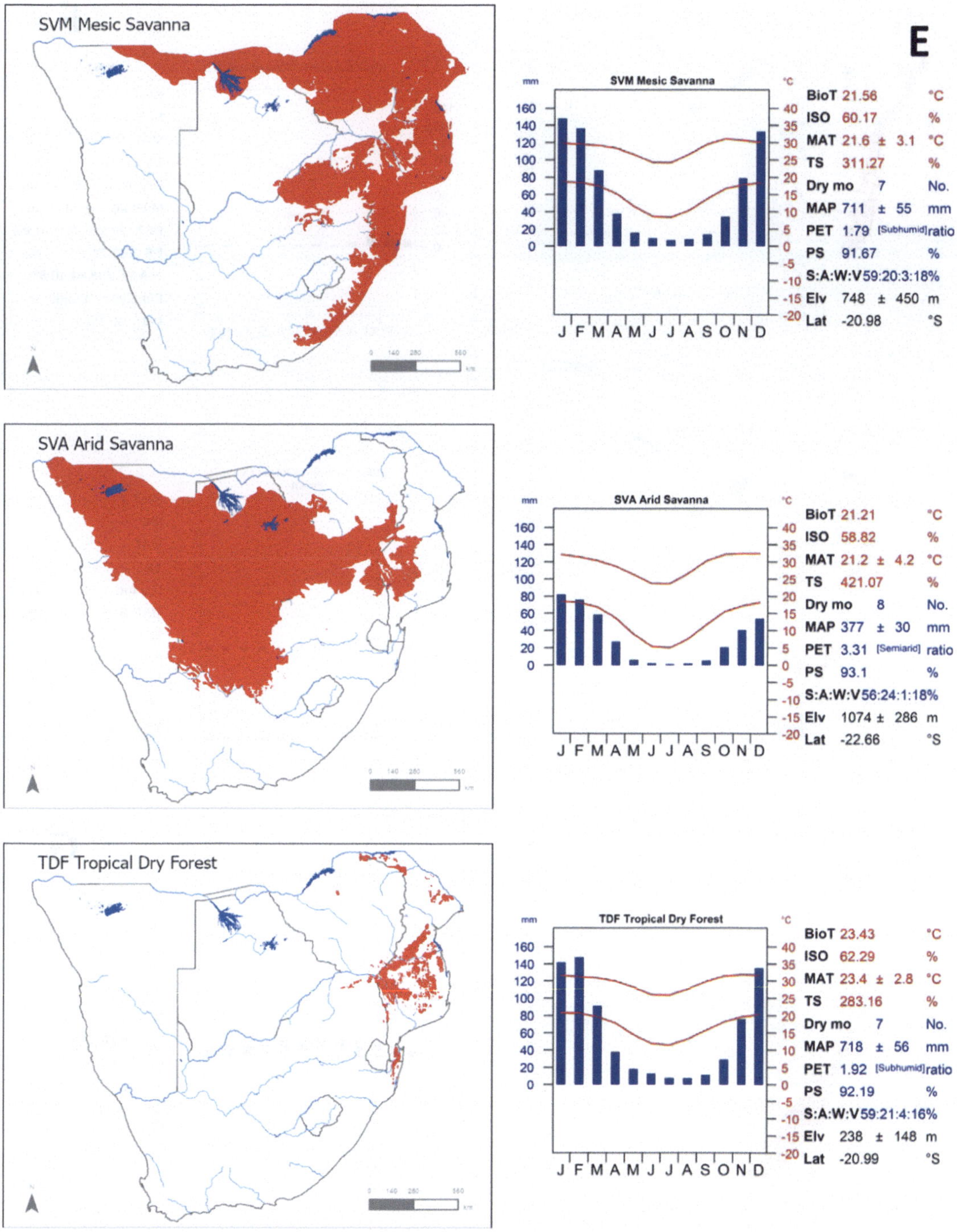

Fig. 2.5 (continued)

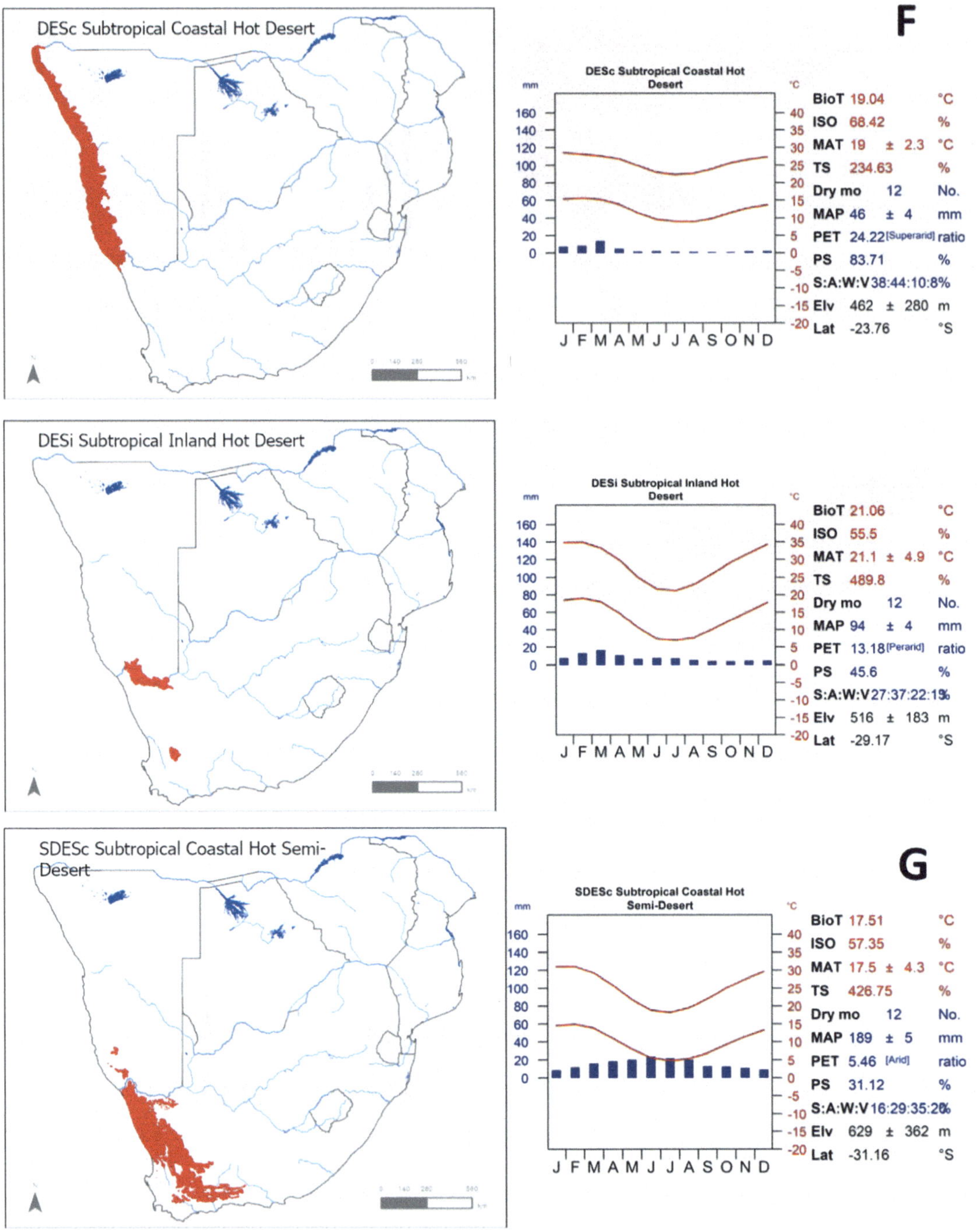

Fig. 2.5 (continued)

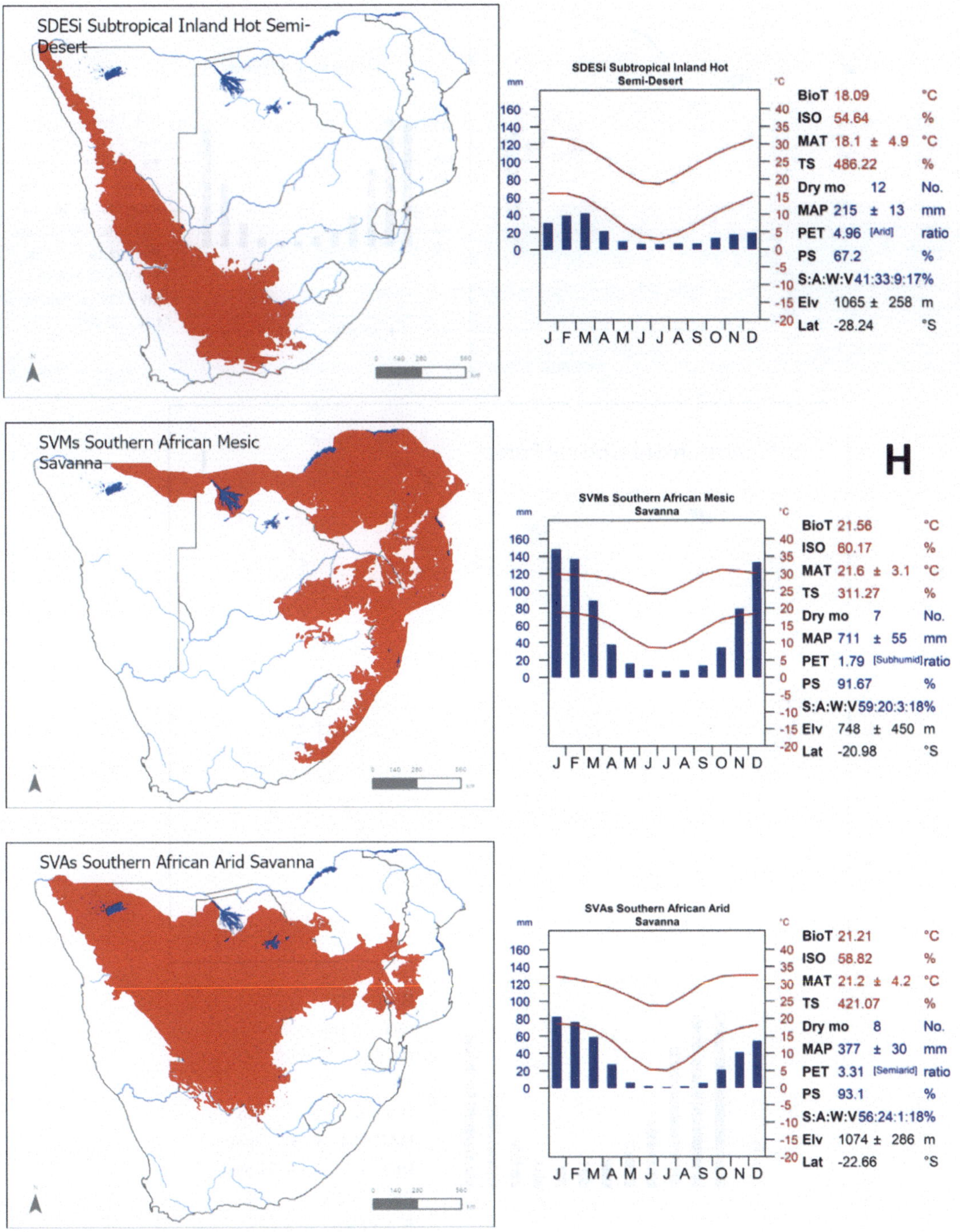

Fig. 2.5 (continued)

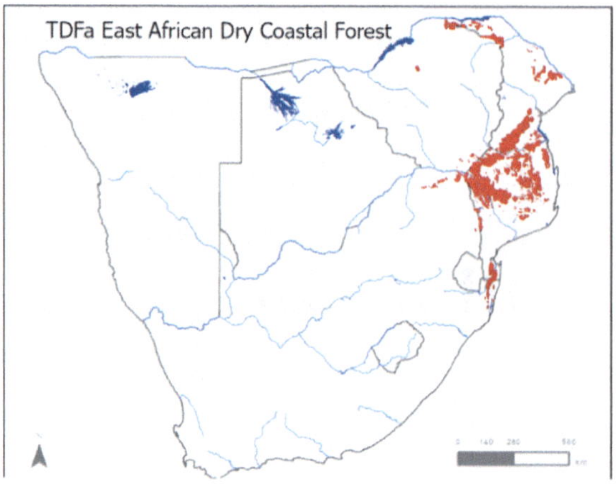

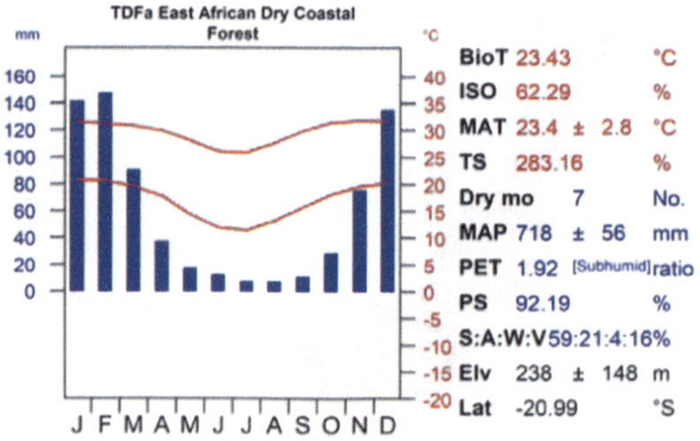

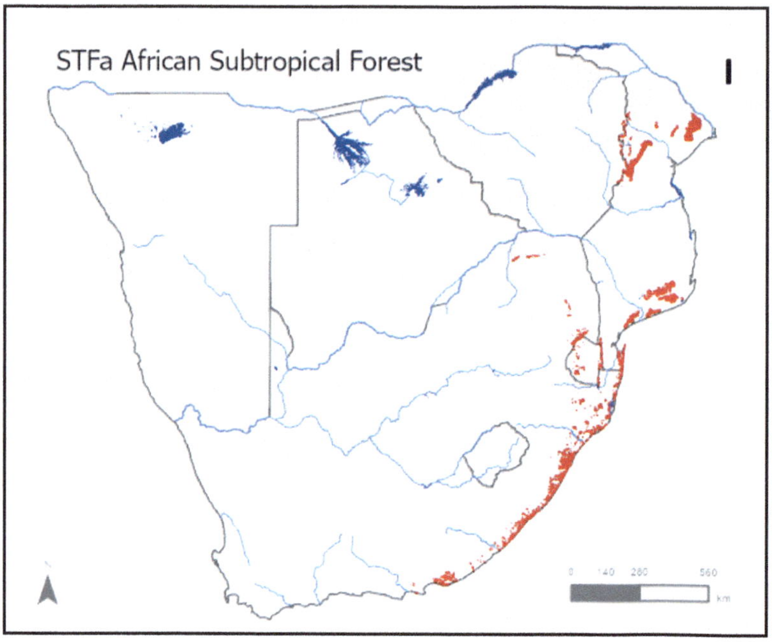

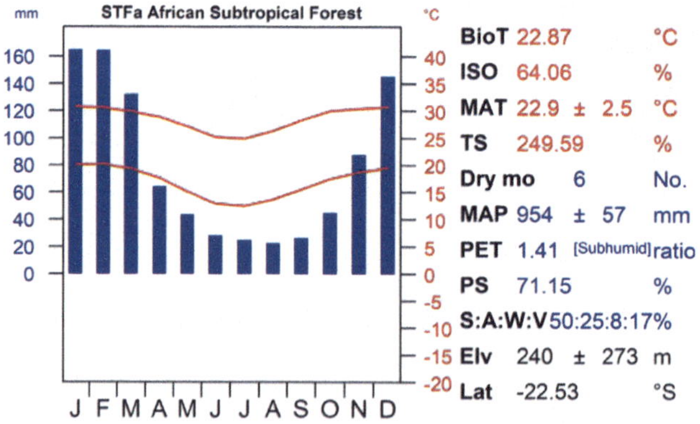

Fig. 2.5 (continued)

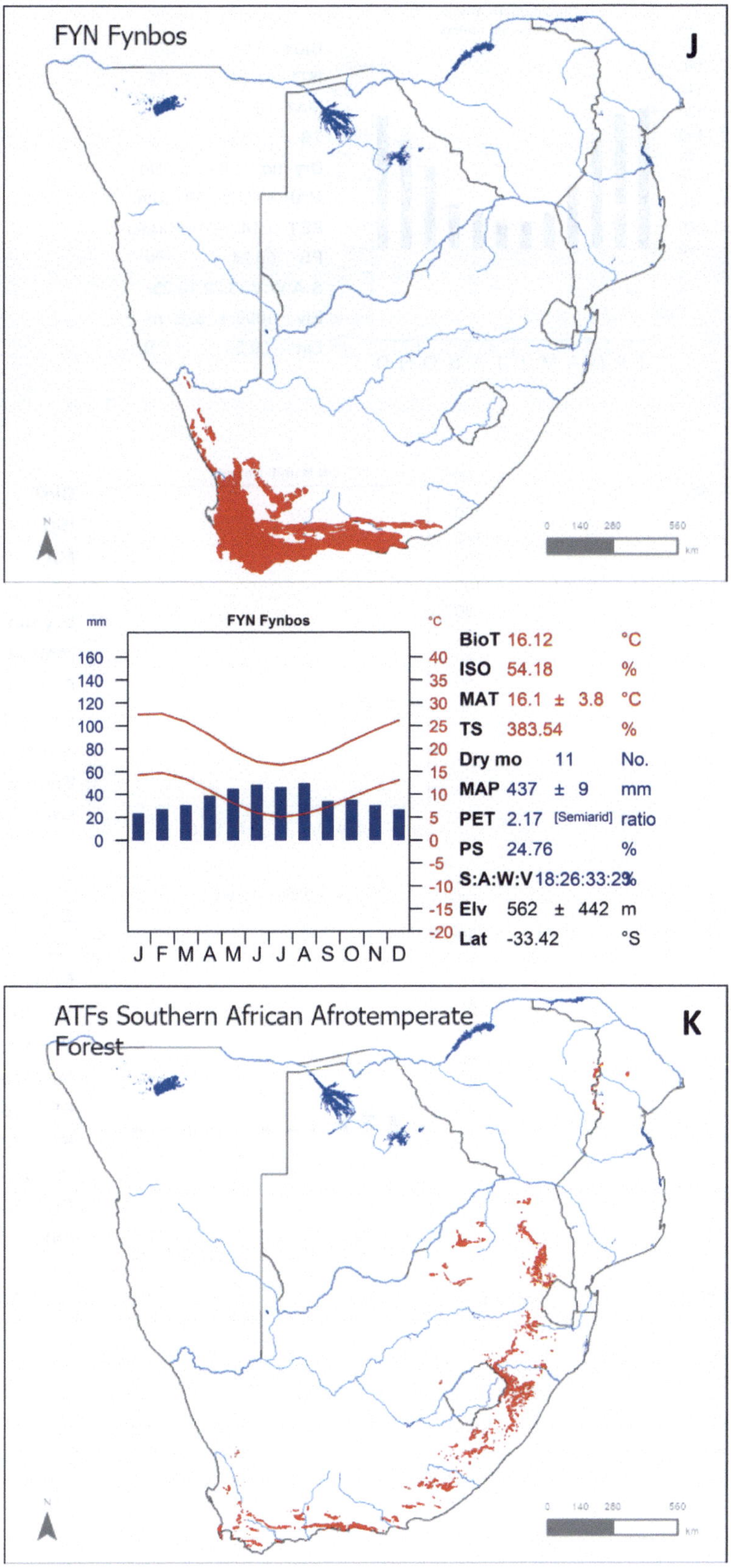

Fig. 2.5 (continued)

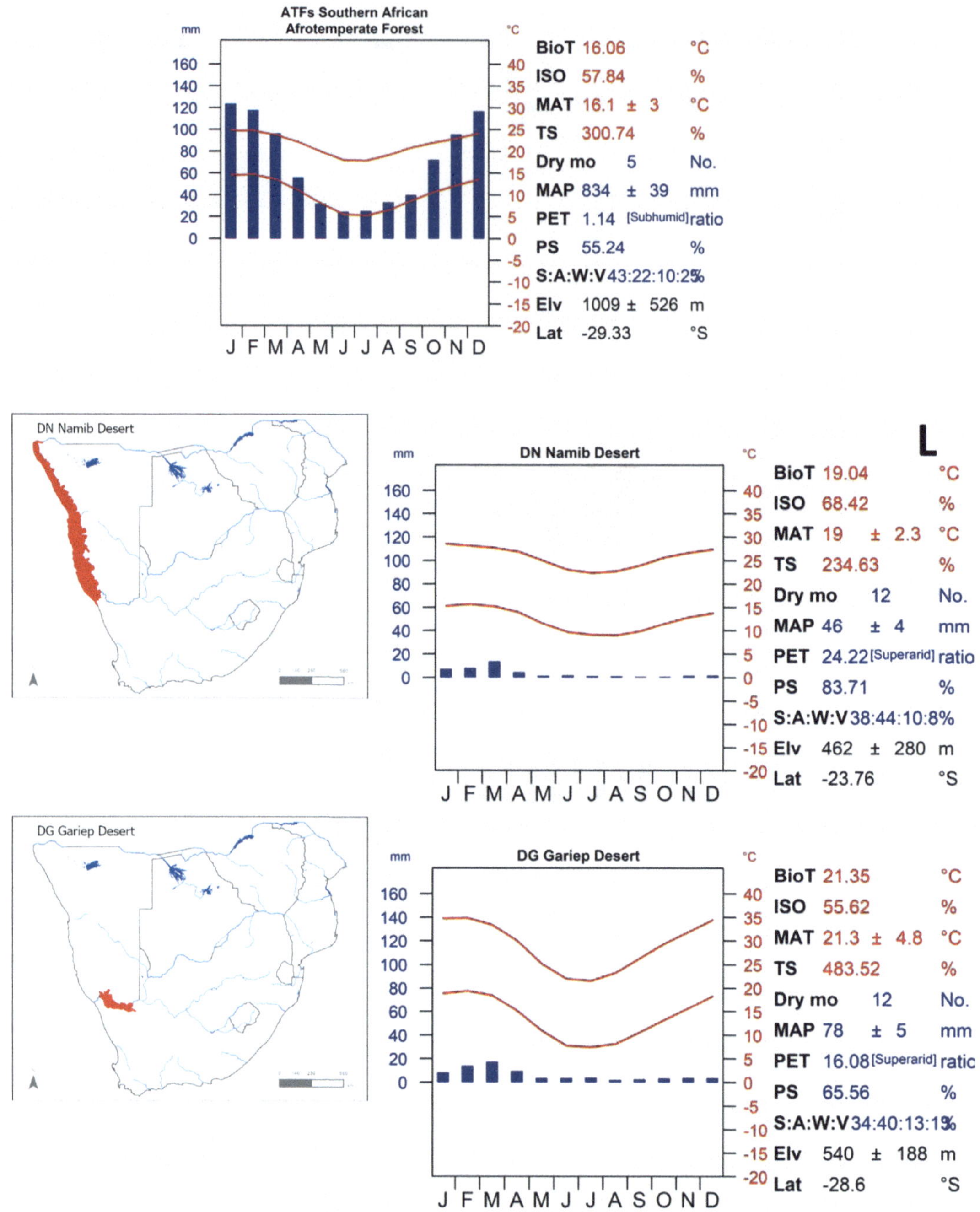

Fig. 2.5 (continued)

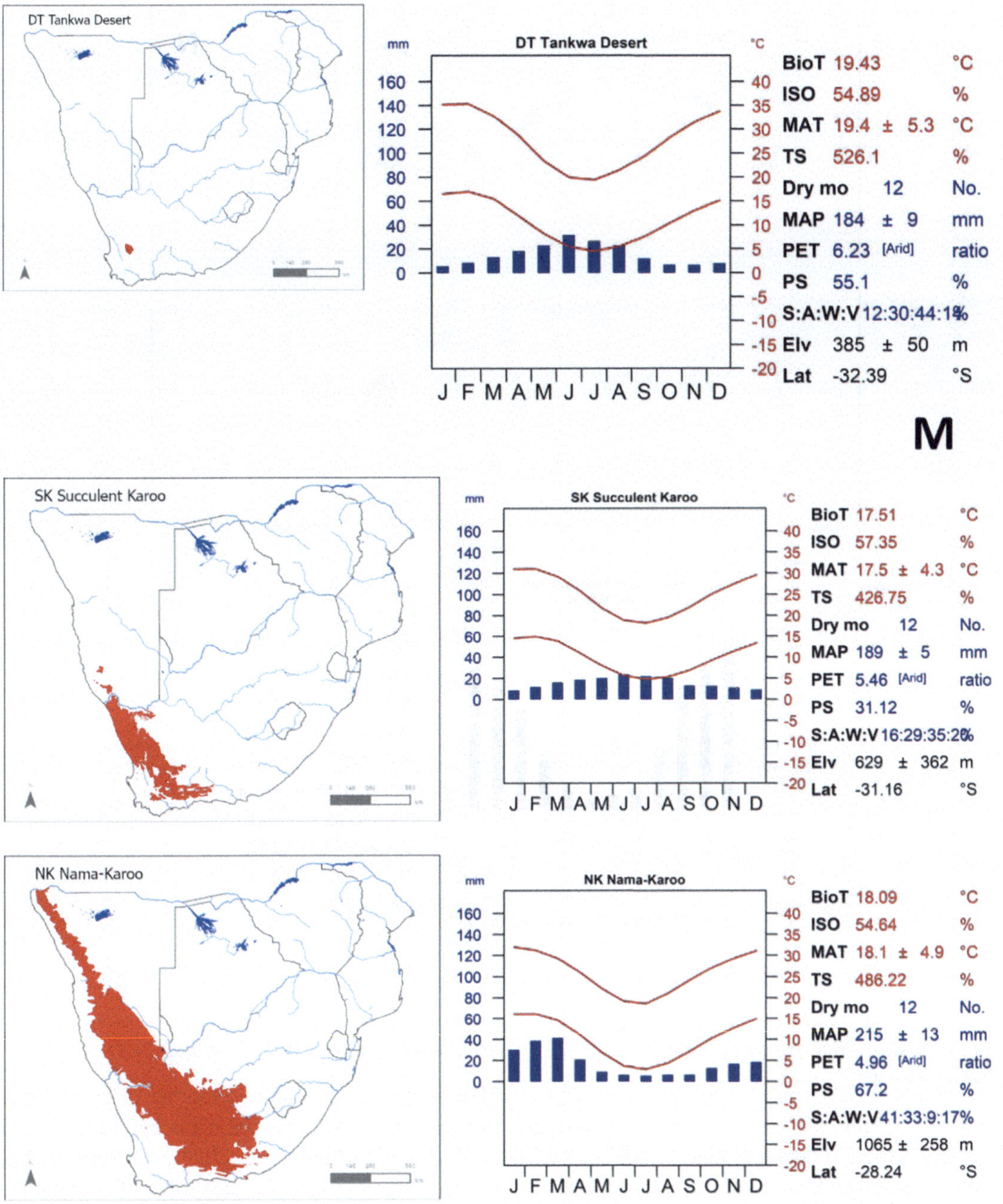

Fig. 2.5 (continued)

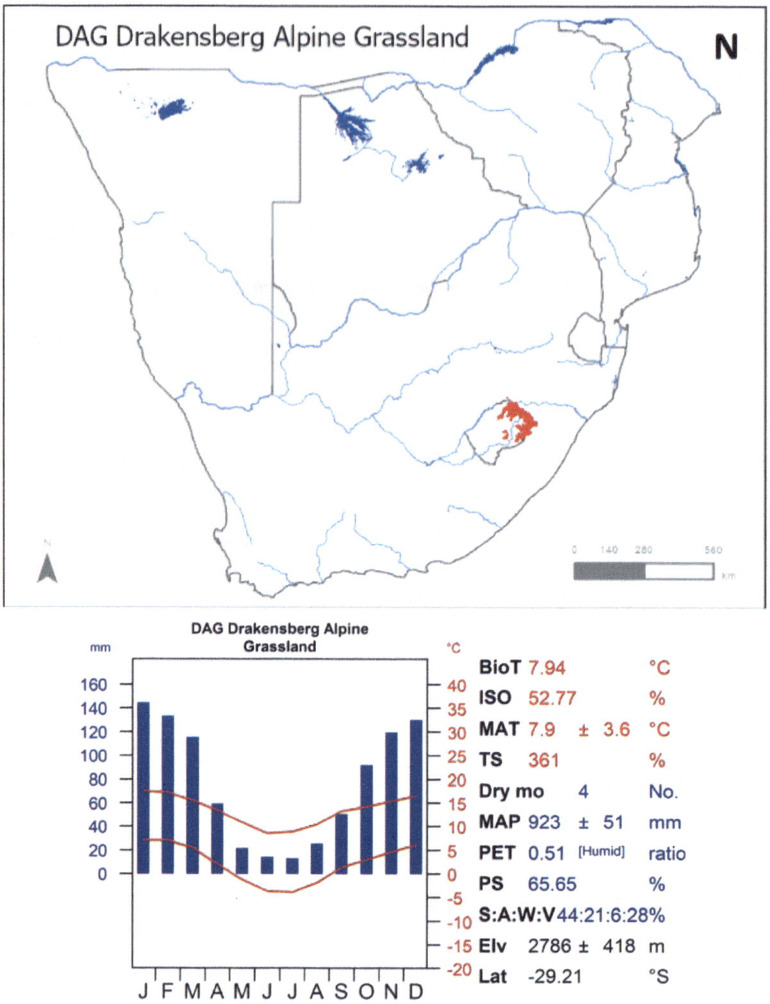

Fig. 2.5 (continued)

Semiarid Warm–Temperate Ecotone S2–E1(sub)–T3

This is a unique zonoecotone endemic to Southern Africa. It accommodates CEB Southern Cape Valley Thicket—a unit spatially embedded within the Fynbos Biome (member of the Zonobiome S1) but bioclimatically positioned between the Warm-Temperate Zone (Zonobiome T3) and the ecotonal Albany Thicket Biome (see above). The Southern Cape Valley Thicket shows a floro-genetic link to the latter.

Desert–Mediterranean Ecotone S2–S1

This ecotone found in the Southern Hemisphere, also in Chile (Mucina 2023; Luebert et al. 2024) and Australia (Mucina 2023), is represented in the MBSA area by a series of vegetation types of the driest sections of the West Coast of South Africa. Formerly, it was known as the Western Strandveld (Rebelo et al. 2006). The corresponding CEB here is the West Coast Arid Strandveld.

Austro-Oromediterranean–Austro-Alpine Ecotone S1–A4

This is a unique zonoecotone endemic to Southern Africa. It has been coined to accommodate the CEB Altimontane Fynbos bioclimatically and floro-genetically, straddling the zonobiomes S1 and A4.

Subtropical–Warm Temperate Wet Ecotone E2(savm)–T3

This zonoecotone is represented in the study area by the warm-temperate mesic (high precipitation) grassland continental ecotonal biome CEB African Montane Grassland dominated by C4 grasses. Sub-Escarpment Grasslands (Mucina et al. 2006) also belong here, as do the minor archipelagos of high-altitude grassland patches on the Great Escarpment north of the Limpopo River (Chimanimani, Nyanga, Gorongosa). These are dominated by C4 grasses and fall (due to high elevation) within the warm-temperate climatic envelope.

Subtropical–Warm-Temperate Dry Ecotone E2(sava)–E2(savm)/T3–S2

This ecotone is found on the interface of the CB Southern African Arid Savanna, Subtropical-Warm Temperate Wet Ecotone, and CB Nama-Karoo. It covers the arid section of the former broadly conceived Grassland Biome (Mucina et al. 2006). The corresponding CEB is the Dry Highveld Grassland.

2.3 Comparison with Previous Biome Classifications in Southern Africa

Probably the most-cited previous works directed specifically at biomes of southern Africa are Rutherford and Westfall (1986, 1994) and its successor, Rutherford et al. (2006a), where the biome scheme was adopted by the National Vegetation Map of South Africa (Mucina and Rutherford 2006). The much-quoted work of Low and Rebelo (1996, 1998) does not provide a rationale for their biomes of southern Africa other than merely quoting Rutherford and Westfall (1986) and some other sources. The work of Rutherford and Westfall (1986) was seeded by a seminar given a year or two earlier by the senior author in Pretoria on the strong distinctions between Succulent Karoo and what Rutherford and Westfall (1986) later named Nama-Karoo. The 1986 work undertook to provide explicit criteria and definitions to substantiate and modify, where necessary, the general biome framework provided by, notably, Huntley (1984) for the area of southern Africa south of 22° S. This area was later extended to include the whole of Namibia and Botswana (Rutherford 1997). The criteria used in 1986 were simple: primarily, the relative dominance of defined plant life forms (i.e., a fundamentally physiognomic scheme) and, secondarily, closely associated climatic factors in which seasonal availability of water was key. Explicit scale requirements deliberately excluded minor occurrences of biomes.

In Rutherford et al. (2006a) more complex criteria were applied which included floristic and additional environmental factors, as well as a relaxation of scale requirements. In this way savanna along the eastern seaboard of South Africa was reinterpreted as the Indian Ocean Coastal Belt and Thicket Biomes and the hyperarid part of the lower Orange River valley, which had been split between a coastal hyperarid part of the Succulent Karoo Biome and interior Desert Biome in Rutherford (1997), were both consigned to Desert.

In Fig. 2.6, we briefly compare the biomes of the most recent of the above accounts, namely Rutherford et al. (2006a), with our current continental biomes, which reflect the closest rank to the 2006 work. The only biome of Rutherford et al. (2006a) which becomes entirely subsumed is the Indian Ocean Coastal Belt. In the current work this much transformed entity reverts to falling within the Mesic Savanna (Savanna Biome of Rutherford and Westfall 1986) and Subtropical Forest Biome. With the exception of the Nama-Karoo Biome, all the other biomes of 2006 are divided more finely at the continental biome level. Some of these splits, such as Desert into Namib Desert and

Gariep Desert, were recognised at the bioregion level in 2006. The division of the Savanna Biome into Mesic and Arid Savannas also corresponded to the groupings of bioregions in 2006. The Dry Highveld Bioregion was re-evaluated as forming the ecotonal Dry Highveld Grassland (see above), and the other Grassland bioregions were suitably adjusted and reinterpreted. On an area basis, the vast majority of the Fynbos Biome of 2006 persists in the current work at the zonal continental biome level, with the newly recognised two important ecotonal continental biomes contributing a very small proportion of the total area. The remaining most significant changes to the 2006 work are the complete reappraisal of the former Grassland and Thicket Biomes, a far more detailed and comprehensive approach to Forests, and the recognition of Tankwa Desert as separate from Succulent Karoo. Of course, this comparison (Fig. 2.6) necessarily ignores a major difference, namely, that the current work also covers over two and a half times the area covered in 2006. The 2006 work had a regional subcontinental context and approach, whereas the current work strives to place all units, large and small, firmly into a global context.

In order to facilitate the correspondence between our biome system and the vegetation units of the VegMap2006 (VM06; Mucina and Rutherford 2006), VegMap2018 (VM18; Dayaram et al. 2019) and the VUs of the Mozambique vegetation map (VegMapMoz; Lötter et al. 2023), we have created a crosswalk available as a supplementary file Ch2 Mucina&Rutherford SUPPL Table S2.2 https://doi.org/10.1007/978-3-031-63731-5_2.

In addition to the earlier subcontinental studies, there are also numerous global classifications of natural terrestrial ecosystems, of which Walter and Box (1976) is to a fair extent conceptually aligned with the current work, but less so in its implementation in southern Africa. Most global maps understandably suffer from a lack of detail or even gross approximations at regional or subcontinental scales. The global zonobiome and zonoecotone map published by Walter and Box (the insert map in 1976) shows (in a very stylistic and simplistic way) four zonobiomes and two zonoecotones for the MBSA:

Zonobiome II (Tropical Seasonal) covers the largest part, despite its dry wing being extracted as a part of the Zonoecotone III–II). The Zonobiome II almost entirely covers Zimbabwe, the whole of southern Mozambique and in South Africa it also encompasses the high-elevation Highveld grasslands. Hence, there is an obvious mismatch (for details, see below) between Walter's and our zonobiome concepts in the study area.

Zonobiome III (Hot Arid) of Walter & Box probably comes the closest to matching our concept of Zonobiome S2 as it encompasses the Namib and Gariep Deserts as well as the entire Succulent Karoo and most of the Nama-Karoo semideserts.

Zonobiome IV (Ethesial) is limited to the westernmost reaches of what is usually considered Fynbos Biome (Rebelo et al. 2006), representative of our Zonobiome S1.

Zonobiome V (Warm-Temperate) is mapped along the seaboards of the East Coast of South Africa, spanning the Eastern Cape and the border region between South Africa and Mozambique. We consider this concept as lacking justification even when Walter's bioclimatic criteria (see Box 2016 for a recent account) are applied. This region is bioclimatically not warm-temperate; it exhibits a subtropical nature (see Chapter 4 in this book).

The first of the two zonoecotonal units suggested by Walter and Box in our study is contentious. Zonoecotone IV–III, which is expected to straddle the Ethesial Zonobiome IV and Hot Arid Zonobiome III, is also recognised in our conceptual framework (as ZE S1–S2). However, its position as mapped by Walter is dubious. In fact, only its westernmost parts, i.e., the narrow stretch along the West Coast of the Atlantic Ocean, qualify as such in our conceptual framework. The remainder belongs either to Zonobiome IV or Zonobiome II (southern reaches of the Nama Karoo). The other zonoecotonal unit, Zonoecotone III–II, largely coincides with our concept of GB Arid Savanna.

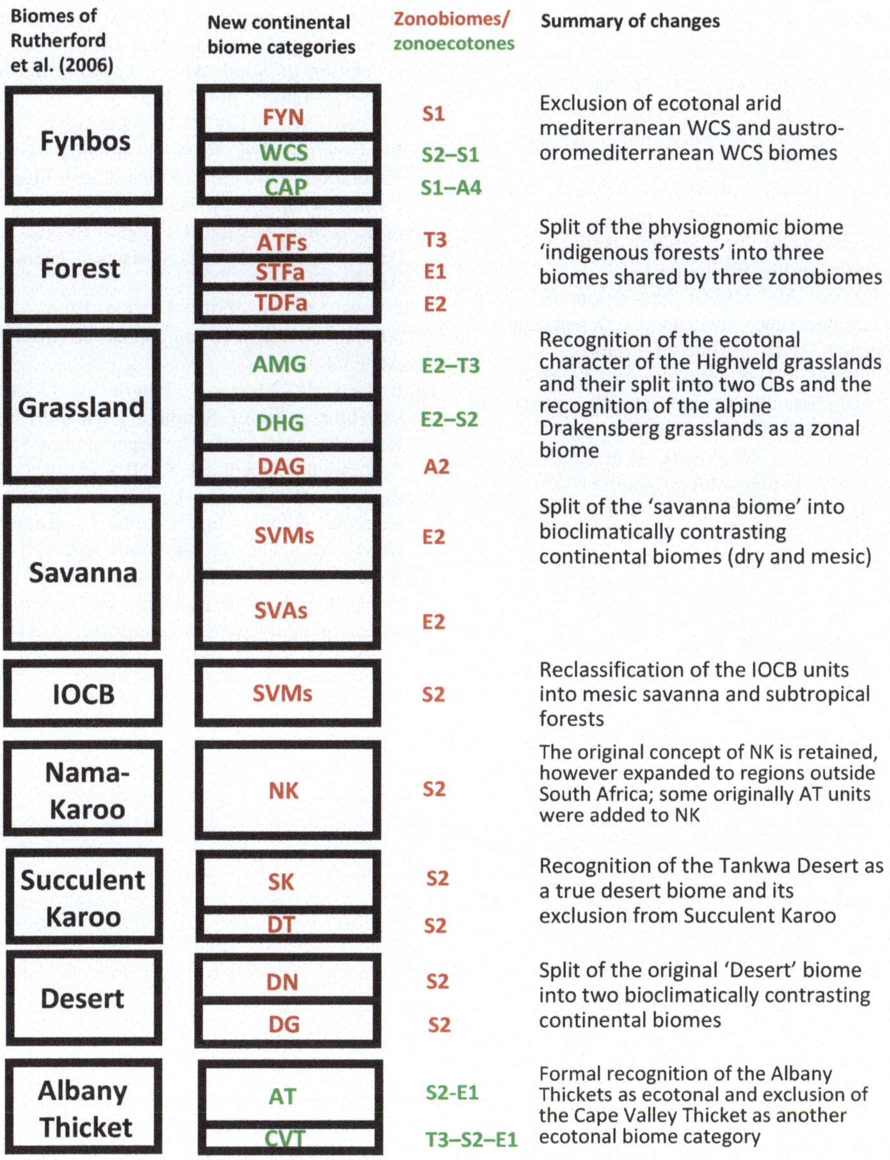

Fig. 2.6 Crosswalk between the biome mapping categories of Mucina and Rutherford (2006) and the new biome system at the CB level

In conclusion, this schematic zonobiome/zonoecotone map of Walter and Box (1976) proved to have limited heuristic value for our purposes. Rutherford and Westfall (1986) also decided against using this 1976 map as a point of departure for their work.

Data Availability The shapefile which served for production of the biome maps is available in the supplementary file called Ch2 Mucina&Rutherford SUPPL Shapefile https://doi.org/10.1007/978-3-031-63731-5_2.

References

Box EO (2016) World bioclimatic zonation. In: Box EO (ed) Vegetation structure and function at multiple spatial, temporal and conceptual scales. Springer International, Cham, pp 3–52

Huntley BJ (1984) Characteristics of South African biomes. In: Booysen PdeV, Tainton NM (eds) Ecological effects of fire in South African ecosystems. Springer-Verlag, Berlin, pp 1–17

Low AB, Rebelo AG (1996) Vegetation of South Africa, Lesotho and Swaziland. Department of Environmental Affairs & Tourism, Pretoria

Low AB, Rebelo AG (1998) Vegetation of South Africa, Lesotho and Swaziland, 2nd ed. Department of Environmental Affairs & Tourism, Pretoria

Luebert F, Pliscoff P, Tsakalos JL et al (2024) Zonal biomes of Chile (in preparation for submission)

Mucina L (2023) Biomes of the Southern Hemisphere. Springer Nature Switzerland, Cham

Mucina L, Hoare DB, Lötter MC et al (2006) Grassland biome. In: Mucina L, Rutherford MC (eds) The vegetation of South Africa, Lesotho and Swaziland. SANBI, Pretoria, pp 348–437

Mucina L, Lötter MC, Rutherford MC et al (2022) Forest biomes of Southern Africa. New Zeal J Bot 60:377–428

Rebelo AG, Boucher C, Helme N et al (2006) Fynbos biome. In: Mucina L, Rutherford MC (eds) The vegetation of South Africa, Lesotho and Swaziland. SANBI, Pretoria, pp 52–219

Rutherford MC (1997) Categorisation of biomes. In: Cowling RM, Richardson DM, Pierce S (eds) Vegetation of southern Africa. Cambridge University Press, Cambridge, pp 91–98

Rutherford MC, Westfall RH (1986) Biomes of southern Africa—an objective categorisation. Mem Bot Surv S Afr 54:1–98

Rutherford MC, Westfall RH (1994) Biomes of southern Africa—an objective categorisation. Mem Bot Surv S Afr 63:1–94

Rutherford MC, Mucina L, Powrie LW (2006a) Biomes and bioregions of Southern Africa. In: Mucina L, Rutherford MC (eds) The vegetation of South Africa, Lesotho and Swaziland. SANBI, Pretoria, pp 30–51

Rutherford MC, Mucina L, Lötter MC et al (2006b) Savanna Biome. In: Mucina L, Rutherford MC (eds) The vegetation of South Africa, Lesotho and Swaziland. SANBI, Pretoria, pp 438–539

Walter H, Box E (1976) Global classification of natural terrestrial ecosystems. Vegetatio 32:75–81

Southern African Savannas and Dry Forests

3

Michael C. Rutherford, Mervyn C. Lötter, Ladislav Mucina, John E. Burrows, James L. Tsakalos and Leslie W. Powrie

Abstract

The biomes of the zonobiome E2 Tropical Seasonal Zone occupy the largest portion of the tropical and subtropical regions of the world. This zone is also the dominant zonobiome of the studied MBSA. Physiognomically, the biomes of this zone are represented by savanna grasslands and woodlands, as well as Tropical Dry Forests (TDF). In bioclimatic terms, these biomes are characterised by alternation of prolonged dry and wet periods. In the study region we recognise Mesic and Arid Savanna at the rank of global biome, each comprising 25 and 9 regional biomes, respectively. While the savanna units are functionally underpinned by the domination of highly productive C4 grasslands, the TDF is characterised by an overall lack of grassy (or shrubby) understorey beneath a (semi)closed canopy of trees of predominantly low stature. There are two regional biomes of TDF recognised within the study area, namely Southern African Dry Forest and Southern African Dry Thicket. This chapter presents descriptions of all regional biomes of the zonobiome E2.

Keywords

Arid savanna · Azonal regional biomes · Bioclimatic thresholds · Kalahari · Lowveld · Mesic savanna · Miombo · Mopane · Seasonal tropical grasslands and woodlands · Thornveld · Tropical dry forest · Zonal regional biomes

M. C. Rutherford (✉)
High Peak, Derbyshire, United Kingdom

Department of Botany & Zoology, Stellenbosch University, Private Bag X1, Matieland, Stellenbosch 7602, South Africa

M. C. Lötter
Mpumalanga Tourism and Park Agency, P/Bag X1088, Lydenburg 1120, South Africa

School of Animal, Plants and Environmental Sciences, University of the Witwatersrand, Private Bag 3, Wits, Johannesburg 2050, South Africa

L. Mucina · J. L. Tsakalos
Harry Butler Institute, Murdoch University, 90 South Street, Murdoch, Perth 6150, Australia

L. Mucina · L. W. Powrie
Department of Geography & Environmental Studies, Stellenbosch University, Private Bag X1, Matieland, Stellenbosch 7602, South Africa

J. E. Burrows
365 St Helena Road, Calitzdorp 6660, Western Cape, South Africa

J. L. Tsakalos
Plant Diversity and Ecosystems Management Unit, School of Biosciences and Veterinary Medicine, University of Camerino, Via Pontoni 5, Camerino 62032, Italy

L. W. Powrie
Kirstenbosch Research Centre, South African National Biodiversity Institute, Private Bag X7 Claremont Cape Town, South Africa

L. Mucina and M. C. Rutherford (eds.), *Zonal Biomes of Southern Africa*, Biome Ecology 2,
https://doi.org/10.1007/978-3-031-63731-5_3

The Southern African Savannas form part of global-level Savanna Biomes which collectively constitute the zonobiome known as the 'Tropical Seasonal Zone' (designated as E2) which is dominated by 'savanna' (or sometimes spelled 'savannah') vegetation. Many authors appropriately limit the use of the term savanna to apply to *tropical and subtropical woodlands* with grassy undergrowth (see Bond 2019 for a recent source). However, this term is misplaced if used to designate structurally similar sparse woodlands in temperate regions of the World (e.g., Volder et al. 2013; see the issue of 'Protea savanna' discussed in Mucina et al. 2006a). Seminal volumes on global savannas, that also cover various ecological aspects of southern African savannas, include Bourliére (1983), Huntley and Walker (1982), Tothill and Mott (1985) and Walker (1987).

Savannas are typically two-layered, comprising a discontinuous tree layer overtopping a continuous or near-continuous grass layer dominated by C4 grasses (Lehmann et al. 2011). An intermediate layer of shrubs or small trees is sometimes present. The savanna climate is characterised by a hot wet season alternated with a warm, dry season. The combination of the near-continuous grass layer and a regular dry season results in the fuel load and flammability needed for the characteristic fires of the savanna. It is consequently unsurprising to find a flora well-adapted to fire and to find, for example, that the protective geoxylic habit in Africa is associated with the savanna biome (Maurin et al. 2014).

Savanna vegetation is largely determined by climate, landscape position, and chronic disturbances, including those arising from fire and megaherbivore grazing (Asner et al. 2009) and human dependencies. One of the best demonstrations in the region of the effect of climate is found along the gradient of rainfall on a relatively uniform sand substrate along the Kalahari transect. This transect effectively captures the increasing structural complexity and species changes from the arid southwest end of the transect in the Northern Cape Province of South Africa to the northeast of Botswana and Caprivi area of Namibia, and outside our mapping area, into Zambia (Scholes et al. 2002). Disturbance is considered natural in savanna and results in strong local variation and changes in canopy vegetation cover and height. Fire and the actions of megaherbivores can have significant effects on vegetation canopy cover and height (Asner and Levick 2012). Local rural inhabitants depend on the savanna resources (Campbell et al. 1996), and the resultant utilisation pressures can strongly affect vegetation structure and composition, with, for example, woodland being transformed into shrubland (Rutherford et al. 2012). The local-level spatial organising processes between tree and grass layers (Scholes 1997) can also affect the final form of the structure of savanna vegetation (also at other scales; see Staver 2018).

Africa hosts the most significant portion of savannas on a global scale. Here, this large spatial extent is found to translate into substantial savanna heterogeneity, both bioclimatically and biogeographically.

In the study region (MBSA), we recognise three global level biomes within the Tropical Seasonal Zonobiome E2, namely GB Mesic Savanna, GB Arid Savanna, and GB Tropical Dry Forest (TDF). The savanna global biomes are represented by one continental biome each—CB Southern African Mesic Savanna and CB Southern African Arid Savanna, respectively. One of the several important bioclimatic distinctions between the Mesic and Arid Savanna in Southern Africa is the dry season's duration, which is, as defined in Chap. 1, at least eight months long for Arid Savanna. The position of the marginal arid savanna regional biomes such as SVAa7 and SVAs8 requires more nuanced consideration of other bioclimatic measures. The arid and moist savanna biomes of Huntley (1982) in Southern Africa correspond relatively closely to our two continental biomes, with his mesic savanna receiving a MAP of more than 650 mm but with a lower limit of less than 600 mm on sands in the south. Sankaran et al. (2005), in their Africa-wide savanna analysis, also indicate a MAP breakpoint of 650 mm between these arid and mesic savanna systems, but they emphasise that this lies within a broad range lying between 514 and 784 mm.

The general distribution and ecology of the TDF, and its distinction from the savanna global biomes, have been handled by Mucina et al. (2022). Figure 3.1 summarizes these aspects in synoptic manner.

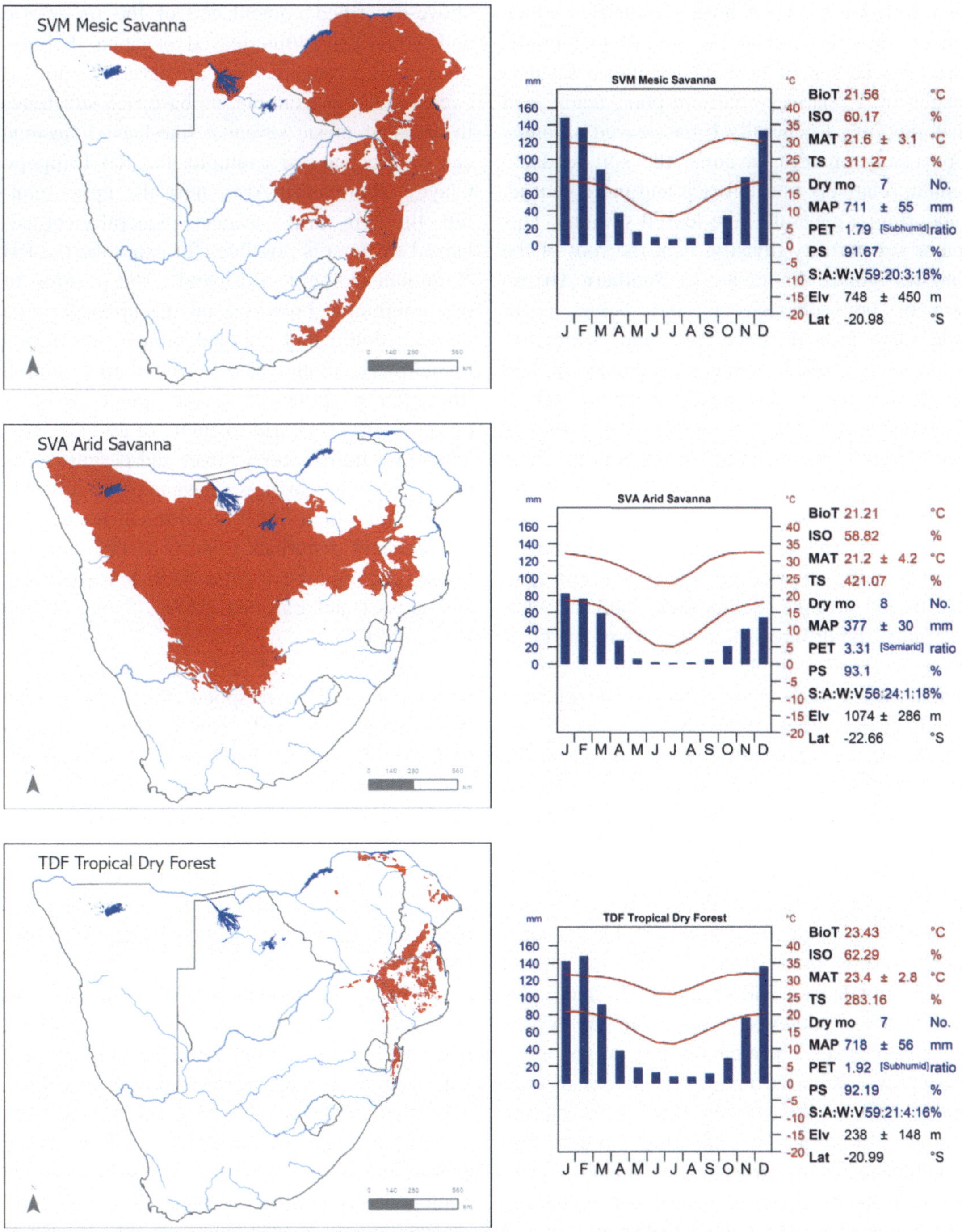

Fig. 3.1 The distribution areas and the climatic diagrams of the global biomes within the zonobiome E2. The construction of the climatic diagrams and the distribution maps of the biomes is described in Chap. 1

Our arid and mesic savanna continental biomes have long been associated with several critical functional differences in Southern Africa and elsewhere. In essence, these are associated with levels of leaching of nutrients which are, in turn, dependent on rainfall magnitude. Greater leaching of nutrients in mesic savanna results in a relatively nutrient-poor, acidic soil that supports a generally broad-leaved savanna. Conversely, in arid savanna, the soil tends to remain relatively nutrient-rich and more neutral, supporting a generally fine-leaved savanna. This major soil fertility division is at the root of the long-recognised distinction by Southern African farmers between 'sweet' and 'sour' bushveld (also termed 'sweetveld' and 'sourveld', respectively) based on the maintenance or loss of grass grazing value through winter, respectively (Huntley 1982). Sweetveld has sufficient leaf protein to be digestible throughout the year, whereas sourveld drops below the digestibility threshold for most of the dry season (Scholes et al. 2002). This explains the greater herbivore numbers and increased grazing pressure in arid, fine-leaved savanna (Scholes and Walker 1993). In reality, the differences in fertility of the grass layer are more complex and involve the relation between carbon assimilation and nutrient supply (Ellery et al. 1995). Several other functional and structural differences between the fine-leaved arid savanna and the broadleaved mesic savanna are summarised in Scholes (1997: Table 11.2), Rutherford et al. (2006: 448) and Geißler et al. (2024: Table 15.1). For example, the rate at which the grass fuel load is produced is higher in broad-leaved mesic savanna, which results in a shorter fire-return interval than in fine-leaved arid savanna. The main tree anti-herbivore defence mechanism in broad-leaved savanna is chemical (mainly polyphenols, especially condensed tannins), whereas in fine-leaved savanna, it is structural-defensive (mainly thorns). In terms of floristic composition, broad-leaved savanna typically includes genera such as *Brachystegia, Burkea, Combretum,* and *Terminalia*, while the fine-leaved savanna includes mainly *Senegalia* and *Vachellia* species (formerly classified as *Acacia*; Kyalangalilwa et al. 2013). In general, Andropogoneae grasses tend to dominate in mesic savanna, whereas Chloridoideae grasses are more important in arid savanna. It should be realised that, although the above-described coincidence of the mesic/arid and broad-leaved/fine-leaved savanna dichotomies applies broadly, there are several important exceptions. For example, on base-rich substrates in areas of mesic savanna, fine-leaved savanna can also occur (for example, the RB Limpopo Clayey Thornveld). Also, near the upper rainfall limit in arid savanna, generally broad-leaved savanna is possible (for example, the RB Namibian Dolomite Bushveld). The position of the enigmatic broad-leaved *Colophospermum mopane* dominated regional biomes is referred to below. It should also be noted that rainfall limitation to potential woody plant cover is more important in arid savanna. In arid savanna, water constrains woody cover and permits grass to coexist, whereas mesic savannas are unstable systems in which MAP is sufficient for canopy closure, and disturbances such as fire and herbivory are required for the coexistence of trees and grass (Sankaran et al. 2005; Sankaran and Ratnam 2013).

Although our arid and mesic savanna continental biomes correspond fairly well with the fine-leaved and broad-leaved savanna units as mapped by Scholes (1997), there are two important conceptual differences. Firstly, Scholes (1997) separates as a third division the mopane-dominated areas associated with the Limpopo, Zambezi, and Cunene River watersheds as broad-leaved, arid, fertile savanna. We prefer to maintain the functional integrity of our arid and mesic savanna continental biomes and place these three mopane areas according to our moisture criteria into our continental biomes, with the highest rainfall Zambezi mopane area placed in mesic savanna. Secondly, he separates a fourth division called 'Mixed savanna' in parts of eastern Southern Africa. Instead, we have broken down this Scholes' unit into its more homogeneous constituent regional biomes and assigned them appropriately to either our CB Southern African Arid Savanna or CB Southern African Mesic Savanna. It should be borne in mind that a particular limitation in the consistent application and mapping of this dichotomy is found, especially where the very common soil catenae of the ancient erosion surfaces create a close, local juxtaposition of the broad-leaved savanna on the ridge crests and slopes of the interfluves and the fine-leaved savanna on the valley bottoms. Indeed, the scale of this soil fertility gradient in savanna can thus vary from regional down to a few meters (Scholes 1990).

3.1 Continental Biome Southern African Mesic Savanna

This generally broad-leaved continental biome covers the higher rainfall parts of the savanna in the sub-continent on a wide range of generally less fertile substrates. In the less water-limited mesic savanna, substrate and soils can be particularly important in determining savannas at the regional biome level. In this work, the lower MAP limit of the continental biome in the eastern parts is roughly 600 mm. However, in northern Namibia and Botswana, the RB Baikiaea Woodland extends into lower rainfall areas with MAP as low as 500 mm on deep, water-storing Kalahari sands, which support a relatively higher woody plant cover (see Sankaran et al. 2005). The CB Southern African Mesic Savanna is mainly located within the Tropics, but it extends into the subtropical zone in northeastern South Africa and southwards along its southeastern seaboards. It also occurs in some parts of southeastern Botswana. Due to this relatively large north–south expanse, there is a southerly decrease in temperature. For example, in the coastal belt, the mean annual temperature (MAT) changes from 24.2 °C for the RB Zambezian Clayey Savanna in the north to 18.0 °C in the RB Sub-Escarpment Savanna in the south. In addition, the temperatures of the savannas of numerous mountains tend to be more variable and cooler than their surroundings.

Southern African mesic savannas have many faces (Fig. 3.2) and contain many more regional biomes than the CB Southern African Arid Savanna. The pride of place in the CB Southern African Mesic Savanna probably goes to the iconic miombo vegetation, which occupies the most mesic parts of the savanna in Zimbabwe and Mozambique. It is often located on well-drained soils, the wettest parts in Central Mozambique and eastern Zimbabwean highlands, with *Brachystegia boehmii, B. microphylla* and *B. utilis* prominent, while in the less wet parts, *B. spiciformis* and *Julbernardia globiflora* are most important. The RB Baikiaea Woodland, referred to above, is found on deep Kalahari sands and is dominated by *Baikiaea plurijuga, Baphia massaiensis, Guibourtia coleosperma,* and *Schinziophyton rautanenii.* On the leached, acidic sands of the northern South African plateau and southeastern Zimbabwean plateau, a broad-leaved savanna is typically found containing, for example, *Burkea africana, Combretum molle, C. zeyheri, Lannea discolor, Ochna pulchra,* and *Terminalia sericea.* Similar conditions develop in the lowveld of mainly Mozambique and South Africa. Here on the sandstones and granites, a sandy savanna occurs dominated by, for example, *Combretum apiculatum, C. zeyheri,* and *Terminalia sericea.* Juxtaposed to these sandy systems, both on the plateaus and the lowveld, are base-rich, clayey soils, which, despite the higher rainfall, as described above, support thornveld savanna woodlands. Here the most prominent elements are *Vachellia hebeclada, V. karroo, V. nilotica, V. rehmanniana,* and *V. tenuispina* on elevated plateaus, and *Senegalia nigrescens, S. senegal, Sclerocarya birrea, V. borleae, V. grandicornuta,* and *V. tortilis* in the lowlands. Main mountain savannas with higher rainfall are often characterised by acidic soils and broader ranges of topography and temperature. Important species include *Combretum apiculatum, Diplorhynchus condylocarpon, Faurea saligna, Heteropyxis natalensis, Parinari curatellifolia, Protea caffra, Pseudolachnostylis maprouneifolia, Senegalia caffra,* and *Vachellia sieberiana.* In contrast to the mountains of the north, the low-lying, incised valleys down the eastern coastal belt of South Africa allow, climatically, for the most southerly extension of savanna vegetation

Fig. 3.2 Physiognomy of selected vegetation types of mesic and arid savanna in southern Africa. **a** Arid Kalahari, South Africa: Kgalagadi Transfrontier Park, partly stabilised dunes and calcrete outcrops fringing dry bed Nossob River. **b** An open savannah woodland dominated by *Senegalia hereroensis* (Fabaceae) in the Daan Viljoen Game Reserve near Windhoek (Namibia), Khomas Hochland. **c** Open mopane (*Colophospermum mopane*; Fabaceae) woodland in Honnet Nature Reserve near Tshipise Resort, Venda, South Africa. **d** Lowveld, South Africa: Kruger National Park, a view from Olifants Rest Camp over plains covered with broad-leaved *Combretum* spp. (Combretaceae) savanna woodlands. **e** Montane miombo woodlands with *Brachystegia boehmii*, *B. spiciformis*, and *B. tamarindoides* subsp. *microphylla* (all Fabaceae), *Uapaca kirkiana* (Phyllanthaceae), and *Vangueriopsis lanciflora* (Rubiaceae) prominent in foreground in a gorge at lower slopes of the Chimanimani Mts. (Zimbabwe). **f** Semi-open *Baikiaea plurijuga* woodland with dense undergrowth of *Schmidtia* and *Eragrostis* grasses on deep Kalahari sands, E of Victoria Falls, Zimbabwe. *Photo credits* **a–d**, **f** L. Mucina, **e** M. C. Lötter

in Southern Africa, in the form of RB Sub-Escarpment Savanna.

Some other specific substrates are important functionally and for their association with relatively high levels of specialisation and endemism in mesic savanna. Particularly important are the savannas on ultramafic, dolomitic, and quartzite substrates. Most prominent of these is the grassy savanna of the ultramafic feature that forms the Great Dyke that runs through Zimbabwe from north to south. Discrete areas of ultramafic sourveld are also found in the vicinity of Barberton in Mpumalanga Province of South Africa. The bands of dolomite in northern South Africa support two regional savanna pedobiomes, one of which (RB Western Dolomite Bushveld) is the least mesic regional biome within the CB Southern African Mesic Savanna (apart from the RB Baikiaea Woodland on deep Kalahari sands). Another dolomite regional biome is found in northern Namibia (RB Namibian Dolomite Bushveld), and is the least arid regional biome within the Southern African Arid Savanna continental biome (see below).

Two distinctive sourveld savannas are found further south. These are on quartzite from southern eSwatini to the northern KwaZulu-Natal Province of South Africa and on sandstone outcrops of the coastal belt from central KwaZulu-Natal to the northern part of the Eastern Cape Province of South Africa. The savannas of these unique substrates are described under their constituent regional pedobiomes below.

The already mentioned mesic position of the mopane-dominated areas associated with the Zambezi River basin is enigmatic in that the MAP is mostly markedly higher than in other mopane regional biomes that lie within the Southern African Arid Savanna continental biome. The mopane-dominated areas in the vicinity of the Okavango River delta in northern Botswana should probably be regarded as transitional between Mesic and Arid Savanna (see discussion under SVMs21 Zambezian Mopane below).

The continental biome Southern African Mesic Savanna consists of 25 regional biomes (Table 3.1).

SVMs01 Limpopo Mountain Bushveld

Gm 20 Leolo Summit Sourveld (Dayaram et al. 2019)
Gm 21 Lydenburg Thornveld (Dayaram et al. 2019)
SVcb 4 Dwarsberg-Swartruggens Mountain Bushveld (Dayaram et al. 2019)
SVcb 5 Pilanesberg Mountain Bushveld (Dayaram et al. 2019)
SVcb 7 Norite Koppies Bushveld (Dayaram et al. 2019)
SVcb 9 Gold Reef Mountain Bushveld (Dayaram et al. 2019)
SVcb 10 Gauteng Shale Mountain Bushveld (Dayaram et al. 2019)
SVcb 11 Andesite Mountain Bushveld (Dayaram et al. 2019)
SVcb 13 Loskop Mountain Bushveld (Dayaram et al. 2019)
SVcb 17 Waterberg Mountain Bushveld (Dayaram et al. 2019)
SVcb 21 Soutpansberg Mountain Bushveld (Dayaram et al. 2019)
SVcb 24 Mamabolo Mountain Bushveld (Dayaram et al. 2019)
SVcb 26 Ohrigstad Mountain Bushveld (Dayaram et al. 2019)
SVcb 28 Sekhukhune Mountain Bushveld (Dayaram et al. 2019)

Distribution Occurs along multiple mountains, ridges, and hills in the northern Free State, North-West, Gauteng, Limpopo, and Mpumalanga Provinces of South Africa and just crossing over into southeastern Botswana. It extends from Parys and Potchefstroom in the south, northwards to the Waterberg and Soutpansberg mountains and as far as the vicinity of Gaborone in the west, and to Lydenburg and Ohrigstad on the west-facing flanks of the Great Escarpment (Fig. 3.3).

Vegetation Physiognomy and Composition This is a very diverse regional biome as it covers a range of altitude and moisture gradients with a strong turnover in diagnostic species across this unit. However, it is usually characterised by broad-leaved deciduous open woodland, occasionally forming a wooded grassland where it abuts grassland units, on rocky mountain slopes. Some of the dominant trees throughout its range include *Albizia tanganyicensis, Aloe marlothii, Berchemia zeyheri, Bridelia mollis, Burkea africana, Combretum apiculatum, C. mollis, Cussonia natalensis, Diplorhynchus condylocarpon, Dombeya rotundifolia, Englerophytum*

Table 3.1 A system of global biomes (GB), continental biomes (CB) and regional biomes (RB) of the zonobiome E2 Tropical Seasonal Zone of Southern Africa

Rank	Code	Biome	Zonality	Azonality type	Climatic diagnosis
ZB	**E2**	**Tropical Seasonal Zone**	**Zonal**		**Tropical-subtropical; mesic to humid, seasonal summer rainfall**
GB	***SVM***	***Mesic Savanna***	***Zonal***		***Tropical-subtropical; mesic to humid, seasonal summer rainfall***
CB	**SVMs**	**Southern African Mesic Savanna**	**Zonal**		**Tropical-subtropical; mesic to humid, seasonal summer rainfall**
RB	SVMs01	Limpopo Mountain Bushveld	Zonal		
RB	SVMs02	Southeastern Limpopo Sandy Bushveld	Zonal		
RB	SVMs03	Limpopo Clayey Thornveld	Zonal		
RB	SVMs04	Zimbabwe Clayey Bushveld	Zonal		
RB	SVMs05	Zimbabwe Sandy Bushveld	Zonal		
RB	SVMs06	Zambezian Wet Miombo	Zonal		
RB	SVMs07	Zambezian Dry Miombo	Zonal		
RB	SVMs08	Zambezian Montane Miombo	Zonal		
RB	SVMs09	Lowveld Mountain Bushveld	Zonal		
RB	SVMs10	Lowveld Sandy Bushveld	Zonal		
RB	SVMs11	Lowveld Clayey Bushveld	Zonal		
RB	SVMs12	Subtropical East Coast Woodland	Zonal		
RB	SVMs13	Tropical East Coast Woodland	Zonal		
RB	SVMs14	Sub-Escarpment Savanna	Zonal		
RB	SVMs15	Zambezian Sandy Savanna	Zonal		
RB	SVMs16	Zambezian Clayey Savanna	Zonal		
RB	SVMs17	Great Dyke Grassland	Intrazonal	Metallobiome	
RB	SVMs18	Western Dolomite Bushveld	Intrazonal	Peinobiome	
RB	SVMs19	Eastern Dolomite Bushveld	Intrazonal	Peinobiome	
RB	SVMs20	Zambezian Inselberg Woodland	Intrazonal	Peinobiome	
RB	SVMs21	Zambezian Mopane	Intrazonal	Argillobiome	
RB	SVMs22	Baikaea Woodland	Intrazonal	Psammobiome	
RB	SVMs23	Bushveld Ultramafic Sourveld	Intrazonal	Metallobiome	
RB	SVMs24	Southern Quartzite Sourveld	Intrazonal	Peinobiome	
RB	SVMs25	Southern Sandstone Sourveld	Intrazonal	Peinobiome	
GB	***SVA***	***Arid Savanna***	***Zonal***		***Tropical-subtropical; subarid low seasonal summer rainfall***
CB	**SVAs**	**Southern African Arid Savanna**	**Zonal**		**Tropical-subtropical; subarid low seasonal summer rainfall**

(continued)

Table 3.1 (continued)

Rank	Code	Biome	Zonality	Azonality type	Climatic diagnosis
RB	SVAs1	Arid Kalahari Mountain Bushveld	Zonal		
RB	SVAs2	Arid Kalahari Sandy Bushveld	Zonal		
RB	SVAs3	Northern Kalahari Bushveld	Zonal		
RB	SVAs4	Namibian Central Upland Bushveld	Zonal		
RB	SVAs5	Kaokoland Mopane	Intrazonal	Peinobiome	
RB	SVAs6	Kalahari Duneveld	Intrazonal	Psammobiome	
RB	SVAs7	Northwestern Limpopo Sandy Bushveld	Zonal		
RB	SVAs8	Limpopo Mopane	Intrazonal	Argillobiome	
RB	SVAs9	Namibian Dolomite Bushveld	Intrazonal	Peinobiome	
GB	***TDF***	***Tropical Dry Forest***	***Zonal***		***Tropical-subtropical; subarid; low seasonal summer rainfall; fire-shy***
CB	**TDFa**	**East African Dry Coastal Forest**	**Zonal**		**Tropical-subtropical; subarid; low seasonal summer rainfall; fire-shy**
RB	TDFa2	Southern African Dry Forest	Extrazonal	Psammobiome	
RB	TDFa3	Southern African Dry Thicket	Extrazonal	Psammobiome	

magalismontanum, Faurea saligna, Heteropyxis natalensis, Hexalobus monopetalus, Kirkia acuminata, K. wilmsii, Maytenus undata, Mimusops zeyheri, Mundulea sericea, Lannea discolor, Ochna pulchra, Ozoroa sphaerocarpa, Protea caffra, Pterocarpus angolensis, Searsia leptodictya, Senegalia caffra, Strychnos cocculoides, Terminalia sericea, Vachellia robusta subsp. *robusta, V. sieberiana,* and *Vitex rehmannii.* Characteristic shrubs include *Carissa edulis, Diospyros lycioides, Elephantorrhiza burkei, Grewia flava, Helichrysum kraussii, Parinari capensis, Salacia rehmannii,* and *Vangueria parvifolia.* The herbaceous layer is dominated by *Abutilon pycnodon, Barleria gueinzii, Hermannia grisea* with the following prominent grasses *Digitaria eriantha, Enneapogon scoparius, Eragrostis racemosa, Loudetia simplex, Setaria sphacelata, Trachypogon spicatus,* and *Tristachya biserata.*

Climate The MAP of 691 mm for this regional biome falls mainly in summer and winters are dry. Summers are hot and winters mild to cool. This regional biome is among those mesic savanna regional biomes with the lowest minimum winter temperature. Occurrence of frost varies from frequent on the highveld of Gauteng to infrequent in the northern and easternmost area (Fig. 3.3).

Topography, Geology and Soils Diverse elevated topography comprises often rocky, low hills and ridges to rugged high mountains and includes inselbergs, plateaus, tablelands, steep faces and cliffs, deeply incised valleys as well as shallow valleys with more moderate slopes. The geological substrates of this regional biome are particularly diverse and include: igneous andesite, basalt, gabbro, granite, norite, pyroxenite, rhyolite, syenite, and tuff; sedimentary conglomerate, sandstone and shale; and metamorphic gneiss and quartzite. This regional biome excludes areas on dolomite. Soils are rocky and gravel lithosols, often shallow to very shallow and skeletal, but deeper soils occur on less steep slopes and valley floors. Soil texture can be sandy or loamy and freely drained, or clayey.

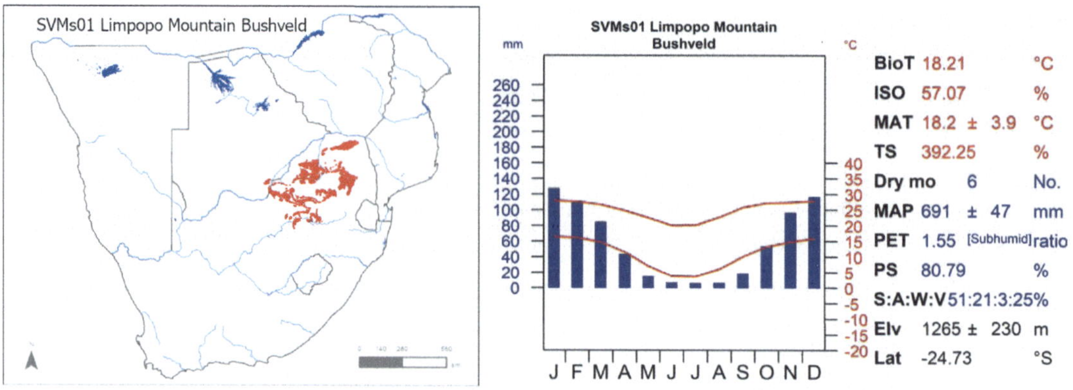

Fig. 3.3 The distribution area and the climatic diagram of the regional biome SVMs01 Limpopo Mountain Bushveld

Sources Acocks (1953, 1988), van Wyk (1959), Wild and Barbosa (1967), van Vuuren and van der Schijff (1970), Coetzee (1972), Theron (1973), Coetzee (1974, 1975), Bredenkamp (1975, 1977), Bredenkamp and Theron (1978), Scholes (1979), van der Meulen (1979), Bredenkamp and Theron (1980), Coetzee et al. (1981, 1993, 1994, 1995), Westfall (1981), Westfall et al. (1983, 1985), Behr and Bredenkamp (1988), Ben-Shahar (1988), Bezuidenhout et al. (1988), du Preez and Venter (1990), van Jaarsveld and Hardy (1991), Bezuidenhout et al. (1994), Zacharias (1994), Panagos (1996), Winter and Hahn (1999), Grobler (2000), Siebert (2001), Siebert et al. (2001, 2002a, b, c, d, 2003a, b, c), Grobler et al. (2002), van Staden (2002), Siebert and van Wyk (2005), van Staden and Bredenkamp (2005), Mucina et al. (2006a), Rutherford et al. (2006), Furniss (2008), Barrett et al. (2010), Mills et al. (2012), Boneschans et al. (2015), van Staden et al. (2017, 2021), Nzuza et al. (2021), Adhikari et al. (2022), Setshedi et al. (2022), Popp and Kalwij (2023), Rapiya et al. (2023), Schroder et al. (2024).

SVMs02 Southeastern Limpopo Sandy Bushveld

SVcb 12 Central Sandy Bushveld (Dayaram et al. 2019)
SVcb 16 Western Sandy Bushveld (Dayaram et al. 2019)

Distribution The unit comprises a broad belt of plains on the central plateau in northern South Africa from Northam eastwards, passing between Pretoria and Hamanskraal to Groblersdal and Jane Furse in the east. Also includes sandy slopes and flats in the vicinity of the Waterberg and Rooiberg, such as the open valleys of Vaalwater and Rankin's Pass, and the wide Matlabas plains which grade northwestwards into the more arid RB Northwestern Limpopo Sandy Bushveld. It includes several broad ridges of sand that tend to bisect the Springbok Flats (Fig. 3.4). Unmapped patches of this regional biome occur on flats near some of the mountains in southeastern Botswana.

Vegetation Physiognomy and Composition Physiognomy is characterised by medium-tall to low, open to moderately closed woodland with variable, but usually well-developed grass layer. Predominantly broad-leaved trees occur in the core areas but with some fine-leaved trees also present, especially in low lying areas or drier parts. Important species include *Burkea africana, Combretum apiculatum, C. imberbe, C. molle, C. zeyheri, Dichrostachys cinerea, Euclea natalensis, Grewia flavescens, Lannea discolor, Ochna pulchra, Ozoroa paniculosa, Peltophorum africanum, Securidaca longepedunculata, Senegalia burkei, Sclerocarya birrea, Strychnos cocculoides, S. pungens, Terminalia sericea, Vachellia luederitzii,* and *Ximenia caffra.* Geoxylic suffrutices can be common and include *Dichapetalum cymosum, Fadogia homblei, Lannea edulis, Parinari capensis,* and *Pygmaeothamnus zeyheri.* Towards the drier NW, the proportion of broad-leaved woody plants declines with the increasing importance of mainly *Senegalia erubescens, S. mellifera, S. nigrescens, Vachellia nilotica,* and *V. tortilis.* Amongst important grasses are *Andropogon schirensis, Anthephora pubescens, Brachiaria nigropedata, Digitaria eriantha* subsp. *eriantha, Diheteropogon amplectans, Eragrostis pallens, E. rigidior, Hyperthelia dissoluta, Panicum maximum, Perotis patens, Rhynchelytrum villosum,* and *Schmidtia pappophoroides.* This regional biome forms a lengthy border with the RB Limpopo Mountain Bushveld, particularly in the Waterberg and Loskop Dam regions, which results in a number of shared high-rainfall floristic elements in places.

Climate Rain falls predominantly in summer; winters are dry. MAP is 597 mm, ranging mainly between 500 and 700 mm, but falling below 500 mm in parts of the northwestern areas. Summers are hot with warm winter days and cold nights with frost. The climate is very similar to that of the RB Limpopo Clayey Thornveld (except for the driest parts), hence suggesting that climate is not an important discriminator between these two interspersed regional biomes, in contrast to their strong edaphic differences (Fig. 3.4).

Topography, Geology and Soils Broad undulating plains and open valleys and sandy flats between mountains are the characteristic features of the topography. The large southern and eastern parts are underlain by granite of the Lebowa Granite Suite and by some granophyre of the Rashoop Granophyre Suite, both of the Bushveld Igneous Complex, as well as by some Archaean granite and gneiss of the Swazian Erathem. Main rocks in the northern parts, in the vicinity of the Waterberg Mountains, are sandstone, conglomerate, siltstone, mudstone, and shale of the Alma and Vaalwater Formations, and Matlabas and Kransberg Subgroups, all of the Waterberg Group. Soils are well-drained, sandy, mesotrophic and dystrophic, sometimes leached but can also be less sandy and eutrophic in parts of the west and lower landscape positions. Well-drained and deep soils on slopes can form a catenary sequence from the Hutton soil form at the top to Clovelly soil form on the lower slopes.

Sources Grunow (1965), Herbst (1973), Coetzee et al. (1976), Rutherford and Kelly (1978), van der Meulen (1979), Rutherford (1980, 1981, 1982, 1983b, 1984), van der Meulen and Westfall (1980), Grossman et al. (1981), Cresswell et al. (1982), Gandar (1982), Huntley and Morris (1982), Morris et al. (1982), Lubke and Thatcher (1983), Lubke et al. (1983), Theron et al. (1984), Whittaker et al. (1984), Cooper and Owen-Smith (1985), Knoop and Walker (1985), Walker et al. (1986), Frost (1987), Yeaton et al. (1988), Blackmore et al. (1990), Peel et al. (1991), Scholes and Walker

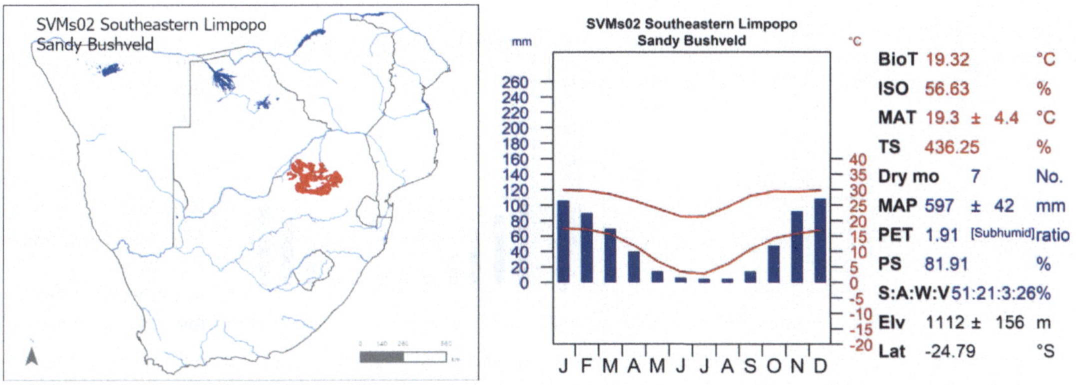

Fig. 3.4 The distribution area and the climatic diagram of the regional biome SVMs02 Southeastern Limpopo Sandy Bushveld

(1993), van Rooyen et al. (1993), Brown and Bredenkamp (1994), Brown et al. (1995, 1996, 1997), Dörgeloh (1998, 1999a, b), Winterbach (1998), Winterbach et al. (2000), Rutherford et al. (2006), le Grange (2010a, b), Tshisikhawe et al. (2012), Whitecross et al. (2017).

SVMs03 Limpopo Clayey Thornveld

SVcb 1 Dwaalboom Thornveld (Dayaram et al. 2019)
SVcb 3 Zeerust Thornveld (Dayaram et al. 2019)
SVcb 6 Marikana Thornveld (Dayaram et al. 2019)
SVcb 8 Moot Plains Bushveld (Dayaram et al. 2019)
SVcb 14 Loskop Thornveld (Dayaram et al. 2019)
SVcb 15 Springbokvlakte Thornveld (Dayaram et al. 2019)
SVcb 27 Sekhukhune Plains Bushveld (Dayaram et al. 2019)

Distribution Comprises large and smaller, low-lying flats in northern South Africa, north of the Highveld and mostly south of the Waterberg-Strydpoort Mountains and west of the Eastern Escarpment, extending to limited parts of southeastern Botswana in the west (Fig. 3.5). From the east, it spans the Sekhukhuneland plains between Burgersfort and Mphahlele as well as plains and valleys between Groblersdal and the areas around the Loskop Dam. More centrally located are the extensive Springbok Flats from Zebediela and Roedtan to Marble Hall, swinging westwards though Pienaars River to Assen. It also extends from Pretoria to the Rustenburg area on the flats of Marikana and the Moot, north of the Magaliesberg and Daspoort Mountain Ranges, respectively. It continues westwards, widening from Rustenburg, Northam, and Thabazimbi, on flats on both sides of the Dwarsberg and Rand Van Tweede Poort Mountain ridges as well as between hills and ridges around Swartruggens and Zeerust, reaching areas near Gaborone, Sikwane, and Lobatse in Botswana.

Vegetation Physiognomy and Composition The vegetation physiognomy is dominated by medium to low height, mostly fine-leaved trees with moderate cover, together with a variable grass layer. Important species include *Boscia foetida, Euclea crispa, E. undulata, Euphorbia tirucalli, Searsia lancea, S. pyroides, Senegalia burkei, S. caffra, S. mellifera, Vachellia erioloba, V. gerrardii, V. karroo, V. luederitzii, V. nilotica, V. tenuispina, V. tortilis,* and *Ziziphus mucronata. Combretum imberbe,* and *Senegalia erubescens* are largely limited to the western sections of this regional biome, with *Searsia engleri* and *Vachellia sieberiana* mainly occurring in the eastern sections. Dominant grasses include *Aristida biparta, Bothriochloa insculpta, Cenchrus ciliaris, Dichanthium annulatum* var. *papillosum, Digitaria argyrograpta, D. eriantha* subsp. *eriantha, Elionurus muticus, Enneapogon cenchroides, Eragrostis lehmanniana, Heteropogon contortus, Ischaemum afrum, Panicum maximum, Setaria incrassata, S. sphacelata, Themeda triandra,* and *Urochloa mosambicensis.*

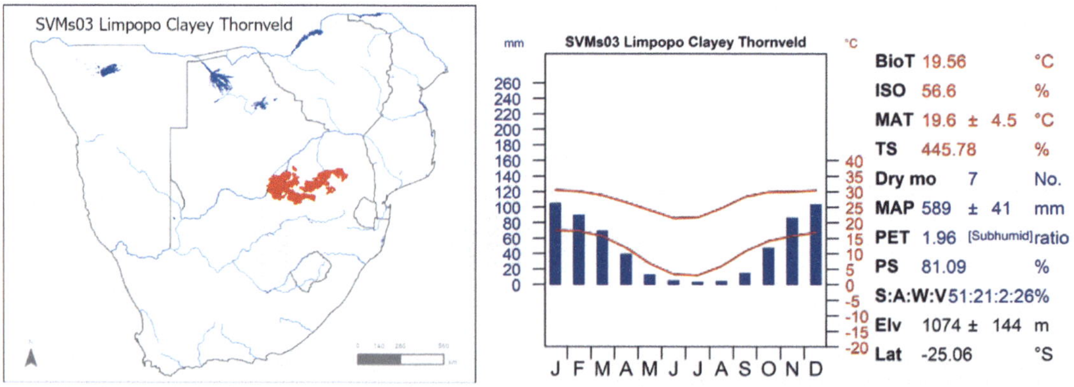

Fig. 3.5 The distribution area and the climatic diagram of the regional biome SVMs03 Limpopo Clayey Thornveld

Climate MAP of 589 mm occurs predominantly in summer, and varies mostly between 500 and 700 mm, but declines below 500 mm in northern parts of the Sekhukhune Plains. Summers are hot, with warm winter days but with cold and frosty nights (but relatively little frost on the Sekhukhune Plains; Fig. 3.5).

Topography, Geology and Soils The topography is characterised by flat to slightly undulating plains and bottomlands as well as open valleys between hills. The geology is complex, with gabbro, norite, bronzite, harzburgite, anorthosite, and pyroxenite, that make up the major mafic rocks of the Rustenburg Layered Suite of the Bushveld Igneous Complex. Further shales, mudstones, carbonates, volcanic rocks, conglomerates, breccias and diamictites of the Rayton, Silverton, Bloempoort and Loskop Formations of the Transvaal Supergroup also occur. Also important are the granite-gneisses of the Swazian Erathem in the NW, and basalts (Letaba Formation), mudstones (Irrigasie Formation), and shales (Ecca Group) of the Karoo Supergroup on the Springbok Flats. Soils are rich in clay, with high base status. They include vertic and melanic black clays (with gilgai soil features in parts of the Rustenburg area), deep colluvial clay-loams, and freely drained, eutrophic, mesotrophic and some dystrophic soils.

Sources Galpin (1926), van Wyk (1959), Coetzee (1971), Morris (1972), Theron (1973), Verster et al. (1973), Coetzee et al. (1976), van der Meulen (1979), van der Meulen and Westfall (1979, 1980), van Rooyen (1983, 1984), Milton (1987), Acocks (1988), Pauw (1988), Rutherford (1993), Breebaart and Deutschländer (1997), Panagos et al. (1998), Winterbach (1998), Siebert (2001), Siebert et al. (2002a, b, d, 2003b), Malan and van Niekerk (2005), Rutherford et al. (2006), la Grange (2010a, b), le Grange (2010a, b), Mochesane (2018), Adhikari et al. (2022), Mndela et al. (2022, 2023).

SVMs04 Zimbabwe Clayey Bushveld

Distribution Occurs on the central plateau of Zimbabwe in the form of variably sized patches stretching from the vicinity of Bulawayo in the southwest to that of Harare in the northeast, with largest patches found between the Harare and Masvingo areas (Fig. 3.6). The patches are generally embedded within either RB Zimbabwe Sandy Bushveld or RB Zambezian Dry Miombo.

Vegetation Physiognomy and Composition Wooded grasslands are a feature of this type, the main grasses being *Loudetia simplex, Themeda triandra, Trachypogon spicatus, Eragrostis hispida*, and *Alloteropsis semialata.* The woody component is composed typically of *Parinari curatellifolia*, often forming monotypic stands, but also with *Faurea rochetiana, F. saligna, Phoenix reclinata, Protea angolensis, P. gaguedi, P. welwitschii, Terminalia sericea*, and the geoxylic suffrutex *Syzygium guineense* subsp. *huillense* frequent in seasonal wetlands or vleis. Patches of *Brachystegia spiciformis–Julbernardia globiflora* miombo woodland may occur within this type, with taxa such as *Pterocarpus angolensis, Pseudolachnostylis maprouneifolia, Ozoroa insignis* subsp. *reticulata, Albizia amara, A. antunesiana*, and *Ekebergia benguelensis* co-occurring. Acacia woodland composed of *Vachellia gerrardii, V. karroo*, and *V. rehmanniana* also occurs on heavier clay soils in lower-lying areas.

Climate This regional biome has a MAP of 701 mm falling mainly in summer and winters are dry. There is a distinct gradient in MAP increasing from around 600 mm in the west in the Matabeleland South Province to over 800 mm in parts of the eastern areas. Summers are hot, winter days are warm and frost is infrequent on winter nights (Fig. 3.6).

Topography, Geology and Soils Occupies flat to undulating plains on the Zimbabwe plateau. Much of the rock types are meta sedimentary and metavolcanic; gneiss is common and basalt occurs locally. Basaltic metavolcanics also occur locally, for example, in the vicinity of the Matopos. Soils are deep to shallow, clay-rich, with poor drainage.

Sources Ingram (1960), Wild and Barbosa (1967), Kennan (1972), Timberlake et al. (1993), Moyo et al. (2011), Mwembe et al. (2013).

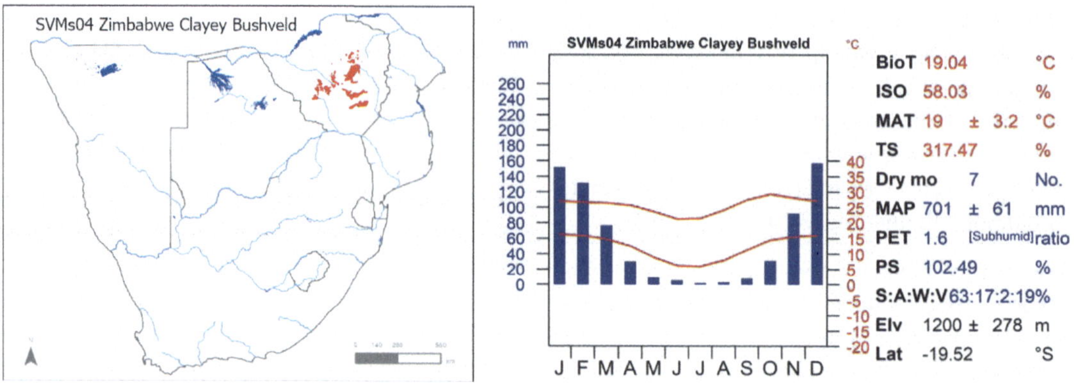

Fig. 3.6 The distribution area and the climatic diagram of the regional biome SVMs04 Zimbabwe Clayey Bushveld

SVMs05 Zimbabwe Sandy Bushveld

Distribution Is found mainly on the plateau above 1000 m of elevation in south-western Zimbabwe, extending along the watershed from Plumtree and Bulawayo to Gweru. The largest area extends south of the watershed in Matabeleland South and includes the Matobo Hills, but also extends northwards to some isolated patches in northern parts of the Midlands Province, as well as slightly in to Botswana north of Francistown (Fig. 3.7).

Vegetation Physiognomy and Composition Physiognomy is mostly open to closed canopy of broad-leaved, medium height trees with a well-developed herbaceous layer. Areas of acacia savanna occur with *Vachellia gerrardii, V. karroo, V. nilotica* subsp. *kraussiana,* and *V. rehmanniana.* Other important taxa include *Burkea africana, Combretum hereroense, C. molle, C. zeyheri, Commiphora mollis, Diplorhynchus condylocarpon, Dombeya rotundifolia, Euphorbia ingens, Erythrina latissima, Ficus burkei, Grewia bicolor, Lannea discolor, Ozoroa reticulata, Peltophorum africanum, Piliostigma thonningii, Pseudolachnostylis maprouneifolia, Pterocarpus angolensis, P. rotundifolius* subsp. *rotundifolius, Sclerocarya birrea* subsp. *caffra, Securidaca longepedunculata,* and *Terminalia sericea. Colophospermum mopane* can extend into the unit along drainage lines from the margins and is also found on localised sodic soils within the mapping unit. The granite hills of Matobo support trees such as *Albizia tanganyicensis, Brachystegia glaucescens, Commiphora marlothii, Croton gratissimus, Euphorbia cooperi, Ficus abutilifolia,* and *F. glumosa.* The dominant grasses are *Themeda triandra* as well as *Heteropogon contortus, Cymbopogon plurinodis,* various species of *Eragrostis*, and *Bothriochloa insculpta.*

Climate The MAP of 588 mm falls mainly in summer and winters are dry. MAP varies from around 500 mm where the regional biome abuts the RB Limpopo Mopane in the south, to around 700 mm in parts of the northeastern areas. Summers are hot, winter days are warm and frost is infrequent (Fig. 3.7).

Topography, Geology and Soils Mostly found on broad undulating plains of the Zimbabwe Plateau, with occasional hills and ridges. Geology constitutes mainly extensive areas of the older gneiss complex in which are embedded numerous large patches of younger intrusive granite. There are lesser occurrences of grits, sandstone and siltstone as well as andesitic and dactic metavolcanics. Soils are sandy with lower clay content than soils of the associated Zimbabwe Clayey Bushveld. Soil depths vary widely and are generally well-drained.

Sources Ingram (1960), Wild and Barbosa (1967), Farrell (1968b), Kennan (1972), Grobler (1974), Timberlake et al. (1993).

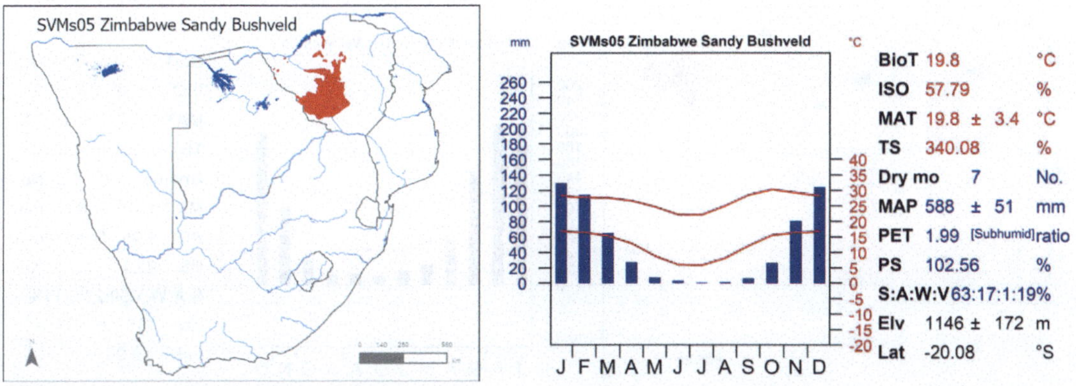

Fig. 3.7 The distribution area and the climatic diagram of the regional biome SVMs05 Zimbabwe Sandy Bushveld

SVMs06 Zambezian Wet Miombo

SVmd2 Barue Plateau Moist Miombo (Lötter et al. 2023)
SVmw5 Cheringoma Coastal Moist Miombo (Lötter et al. 2023)
SVmw6 Cheringoma Escarpment Moist Miombo (Lötter et al. 2023)
SVmw7 Cheringoma Plateau Moist Miombo (Lötter et al. 2023)
SVmw9 Chimoio Moist Miombo (Lötter et al. 2023)
SVmw10 Choa Escarpment Moist Miombo (Lötter et al. 2023)
SVmw12 Dombe Escarpment Miombo (Lötter et al. 2023)
SVmw13 Gorongosa Escarpment Moist Miombo (Lötter et al. 2023)
SVmw14 Gorongosa Foothills Moist Miombo (Lötter et al. 2023)

Distribution In Mozambique, Zambezian Wet Miombo covers most of the central parts of the Manica Province and neighbouring parts of Sofala Province, also occurring on the isolated Cheringoma Plateau between Beira and the Zambezi River. In addition, it straddles the Zimbabwe–Mozambique border from the vicinity of Chipinge and the Mossuriza District in the south to the slopes of the Nyanga Mountains in the north (Fig. 3.8). Along this mountainous border region it remains at elevations below that of RB Eastern Highlands Grassland or RB Zambezian Montane Miombo.

Vegetation Physiognomy and Composition The vegetation physiognomy is an open to closed woodland varying in height from 8 to 15 m tall, mainly dominated by the miombo pairing of *Brachystegia spiciformis* throughout and *Julbernardia globiflora* coming in at lower altitudes, as well as *Brachystegia utilis* and *B. boehmii* in areas of higher rainfall. *Parinari curatellifolia* is the most characteristic non-miombo species throughout but other typical trees include *Antidesma venosum*, *Bridelia micrantha*, *Combretum molle*, *C. zeyheri*, *Cussonia arborea*, *Faurea saligna*, *F. rochetiana*, *Heteropyxis dehniae*, *Pericopsis angolensis*, *Protea gaguedi*, *Syzygium guineense* subsp. *guineense*, *Terminalia gazensis*, *Uapaca kirkiana*, *Vachellia amythethophylla*, *V. sieberiana* var. *woodii*, and *Vitex payos* var. *glabrescens*. Groves of *Vachellia abyssinica* are characteristic of wetter soils in open areas at higher altitudes within this biome. Grasses include *Themeda triandra*, *Loudetia simplex*, *Cymbopogon validus*, *Setaria* spp., *Pennisetum unisetum*, *Hyparrhenia filipendula*, and *Hyperthelia dissoluta*. The disjunct region of wet miombo woodland occurring on the Cheringoma Plateau in Mozambique represents a distinct subtype of this regional biome. It shares the dominants of *Brachystegia spiciformis*, *Julbernardia globiflora*, and *Parinari curatellifolia* but many of the associated trees and shrubs differ markedly; namely and most typically *Albizia adianthifolia*, *A. versicolor*, *Amblygonocarpus andongensis*, *Hirtella zanguebarica*, *Millettia stuhlmannii*, *Terminalia myrtifolia*, *Uapaca nitida*, *U. sansibarica*, and

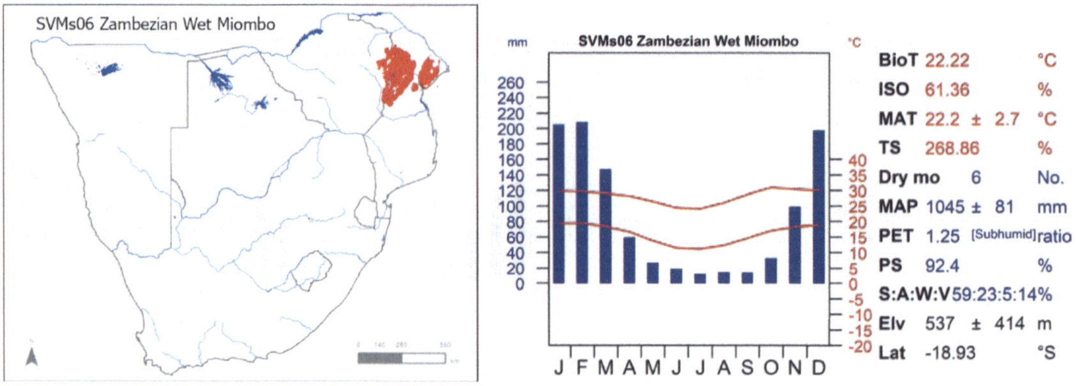

Fig. 3.8 The distribution area and the climatic diagram of the regional biome SVMs06 Zambezian Wet Miombo

Vitex doniana. The grass layer is sparse, often with dense stands of wild gingers, *Aframomum* spp., and the canopy is typically rich in epiphytes, mainly orchids and ferns, especially *Platycerium* spp. It may be noted that this regional biome lacks the additional miombo component of *Isoberlinia angolensis* that is typically found in some wet miombo to the north outside the MBSA, often with a greater diversity of *Brachystegia* species, for example in northern Zambia.

Climate With a MAP of 1045 mm this biome is one of the most mesic savanna regional biomes; only two other regional biomes are similar and have a MAP that exceeds 1000 mm. Rain falls predominantly in summer, and although winters are relatively dry, an average of about 50 mm falls in the winter months combined. Summers are hot and humid whereas winters are warm, but conditions are cooler at higher altitude near the mountains (Fig. 3.8).

Topography, Geology and Soils The landscape varies from extensive, undulating plains and rolling hills to often rugged, rocky slopes and lower ridges associated with the Chimanimani-Nyanga Mountain axis. Geological substrates comprise metamorphic rocks, especially granulitic gneiss, including migmatite, as well as micaschist, granite and, in lower lying areas, recent fluvial deposits are found. Localised occurrences of limestone, shale and quartzite also occur. The soils are mainly fairly acidic sandy clay loams and sandy loams.

Sources Pedro and Barbosa (1955), Goodier and Phipps (1963), Wild and Barbosa (1967), Frost (1996), Smith and Allen (2004), Ribeiro et al. (2020), Cizek (2021), Lötter et al. (2023).

SVMs07 Zambezian Dry Miombo

SVcb 22 VhaVenda Miombo (Dayaram et al. 2019)
SVmd2 Barue Escarpment Miombo (Lötter et al. 2023)
SVmd3 Inhambane Coastal Miombo (Lötter et al. 2023)
SVmw30 Mossurize Escarpment Miombo (Lötter et al. 2023)
SVmd9 Pangue Dry Miombo (Lötter et al. 2023)
SVmd10 Save Coastal Miombo (Lötter et al. 2023)
SVmd12 Vilanculos Coastal Miombo (Lötter et al. 2023)

Distribution Occurs on much of the central and northeastern plateau of Zimbabwe tending to become more dissected and patchy closer to the Zambezi River in the north, as well as towards the Save River in the south. It occurs to a lesser extent along inland parts of the Mozambique lowland plain in the Zambezi River valley but is found as an extensive, broad southern coastal belt between the Save River in the north and the Limpopo River in the south of Mozambique (Fig. 3.9). It remains inland of the subtropical forests in the higher rainfall zone along the (northern) coast in Mozambique. A very small relict patch of miombo occurs in northeastern South Africa where *Brachystegia spiciformis* is the sole miombo element.

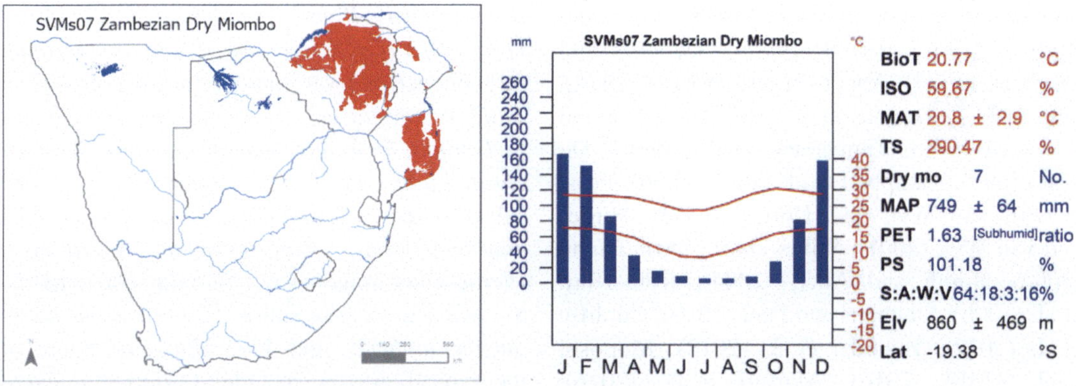

Fig. 3.9 The distribution area and the climatic diagram of the regional biome SVMs07 Zambezian Dry Miombo

Vegetation Physiognomy and Composition The physiognomy is often a medium to tall closed tree layer dominated by *Brachystegia spiciformis* and *Julbernardia globiflora* and a poorly developed herbaceous layer although the grass cover increases rapidly with some degree of opening of the tree canopy. Tree height appears related to rainfall and soil depth but in general few trees grow taller than 20 m and in many areas a tree height of 12–15 m is more common. Other fairly common or diagnostic trees and shrubs are *Albizia antunesiana*, *Azanza garckeana*, *Brachystegia boehmii*, *Burkea africana*, *Combretum molle*, *Cussonia arborea*, *Diplorhynchus condylocarpon*, *Faurea rochetiana*, *Flacourtia indica*, *Lannea discolor*, *Monotes glaber*, *Ozoroa insignis*, *Parinari curatellifolia*, *Pericopsis angolensis*, *Protea angolensis*, *Pseudolachnostylis maprouneifolia*, *Pterocarpus angolensis*, *P. rotundifolius* subsp. *rotundifolius*, *Strychnos cocculoides*, and *Uapaca kirkiana*. The large spreading tree, *Brachystegia glaucescens*, typifies the numerous granite hills scattered throughout much of this region, while the gregarious *Senegalia polyacantha* subsp. *campylacantha* is characteristic of lower-lying heavier soils along drainage lines. Low shrubs include species of *Dolichos*, *Eriosema*, *Indigofera*, *Kotschya*, and *Sphenostylis*, and saplings of canopy trees. Important grasses include *Andropogon gayanus*, *Bothriochloa glabra*, *Brachiaria brizantha*, *Digitaria diagonalis*, *D. gazensis*, *D. milanjiana*, *Hyparrhenia filipendula*, *Hyperthelia dissoluta*, *Loudetia flavida*, *L. simplex*, *Panicum maximum*, *Schizachyrium jeffreysii*, *Themeda triandra*, and *Tristachya nodiglumis*. Despite the usual sparseness of the grass layer, the accumulation of biomass production of this layer (Rutherford 1978) is sufficient to support fire at least every few years.

Climate This regional biome has a MAP of 749 mm ranging from about 600 mm or 650 mm in southern Zimbabwe and southern Mozambique where it borders RB Limpopo Mopane to approximately 850 mm or 900 mm on parts of the northeastern Zimbabwe Plateau. Summers are hot and winters are mild to warm (Fig. 3.9).

Topography, Geology and Soils This regional biome is positioned on extensive level to undulating plains, valleys, ridges and some rocky outcrops. Geology on the plateau comprises metavolcanics and meta-sediments of the Basement Complex as well as intrusive granites and granitic gneisses. It also extends on to sands, including dunes, in Mozambique, and marginally on to aeolian Kalahari sands in parts of western Zimbabwe. Also included are some more restricted occurrences of grits, phyllite, quartzite, sandstone, shale and siltstone. Soils vary from shallow stony soils to deeper sand and loams, and most soils are freely drained, although drainage can be restricted locally by, for example, clay subsoils.

Sources Pedro and Barbosa (1955), Wild and Barbosa (1967), Farrell (1968b), Jacobsen and Jacobsen (1973), Anderson and Walker (1974), Guy (1977), Thomas et al. (1977). Werger and Coetzee (1978), Timberlake et al. (1993), Du Toit (1993), Campbell et al. (1995, 1996), Frost (1996), van Wyk and Hurter (2000), Hurter and van Wyk (2001), Willis et al. (2001), Hahn (2002), Smith and Allen (2004), Rutherford et al. (2006), Stalmans and Peel (2010), Cunliffe et al. (2012), Zimudzi et al. (2013), Muposhi et al. (2014, 2016), Martini et al. (2016), Tafangenyasha et al. (2016), Zimudzi and Chapano (2016), Nyazika et al. (2017), Ribeiro et al. (2020), Cizek (2021), Lötter al. (2023).

SVMs08 Zambezian Montane Miombo

SVmw8 Chimanimani Montane Miombo (Lötter et al. 2023)

Distribution The distribution of this regional biome comprises relatively limited areas, mostly within Mozambique along the border with Zimbabwe, on mountain slopes of the Chimanimani region, from around Espungabera in the south to the vicinity of Manica in the north (Fig. 3.10).

Vegetation Physiognomy and Composition This is a relatively short miombo woodland typically 10–15 m high, the miombo component composed of *Brachystegia microphylla* (dominant, especially on quartzite ridges), *B. boehmii*, *B. spiciformis*, *B. utilis*, and *Julbernardia globiflora*. Other typical trees are *Albizia antunesiana*, *A. schimperiana* var. *amaniensis*, *A. versicolor*, *Antidesma venosum*, *Cussonia spicata*, *Faurea rochetiana*, *Ficus sur*, *Monotes engleri*, *Parinari curatellifolia*, *Pericopsis angolensis*, *P. caffra* subsp. *gazensis*, *P. petiolaris*, *Pseudolachnostylis maprouneifolia*, *Psorospermum febrifugum*, *Pterocarpus angolensis*, *Searsia chirindensis*, *Syzygium guineense* subsp. *guineense*, *Vachellia amythethophylla*, and *Vitex doniana*. Some of the typical grasses include *Andropogon eucomus* subsp. *huillensis*, *Digitaria gazensis*, *Hyparrhenia filipendula*, *H. dichroa*, *Loudetia simplex*, *Panicum maximum*, and *Themeda triandra*. *Pteridium aquilinum* and the thorny climber *Smilax anceps* are characteristic species in moist grassy communities of this biome.

Climate With a MAP of 1026 mm this biome is one of the three most mesic savanna regional biomes in the MBSA area. Rain falls predominantly in summer and winters are dry. Summers are warm to hot and winters cool to mild. It has a MAT of 19.9 °C and, owing to its increased elevation, it has the lowest summer temperature maxima of regional biomes at similar latitude (Fig. 3.10).

Topography, Geology and Soils The topography comprises broken and often rugged terrain and ridges. Soils and geology are variable. Most important rocks include gneissic and other

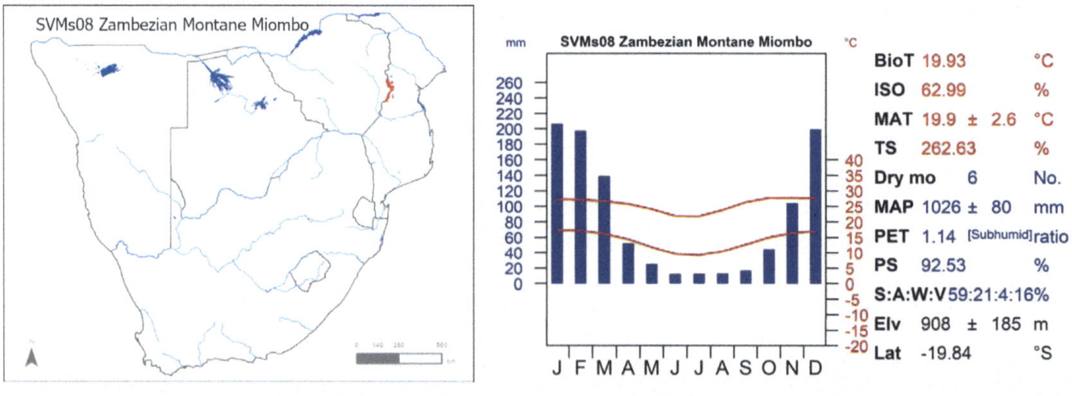

Fig. 3.10 The distribution area and the climatic diagram of the regional biome SVMs08 Zambezian Montane Miombo

granitoids, dolerite, pyroxenite gabbros, siltstones, micaschists, and quartzites. Soils are mainly fairly acidic sandy clay loams.

Sources Pedro and Barbosa (1955), Goodier and Phipps (1963), Wild and Barbosa (1967), Lötter et al. (2023).

SVMs09 Lowveld Mountain Bushveld

SVl 7 Gravelotte Rocky Bushveld (Dayaram et al. 2019)
SVl 8 Tzaneen Sour Bushveld (Dayaram et al. 2019)
SVl 9 Legogote Sour Bushveld (Dayaram et al. 2019)
SVl 11 Malelane Mountain Bushveld (Dayaram et al. 2019)
SVl 12 Kaalrug Mountain Bushveld (Dayaram et al. 2019)
SVl 15 Northern Lebombo Bushveld (Dayaram et al. 2019)
SVl 16 Southern Lebombo Bushveld (Dayaram et al. 2019)
SVl 17 Lebombo Summit Sourveld (Dayaram et al. 2019)
SVl 27 Crocodile Gorge Mountain Bushveld (Dayaram et al. 2019)
SVsl2 Lebombo Summit Sourveld (Lötter et al. 2023)
SVsl3 Northern Lebombo Bushveld (Lötter et al. 2023)
SVsl4 Southern Lebombo Bushveld (Lötter et al. 2023)

Distribution The two parallel ridges of this regional biome runs, firstly, from Makhado (Louis Trichardt) in the north, southwards along the base of the northern Great Escarpment to the foothills of Barberton in Mpumalanga. The second ridge runs along the Lebombo mountains, from Shingwedzi in the north and southwards to Mkuzi in the south (Fig. 3.11).

Vegetation Physiognomy and Composition Predominantly open woodlands on steep slopes and shallow soils with higher than average rainfall. Vegetation of this regional biome covers a range in altitude which results in a diverse composition. Some of the dominant trees throughout its range include *Dombeya rotundifolia, Heteropyxis natalensis, Combretum apiculatum, C. zeyheri, Englerophytum magalismontanum, Erythrina lysistemon, E. latissima, Euphorbia confinalis, E. ingens, Faurea rochetiana, F. saligna, Ficus burkei, F. ingens, Lannea discolor, Mystroxylon aethiopicum, Mundulea sericea, Parinari curatellifolia, Peltophorum africanum, Pterocarpus angolensis, Senegalia ataxacantha, S. burkei, S. erubescens, Terminalia sericea, Vachellia caffra, V. davyi, V. sieberiana,* and *Vitex obovata.* Large shrubs or scramblers are *Bauhinia galpinii, Diospyros galpinii, D. lycioides* subsp. *sericea, Erythroxylon delagoense, Flemingia grahamiana, Grewia bicolor, Gymnanthemum amygdalinum,* and *Pseudarthria hookeri.* Herbaceous and succulent plants are characterised by *Aloe affinis, A. marlothii, A. petricola, A. spicata, Barleria affinis, Euphorbia vandermerwei, Gerbera ambigua, G. viridifolia, Hemizygia persimilis, Hibiscus sidiformis, Kalanchoe rotundifolia, Ocimum gratissimum, Waltheria indica.* Dominant graminoids are *Andropogon schirensis, Aristida congesta, Cymbopogon caesius, C. nardus, Hyparrhenia cymbaria, Hyperthelia dissoluta, Panicum maximum,* and *Themeda triandra.*

Climate This regional biome experiences summer rainfall, with about half of the MAP of 743 mm falling in the three summer months, and winters are relatively dry. The MAP generally increases from the foot slopes to higher elevations of the mountains. MAT is 20.7 °C and summers are hot and winters are relatively mild. Frequency of frost varies from very infrequent to occasional at higher elevations. The regional biome has higher rainfall and temperatures than that of the closely related RB Limpopo Mountain Bushveld (Fig. 3.11).

Topography, Geology and Soils The topography is formed by low to high mountains often with rocky, broken slopes and ridges but also with gentle lower slopes and undulating plains near mountain bases or on pediments. Geology is varied including igneous basalt, dolerite, granite, granodiorite, granophyre, rhyolite and tonalite as well as sedimentary sandstone and shale, and metamorphic amphibolite, gneiss, migmatite, quartzite and schist. Soils are generally stony shallow lithosols, sandy or gravelly and well drained, but can include some clayey soils at lower pediment edges.

Sources van der Schijff and Schoonraad (1971), Scheepers (1977), Moll (1978), van Rooyen (1978), Coetzee (1983), Gertenbach (1983b), Deall (1985), Acocks (1988), Deall et al. (1989),

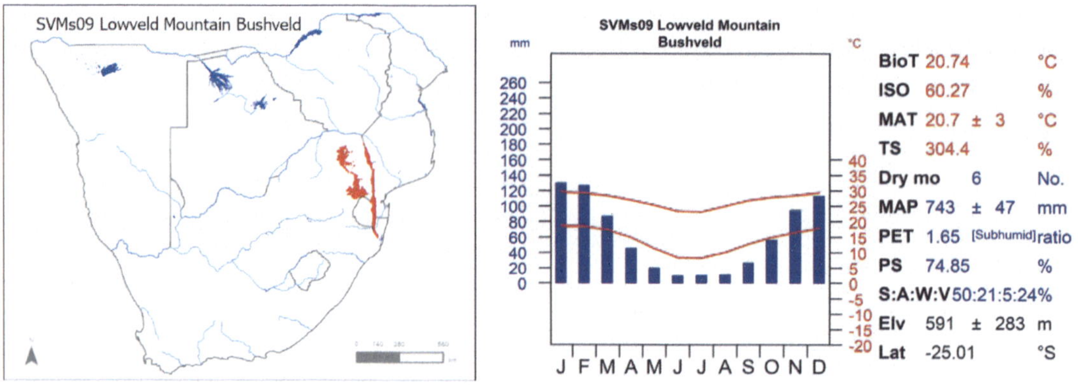

Fig. 3.11 The distribution area and the climatic diagram of the regional biome SVMs09 Lowveld Mountain Bushveld

Stalmans (1990), Bredenkamp and Matthews (1991), Camp (1999c), de Frey (1999), Smith (2001), van Wyk and Smith (2001), Rutherford et al. (2006).

SVMs10 Lowveld Sandy Bushveld

SVl 1 Makuleke Sandy Bushveld (Dayaram et al. 2019)
SVl 2 Nwambyia-Pumbe Sandy Bushveld (Dayaram et al. 2019)
SVl 3 Granite Lowveld (Dayaram et al. 2019)
SVl 10 Pretoriuskop Sour Bushveld (Dayaram et al. 2019)
SVl 18 Tembe Sandy Bushveld (Dayaram et al. 2019)
SVl 19 Western Maputaland Sandy Bushveld (Dayaram et al. 2019)
SVl 25 Maputaland Pallid Sandy Bushveld (Dayaram et al. 2019)
SVl 26 Muzi Palm Veld and Wooded Grassland (Dayaram et al. 2019)
SVsl1 Gaza Sandy Guibourtia Woodland (Lötter et al. 2023)
SVsl6 Nwambiya-Pumbe Sandy Bushveld (Lötter et al. 2023)
SVsl7 Tembe Sandy Bushveld (Lötter et al. 2023)
SVsl8 Urronga Lowland Dry Woodland (Lötter et al. 2023)

Distribution Situated mainly on the more sandy low-lying inland plains lying between the Great Escarpment in South Africa and the coast in Mozambique and northern KwaZulu-Natal (Fig. 3.12). It stretches discontinuously from northeastern KwaZulu-Natal in the south, through to the Gaza and Inhambane Provinces in Mozambique and the Save River in the north. It also extends from Eswatini northwards through the lowlands of the Mpumalanga and Limpopo Provinces as far as the eastern Soutpansberg and includes a small part of the Zimbabwean lowveld in the Gonarezhou National Park.

Vegetation Physiognomy and Composition The vegetation of the biome is physiognomically open to closed woodland. The most dominant trees are *Vachellia nigrescens, Terminalia sericea, Afzelia quanzensis, Albizia forbesii, A. versicolor, Annona senegalensis, Balanites maughamii, Cassia abbreviata, Cleistanthus schlechteri, Combretum apiculatum, C. collinum, C. zeyheri, Crossopteryx febrifuga, Erythrophleum lasianthum, Guibourtia conjugata, Hymenocardia ulmoides, Lannea antiscorbutica, Lannea schweinfurthii* var. *stuhlmannii, Peltophorum africanum, Pseudolachnostylis maprouneifolia, Terminalia myrtifolia, Pterocarpus lucens, Pterocarpus rotundifolius, Pterocarpus angolensis, Strychnos madagascariense, S. spinosa,* and *Ozoroa sphaerocarpa*. Towards the northeast *Guibourtia conjugata, Millettia stuhlmannii,* and *Philenoptera bussei* may become dominant. Shrubs include *Alchornea laxiflora, Albertisia delagoensis, Antidesma membranaceum, A. rufescens, Coptosperma zygoon, Croton pseudopulchellus, Dalbergia nitidula, Ehretia amoena, Erythrina humeana, Erythrococca menyharthii, Grewia bicolor, G. gracillima, G. microthyrsa, Gymnanthemum coloratum* subsp. *coloratum, Pavetta catophylla, P. vanwykiana,*

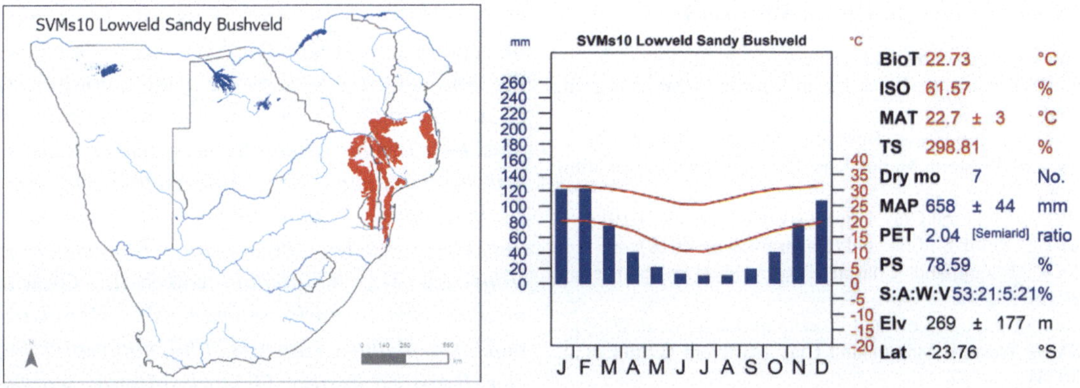

Fig. 3.12 The distribution area and the climatic diagram of the regional biome SVMs10 Lowveld Sandy Bushveld

Senna petersiana, Uvaria gracilipes, Vepris bremekampii, Vitex ferruginea, and *Xylotheca kraussianum.* The herbaceous layer is composed of *Alloteropsis papillosa, Aristida congesta* subsp. *barbicollis, A. stipitata subsp. gracilifolia, Brachiaria deflexa, B. nigropedata, Cenchrus biflorus, C. incertus, Chloris gayana, C. virgata, Cynodon dactylon, Cyperus distans, Cymbopogon pospischilii, Digitaria eriantha, Dicerocaryum senecioides, Enneapogon cenchroides, Eragrostis pallens, E. superba, Hemizygia bracteosa, Heliotropium indicum, H. steudneri, Indigofera filipes, Panicum maximum, Perotis patens, Phyllanthus parvulus, Pogonarthria squarrosa, Schmidtia pappophoroides, Urochloa mossambicensis,* and *Vigna unguiculata.*

Climate The MAP of 658 mm varies from below 500 mm in the northern parts of both the South African lowveld and Gaza Province of Mozambique to over 800 mm towards the coast in Inhambane Province in Mozambique and towards the escarpment in South Africa and Eswatini. Winters are dry. Summers are hot and winters are usually warm. The biome is generally frost free, but infrequent frost can occur near higher elevation margins near the escarpment in the west (Fig. 3.12).

Topography, Geology and Soils The landscape is made up largely of extensive undulating plains and flats. Geological substrate in South Africa is mainly granite and sandstone, but also includes gneiss, conglomerate and migmatite. Shallow to deep, freely drained sandy and sandy loam soils, and dune sands are dominant in Mozambique and northeastern KwaZulu-Natal. Also important in Mozambique is reef limestone, sandy limestone, gritty limestone and gritty conglomerate. On the broad granitic plains in South Africa, localized clayey soils typically occur near the bottom of the catena created by the low-lying incised drainage lines, in a landscape dominated by sandy soils.

Sources de Moor et al. (1977), Moll (1978), van Rooyen (1978), van Rooyen et al. (1981a), Bredenkamp (1982, 1985, 1986a, 1987), Bredenkamp et al. (1983, 1993), Coetzee (1983), Gertenbach (1983b), Bredenkamp and Theron (1985, 1988, 1990, 1991), Goodman (1990), Witkowski and O'Connor (1996), Trollope et al. (1998), Camp (1999e), Higgins et al. (1999), Matthews et al. (2001), Scholes et al. (2001), Gaugris et al. (2004), Rutherford et al. (2006), Peel et al. (2007), Cronje et al. (2008), Siebert and Eckhardt (2008), February and Higgins (2010), Smit et al. (2010), Shackleton and Scholes (2011), Shannon et al. (2011), Smit et al. (2013), Wigley et al. (2014), Combrink (2017), Bouwer et al. (2020), Janecke (2020), Janecke et al. (2020), Theron et al. (2020), Urban et al. (2020), van Aardt et al. (2020), LeClare et al. (2020), van Coller et al. (2021), Lenfers et al. (2022), Coetsee et al. (2023a, b), Cook et al. (2023), Gordon et al. (2023).

SVMs11 Lowveld Clayey Bushveld

SVl 4 Delagoa Lowveld (Dayaram et al. 2019)
SVl 5 Tshokwane-Hlane Basalt Lowveld (Dayaram et al. 2019)
SVl 6 Gabbro Grassy Bushveld (Dayaram et al. 2019)
SVl 20 Western Maputaland Clay Bushveld (Dayaram et al. 2019)
SVl 21 Makatini Clay Thicket (Dayaram et al. 2019)
SVl 23 Zululand Lowveld (Dayaram et al. 2019)
SVl 24 Zululand Coastal Thornveld (Dayaram et al. 2019)
SVsl5 Limpopo Lowland Woodland (Lötter et al. 2023)
SVsl9 Western Maputaland Clay Bushveld (Lötter et al. 2023)

Distribution Occurs mainly along the more clayey low-lying inland plains between the Great Escarpment of South Africa and the coast of southern Mozambique (Fig. 3.13). It stretches fairly continuously from Ulundi and Empangeni in northeastern KwaZulu-Natal in the south, forming a N–S orientated belt on the plains west of the base of the Lebombo Mountains through Eswatini and northeastern South Africa into the Kruger National Park at the Olifants River. A roughly parallel belt lies E of the base of the Lebombo Mountains from the Mkuzi Game Reserve in KwaZulu-Natal in the south and extending northwards into the Maputo and Gaza Provinces of Mozambique to the vicinity of the lower Limpopo River and Banhine National Park.

Vegetation Physiognomy and Composition The biome supports a dry, mixed deciduous open woodland, or wooded grassland on deep clay soils. The vegetation is usually comprised of open or closed woodland on clay soils where the dominant tree species are usually comprised of *Albizia anthelmintica, A. harveyi, A. petersiana, Berchemia zeyheri, Bolusanthus speciosus, Boscia mossambicensis, Combretum imberbe, C. hereroense, Cordia ovalis, Dalbergia melanoxylon, Euclea divinorum, Elaeodendron transvaalensis, Gardenia volkensii, Grewia bicolor, Gymnosporia senegalensis, Manilkara mochisia, Pappea capensis, Philenoptera violacea, Pyrostria hystrix, Olax dissitiflora, Searsia gueinzii, Scotia brachypetala, Senegalia senegal* var. *rostrata, Terminalia prunioides, Vachellia grandicornuta, V. luederitzii* var. *retinens, V. nigrescens, V. nilotica* subsp. *kraussiana,* and *Ziziphus mucronata.* The shrubs and climbers are dominated by *Azima tetracantha, Cadaba termitaria, Capparis tomentosa, Cissus quadrangularis, C. rotundifolia, Cocculus hirsutus, Croton menyhartii, Euphorbia grandicornis, Flueggea virosa, Gossypium herbaceum, Grewia villosa, Maerua juncea, M. parvifolia, Mystroxylon aethiopicum, Phyllanthus reticulatus,* and *Schotia capitata.* Subshrubs and herbaceous species include *Abutilon indicum* subsp. *guineense, Adenium multiflorum, A. swazicum, Aloe suffulta, Corbichonia decumbens, Cotyledon barbeyi, Crotalaria schlechteri, Ecbolium glabratum, Gladiolus brachyphyllus, Euphorbia knuthii, E. lugardiae, Gnidia capitata, Gossypium herbaceum* var. *africanum, Hibiscus trionum, Hilliardiella oligocephala,*

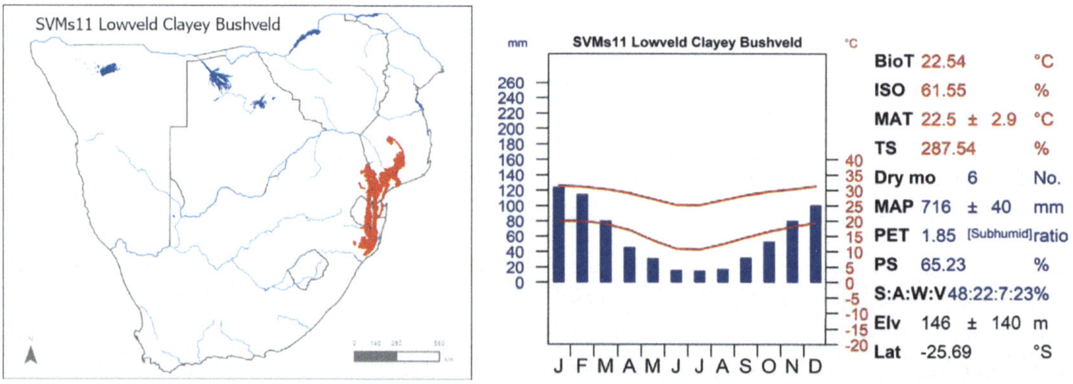

Fig. 3.13 The distribution area and the climatic diagram of the regional biome SVMs11 Lowveld Clayey Bushveld

Indigofera lupatana, I. schimperi var. *schimperi, Jatropha variifolia, Kalanchoe lanceolata, Merremia palmata, Neorautanenia mitis, Polygala senensis, Rhynchosia albissima, Sesbania sesban, S. tetraptera* subsp. *tetraptera, Sida hoepfneri, Solanum torreanum, Tephrosia purpurea* var. *pubescens*, and *Turbina oblongata*. Grasses are often dominant, some of which include *Alloteropsis cimicina, Andropogon gayanus* var. *polycladus, Aristida congesta* subsp. *barbicollis, Bothriochloa insculpta, Brachiaria eruciformis, Cenchrus ciliaris, Dinebra retroflexa* var. *condensata, Enneapogon cenchroides, Eragrostis barbinodis, E. cilianensis, E. cylindriflora, E. superba, Heteropogon contortus, Hyperthelia dissoluta, Leptochloa eleusine, L. panicea, Panicum coloratum, P. deustum, P. maximum, Perotis patens, Schoenefeldia transiens, Sehima galpinii, Setaria incrassata, Sorghum versicolor, Sporobolus pyramidalis, Themeda triandra,* and *Urochloa mossambicensis.*

Climate The MAP of 716 mm varies from less than 550 mm in the central parts of the Kruger National Park in South Africa and the Banhine National Park area in Mozambique, to over 800 mm in the southernmost parts in KwaZulu-Natal in South Africa and areas near the coast north of Maputo. Although rain falls predominantly in summer, slightly less than half falls in the summer months. Notwithstanding the relatively dry winters, some winter rains occur, especially in those parts nearer the coast in KwaZulu-Natal and Mozambique. Summers are hot and winters are warm. The regional biome is generally frost free (Fig. 3.13).

Topography, Geology and Soils The biome comprises extensive flat plains or gently undulating terrain, including various alluvial and bottomland areas. Most common underlying rock is basalt, although shale, siltstone, conglomerate, dolerite and gabbro also occur. Extensive areas of clay-rich fluvial deposits are found in Mozambique. Poorly drained black, brown or red clayey soils dominate, including clay loams, sandy clay loams and sandy clays. Dark vertic or melanic clay soils develop in low lying areas.

Sources Ward (1962), Downing (1972), Coetzee et al. (1977), de Moor et al. (1977), Gertenbach (1978, 1983b), Gertenbach and Potgieter (1978), Moll (1978), Downing (1980), Macdonald (1980, 1981), Coetzee (1983), Watson and Macdonald (1983), Whateley and Porter (1983), Goodman (1990), Bredenkamp and Deutschländer (1994, 1995), Trollope et al. (1998), Camp (1999a, e), Skowno et al. (1999), van der Linden (2004), van der Linden et al. (2005, 2008), Rutherford et al. (2006), Waldram et al. (2008), February and Higgins (2010), Smit et al. (2010), Wigley et al. (2014), Veldhuis et al. (2016), Harris et al. (2019).

SVMs12 Subtropical East Coast Woodland

CB 1 Maputaland Coastal Belt (Dayaram et al. 2019)
CB 2 Maputaland Wooded Grassland (Dayaram et al. 2019)
CB 3 KwaZulu-Natal Coastal Belt Grassland (Dayaram et al. 2019)
CB 5 Transkei Coastal Belt (Dayaram et al. 2019)
CB 6 KwaZulu-Natal Coastal Belt Thornveld (Dayaram et al. 2019)
SVcs1 Maputaland Coastal Wooded Grassland (Lötter et al. 2023)

Distribution It forms a belt along the eastern coast from the vicinity of Haga-Haga in the Eastern Cape of South Africa in the south, stretching irregularly northwards along the coast of Transkei to around Mboytyi (where it is interrupted by the coastal belt of the RB Southern Sandstone Sourveld). It continues as a broad coastal belt from around Port Edward in KZN northwards through Durban, Richards Bay and coastal Maputaland, into Mozambique and the areas south of Maputo Bay. It continues along the coast north of Maputo to the vicinity of Bilene in southern Mozambique (Fig. 3.14).

Vegetation Physiognomy and Composition Forms an open semi-deciduous woodland, or more commonly a wooded grassland on coastal sandy soils. It is dominated mainly by the palm *Hyphaene coriacea* and the trees *Garcinia livingstonei, Strychnos madagascariensis, Strychnos spinosa*, and *Syzygium cordatum*. Other frequent species are *Vachellia natalitia, Annona senegalensis, Antidesma venosum, Bridelia* *carthartica* subsp. *cathartica, Phoenix reclinata, Psydrax moggii, Sclerocarya birrea* subsp. *caffra, Sclerocroton integerrimum, Trichilia emetica, Vangueria infausta* and *Xylotheca kraussiana* var. *glabrifolia.* Trees and woody shrubs occasionally present are *Vachellia gerrardii, Afzelia quanzensis, Albizia adianthifolia, Catunaregam obovata, Coddia rudis, Commiphora schlechteri, Dialium schlechteri, Empogona coriacea, Ficus burtt-davyi, Gymnosporia arenicola, Kraussia floribunda, Lagynias monteiroi, Psorospermum febrifugum, Psydrax locuples, Searsia natalensis, Terminalia sericea, Tricalysia delagoensis,* and *Turraea nilotica.* The shrub *Helichrysum kraussii* and scrambler *Smilax anceps* are often dominant but other shrubs and herbaceous species include *Agathisanthemum bojeri, Alysicarpus vaginalis, Chamaecrista plumosa, Chrysocoma ciliata, Clerodendrum ternatum, Crotalaria monteiroi, Desmodium dregeanum, Dicerocaryum senecioides, Disa woodii, Eriosema parviflorum, Gazania krebsiana* subsp. *serrulata, Gladiolus crassifolius, Gloriosa superba, Helichrysum adenocarpum* subsp. *ammophilum, H. candolleanum, H. longifolium, H. silvaticum, Indigofera inhambanensis, Merremia tridentata* subsp. *angustifolia, Oocephala centaureoides, Pentarrhinum insipidum, Phyllanthus delagoensis, Polygala capillaris, Trachyandra saltii,* and *Tritonia moggii.* The fern *Pteridium aquilinum* subsp. *capense* may form extensive colonies in moist areas. The coastal grasslands are characterised by an abundance of geoxylic suffrutices or geoxyles (underground trees), represented by *Albertisia delagoensis, Eugenia capensis* subsp. *multiflora, E. mossambicensis, Diospyros villosa* (geoxyle form), *Gymnosporia markwardii, Ochna natalitia* (geoxyle form), *Parinari capensis* subsp. *incohata, Salacia rehmannii*, and particularly common or abundant, *Salacia kraussii.* Some of the grasses and sedges recorded are *Andropogon eucomus* subsp. *huillensis, Bewsia biflora, Brachiaria chusqueoides, Brachychloa schiemanniana, Cymbopogon caesius, Cynodon dactylon, Digitaria argyrotricha, D. milanjiana, Diheteropogon amplectens, Elionurus argenteus, Eragrostis ciliaris, E. inamoena, Fimbristylis cymosa, Hyparrhenia dissoluta, Ischaemum arcuatum, I. fasciculatum, Melinis nerviglumis, Mnesithea laevis, Panicum pleianthum, Perotis patens, Pogonarthria squarrosa, Rhynchospora brownii, Trachypogon spicatus,* and *Urelytrum squarrosum.*

Climate A MAP of 983 mm places this biome among the higher rainfall mesic regional biomes. Although only 37% of MAP falls in the summer months, rainfall is clearly concentrated in the warm season. Highest rain occurs generally nearer the coast. The winters are relatively dry although a not insignificant 11% of MAP falls in the winter months. Temperatures are relatively equitable with warm to fairly hot summers and warm to fairly cool winters. Maximum temperatures in summer are among the lowest compared to those in the other savanna regional biomes (Fig. 3.14).

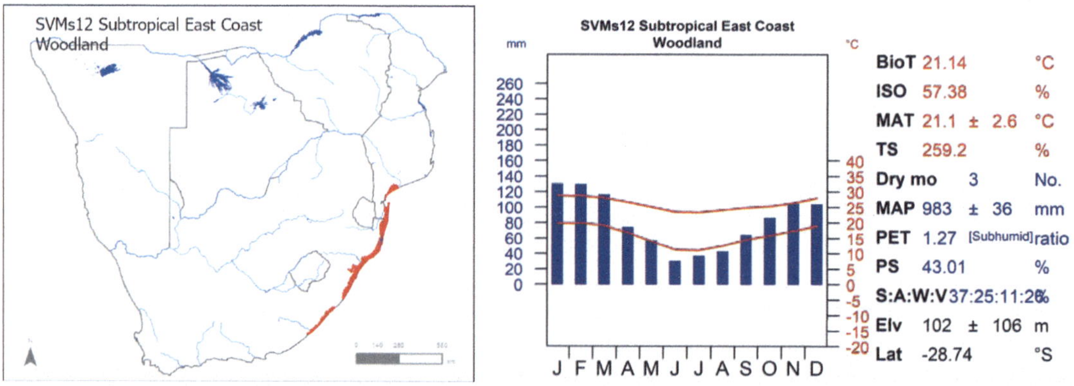

Fig. 3.14 The distribution area and the climatic diagram of the regional biome SVMs12 Subtropical East Coast Woodland

Topography, Geology and Soils The landscape comprises flat to undulating coastal plains, including fixed dunes and redistributed sands as well as recent alluvium. Towards the south the terrain includes highly dissected undulating plains, becoming hilly with steep slopes and coastal ridges. Here the rock types include sandstone, shale, mudstone and tillite. The sandy soils are nutrient poor except in some interdune depressions where organic-rich soils are sometimes found.

Sources Acocks (1953), Pedro and Barbosa (1955), Myre (1964, 1971), Edwards (1967), Wild and Barbosa (1967), Moll (1972), Venter (1972), White (1976), Moll (1978, 1980), Moll and White (1978), Weisser (1978), Weisser and Marques (1979), Ward (1980), Hoffman (1983), White (1983), Weisser (1987), Acocks (1988), Roberts (1993), van Wyk (1994, 1996), Lubbe (1997), Camp (1999a), Matthews et al. (1999), Smith (2001), van Wyk and Smith (2001), Felton (2002), Heijnis (2004), van der Linden et al. (2005), Mucina et al. (2006c), Siebert et al. (2011), Stewart et al. (2020), Lötter et al. (2023), Starke et al. (2023).

SVMs13 Tropical East Coast Woodland

SVct1 Cheringoma Coastal Palm Savanna (Lötter et al. 2023)
SVct2 Inharrime Coastal Palm Savanna (Lötter et al. 2023)
SVct5 Save Coastal Palm Savanna (Lötter et al. 2023)

Distribution Falling entirely within Mozambique, it occurs in the south as a broad, rather fragmented belt on the low-lying inland coastal plain between the vicinity of Xai-Xai and Inhambane. Further north, it is found as a coastal belt between the Vilanculos region and an area south of Beira, as well as from Beira to the lowest reaches of the Zambezi River (Fig. 3.15); it also extends further north into northern Mozambique which lies outside the MBSA.

Vegetation Physiognomy and Composition It forms an open woodland interspersed with coastal grasslands and open grassy wetlands. This biome is typified by the presence of some, or all of three palm species *Hyphaene coriacea*, *Phoenix reclinata,* and *Borassus aethiopum*. The tree component is typically characterised by *Annona senegalensis*, *Euphorbia lividiflora*, *Garcinia livingstonei*, *Lannea schweinfurthii*, *Ozoroa obovata*, *Piliostigma thonningii*, *Pseudolachnostylis maprouneifolia*, *Sclerocarya birrea* subsp. *caffra*, *Sideroxylon inerme*, *Strychnos madagascariensis*, *S. spinosa*, *Uapaca nitida*, *Vachellia gerrardii,* and *Vitex doniana*. Shrubby species include *Ancylobotrys petersianus*, *Cordia sinensis*, *Croton madandensis*, *C. megalocarpus*, *Dichrostachys cinerea* subsp. *plurijuga*, *Grewia transzambesiaca*, *Gymnosporia buxifolia*, *Morella spathulata*, *Polysphaeria lanceolata*, and *Xylotheca kraussiana*. The grassy component includes *Andropogon appendiculatus*, *A. chinensis*, *Bothriochloa bladhii*, *Cynodon dactylon*, *Digitaria diagonalis*, *D. milanjiana*, *Eragrostis chapelieri*, *Heteropogon contortus*, *Hyperthelia dissoluta*, *Imperata cylindrica*, *Oryza longistaminata*, *Pogonarthria squarrosa*, *Panicum maximum,* and *P. subflabellatum.*

Climate This biome has a MAP of 968 mm falling mainly in summer with highest rainfall generally nearer the coast. Winters are relatively dry but some rain does occur. Temperatures are high and fairly equitable with hot summers and warm winters. Its MAT of 24.4 °C is among the highest of the savanna regional biomes (Fig. 3.15).

Topography, Geology and Soils This biome comprises extensive coastal flats which include fixed and consolidated dunes as well as remobilized longitudinal dunes. Recent alluvia are widespread and with fluvial overflow, and the dominant soils are sandy loams and sandy clay loams.

Sources Pedro and Barbosa (1955), Wild and Barbosa (1967), Lötter et al. (2023).

SVMs14 Sub-Escarpment Savanna

Gs 20 Moist Coast Hinterland Grassland (Dayaram et al. 2019)
SVl 14 Swaziland Sour Bushveld (Dayaram et al. 2019)
SVs 1 Thukela Valley Bushveld (Dayaram et al. 2019)
SVs 2 Thukela Thornveld (Dayaram et al. 2019)
SVs 3 KwaZulu-Natal Hinterland Thornveld (Dayaram et al. 2019)
SVs 4 Ngongoni Veld (Mucina and Rutherford 2006)
SVs 6 Eastern Valley Bushveld (Dayaram et al. 2019)
SVs 7 Bhisho Thornveld (Dayaram et al. 2019)
SVs 8 South Eastern Coastal Thornveld (Dayaram et al. 2019)

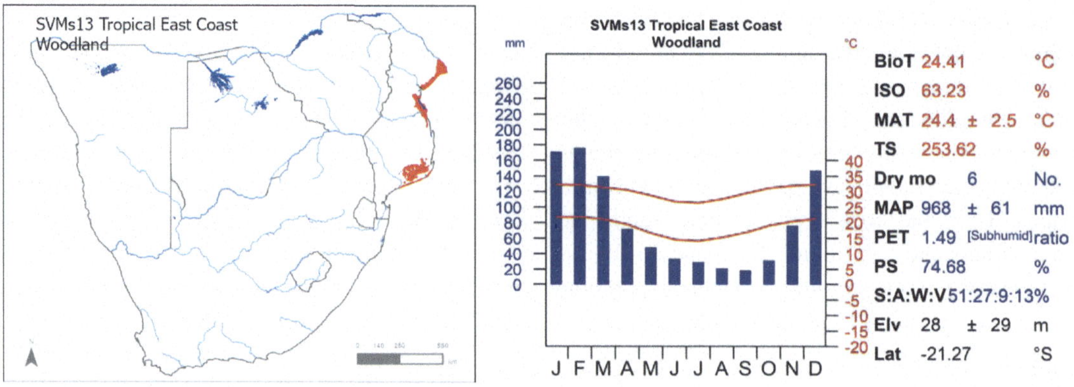

Fig. 3.15 The distribution area and the climatic diagram of the regional biome SVMs13 Tropical East Coast Woodland

Distribution The biome occurs as a relatively narrow belt and series of patches for the most part immediately inland of the RB Subtropical East Coast Woodland from the Bisho area, inland of East London in the Eastern Cape Province, northwards through the Transkei and KwaZulu-Natal extending further between the lowveld and grassland areas in Eswatini to the Badplaas area in Mpumalanga. It extends further inland along the lower reaches of several river valleys including those of the Kei River in the south and the Tugela River in the north (Fig. 3.16).

Vegetation Physiognomy and Composition The vegetation physiognomy varies from dense bush clumps at least 3–5 m high, to scattered low trees and occasionally taller trees. Open to dense succulent vegetation is found on the hottest and driest slopes in the north and more widely further S. Markedly more grassy vegetation occurs in the upland areas. This regional biome extends over a lengthy N–S gradient and species composition varies considerably across this gradient. Important tree and shrub species include *Coddia rudis, Combretum apiculatum, C. molle, Dichrostachys cinerea, Euclea crispa* subsp. *crispa, Erythrina lysistemon, Grewia occidentalis, Pappea capensis, Schotia brachypetala, Spirostachys africana, Vachellia davyi, V. natalitia, V. nilotica, V. tortilis* subsp. *heteracantha, V. sieberiana,* and *Ziziphus mucronata.* Succulent plants include *Aloe arborescens, A. marlothii* subsp. *marlothii, A. rupestris, Euphorbia grandidens, E. tirucalli, E. ingens,* and *E. triangularis*. Typically occurring grasses include *Aristida congesta, A. junciformis* subsp. *junciformis, Eragrostis curvula, E. plana, Heteropogon contortus, Hyparrhenia hirta, Melinis nerviglumis, M. repens, Panicum maximum, Sporobolus africanus, Themeda triandra,* and *Tristachya leucothrix.*

Climate The MAP of 837 mm falls mainly in summer, winters are relatively dry but some rain occurs. With a MAT of 18 °C this regional biome is among the coolest mesic savanna regional biomes. Summers are warm to hot and winters are mild to cool. Arguably, at least the southern part of this unit has the least 'typical' savanna climate within the CB Southern African Mesic Savanna (Fig. 3.16).

Topography, Geology and Soils As may be expected from the highly dissected landscapes of this regional biome, a great variety of soils is found. In the main valleys there are base-rich clay soils developed along the lower valley sides, and soils developed from colluvial material, mainly originating from sandstone in the northern areas. Many of the soils are highly erodible. Ferralitic soils are found on the well-drained upland parts.

Sources West (1951), Killick (1958), Edwards (1967), Morris (1967, 1969), Grunow and Morris (1969), Moll (1976), Coetzee and

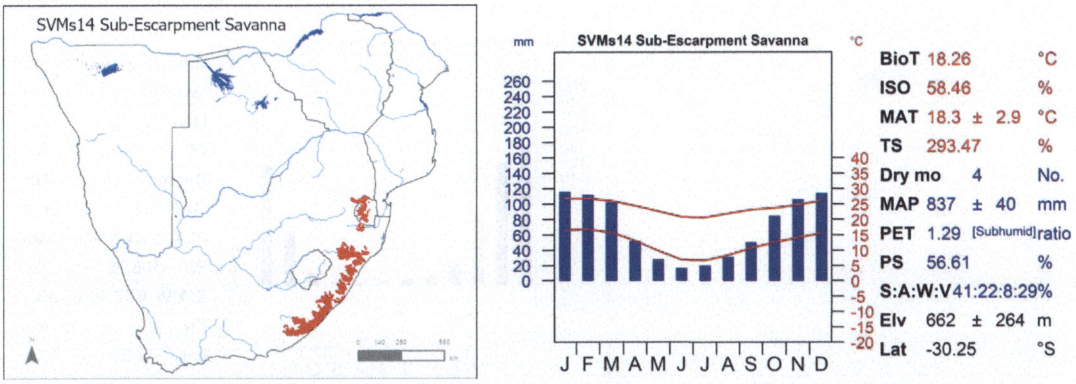

Fig. 3.16 The distribution area and the climatic diagram of the regional biome SVMs14 Sub-Escarpment Savanna

Nel (1978), Acocks (1988), Perkins (1997), Camp (1999a, d, e), De Frey (1999), Perkins et al. (1999), Stalmans et al. (1999), Breebaart et al. (2001), Camp (2001), Stalmans (2002), Wiseman et al. (2004), Rutherford et al. (2006), Mills et al. (2012), Nicolay et al. (2023).

SVMs15 Zambezian Sandy Savanna

SVsz4 Guro Dry Woodland (Lötter et al. 2023)
SVsz6 Madanda Sandstone Dry Woodland (Lötter et al. 2023)
SVsz7 Maringue Sandstone Dry Woodland (Lötter et al. 2023)

Distribution Divided between Mozambique and Zimbabwe, this regional biome falls almost entirely within the catchment area of the Zambezi River. In Mozambique it covers a broad lowland area from the vicinity of Tete and the Nyamapanda border post, tapering southwards to a point at the Gorongosa National Park. It extends into northeastern Mashonaland in Zimbabwe, where it occupies parts of the low-lying areas including those around the Umfurudzi Safari Park and Nyamapanda as well as valleys northwest of the Nyanga Mountains (Fig. 3.17).

Vegetation Physiognomy and Composition It forms a mixed deciduous woodland rarely with any dominance except in areas of *Colophospermum mopane*. The miombo tree *Julbernardia globiflora* is a typical Zambezian element that occurs sporadically or frequently throughout. Acacias are often dominant or frequent, especially *Senegalia nigrescens*, *Vachellia gerrardii*, *V. nilotica* subsp. *kraussiana*, and *V. tortilis*. The typical woody species include *Adansonia digitata*, *Afzelia quanzensis*, *Albizia anthelmintica*, *A. versicolor*, *Combretum adenogonium*, *C. apiculatum*, *C. hereroense*, *C. imberbe*, *C. molle*, *Crossopteryx febrifugum*, *Diospyros mespiliformis*, *Guibourtia conjugata*, *Lannea schweinfurthii*, *Markhamia sansibarica*, *Philenoptera violacea*, *Piliostigma thonningii*, *Pterocarpus angolensis*, *Sclerocarya birrea* subsp. *caffra*, *Strychnos madagascariensis*, *Terminalia myrtifolia*, *T. sericea*, and *Xeroderris stuhlmannii*. Some shrubs and lianas include *Artabotrys brachypetala*, *Capparis tomentosa*, *Combretum mossambicense*, *Dalbergia melanoxylon*, *Grewia* spp., and *Landolphia kirkii*. Soft shrubs and herbaceous include *Alchornea laxiflora*, *Cissus cornifolia*, *Commelina forskaolii*, *Crotalaria anthyllopsis*, *C. hyssopifolia*, *C. laburnifolia*, *Cyphostemma congestum*, *Hermannia glanduligera*, *H. kirkii*, *Hibiscus caesius*, *H. meeusei*, *Indigofera trita* var. *scabra*, *Melhania acuminata*, *M. forbesii*, *Neorautanenia mitis*, *Scutellaria schweinfurthii* subsp. *paucifolia*, *Sphaeranthus angolensis*, *Sphenostylis erecta*, *Tacazzea apiculata*, *Tephrosia villosa* subsp. *ehrenbergiana*, *Thunbergia petersiana*, and *Xerophyta humilis*. Some grasses include *Alloteropsis semialata*, *Cenchrus purpureus*, *Enteropogon macrostachyus*, *Hyparrhenia*

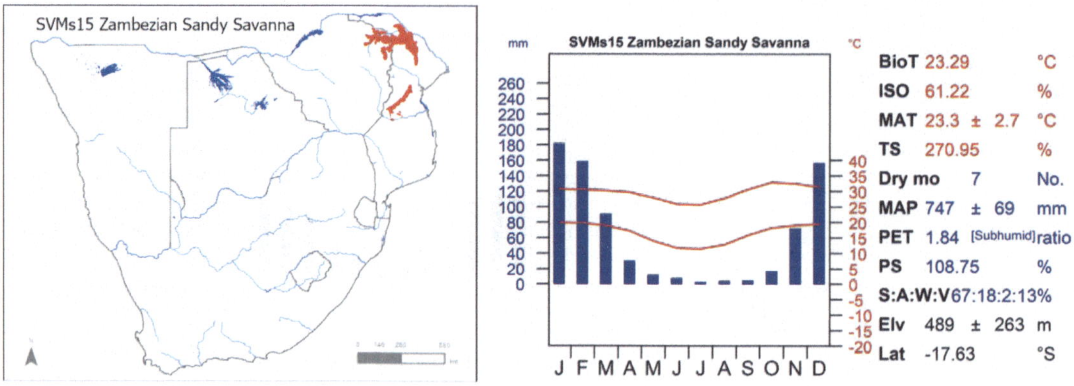

Fig. 3.17 The distribution area and the climatic diagram of the regional biome SVMs15 Zambezian Sandy Savanna

finitima, Hyperthelia dissoluta, Leptocarydion vulpiastrum, Schmidtia pappophoroides, and *Tricholaena monachne*.

Climate The MAP of 747 mm falls mainly in summer and winters are dry. With a MAT of 23.3 °C summers are hot and winters are warm. Day temperature maxima clearly peak in late spring and early summer before onset of the main summer rains (Fig. 3.17).

Topography, Geology and Soils The biome is found on extensive undulating plains and valleys with broad alluvial deposits as well as rocky hills. Rock types include gritty conglomerate, sandstone, banded gneiss, granulitic gneiss, hornblende gneiss, and granulite. Sandy loam soils are widespread.

Sources Pedro and Barbosa (1955), Wild and Barbosa (1967), Stalmans and Wishart (2005), Stalmans and Peel (2010), Lötter et al. (2023).

SVMs16 Zambezian Clayey Savanna

SVsz1 Bangomatete Rhyolite Dry Woodland (Lötter et al. 2023)
SVsz2 Canxixe Lowland Dry Woodland (Lötter et al. 2023)
SVsz3 Dombe Basalt Dry Woodland (Lötter et al. 2023)
SVsz9 Rift Valley Lowland Woodland (Lötter et al. 2023)
SVsz10 Save Lowland Dry Woodland (Lötter et al. 2023)
SVsz12 Stormberg Dry Woodland (Lötter et al. 2023)

Distribution Occurs mainly on the lowlands of Mozambique, with several patches east of Tete and a major broad inland belt running from the Chiramba and Caia areas on the lower Zambezi River, southwards through the Gorongosa National Park before bifurcating near Beira, with both resulting arms of the belt terminating in the south near the Save River around Massangena and Save. The regional biome also enters Zimbabwe in the northeastern Mashonaland lowlands in the vicinity of the lower Mazoe and Ruenya Rivers (Fig. 3.18).

Vegetation Physiognomy and Composition A mixed deciduous woodland with little dominance except for *Senegalia nigrescens* which may be frequent over large areas. Both *Colophospermum mopane* and *Androstachys johnsonii* may form pure stands in places, the latter species mainly in the south of this biome. The miombo element *Julbernardia globiflora* may occur scattered in areas of higher ground. *Acacia* woodland is composed of *Senegalia welwitschii* subsp. *delagoensis, Vachellia gerrardii, V. karroo, V. nilotica* subsp. *kraussiana, V. tortilis* and *V. xanthophloea*, the latter gregarious and often frequent in low-lying areas. The typical woody species include *Adansonia digitata, Albizia harveyi, A. versicolor, Combretum adenogonium, C. apiculatum, C. imberbe, Diospyros kirkii, Diplorhynchus condylocarpon, Lannea schweinfurthii, Philenoptera violacea, Piliostigma thonningii, Pterocarpus brenanii, Sclerocarya birrea* subsp. *caffra, Sterculia appendiculata, S. africana, Terminalia prunioides, Xeroderris*

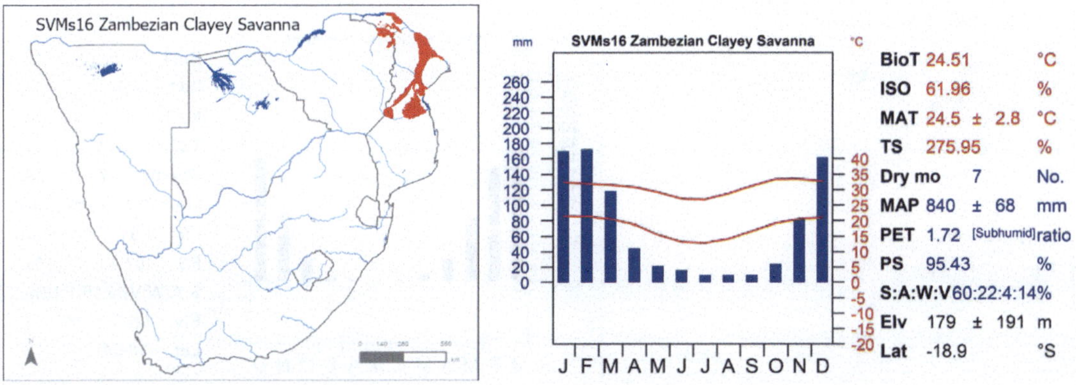

Fig. 3.18 The distribution area and the climatic diagram of the regional biome SVMs16 Zambezian Clayey Savanna

stuhlmannii, and *Ziziphus mucronata.* Some shrubs and lianas include *Artabotrys brachypetala, Capparis tomentosa, Cleistochlamys kirkii, Combretum mossambicense, Dalbergia melanoxylon, Grewia* spp., and *Maclura africana*. Open grassy areas are dominated by *Heteropogon contortus, Themeda triandra*, as well as the species of the genera *Setaria, Panicum,* and *Eragrostis*. The palms *Hyphaene coriacea* and *H. petersiana* are often characteristic elements within this biome.

Climate The MAP of 840 mm falls mainly in summer and winters are dry. Summers are hot and winters are warm. The MAT of 24.5 °C is one of the highest compared to that of other savanna regional biomes, which is commensurate with its position at low elevations and latitude (Fig. 3.18).

Topography, Geology and Soils The regional biome occurs on extensive flats and undulating plains comprising recent alluvia, gritty conglomerate and some tholeitic basalt, paragneiss, gneiss, limestone and rhyolite that overlays basalt. Soils are mainly sandy clay loams, loams, sandy loams, and gritty loams.

Sources Pedro and Barbosa (1955), Wild and Barbosa (1967), Lötter et al. (2023).

SVMs17 Great Dyke Grassland

Distribution Limited to Zimbabwe, it comprises a narrow N–S trending linear strip of ranges from the Mulingura Hills in the Mashonaland Central Province in the north to the Doro Range in the Matabeleland South Province in the south (Fig. 3.19). It is fully embedded in the surrounding RB Zambezian Dry Miombo and RB Zimbabwe Sandy Bushveld.

Vegetation Physiognomy and Composition Vegetation physiognomy is dominated by grasslands, mainly composed of *Loudetia simplex*, but also the other C4 grasses of the genera *Andropogon, Brachiaria, Digitaria, Diheteropogon, Eragrostis, Hyparrhenia, Loudetia, Setaria*, and *Tripogon*. Typical woody species are *Combretum molle, Diplorhynchus condylocarpon, Lannea discolor, Ozoroa longipetiolata, O. reticulata, Protea angolensis, P. gaguedi*, and *Tapiphyllum velutinum.* Rare patches of miombo woodland may occur including *Brachystegia glaucescens, B. spiciformis,* and *Julbernardia globiflora*, growing with *Pterocarpus angolensis.* Due to specific geology (phytotoxic ultramafics) the unit supports a number of endemic species, making it a regional centre of endemism. Endemics include *Aloe ortholopha, Dicoma niccolifera, Euphorbia memoralis, E. wildii, Leucas aggerestris, L. hephaestis, Lotononis serpenticola, Pearsonia metallifera*, and several others.

Climate The MAP of 733 mm ranges from less than 500 mm in the south in the Matabeleland South Province to over 800 mm in the north in the Mashonaland Central Province. Rain falls

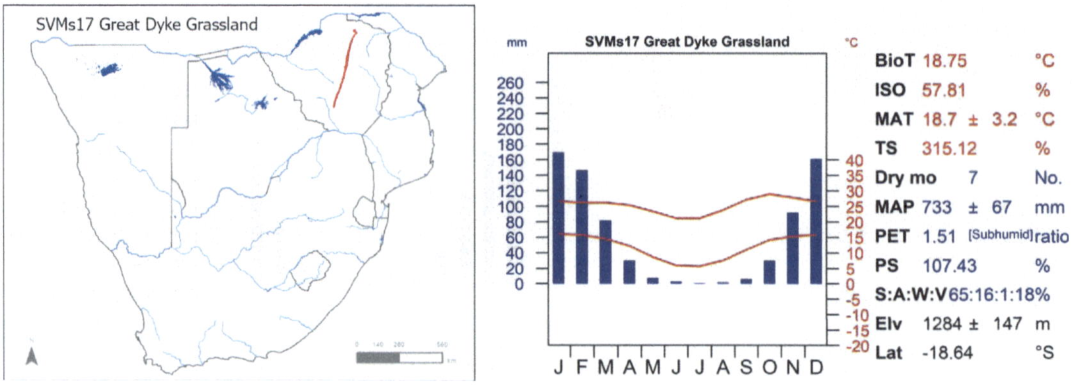

Fig. 3.19 The distribution area and the climatic diagram of the regional biome SVMs17 Great Dyke Grassland

predominantly in summer and winters are dry. Summers are hot and winters are warm to mild (Fig. 3.19).

Topography, Geology and Soils The landscape of this regional biome is characterised by a long sequence of undulating ridges and hills elevated above the surrounding plateaus. Invariably, the ultramafic geology dominates and includes serpentinite, pyroxenite, norite, gabbro, and dunite that all support shallow soils with high concentration of heavy metals (Mg, Cu, Ni, Cr and some more).

Sources Barclay-Smith (1963), Wild (1965, 1978), Proctor and Craig (1973), Werger et al. (1978a, b), Brooks and Yang (1984), Proctor and Cole (1992), van Wyk and Smith (2001).

SVMs18 Western Dolomite Bushveld

SVcb 2 Madikwe Dolomite Bushveld (Dayaram et al. 2019)

Distribution In northern South Africa, this biome unit extends as a mainly linear feature, from the hills and mountain ridges west of Thabazimbi in the east, westwards along the ridges of Witfonteinrant, Makane, and Rant Van Tweede Poort. It marginally enters Botswana around Ramotswa and Lobatse in the west. It also occurs as a few small isolated outcrops to the east of Northam in the Crocodile River valley (Fig. 3.20).

Vegetation Physiognomy and Composition Vegetation physiognomy comprises an open savanna with a prominent, well-developed continuous grass layer, with scattered or occasional tall trees and moderate cover of medium-height to low trees and shrubs. Important woody species include *Combretum apiculatum, C. imberbe, C. molle, Dichrostachys cinerea, Dombeya rotundifolia, Grewia flava, G. monticola, Kirkia wilmsii, Ozoroa paniculata, Pappea capensis, Searsia lancea, S. leptodictya, Sclerocarya birrea, Senegalia caffra, Tarchonanthus parvicapitulatus,* and *Vitex zeyheri*. The grass layer consists mainly of *Aristida canescens, A. congesta, A. diffusa, Brachiaria deflexa, Chrysopogon serrulatus, Digitaria eriantha, Enneapogon cenchroides, E. scoparius, Eragrostis rigidior, Heteropogon contortus,* and *Melinis repens*.

Climate The MAP of 574 mm is one of the lowest in mesic savanna falling mainly in summer and winters are dry. MAP is higher in the east and lower in the west of the regional biome. Summers are hot and winter days are warm with cold nights and frost (Fig. 3.20).

Topography, Geology and Soils The dominant topography is made up of low ridges, hills and pediments underlain mainly by dolomite, subordinate chert, minor carbonaceous shale, limestone and quartzite of the Malmani Subgroup (Chuniespoort Group, Transvaal Supergroup).

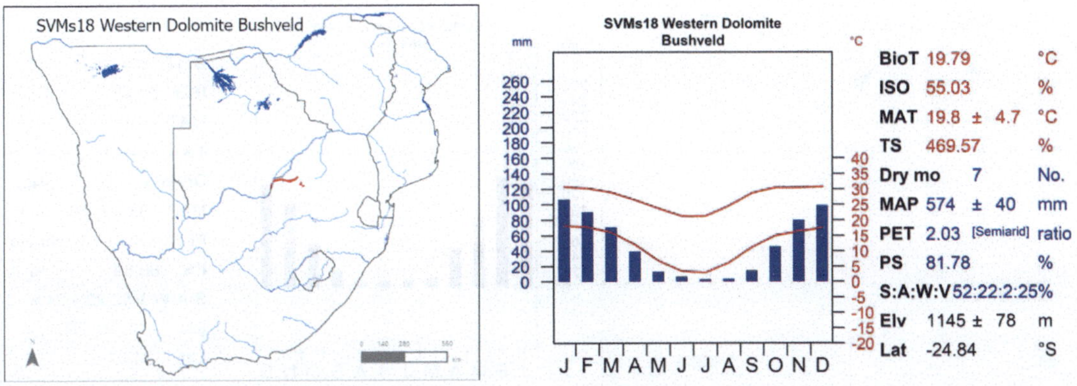

Fig. 3.20 The distribution area and the climatic diagram of the regional biome SVMs18 Western Dolomite Bushveld

Soils are predominantly shallow, stoney sands and sandy loams which become calcareous with high base status and high levels of calcium and magnesium on rocky outcrops.

Sources Zacharias (1994), Hudak and Wessman (2001), Rutherford et al. (2006), la Grange (2010a, b).

SVMs19 Eastern Dolomite Bushveld

SVcb 25 Poung Dolomite Mountain Bushveld (Dayaram et al. 2019)

Distribution Lies almost entirely in the Limpopo Province of South Africa along mountain slopes from the area south of the Abel Erasmus Pass, more or less continuously northwards along the western rain shadow side of the escarpment, including Poung Mountain near Penge becoming discontinuous towards the Wolkberg and westwards, away from the escarpment, along the Strydpoort Mountains to Chuniespoort and Mokopane (Fig. 3.21).

Vegetation Physiognomy and Composition This regional biome forms an open to closed woodland with well-developed shrub layers. Dominant trees include *Euphorbia tirucalli, Hippobromus pauciflorus, Kirkia wilmsii, Lydenburgia cassinoides* and *Seemannaralia gerrardii* and also important are *Boscia albitrunca, Combretum hereroense, Croton gratissimus, Dombeya autumnalis,* and *Vitex obovata* subsp. *wilmsii*. Important shrubs include *Barleria rotundifolia, Euchaetis schlechteri, Plectranthus xerophilus, Pouzolzia mixta, Senna petersiana, Rhynchosia nitens,* and the geoxylic suffrutex *Ozoroa albicans*. Important grasses include *Bewsia biflora, Brachiaria serrata, Enneapogon scoparius, Eragrostis lehmanniana, Heteropogon contortus, Loudetia simplex, Melinis nerviglumis, M. repens, Panicum deustum, P. maximum, Themeda triandra,* and *Tragus berteronianus*. Also characteristic is the megaherb *Dracaena transvaalensis*.

Climate The MAP of 783 mm falls mainly in summer and winters are dry. MAP is lowest in the escarpment rain shadow and highest at upper elevations where the biome meets the RB Eastern Dolomite Grassland. Summers are hot, especially at lower elevations, and winters are warm to mild. Mist is absent and frost is generally infrequent (Fig. 3.21).

Topography, Geology and Soils Positioned on low to high mountain slopes, including steep valleys and ridges. Geology is almost entirely Malmani Formation dolomite of the Transvaal Supergroup with chert always interlayered. Soils are usually shallow (Mispah soil form) and occasionally deep (Hutton and Griffin soil forms) with high pH, rich in calcium and magnesium, and with low levels of phosphorus.

Sources Stalmans (1990), Matthews (1991), Stalmans and de Klerk (1991, 1992), van Wyk and Smith (2001), Rutherford et al. (2006), Mills et al. (2012).

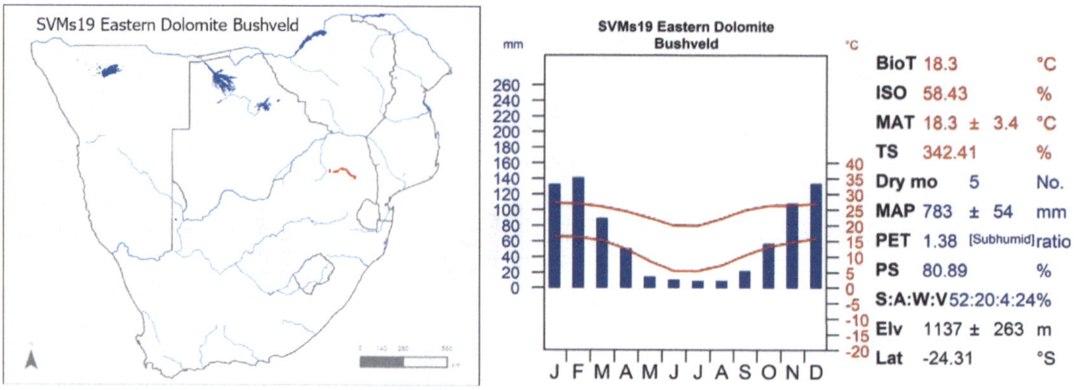

Fig. 3.21 The distribution area and the climatic diagram of the regional biome SVMs19 Eastern Dolomite Bushveld

SVMs20 Zambezian Inselberg Woodland

SVin2 Southern Inselberg Woodland (Lötter et al. 2023)

Distribution This regional biome forms a mosaic of granite or gneiss inselbergs and rocky hills, embedded within other woodland biomes, that are characteristic of northern Mozambique, extending southwards into Manica Province, and westwards in eastern Zimbabwe (Fig. 3.22). So far, not all the known occurrences have been mapped in Zimbabwe.

Vegetation Physiognomy and Composition This regional biome is characterised by open mixed deciduous woodland that is largely bare of leaves in the dry season, interspersed with bare rock and sheetrock communities. Although there is the expected species turnover from west to east, the following are generally present throughout the range of this regional biome. Trees include *Afzelia quanzensis, Albizia tanganyikensis, A. versicolor, Brachystegia glaucescens, B. microphylla, B. spiciformis, B. torrei, Commiphora africana, C. marlothii, Croton pseudopulchellus, D. natalensis* subsp *nummularia, Euphorbia cooperi, E. griseola* subsp. *mashonica, E. matebelensis, E. tirucallii, Ficus abutilifolia, F. glumosa, F. ingenis, F. natalensis, Garcinia buchananii, Hymenodictyon floribundum, Kirkia acuminata, Rothmannia fischeri* subsp. *fischeri, Schrebera alata, Steganotaenia araliacea, Sterculia africana,* and *S. quinqueloba*. Shrubs and small trees include *Clerodendron myricoides, Coptosperma neurophyllum, C. supra-axillare, Cynanchum viminale, Grewia micrantha, Monanthotaxis obovata, Mundulea sericea, Polysphaeria lanceolata, Pavetta gardenifolia, Pouzolzia mixta, Psydrax livida, Strophanthus petersianus, Tetradenia riparia,* and *Vepris reflexa*. Grasses are few and the graminoid component is dominated by the sedge *Coleochloa* spp. A number of range restricted species of *Encephalartos* (Zamiaceae) are confined to this regional biome in Mozambique and the border areas of eastern Zimbabwe.

Climate The MAP of 1108 mm is by far the highest of the savanna regional biomes, possibly positively biased by the majority of the mapped occurrences of this regional biome being embedded within the region dominated by the RB Zambezian Wet Miombo. Rain falls predominantly in summer and winters are relatively dry. Summers are hot and winters are mild (Fig. 3.22).

Topography, Geology and Soils Generally prominent, sometimes steep, rock domes as well as other assemblages of usually massive rocks including many with eroded summits. Predominant rocks are mica schists, gneissic granitoids, aluminous quartzite and granodiorite. Soils can be skeletal but are better developed on the lower slopes where moderately acidic sandy clay loams occur.

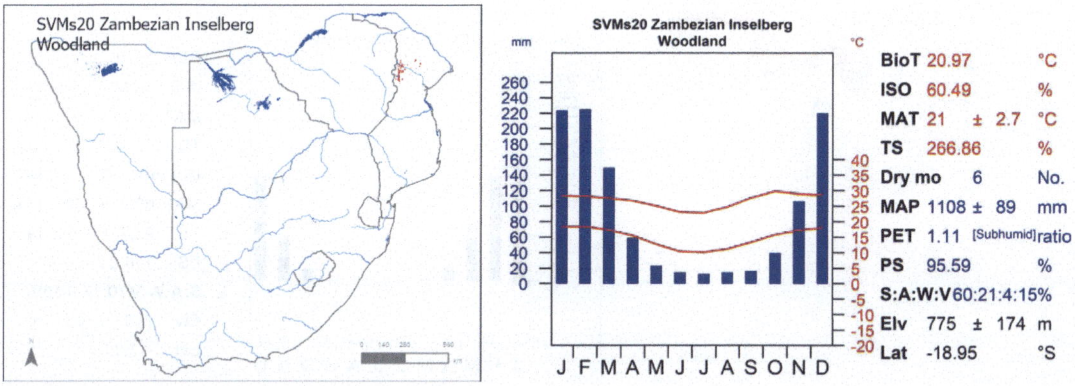

Fig. 3.22 The distribution area and the climatic diagram of the regional biome SVMs20 Zambezian Inselberg Woodland

Sources Pedro and Barbosa (1955), Wild and Barbosa (1967), Seine et al. (1998), Lötter et al. (2023).

SVMs21 Zambezian Mopane

SVmwz1 Luia Basalt Mopane Woodland (Lötter et al. 2023)
SVmwz2 Mágoè Sandstone Mopane Woodland (Lötter et al. 2023)
SVmwz3 Zambezi Valley Mopane Woodland (Lötter et al. 2023)

Distribution Mainly located on the Zambezi Valley lowlands of northern Zimbabwe and the adjacent Tete Province of Mozambique, extending southwards up valleys of mainly the Sengwa and Gwayi Rivers in Zimbabwe and in the form of a few isolated patches located south of Tete towards Gorongosa in Mozambique (Fig. 3.23). This regional biome also extends westward at higher elevations, mostly above 1000 m, from the Hwange area in northwestern Zimbabwe into the eastern Caprivi part of Namibia as well as into the peripheral area of the Okavango River Delta, north of Maun in northern Botswana.

Vegetation Physiognomy and Composition The typical vegetation of this variable regional biome are deciduous woodlands often comprising monospecific *Colophospermum* woodland but also with mixed dry woodlands on rocky ridges where mopane is more sporadic. *Colophospermum* mopane is characteristic and often dominant, usually most abundant on clay soils. *Adansonia digitata* occurs as scattered emergent while the main trees include *Afzelia quanzensis, Albizia harveyi, Balanites aegyptiaca, Combretum apiculatum, C. fragrans, Commiphora* spp., the family Capparaceae (*Boscia, Cadaba, Capparis, Courbonia, Maerua, Thilachium*), *Dalbergia boehmii, D. melanoxylon, Diospyros quiloensis, Erythroxylum zambesiacum, Grewia* spp., *Kirkia acuminata, Lannea schweinfurthii, Philenoptera violacea, Senegalia nigrescens, Sterculia africana, Terminalia prunioides, T. stuhlmannii, Vachellia nilotica* subsp. *kraussiana, V. tortilis,* and *Xeroderris stuhlmannii.* Grasses include *Andropogon gayanus, Aristida* spp., *Dactyloctenium giganteum, Eragrostis* spp., *Heteropogon contortus, Hyperthelia dissoluta, Hyparrhenia rufa, Panicum maximum, Schmidtia pappophoroides, Sporobolus cordofanus, S. ioclados,* and *Urochloa mossambicensis.* The generally taller and denser woodland physiognomy of this regional biome contrasts with the usually lower and more open savanna of the RB Limpopo Mopane and RB Kaokoland Mopane of the CB Southern African Arid Savanna. Whether the reduced dominance of mopane observed in parts of the mesic RB Zambezi Mopane is linked to the higher MAP of the regional biome requires further evaluation.

Climate Rainfall is strongly seasonal with most falling in summer, with remarkably little falling in winter. The MAP of 608 mm varies

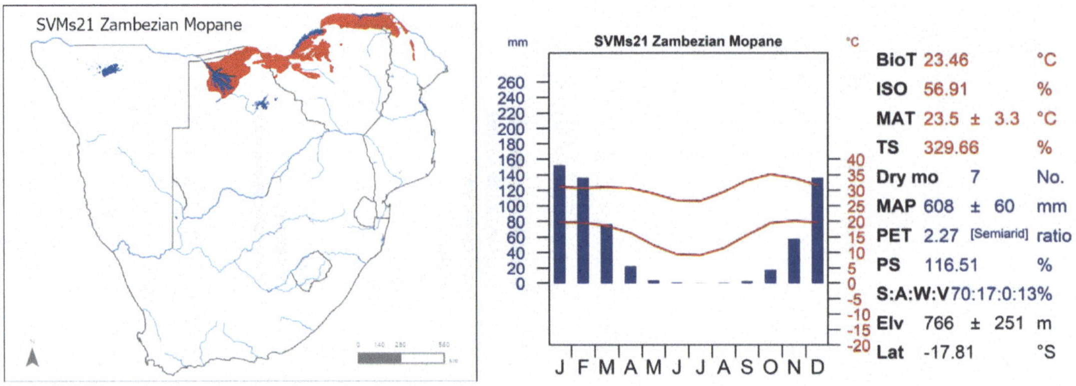

Fig. 3.23 The distribution area and the climatic diagram of the regional biome SVMs21 Zambezian Mopane

considerably within the regional biome dropping to a MAP of less than 500 mm around Maun in Botswana but exceeding 700 mm in most of the Zambezi River valley of Zimbabwe and Mozambique, for instance at Victoria Falls, Chirundu, and Songo. Late spring and summers are hot to very hot and winters are mild and frost is rare. It may be noted that the MAP of 608 mm for this regional biome is considerably higher than the MAP for the two other mopane savanna regional biomes of the Arid Savanna (Kaokoland Mopane 342 mm; Limpopo Mopane 454 mm) and, importantly, that this contrast becomes considerably enhanced if the drier, more transitional area of RB Zambezi Mopane in Botswana is excluded from this regional biome (Fig. 3.23).

Topography, Geology and Soils This regional biome occupies very varied landscapes from wide open valleys to narrow gorges, and includes hills and ridges, as well as undulating or alluvial plains, and some areas of Kalahari sands on plains in the west. Important geological substrates are grits, sandstones, and siltstones, as well as basalt, conglomerate, gneiss, mudstone, schist and calcareous lacustrine sediments. Soils are typically moderately deep, fairly neutral, sandy clay loams or clay loams. In the west, the aeolian Kalahari sands can, in places, conceal a deeper subsurface layer with higher clay content (K. L. Tinley personal communication). The occurrence of this regional biome around the major endorheic basin of the Okavango Delta on the extensive sand plains of the Kalahari, is probably due to the past, broad-scale, accumulation of finer fluvial particles which results in the formation of heavier soils, sufficient to support *C. mopane* dominance.

Sources Pedro and Barbosa (1955), Rattray (1962), Tinley (1966), Wild and Barbosa (1967), Child (1968), Farrell (1968b), Anderson and Walker (1974), Guy (1977), Werger and Coetzee (1978), Craig (1983), Timberlake et al. (1993), du Toit (1993), Mhlanga and Mapaure (1999), Kalwij et al. (2010), Poilecot and Gaidet (2011), Revermann and Finckh (2013), Stellmes et al. (2013a, b), Tedder et al. (2013), Revermann (2016), Revermann et al. (2016a, b), Tafangenyasha et al. (2016), Moses et al. (2017), Arraut et al. (2018), Teketay et al. (2018), Lötter et al. (2023).

SVMs22 Baikiaea Woodland

Distribution The Baikiaea woodlands are found in the Matabeleland North Province in the northwest of Zimbabwe and the adjacent northeastern part of the Chobe District in Botswana. They extend westwards along the northern limit of our study area, including most of the Caprivi Strip in Namibia and the adjoining parts in far-northern Botswana, and the Kavango and Ohangwena Regions in northern Namibia (Fig. 3.24). This regional biome also extends beyond our study area into southern Angola and southwestern Zambia.

Vegetation Physiognomy and Composition It is best developed in the east (and especially north of our mapping area in Zambia) where closed tall woodland can be common. Further west the woodlands tend to be more open, shorter and often restricted to dune crest areas and the mapped zone of this regional biome can lack *Baikiaea plurijuga* in places. Some of these absences may have been exacerbated by heavy utilization of this valuable timber tree. Beside the dominant *Baikiaea*, the most important associated trees are *Brachystegia spiciformis*, *Burkea africana*, *Combretum apiculatum*, *C. collinum*, *C. molle*, *C. zeyheri*, *Commiphora mossambicensis*, *Dialium englerianum*, *Diplorhynchus condylocarpon*, *Erythrophleum africanum*, *Guibourtia coleosperma*, *Ochna pulchra*, *Pseudolachnostylis maprouneifolia*, *Pterocarpus angolensis*, *Schinziophyton rautanenii*, *Strychnos pungens*, and *Terminalia sericea*. Small trees and shrubs include *Baphia massaiensis*, *Bauhinia macrantha*, *Dichapetalum rhodesicum*, *Euphorbia matabelensis*, *Grewia bicolor*, *G. monticola*, *Monanthotaxis obovata*, *Paropsia brazzeana*, *Pavetta schumanniana*, *Psydrax livida*, *Searsia tenuinervis*, and *Vangueria infausta*. Grasses are few and sparse and include *Eragrostis nindensis*, *Heteropogon melanocarpus*, *Leptocarydion vulpiastrum*, *Panicum maximum*, *Perotis patens*, *Pogonarthria squarrosa*, *Setaria pumila*, and *Sporobolus panicoides*.

Climate This regional biome's MAP of 554 mm shows relatively little variation over the spatial range of the regional biome and represents the driest of the Mesic Savanna regional biomes. Rainfall is strongly seasonal, occurring predominantly in summer with remarkably dry winters. It is one of the warmer savanna regional biomes with an MAT of 22.3 °C with hot summers and warm winters. Frost is infrequent but severe frost can occur at irregular intervals of several years (Fig. 3.24).

Topography, Geology and Soils Deep Kalahari sands are the typical substrate supporting the communities of this regional biome. *Baikiaea plurijuga* is often confined to the deepest sands on and near the crest of stable parallel dunes, notably in Hwange National Park (Zimbabwe), Western Caprivi and in parts of the Kavango region in Namibia. This regional biome broadly corresponds to the Barotse Centre of Endemism. It has been suggested that difficulty in drawing the boundary between the RB Baikiaea Woodland and the surrounding vegetation is because *Baikiaea* has been exploited rigorously as it provides valuable timber.

Sources Martin (1939), Mitchell (1962), Wild and Barbosa (1967), Child (1968), Weare and Yalala (1971), Simpson (1975), Werger and Coetzee (1978), Childes and Walker (1987), Rogers (1993), Timberlake et al. (1993), Mendelsohn and Roberts (1997), Mosugelo

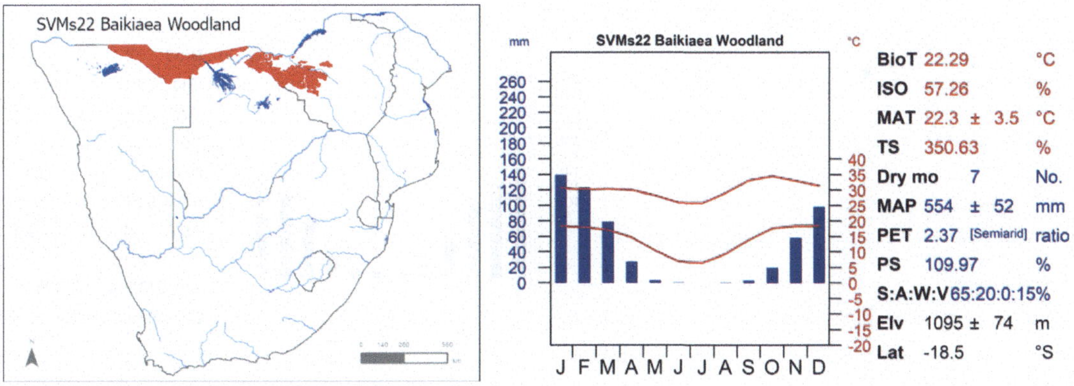

Fig. 3.24 The distribution area and the climatic diagram of the regional biome SVMs22 Baikiaea Woodland

et al. (2002), Ringrose et al. (2003), Timberlake and Childes (2004), Holdo (2006, 2007), Jürgens et al. (2010), Kalwij et al. (2010), Aarrestad et al. (2011), Revermann and Finckh (2013), Stellmes et al. (2013a, b), Wanke et al. (2013), De Cauwer et al. (2016), Revermann (2016), Revermann et al. (2016a, b), Sianga and Fynn (2017), Arraut et al. (2018), Mandinyenya et al. (2020), Vittoz et al. (2020), Wilson et al. (2021).

SVMs23 Bushveld Ultramafic Sourveld

SVl 13 Barberton Serpentine Sourveld (Dayaram et al. 2019)

Distribution Located in the eastern Mpumalanga Province of South Africa and occurs in the form of relatively isolated patches on the exposed ultramafic substrates scattered in the broad vicinity of Barberton from Malelane in the east, to the Songimvelo Nature Reserve area in the south, Badplaas in the west, and to west of Mmombelo in the north (Fig. 3.25).

Vegetation Physiognomy and Composition Herbaceous grasslands with stunted woody vegetation occur on the southern ultramafic outcrops, with more woody vegetation apparent on the lower-lying Noordkaap area and towards Malelane. Important trees include *Senegalia caffra, A. davyi, Aloe marlothii* subsp. *marlothii, Faurea rochetiana, Pavetta edentula,* and *Searsia pondoensis*. Shrubs are *Erythroxylum delagoense, Gnidia sericocephala, Helichrysum kraussii, Jamesbrittenia grandiflora, Jatropha latifolia* var. *latifolia, Lasiosiphon caffrer, Pearsonia sessilifolia, Searsia rogersii, Tephrosia elongata,* and *Thunbergia atriplicifolia.* Grasses include *Andropogon schirensis, Bewsia biflora, Cymbopogon caesius, Diheteropogon amplectens, Eragrostis capensis, E. racemosa, Heteropogon contortus, Loudetia simplex, Panicum natalense,* and *Themeda triandra.* Herbs comprise *Berkheya coddii, Cheilanthes involuta, Helichrysum nudifolium* var. *pilosellum, Hemizygia persimilis, Rabdosiella calycina, Rhynchosia minima, R. monophylla, R. totta, Senecio coronatus, S. venosus, Vernonia natalensis,* and *V. sutherlandii.* It may be noted that there are many edaphic specialist plant species in this regional biome.

Climate MAP of 815 mm is concentrated in summer and winters are dry. Summers are hot and winters warm to mild. Frost is infrequent (Fig. 3.25).

Topography, Geology and Soils The terrain is often hilly, but is very varied. The geological substrates comprise ultramafic lavas (including komatiites and serpentinites), predominantly of the Onverwacht Group of the Barberton Supergroup (Barberton Greenstone Belt). The ultramafic geology gives rise to soils

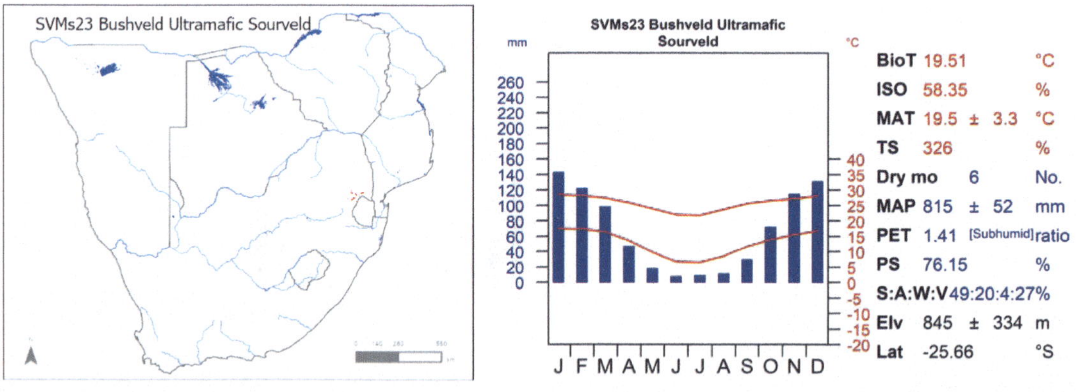

Fig. 3.25 The distribution area and the climatic diagram of the regional biome SVMs23 Bushveld Ultramafic Sourveld

with unusually high magnesium: calcium ratios. These soils are also associated with high concentrations of heavy metals such as Ni and Cr, which are generally toxic to most plants.

Sources Morrey et al. (1989, 1992), Balkwill et al. (1997), Smith et al. (2001), Rutherford et al. (2006).

SVMs24 Southern Quartzite Sourveld

Gs 2 Ithala Quartzite Sourveld (Dayaram et al. 2019)
SVl 22 Northern Zululand Sourveld (Dayaram et al. 2019)

Distribution This azonal (peinobiome) regional biome is restricted to northern KwaZulu-Natal Province of South Africa, with small pockets occurring also in SE Mpumalanga and southern eSwatini (Fig. 3.26). The detailed breakdown of its geographical coverage is given by Rutherford et al. (2006: 505, for SV l22 Northern Zululand Sourveld and Gs 2 Ithala Quartzite Sourveld). Mucina et al. (2006a) classified part of this regional biome as a member of the Grassland Biome due to its overall grassy physiognomy.

Vegetation Physiognomy and Composition The dominant physiognomy is open, grass-dominated vegetation, supporting scattered trees (various species of *Vachellia* and *Senegalia,* and *Syzygium legatii*), succulent tree-like megaherbs (*Aloidendron*), and plethora of shrub species. Dense bushveld thickets occur rarely in this unit. Vegetation structure varies according to altitude and rockiness, but the basal density of the grass sward remains relatively low. This unit occurs in the zone between grasslands and savanna woodlands where the dominant grassland gives way to woodland as elevation decreases. The grasslands are species-rich covering a variety of upper altitudes but sharing common species unique to the dystrophic quartzite geology. The dominant woody species include *Combretum molle, Cassipourea swaziensis, Englerophytum magalismontanum, Gymnosporia tenuispina, Morella pilulifera, Pavetta gracilifolia,* and *Syzygium legatii.* The most prominent grasses are all C4 and include *Heteropogon contortus, Hyperthelia dissoluta, Loudetia simplex, Monocymbium ceresiiforme, Panicum natalense, Themeda triandra,* and *Trachypogon spicatus.*

Climate Rainfall is less strongly seasonal than many other of the savanna regional biomes, but nevertheless peaks in midsummer, with a little rain in winter. The MAP of 852 mm varies from over 1000 mm in the west to less than 600 mm in the east. Frost is infrequent to occasional. The MAT 19.2 °C indicates a subtropical thermal climate. Summers are hot and winters mild to warm (Fig. 3.26).

Topography, Geology and Soils The topography is characterised by low, undulating and sometimes highly dissected mountains. The geological substrate varies. The most prominent are quartzitic sandstones of the Madzaringwe and Pietermaritzburg Formations with sedimentary

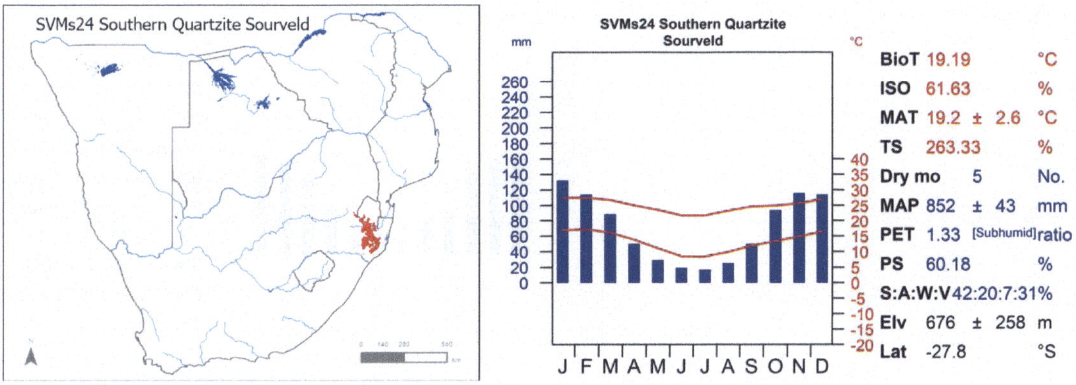

Fig. 3.26 The distribution area and the climatic diagram of the regional biome SVMs24 Southern Quartzite Sourveld

(often metamorphosed) Dwyka diamictites, shales, and siltstones as well as Archaean granite and gneiss also significant. The Glenrosa and Mispah soils are well-drained, shallow, and nutrient poor.

Sources Acocks (1953, 1988), Camp (1999c, b), Wiseman et al. (2004), Mucina et al. (2006a), Rutherford et al. (2006), Gordijn (2010).

SVMs25 Southern Sandstone Sourveld

CB 4 Pondoland-Ugu Sandstone Coastal Sourveld (Dayaram et al. 2019)
SVs 5 KwaZulu-Natal Sandstone Sourveld (Dayaram et al. 2019)

Distribution The largest patch of this regional biome forms a coastal belt north of Port St Johns in Pondoland (northeastern Transkei, Eastern Cape Province of South Africa). It extends further north from the surrounds of Port Shepstone (Ugu District of KwaZulu-Natal) as a series of isolated inland elevated sandstone plateaus via St Faith's, Highflats, Hillcrest and Table Mountain to Mapumulo near Kranskop in central KwaZulu-Natal (Fig. 3.27). The elevation of the biome rises in a northeasterly direction.

Vegetation Physiognomy and Composition This regional biome is characterised by natural, species-rich grasslands punctuated with scattered geoxylic suffrutices, low shrubs and small trees including *Faurea saligna, Protea caffra, P. roupelliae,* and *Leucadendron spissifolium.* Grasses such as *Alloteropsis semialata* subsp. *eckloniana, Aristida junciformis* subsp. *galpinii* and subsp. *junciformis, Cymbopogon nardus, Heteropogon contortus, Themeda triandra, Trachypogon spicatus,* and *Tristachya leucothrix* are the most prominent in these edaphic grasslands. Herbs are common and comprise, for example, *Chaetacanthus burchellii, Cyanotis speciosa, Helichrysum allioides, H. appendiculatum, H. krebsianum, H, spiralepis, Pentanisia angustifolia, Rhynchosia totta,* and *Tephrosia macropoda.* The vegetation of this regional biome is rich in local endemics.

Climate The MAP of 983 mm is not as strongly seasonally distributed compared to most other savanna regional biomes and rain can peak from late spring to early autumn and, although winters are relatively dry, 88 mm of rain occurs in winter on average. Mist is common in some inland localities in the N section of the unit. With a MAT of 18.3 °C the regional biome is one of the coolest savanna regional biomes; daily maximum temperatures in summer are particularly low relative to those in most other savanna regional biomes. However, winter frost does not occur or is generally infrequent (Fig. 3.27).

Topography, Geology and Soils The southern section of the unit is a coastal peneplain, dissected by deep gorges harbouring subtropical

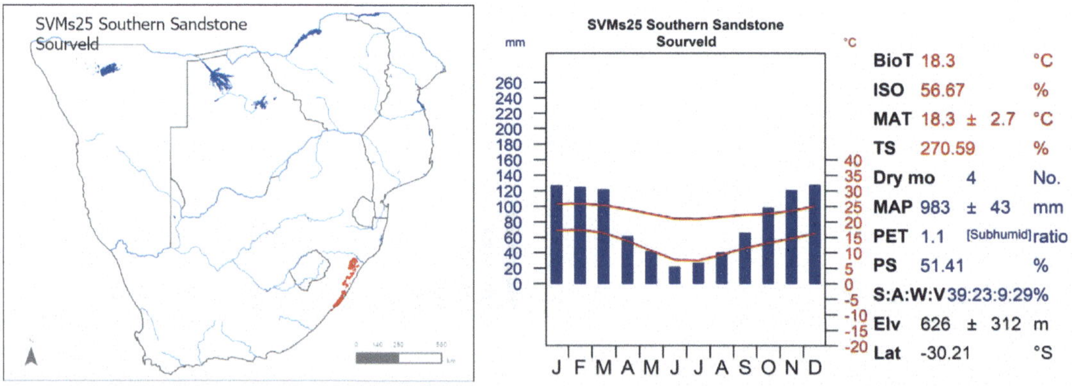

Fig. 3.27 The distribution area and the climatic diagram of the regional biome SVMs25 Southern Sandstone Sourveld

forests. The rest of the area of this regional biome is characterised by undulating hills with flat tablelands. These landscapes are built of hard, white, coarse-grained, siliceous quartz arenites (sandstones) of the Msikaba Formation of the Devonian (Thomas et al. 1992) in the south, while Ordovician Natal Group sandstones are dominant in the north. Both give rise to shallow, nutrient-poor (highly leached), skeletal, acidic sandy soils, freely drained and classified mainly as Glenrosa and Mispah forms.

Sources Moll and White (1978), Midgley (1986), Shackleton (1989, 1990), van Wyk (1990a, b), Shackleton et al. (1991), Shackleton (1992), Abbott (1993), Shackleton and Shackleton (1994), le Roux (1995), Glen (1996), Camp (1999a, b), Abbott et al. (2000), Carbutt and Edwards (2001), van Wyk and Smith (2001), Carbutt (2005), Mucina et al. (2006c), Rutherford et al. (2006), Ngubane et al. (2014), Boon et al. (2016), Drury et al. (2016), Odindi et al. (2016), Morris and Scott-Shaw (2019).

3.2 Continental Biome Southern African Arid Savanna

This continental biome covers more than half the savanna area of Southern Africa but is considerably less diverse than the Southern African Mesic Savanna continental biome. It extends from the western Great Escarpment in northern Namibia eastwards to the arid lowlands associated with the Limpopo River catchment in southern Mozambique and the northern Lowveld of South Africa. The western part comprises the relatively high elevation upper Escarpment and associated hinterland of mopane-dominated vegetation of Kaokoland in the north of Namibia, to the thornveld of the Khomas Hochland further south. In the east, the mopane-dominated vegetation of the Limpopo lowlands is joined by other relatively low-lying thornveld of the northwestern regions of the Limpopo Province of South Africa and adjacent southeastern Botswana. Between these two flanks lies the broad expanse of deep Kalahari sands, which includes the extensive dune field in the SW. The main expanse of Kalahari sand extends along a gradually increasing rainfall gradient northwards from northern South Africa into Botswana and Namibia. Thornveld (e.g., *Senegalia mellifera* and *Vachellia erioloba*) predominates. However, the proportion of broadleaved trees (e.g., *Burkea africana* and *Combretum collinum*) increases northwards within the RB Northern Kalahari Bushveld, which meets the RB Baikiaea Woodland of the CB Southern African Mesic Savanna in the far north of Namibia and Botswana. While the RB Namibian Dolomite Bushveld of the eastern, higher-rainfall Karstveld of Namibia contains several species typical of mesic savanna (e.g., *Lannea discolor*, *Peltophorum africanum*, *Sclerocarya birrea*, and *Spirostachys africana*), it also hosts many other species that are more common in arid savanna (e.g., *Cissus nymphaeifolia*, *Commiphora glaucescens*, *Danthoniopsis dinteri*, and *Pachypodium lealii*). Although this regional biome has a higher MAP than the other arid savanna regional biomes, and much of it shares a similar long-term NVDI with that of mesic Baikiaea Woodland to the north (Westinga et al. 2020), on overall bioclimatic grounds, Namibian Dolomite Bushveld qualifies, albeit marginally, for placement within the CB Southern African Arid Savanna.

The CB Southern African Arid Savanna consists of nine regional biomes (Table 3.1), namely:

SVAs1 Arid Kalahari Mountain Bushveld

SVk 10 Kuruman Mountain Bushveld (Dayaram et al. 2019)
SVk 15 Koranna-Langeberg Mountain Bushveld (Dayaram et al. 2019)

Distribution This regional biome occurs north of the Gariep River in the Northern Cape Province of South Africa, in the form of two main parallel mountain ranges, as well as some minor associated ridges. The western range comprises the Langeberg which runs northwards past Olifantshoek, extending further north as the Koranna Mts and terminating around the Tswalu Kalahari Reserve. The eastern range stretches

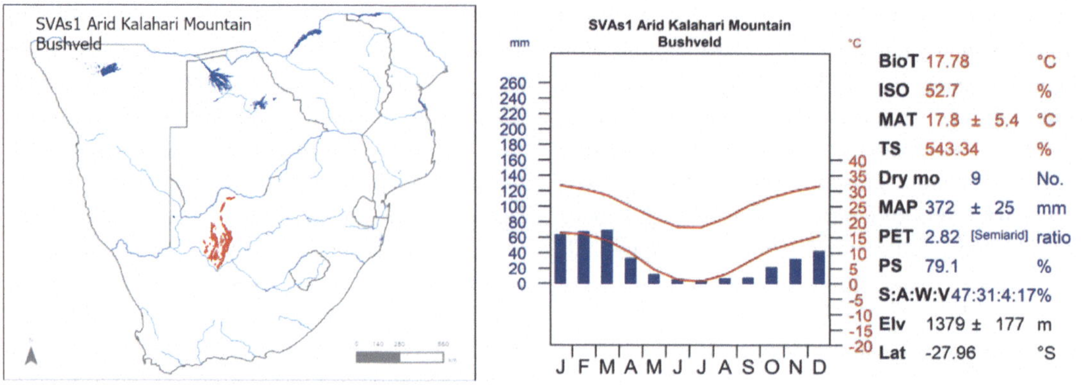

Fig. 3.28 The distribution area and the climatic diagram of the regional biome SVAs1 Arid Kalahari Mountain Bushveld

from near Niekerkshoop in the south, along the Asbestos Mountains to Danielskuil, running further northward past Kuruman in the form of the Kuruman Hills, and re-emerging as isolated hills as far as the vicinity of Pomfret in the North West Province (Fig. 3.28).

Vegetation Physiognomy and Composition The vegetation physiognomy is open, grassy, tall to medium shrubland with occasional low trees, becoming more sparse southwards. Important grasses include *Andropogon chinensis, Anthephora pubescens, Aristida congesta, A, diffusa, Brachiaria nigropedata, Digitaria eriantha, D. polyphylla, Eragrostis curvula, Fingerhuthia africana, Heteropogon contortus,* and *Triraphis andropogonoides*. Prominent shrubs include *Boscia albitrunca, Euclea crispa, E. undulata, Grewia flava, Rhigozum obovatum, Searsia burchellii, S. ciliata, Senegalia mellifera,* and *Tarchonanthus camphoratus*. *Croton gratissimus* and *Euphorbia avasmontana* are particularly prominent on the Koranna Mts while *Calobota cuspidosa* and *Olea europaea* subsp. *africana* are prominent on the Asbestos Mts. The latter mountain ranges are recorded as particularly species rich (302 taxa; the highest species count of any Acocks 1988 recording plots in South Africa). The RB Arid Kalahari Mountain Bushveld forms a major part of the Griqualand West Centre of Plant Endemism (van Wyk and Smith 2001).

Climate MAP of 372 mm falls mainly in summer and early autumn with relatively little rain in winter; there is a distinct declining rainfall gradient from the east to the west so that MAP is considerably lower for the Koranna Mts (< 300 mm) than that of the Asbestos Mts. There is also a N–S decline in MAP so that the Kuruman Hills (part of the Asbestos Mountains) represent the least arid parts of this regional biome. Summer temperatures are hot whereas winter nights are cold with frost. The MAT of 17.8 °C is the lowest of any of the arid or mesic savanna regional biomes (Fig. 3.28).

Topography, Geology and Soils The two main ranges are made up of low mountains and hills rising typically 100–400 m above the surrounding plains while attaining a maximum elevation of about 1800 m. The geology of the two ranges are distinct. The Koranna Mts consist of quartzite, greywacke and lenses of hematite of the Olifantshoek Supergroup. These mountains are generally more rugged and broken with steep slopes (mostly 10–50°) and exposed boulders and rocks covering more than 80% of the landscape. Despite the high surface rock cover of the quartzitic Koranna Mts, extensive subsurface fissuring of the rock allows for deeper soil penetration and root growth. By contrast, Asbestos Mts have often gentler slopes and smoother surface profiles, and consist of banded iron formation, with jaspilite, chert and riebeckite-asbestos of the Asbestos Hills Subgroup of the Griqualand West Supergroup. Soils are shallow, of the Hutton Form.

Sources Rutherford (1983a), Acocks (1988), Rutherford et al. (2006), Davis et al. (2010), van Staden et al. (2020).

SVAs2 Arid Kalahari Sandy Bushveld

SVk 1 Mafikeng Bushveld (Dayaram et al. 2019)
SVk 2 Stella Bushveld (Dayaram et al. 2019)
SVk 3 Schweizer-Reneke Bushveld (Dayaram et al. 2019)
SVk 4 Kimberley Thornveld (Dayaram et al. 2019)
SVk 5 Vaalbos Rocky Shrubland (Dayaram et al. 2019)
SVk 6 Schmidtsdrif Thornveld (Dayaram et al. 2019)
SVk 7 Ghaap Plateau Vaalbosveld (Dayaram et al. 2019)
SVk 8 Kuruman Vaalbosveld (Dayaram et al. 2019)
SVk 9 Kuruman Thornveld (Dayaram et al. 2019)
SVk 11 Molopo Bushveld (Dayaram et al. 2019)
SVk 12 Kathu Bushveld (Dayaram et al. 2019)
SVk 13 Olifantshoek Plains Thornveld (Dayaram et al. 2019)
SVk 14 Postmasburg Thornveld (Dayaram et al. 2019)
SVk 16 Gordonia Plains Shrubland (Dayaram et al. 2019)

Distribution This is the largest regional biome comprising the southern section of the sandy Kalahari Basin from the area north of the Gariep River in the Northern Cape Province of South Africa northwards as far as south of the Ghanzi Ridge and the Makgadikgadi Pans area in Botswana (Fig. 3.29). It is situated mostly to the east of the RB Kalahari Duneveld but extends narrowly westward between the RB Kalahari Duneveld and RB Namibia Central Upland Bushveld south of Rehoboth. Included is an isolated, extensive patch north of Otjovasando in the western Etosha National Park which extends northwards in the Omusati Region in Namibia. On the eastern flank it is bordered by the RB Northwestern Limpopo Sandy Bushveld in southeastern Botswana. In the southeast, it borders dry highveld grassland regional biomes in the North West and Free State Provinces of South Africa.

Vegetation Physiognomy and Composition The shrub layer is usually well developed, with generally scattered (or forming small groups) low trees. The herbaceous layer is dominated by grasses and is well developed. The common shrubs include *Boscia albitrunca, Senegalia mellifera,* and *Vachellia erioloba* (generally the tallest tree) and *V. hebeclada* (can be very common throughout most of the regional biome). In higher rainfall areas, further to the north and the east, trees such as *Terminalia sericea* and *Vachellia luederitzii,* and shrubs such as *Grewia flava* become more important. Calcretes in the southeast are often dominated by *Tarchonanthus camphoratus* while the calcrete outcrops in the north are often dominated by *Catophractes alexandri.* There is no sharp border in the north with the RB Northern Kalahari Bushveld, and a number of floristic elements typical of RB Northern Kalahari Bushveld (for example, *Dichrostachys cinerea, Philenoptera nelsii,* and *Senegalia ataxacantha*) are encountered increasingly along this rainfall gradient towards the north of the

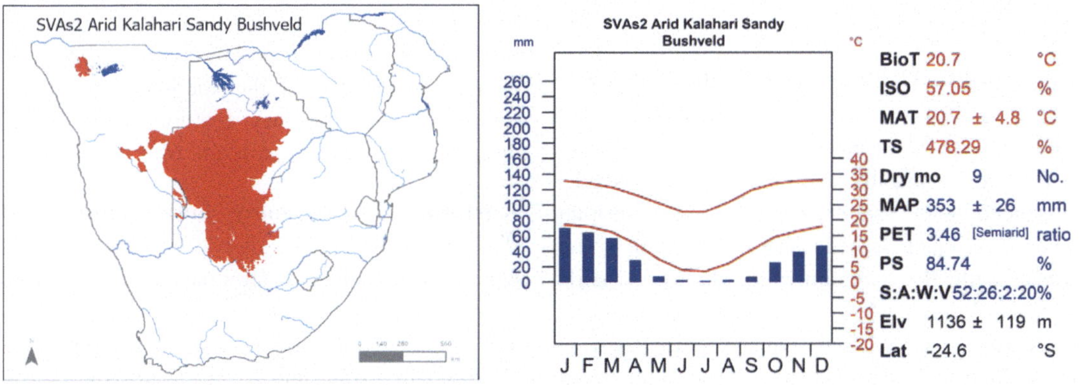

Fig. 3.29 The distribution area and the climatic diagram of the regional biome SVAs2 Arid Kalahari Sandy Bushveld

Arid Kalahari Sandy Bushveld. The other major increasing rainfall gradient from the southwest towards the dry Highveld grassland regional biomes in South Africa is associated with an increase in species typical of less arid areas, for instance *Olea europaea* subsp. *africana* and *Vachellia tortilis*. It is especially in the southeastern corner of this regional biome (roughly southeast of the Stella–Kuruman–Griekwastad line) that the facultative calcicole *Tarchonanthus camphoratus* can be overwhelmingly dominant but it tends to be far less important, even on calcrete, in the majority of this regional biome, particularly further north. *Tarchonanthus* is dominant, together with *Olea europaea* subsp. *africana*, on the dolomite and limestone of the Ghaap Plateau. Dominant grasses in this regional biome include *Anthephora pubescens, Aristida congesta, A. meridionalis, Cenchrus ciliaris, Centropodia glauca, Enneapogon cenchroides, Eragrostis lehmanniana, Schmidtia kalahariensis, S. pappophoroides, Stipagrostis ciliata,* and *S. uniplumis.*

Climate MAP ranges mostly between about 250 mm where it meets the Kalahari Duneveld regional biome in the southwest, and 400 mm in the north but reaches about 500 mm in the southeast, in the belt between Kimberly and Mafikeng, close to the border with the RB Dry Highveld Sandy Grassland and RB Dry Clayey Grassland. Rain falls mainly in summer and winters are usually dry. MAT is 20.7 °C, with hot summers strongly contrasted with very cold nights, with frequent frost, in winter. The TS of 478 is one of the greatest in savanna, after the RB Kalahari Duneveld, indicating a continental climate (Fig. 3.29).

Topography, Geology and Soils Flat landscapes with very few dunes characterise the landscape of this regional biome. The substrate is formed by usually deep to very deep sands, alternated with scattered sub-surface hardpan layers and calcrete beds and outcrops in some areas, with the calcium-rich, shallower soils of the dolomitic Ghaap Plateau being the most distinctive. Tree roots of some species such as *Vachellia erioloba*, have been found at great depths of sand during borehole drilling.

Sources Weare and Yalala (1971), Skarpe (1980, 1986, 1990), Bekker and de Wit (1991), van Rooyen and van Rooyen (1998), Mendelsohn et al. (2000), van Rooyen (2000), Ringrose et al. (2003), Rutherford et al. (2006), Bezuidenhout (2009), Mills et al. (2012), Dingaan et al. (2013, 2016, 2017), Mishra et al. (2015), Dreber et al. (2018), Ravhuhali (2018), Dingaan and Tsubo (2019), Lori et al. (2019), Strohbach et al. (2019), Makhabu et al. (2021), Tiawoun et al. (2022), Harmse et al. (2023), Mogonong et al. (2023).

SVAs3 Northern Kalahari Bushveld

Distribution Extends from the vicinity of Otjiwarongo and the northeast corner of the Etosha National Park in northern Namibia eastwards across the Kalahari sandveld plains of the Otjozondjupa and northern Omaheke Districts. Further eastwards, it extends into Botswana where it occupies the northern Kalahari plains north of the Ghanzi Ridge, remaining mostly south of the Okavango Delta and west of the Makgadikgadi Pans, and thus covers much of the southern Ngamiland District. It also extends into northern parts of the Central Kalahari Game Reserve. It remains immediately south of the belt of the RB Baikiaea Woodland in Namibia and Botswana. A widely separated sandy patch of RB Northern Kalahari Bushveld occurs immediately south of Ruacana on the border between Namibia and Angola (Fig. 3.30). Mountains and hills are rare in this regional biome, but these include the Waterberg in Namibia and the Tsodilo Hills in Botswana. This is the only regional biome in our study area that lies almost wholly within endorheic basins, mostly that of the southern Okavango, associated with drainage features such as the Omatako, Epukiro, Ngami, and Makgadikgadi.

Vegetation Physiognomy and Composition This regional biome shows a very variable vegetation structure from open to closed, low to medium height canopies of microphyllous and broadleaved trees, sometimes separated but often intermingling to different extents and scale. The proportion of broadleaved trees tends to increase further northwards. The grass layer is usually well developed. The most important woody species include *Bauhinia petersiana, Burkea africana, Combretum collinum, C. hereroense, C. psidioides, Commiphora angolensis, C. africana, Croton gratissimus, Dichrostachys cinerea, Grewia avellana, G. flavescens, G. retinervis, Mundulea sericea, Ochna pulchra, Philenoptera nelsii, Rhigozum brevispinosum, Securidaca longepedunculata, Senegalia ataxacantha, S. fleckii, S. mellifera, Terminalia sericea, Vachellia erioloba,* and *V. luederitzii.* Surface calcrete supports species such as *Boscia albitrunca, Catophractes alexandri, Combretum imberbe, Senegalia mellifera,* and *Ziziphus mucronata.* Dominant grasses include *Anthephora pubescens, Aristida meridionalis, A. rhiniochloa, A. stipitata, Brachiaria nigropedata, Digitaria eriantha, D. seriata, Enneapogon cenchroides, Eragrostis lehmanniana, E. pallens, E. porosa, E. rigidior, E. trichophora, Heteropogon contortus, Megaloprotachne albescens, Melinis repens, Panicum coloratum, P. kalaharense, Schmidtia pappophoroides, Stipagrostis hirtigluma, S. uniplumis, Triraphis schinzii,* and *Urochloa brachyura.* Tsodilo Hills lie near the southern boundary of the RB Baikiaea Woodland and occasional isolated *Baikiaea plurijuga* occurs, for instance on a localised sand plateau on top of Female Hill (M. C. Rutherford, personal observation). The sandy patch south of Ruacana also contains some localised patches of *Baikiaea plurijuga*, at least in part associated with sandy water drainage channels or oshanas from the adjacent Cuvelai System, but is too dry to support the full complement of species typical of the RB Baikiaea Woodland.

Climate MAP is 411 mm but ranges from approximately 380 mm in the south to about 500 mm in the north. Rain falls mainly in summer and winters are dry. Winter frosts are common, especially in the southern parts of this regional biome (Fig. 3.30).

Topography, Geology and Soils This regional biome occurs on mainly flat to slightly undulating plains of deep Kalahari sands (ferralic arenosols) on the central plateau of Southern Africa. Fixed dunes tend to be far less common than in the RB Baikiaea Woodland to the north. Calcrete is often found to a varying extent and depths below the sands. In the vicinity of Tsumkwe, near the eastern border of Namibia with Botswana, surface calcrete is widespread on petric calcicoles. Sands are well drained, fairly acidic (pH 5 to 7), with a low carbon content. The deep sands on the western, most elevated part of the Waterberg Plateau, that supports a broadleaved *Burkea africana* savanna, can be very acidic (pH 3.8) which, also in other respects, shows an affinity with the mesic RB Southeastern Limpopo Sandy Bushveld on better leached sands in South Africa.

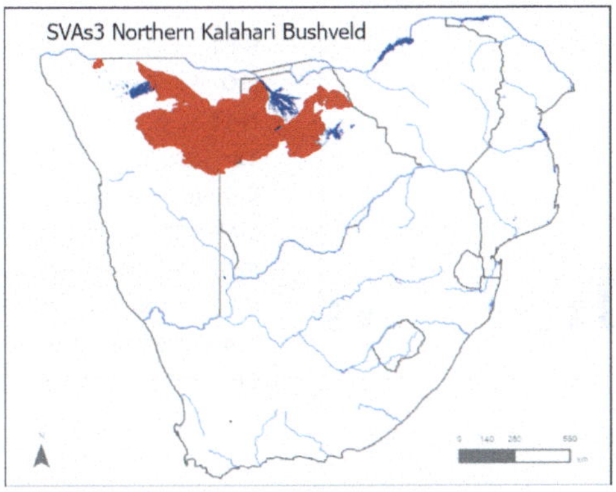

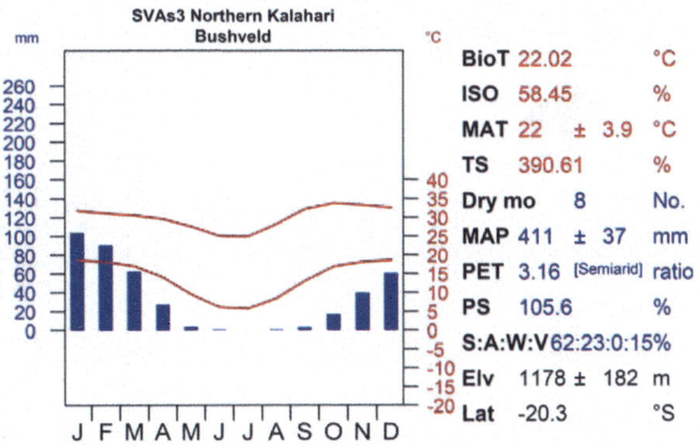

Fig. 3.30 The distribution area and the climatic diagram of the regional biome SVAs3 Northern Kalahari Bushveld

Sources Child (1968), Weare and Yalala (1971), Leser (1972), Rutherford (1972, 1975), Cole and Brown (1976) Jankowitz and Venter (1987), le Roux et al. (1988), Moore and Attwell (1999), Mendelsohn et al. (2000), Hüttich et al. (2009, 2011), Jürgens et al. (2010), Revermann and Finckh (2013), Stellmes et al. (2013a, b), Strohbach (2013, 2014), Tedder et al. (2013), Wanke et al. (2013), Kaholongo and Mapaure (2014), Strohbach and Katuahuripa (2014), Revermann (2016), Revermann et al. (2016a, b), Hauwanga et al. (2018), Geppert et al. (2021), Haindongo et al. (2021), Strohbach and Strohbach (2023).

SVAs4 Namibian Central Upland Bushveld

Distribution This regional biome dominates the Khomas Highlands around Windhoek in Central Namibia, extending north-westward along the broad, high-lying upper watershed between the Omaruru and Omatako River catchments to the vicinity of Kalkfeld (Fig. 3.31). It remains W of the deep ferralic arenosols of the neighbouring RB Northern Kalahari Bushveld. The area of the RB Namibian Central Upland Bushveld also includes several high mountains, such as the Auas and Erongo Mountains, reaching elevations of 2479 m and 2305 m, respectively.

Vegetation Physiognomy and Composition The vegetation physiognomy is determined by well-developed shrub layer grading into open to closed low tree savanna. The dominant species include *Combretum apiculatum* and *Senegalia hereroensis* in the Highlands and *Senegalia mellifera* and *Vachellia reficiens* towards the north-west. Other important species include *Albizia anthelmintica, Catophractes alexandri, Croton gratissimus, Dichrostachys cinerea, Dombeya rotundifolia, Elephantorrhiza suffruticosa, Euclea undulata, Grewia flava, Manuleopsis dinteri, Searsia marlothii, Senegalia erubescens, Vachellia tortilis,* and *Ziziphus mucronata.* The most prominent grasses are *Anthephora pubescens, Aristida adscensionis, Brachiaria nigropedata, Cenchrus ciliaris, Danthoniopsis ramosa, Digitaria eriantha, Enneapogon cenchroides, Eragrostis nindensis, E. scopelophila, Heteropogon contortus, Microchloa caffra, Pennisetum foermerianum, Stipagrostis uniplumis,* and *Triraphis ramosissima. Cheilanthes dinteri* is a common fern.

Climate MAP is about 330 mm, ranging from approximately 200 mm in the west to above 400 mm on some of the highest points in the Highlands and in the northeast. Rainfall occurs predominantly in summer while winters are dry. Frequent frosts occur in winter in the Khomas Highlands (Fig. 3.31).

Topography, Geology and Soils This regional biome ranges from sharply undulating hilly and mountainous landscapes to high-lying plains

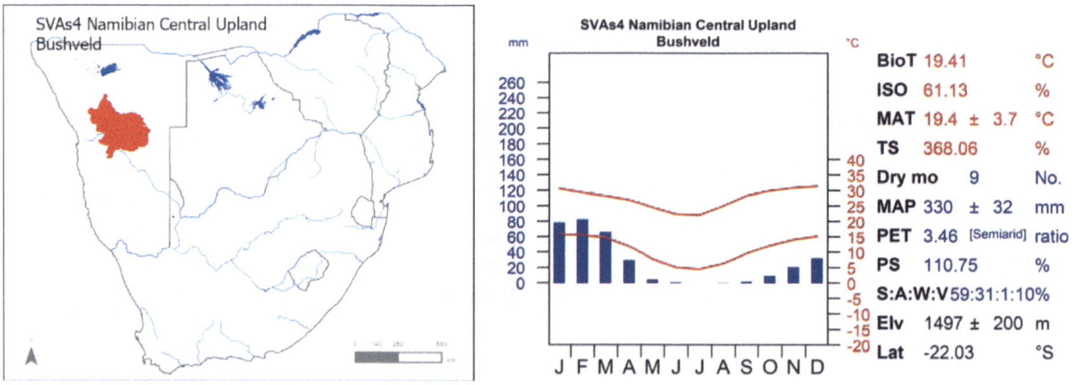

Fig. 3.31 The distribution area and the climatic diagram of the regional biome SVAs4 Namibian Central Upland Bushveld

dotted with hills and mountains in places. The soils are mostly relatively shallow leptosols and regosols, as well as on some cambisols. The geology is complex, but schists, granites and sandstones are most common.

Sources Giess (1971), Volk and Leippert (1971), Mendelsohn et al. (2002), Burke and Wittneben (2008), Jürgens et al. (2010), Strohbach (2012, 2017, 2019, 2021), Haimbili (2015), Strohbach et al. (2019).

SVAs5 Kaokoland Mopane

Distribution This mopane unit occupies most of the plateau area in northwestern Namibia, constituting a major part of Kaokoland (Fig. 3.32). It extends from the northwestern Escarpment, above the belt of the RB Kaokoland Pediment Karoo in the west, to the surrounds of the Etosha Pan in the east. In the south, it is found from the upper Ugab River valley near Outjo, northwards to the Cunene River where it continues beyond the study area into southwestern Angola. In contrast to the sometimes remarkably sharp boundary of this regional biome in the south, its boundary with the RB Kaokoland Pediment Karoo in the west is more diffuse.

Vegetation Physiognomy and Composition The vegetation of this regional biome is characterised by very open, low tree and shrub savanna in the west to open to closed, low to medium height tree savanna in parts of the east. It is overwhelmingly dominated by *Colophospermum mopane*, but with *Combretum apiculatum* and *Terminalia prunioides* also widespread. Other common woody species include *Boscia albitrunca, Catophractes alexandri, Combretum imberbe, Croton gratissimus, Commiphora africana, C. glandulosa, C. glaucescens, C. mollis, C. pyracanthoides, Dichrostachys cinerea, Grewia bicolor, G. tenax, Kirkia acuminata, Mundulea sericea, Pachypodium lealii, Phyllogeiton discolor, Senegalia mellifera,* and *Vachellia reficiens*. Other species closely associated with the escarpment area include *Commiphora crenato-serrata, C. multijuga, Sesamothamnus guerichii,* and *Sterculia quinqueloba,* together with a number of rarer species such as *Albizia brevifolia, Combretum oxystachyum, Commiphora anacardiifolia, C. discolor, Ficus glumosa, Fockea multiflora, Portulacaria longipedunculata,* and *Ptaeroxylon obliquum.* Main grasses are *Anthephora pubescens, A. schinzii, Aristida adscensionis, A. effusa, A. hordeacea, A. rhiniochloa, Cenchrus ciliaris, Enneapogon cenchroides, E. desvauxii, Eragrostis annulata, E. nindensis, Fingerhuthia africana, Monelytrum luederitzianum, Pogonarthria fleckii, Stipagrostis hirtigluma, S. hochstetteriana, S. uniplumis,* and *Triraphis ramosissima.* Several Kaokoland endemic species (e.g., several species of *Commiphora*) occur primarily in the adjacent lower lying, drier RB Kaokoland Pediment, but with some of these overlapping along the western edge of RB Kaokoland Mopane. Despite the abovementioned W–E gradient in height of vegetation, there is considerable variation in the height of mopane on the eastern plains, from extensive areas of low shrubby mopane (about 2.5 m tall) northwest of the Etosha Pan, to areas with relatively tall mopane southeast of the Etosha Pan.

Climate This regional biome has a MAP of 342 mm, ranging from as low as 180 mm along parts of the western escarpment to about 450 mm on plains in the east. Rain falls mainly in summer; winters are very dry. The MAP of this unit is lower than that for the other arid savanna mopane—the RB Limpopo Mopane (Fig. 3.32).

Topography, Geology and Soils Comprises the upper parts of the rugged western escarpment and the extensive plains of the interior plateau. Its elevation (mean of 1166 m) is considerably higher than that of the other two mopane regional biomes (RB Limpopo Mopane and RB Zambezi Mopane). Geology is highly varied, and includes limestone, calcrete, granite, rhyolite, sandstone, schists as well as rocks of the Epupa and Huab metamorphic complexes and the Cunene Complex. Rocky outcrops and lithic leptosols are common while other soils include eutric regosols, calcisols, and cambisols. High levels of salinity are common. In the east, this regional biome also occurs on shallow, often

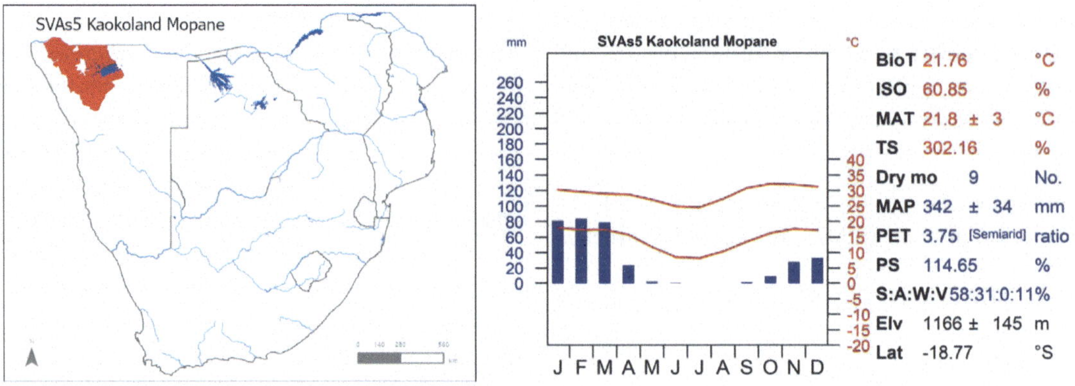

Fig. 3.32 The distribution area and the climatic diagram of the regional biome SVAs5 Kaokoland Mopane

loamy sands. On deeper, less loamy sands as, for instance in parts of the Omusati Region, including an area of western Etosha National Park, the RB Kaokoland Mopane is replaced by the RB Arid Kalahari Sandy Bushveld or the RB Northern Kalahari Bushveld.

Sources Giess (1971), Joubert (1971), le Roux et al. (1988), Becker (2000), Mendelsohn et al. (2000), Leggett et al. (2003), Siebert et al. (2003a), Kangombe (2010), Kaholongo and Mapaure (2014), Makhado et al. (2014), Inman et al. (2020), De Cauwer et al. (2023), Marufu (2023).

SVAs6 Kalahari Duneveld

SVkd 1 Gordonia Duneveld (Dayaram et al. 2019)
SVkd 2 Gordonia Kameeldoring Bushveld (Dayaram et al. 2019)
SVkd 3 Auob Duneveld (Dayaram et al. 2019)
SVkd 4 Nossob Bushveld (Dayaram et al. 2019)

Distribution The RB Kalahari Duneveld extends in a broad band from around Upington and Vanzylsrus in South Africa in the south, northwards into Namibia, remaining east of Koes and Mariental and south of Rehoboth and Gobabis (Fig. 3.33). A small part fringes the Nossob River in Botswana, also covering the larger area northeast of the confluence of the Molopo and Nossob Rivers in the extreme southwestern Botswana.

Vegetation Physiognomy and Composition In the southern dunefields, the grass *Stipagrostis amabilis* dominates the dune crests, with the low tree *Vachellia haematoxylon* scattered on the slopes. *Senegalia mellifera* is common on lower slopes and the shrub *Rhigozum trichotomum* dominates the interdune habitats. On the dunes in the less arid northern parts of the dune field in Namibia, *Vachellia haematoxylon* tends to be replaced by *Vachellia erioloba* and *Terminalia sericea*, while *Stipagrostis amabilis* can be replaced by *Centropodia glauca* on the northern dunes crests. Other important trees or shrubs include *Albizia anthelmintica, Boscia albitrunca, Crotalaria spartioides, Ehretia rigida, Grewia flava, Hermannia tomentosa, Lebeckia linearifolia, Requienia sphaerosperma, Searsia tenuinervis,* and *Vachellia luederitzii.* Other important grasses are *Aristida meridionalis, Brachiaria glomerata, Eragrostis lehmanniana, E. porosa, E. trichophora, Schmidtia kalahariensis, Stipagrostis ciliata, S. obtusa,* and *S. uniplumis.* Prostrate annuals can be common, for example, *Limeum arenicolum* and *Tribulus zeyheri.*

Climate MAP is 230 mm, ranging from about 180 mm in the southwest to 250 in the north, but reaching almost 300 mm in the extreme northeast. Summers can be very hot but winter nights are cold and heavy frost can also occur especially in the south (Fig. 3.33).

Topography, Geology and Soils The topography is shaped by a vast system of parallel dunes trending NW–SE. The dunes of between 3 and

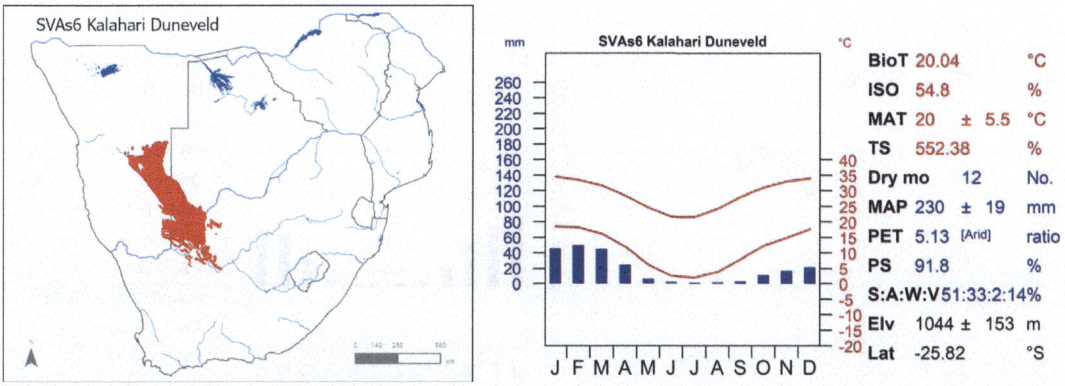

Fig. 3.33 The distribution area and the climatic diagram of the regional biome SVAs6 Kalahari Duneveld

8 m tall rise above the generally flat plains, with higher clay content sands and calcrete in the intervening parallel flat corridors between the dunes. The dunes are normally fixed by vegetation but can become mobile after relatively severe disturbance impacts, such as heavy grazing.

Sources Leistner (1967), Werger (1978), van Rooyen et al. (1984, 1990, 1991a, b, 1994, 2008), van Rooyen and van Rooyen (1998), van Rooyen (2000), Rutherford et al. (2006), Jürgens et al. (2010), Rutherford and Powrie (2010), Strohbach et al. (2019), van der Merwe et al. (2019), Schmiedel et al. (2021).

SVAs7 Northwestern Limpopo Sandy Bushveld

SVcb 18 Roodeberg Bushveld (Dayaram et al. 2019)
SVcb 19 Limpopo Sweet Bushveld (Dayaram et al. 2019)
SVcb 20 Makhado Sweet Bushveld (Dayaram et al. 2019)
SVcb 23 Polokwane Plateau Bushveld (Dayaram et al. 2019)

Distribution This regional biome is mainly associated with the upper Limpopo River Valley in South Africa and Botswana, positioned east of the deep sands of the Kalahari Basin (RB Arid Kalahari Sandy Bushveld) and south of the RB Limpopo Mopane of the lower Limpopo River Valley (Fig. 3.34). It is centred on the Limpopo Sweet Bushveld vegetation type in northwestern South Africa and its extension into southeastern Botswana. It encompasses the plains around Kanye, Gaborone, and northwards to Mahalapye in Botswana, as well as the plains around Lephalale, Swartwater, Rebone (Steiloopbrug), Vivo, and Mogwadi (Dendron), including much of the Polokwane Plateau and extending marginally to Makhado (Louis Trichardt), in South Africa.

Vegetation Physiognomy and Composition The vegetation typical of this regional biome is open to moderately dense bushveld with occasional to scattered medium-tall trees, with a well developed grass layer on the Polokwane Plateau. Important species include *Boscia albitrunca, Combretum apiculatum, C. hereroense, Commiphora africana, C. pyracanthoides, Dichrostachys cinerea, Euclea undulata, Grewia bicolor, G. flava, G. retinervis, Gymnosporia senegalensis, Kirkia acuminata, Peltophorum africanum, Rhigozum brevispinosum, Senegalia erubescens, S. fleckii, S. mellifera, S. nigrescens, S. senegal, Terminalia sericea, Vachellia nilotica, V. permixta, V. rehmanniana, V. robusta,* and *V. tortilis*. In addition, nearer the border with the neighbouring mesic savanna regional biomes, RB Limpopo Mountain Bushveld and RB Southeastern Limpopo Sandy Bushveld, taxa such as *Burkea africana, Combretum zeyheri,* and *Lannea discolor* can occur. Main grasses include *Aristida canescens, A. congesta, A. stipitata* subsp. *graciliflora, Cenchrus ciliaris, Chloris virgata, Digitaria eriantha* subsp. *eriantha, Enneapogon cenchroides, E. scoparius, Eragrostis curvula, E. lehmanniana, E. rigidior, E. trichophora, Melinis repens,*

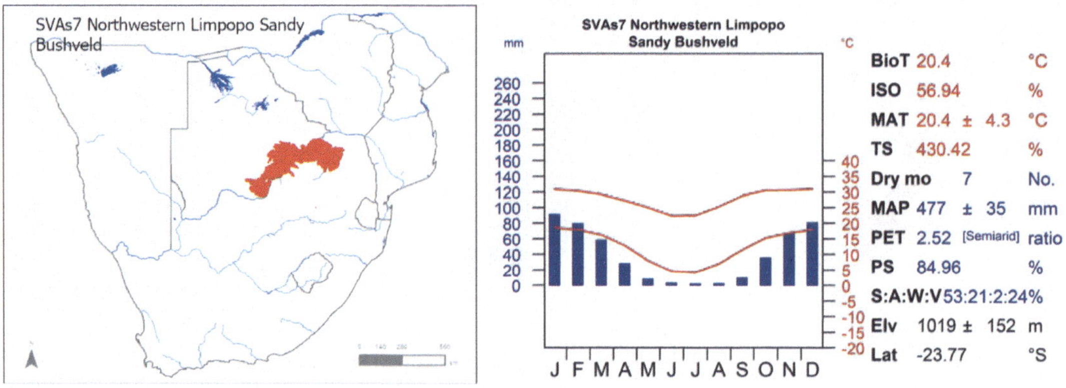

Fig. 3.34 The distribution area and the climatic diagram of the regional biome SVAs7 Northwestern Limpopo Sandy Bushveld

Panicum coloratum, P. maximum, Pogonarthria squarrosa, Schmidtia pappophoroides, Tragus berteronianus, and *Urochloa mosambicensis.* The northwestern extremities of the vegetation type Western Sandy Bushveld (part of the RB Southeastern Limpopo Sandy Bushveld) are transitional to this regional biome.

Climate MAP is 477 mm, ranging mostly from about 400 to 550 mm, and falling mainly in summer. Summers are hot with cool winter nights with occasional frost becoming infrequent in lower lying, northern parts of the Limpopo River valley. In a number of respects, the climate of this regional biome finds parallels with that of the RB Northern Kalahari Bushveld (Fig. 3.34).

Topography, Geology and Soils The topography is characterised by slightly to moderately undulating plains, built of, primarily, granite, gneiss and migmatite with sandstone, conglomerate, siltstone, mudstone and shale being most common on the plains directly north of the Waterberg. Soils are variable but predominantly freely drained, of high base status, usually sandy, but also loamy in places on the plains and clayey in bottomlands. The Botswana part of this regional biome is referred to locally as 'hardeveld', in contrast to the 'sandveld' of the extensive deep Kalahari sands on its western border. However, this hardeveld includes the occasional rocky hills of southeastern Botswana which form part of the RB Limpopo Mountain Bushveld. The RB Northwestern Limpopo Sandy Bushveld here constitutes the plains and pediments between such hills with generally sandy surface horizons but with more variable, yet usually well-drained, subsoils, and also including some loam and clay sections as well as some deep Kalahari sands.

Sources Weare and Yalala (1971), Bredenkamp and Van Vuuren (1977), van Vegten (1984) Bredenkamp (1986b), Panagos et al. (1986), de Wit and Bekker (1990), Peel (1990), Bekker and de Wit (1991), van Staden (1991), Schmidt et al. (1993), Schulze et al. (1994), Skarpe et al. (2000), Käller (2003), Rutherford et al. (2006), Teketay et al. (2016), Potgieter et al. (2017), Tshireletso et al. (2018), Makhabu et al. (2021).

SVAs8 Limpopo Mopane

SVmp 1 Musina Mopane Bushveld (Dayaram et al. 2019)
SVmp 2 Limpopo Ridge Bushveld (Dayaram et al. 2019)
SVmp 3 Cathedral Mopane Bushveld (Dayaram et al. 2019)
SVmp 4 Mopane Basalt Shrubland (Dayaram et al. 2019)
SVmp 5 Tsende Mopaneveld (Dayaram et al. 2019)
SVmp 6 Lowveld Rugged Mopaneveld (Dayaram et al. 2019)
SVmp 7 Phalaborwa-Timbavati Mopaneveld (Dayaram et al. 2019)
SVmp 8 Mopane Gabbro Shrubland (Dayaram et al. 2019)
SVmwl1 Limpopo Ridge Mopane Woodland (Lötter et al. 2023)
SVmwl2 Southern Mopane Woodland (Lötter et al. 2023)

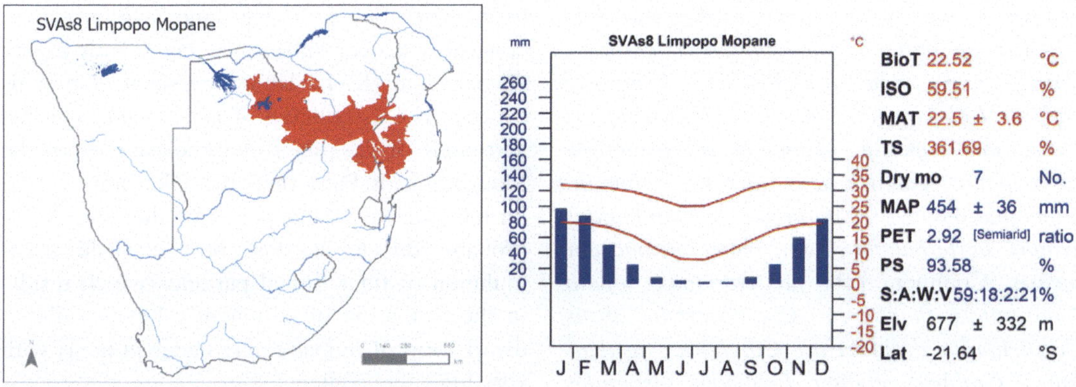

Fig. 3.35 The distribution area and the climatic diagram of the regional biome SVAs8 Limpopo Mopane

Distribution This regional biome encompasses a large, almost contiguous area, mostly within the lower lying parts of the mid-Limpopo river catchment in Botswana, Zimbabwe, Mozambique, and South Africa (Fig. 3.35). It stretches from areas around the Makgadikgadi Pans, Francistown, Palapye, and the Tuli Block in Botswana, entering Zimbabwe in the southernmost part of Hwange National Park and areas northwest of Bulawayo. The main extent in Zimbabwe is in the southern and eastern lowveld including the areas of the Shashi River valley, Beitbridge, Rutenga, Chiredzi, and lower Sabi River valley. It continues into the South African Lowveld south of the Limpopo River, remaining north of the Soutpansberg Range including the Alldays and Musina areas, but extending further south to include the northern half of the Kruger National Park and adjacent parts of the northeastern Lowveld of South Africa. It extends finally into mostly the Gaza Province of Mozambique where it reaches the southern limit of mopane vegetation a little south of the Limpopo National Park.

Vegetation Physiognomy and Composition The physiognomy is characterised by usually open to moderately closed, medium-height savanna, but varies from low shrubland, usually on heavy clay soils, to occasional, relatively closed, tall woodland (cathedral mopane). The herbaceous layer is usually poorly developed. The species *Colophospermum mopane* can often appear as the sole dominant of the vegetation. Other common species, usually on better drained soils, can include *Senegalia nigrescens, Aloe littoralis, Allophulus alnifolius, Boscia albitrunca, Catophractes alexandri, Combretum apiculatum, Combretum hereroense, C. imberbe, C. mossambicense, Commiphora africana, C. glandulosa, C. mollis, C. pyracanthoides, C. tenuipetiolata, Dichrostachys cinerea subsp. africana, Euclea divinorum, Grewia bicolor, G. flavescens, G. monticola, G. villosa, Kirkia acuminata, Rhigozum zambesiacum, Senegalia nigrescens, S. senegal* subsp. *rostrata, Sesamothamnus lugardii, Sclerocarya birrea, Terminalia prunioides, Vachellia exuvialis,* and *V. nilotica*. Other species that are characteristic of this biome are *Asparagus nodulosus, Barleria affinis, Blepharis diversispina, Catophractes alexandri, Cissus cornifolia, Craterostigma plantagineum, Pavetta harborii, Selaginella nivea*, and *Xerophyta humilis*. *Adansonia digitata* is prominent in places. Several important woody species in eastern areas of mopane veld are shared with the Lowveld Savanna, for example, *Cassia abbreviata, Dalbergia melanoxylon, Sclerocarya birrea, Flueggea virosa, Philenoptera violacea,* and *Sterculia africana.* The herbaceous layer can appear sparse in winter but a high-cover, species-rich annual-forb layer together with grasses is present in the summer of good rainfall years, especially in eastern areas. Grasses here include *Aristida adscensionis, A. congesta, Brachiaria* *deflexa, B. eruciformis, Bothriochloa radicans, Digitaria milanjiana, Enneapogon cenchroides, Enteropogon macrostachyus, Eragrostis rigidior, Heteropogon contortus, Melinis repens, Oropetium capense, Panicum coloratum, P. maximum, Pogonarthria squarrosa, Schmidtia pappophoroides, Sporobolus panicoides, Tragus berteronianus,* and *Urochloa mosambicesis.* Common forbs are *Acrotome inflata, Amaranthus thunbergii, Chamaecrista absus, Corbichonia decumbens, Corchorus asplenifolius, Crabbea velutina, Evolvulus alsinoides, Harpagophytum procumbens, Hermbstaedtia odorata, Hibiscus micranthus, H. sidiformis, Ipomoea magnusiana, Justicia flava, Kohautia virgata, Pupalia lappacea, Melhania acuminata, Ruellia patula, Seddera suffuticosa, Tephrosia purpurea,* and *T. villosa.*

Climate The MAP of the entire regional biome is 454 mm but it drops to about 350 mm in the Beitbridge to Pontdrif area. Most of the annual rainfall occurs in summer and winters are dry. Summers are hot with warm days and mild nights in winter. Frost is relatively rare. Colder temperatures have been identified as a key limitation in preventing *C. mopane* from extending further south in the South African Lowveld (Fig. 3.35).

Topography, Geology and Soils Flats, slopes, and extensive undulating plains, with ridges and rocky outcrops in places are the most characteristic features of the landscapes supporting this regional biome. The elevation is concentrated mainly below 1000 m. Rocks of the Archaean Complexes and Stormberg lavas and sediments (Karoo Supergroup) are the most widespread, with post-Stormberg intrusions, Ecca sediments, ultramafic metavolcanics and other geological units of lesser occurrence. Rock types thus vary considerably and include gneiss, granite, basalt, granophyre, sandstone, shale, mudstone, gabbro, calcrete, colluvium and alluvial sediments. This biome occurs on a relatively wide range of soil types, ranging from moderately deep to deep red, brown and dark heavy clay soils, through better drained sandy clay loams and sandy loams, calcareous soils, sands and gravel, to shallow stony soils. The sands can, in places, conceal a deeper subsurface layer with higher clay content (K. L. Tinley, personal communication). The association of *C. mopane* with the surrounds of the major endorheic basins of the Makgadikgadi Pans (and the Okavango Delta) on the extensive sand plains of the Kalahari is probably due to the past, broader scale, accumulation of finer fluvial particles which results in the formation of sufficiently heavy soils. A dwarf form of mopane may occur in areas with very high clay content.

Sources Pedro and Barbosa (1955), Ingram (1960), Rattray (1962), Wild and Barbosa (1967), Child (1968), Farrell (1968a), Louw (1970), Kelly and Walker (1976), Gertenbach (1978, 1983a, b), Allen (1978), van Rooyen (1978), Gertenbach and Potgieter (1979), van Rooyen et al. (1981b), O'Connor and Campbell (1986), Fraser et al. (1987), O'Connor (1992), Rogers (1993), Dekker and van Rooyen (1995), Visser et al. (1996), Dahlberg (2000), du Plessis (2001), Götze (2002), Straub (2002), Jordaan et al. (2004), Stalmans and Wishart (2005), Rutherford et al. (2006, 2012), Peel et al. (2007), Götze et al. (2008), Gandiwa and Kativu (2009), Siebert et al. (2010, 2021), Smit et al. (2010), Stalmans and Peel (2010), Clegg and O'Connor (2012), Cunliffe et al. (2012), Joubert (2013), Stevens et al. (2013), Gandiwa et al. (2014), Makhado et al. (2014), Wigley et al. (2014), O'Connor (2015), Martini et al. (2016), O'Connor and Goodall (2017), Akinyemi and Kgomo (2019), Pais et al. (2021), Nkosi et al. (2022), Aldworth et al. (2023), Cook et al. (2023), Lötter et al. (2023), Wedel (2024).

SVAs9 Namibian Dolomite Bushveld

Distribution The Namibian Dolomite Bushveld comprises the eastern Karstveld in northern Namibia, extending from northwest of Outjo along the Otavi Hills to the Tsumeb area, and including the areas between Otavi and Grootfontein (Fig. 3.36). The very separate and quite extensive dolomite area around the Aha Hills, on the Namibia–Botswana border, is largely overlain by Kalahari sand and bears the RB Northern Kalahari Bushveld. The much smaller and dissected areas of the dolomitic Aha Hills lie at the lower MAP limit of this regional biome and, as with other more arid dolomite areas, are not included in the Namibian Dolomite Bushveld.

Vegetation Physiognomy and Composition Physiognomy is moderately open to closed, medium-tall to low tree savanna also including short to tall, moderately closed thicket. Important taxa on the dolomite hills are *Cissus nymphaeifolia, Combretum apiculatum, Commiphora glaucescens, C. mollis, Croton gratissimus, Dombeya rotundifolia, Euphorbia guerichiana, Flueggea virosa, Grewia flavescens, Gymnosporia senegalensis, Gyrocarpus americanus* subsp. *africanus, Mundulea sericea, Kirkia acuminata, Lannea discolor, Pachypodium lealii, Peltophorum africanum, Sclerocarya birrea, Searsia marlothii, Steganotaenia araliacea, Vangueria infausta, Ximenia caffra,* and *X. americana.* Prominent on the intervening and surrounding Karstveld plains are *Combretum apiculatum, C. imberbe, Commiphora glandulosa, Croton menyharthii, Dichrostachys cinerea, Euclea undulata, Grewia bicolor, Senegalia mellifera, Spirostachys africana, Terminalia prunioides, Vachellia luederitzii, V. nilotica,* and *Ziziphus mucronata.* Included, in the west of this regional biome, is a relatively high rainfall area of generally taller *Colophospermum mopane* trees compared with the average in the Kaokoland Mopane. In the eastern part of this regional biome, scattered, low to tall *Hyphaene petersiana* form a distinctive palmveld on the plains. Important grasses in this regional biome include *Andropogon chinensis, Aristida adscensionis, Brachiaria deflexa, B. malacodes, Cenchrus ciliaris, Danthoniopsis dinteri, Digitaria seriata, Enneapogon cenchroides, E. desvauxii, E. scoparius, Eragrostis lehmanniana, E. nindensis, E. porosa, E. scopelophila, Heteropogon contortus, Melinis repens, Panicum maximum, Pogonarthria fleckii, Schizachyrium exile, Tragus berteronianus,* and *Urochloa brachyura.*

Climate This regional biome represents a relatively mesic dolomitic unit, encompassing the highest rainfall parts of the Karstveld of Namibia, and is the least dry of all regional biomes within the CB Southern African Arid Savanna. MAP is 509 mm, ranging from about 450 mm to over 550 mm on some of the higher eastern mountains. Rain falls mainly in summer and winters are very dry (Fig. 3.36).

Topography, Geology and Soils The topography comprises steep to moderately steep slopes on hills, rising up to about 500 m above the surrounding plains, as well as low hills and ridges, surrounded by generally flat plains, and plains interspersed between the hills. The hills are largely dolomite of the Otavi Group and quartzites, limestones and conglomerates also occur. Soils are stoney, rocky and gravelly on the hills

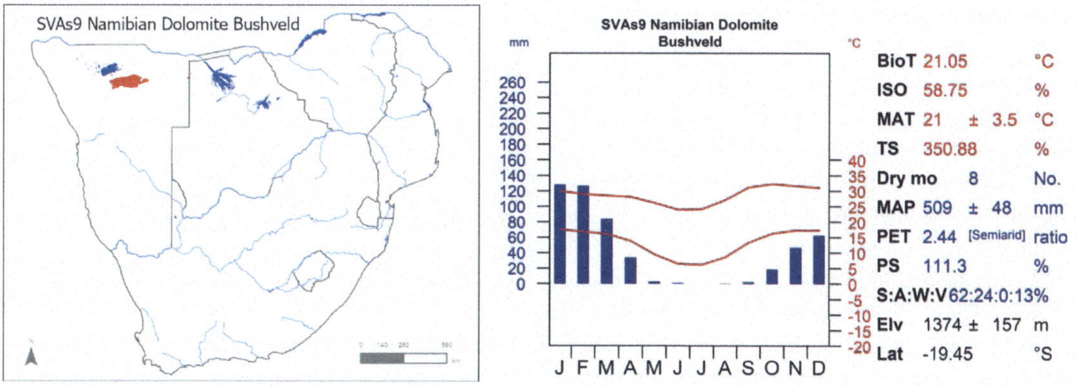

Fig. 3.36 The distribution area and the climatic diagram of the regional biome SVAs9 Namibian Dolomite Bushveld

and generally shallow both on the hills and on the plains. On the plains these shallow soils often overlie calcrete, and vary from calcareous sandy loams to clay loams usually in the lower lying areas. Pockets of deeper dolomite sands with rapid water drainage occur around or below the hills.

Sources Giess (1971), le Roux et al. (1988), Mendelsohn et al. (2000, 2002), Pickford and Senut (2010), Strohbach and Strohbach (2023).

3.3 Southern African Tropical Dry Forest

The following account leans heavily upon a detailed account of the tropical dry forests in Southern Africa published recently by Mucina et al. (2022) where more detail on ecology and evolutionary assembly of this enigmatic vegetation type can be found. The zonobiome S2 (also called Zonobiome II in the Walterian terminology) is represented in Southern Africa, not only by the very extensive and diverse savanna biomes, but also by a small portion of the global Tropical Dry Forest (TDF)—a biome which has been, until recently, somewhat neglected and mainly misunderstood in Africa. Its occurrence in the MBSA has already been hinted at in VegMap2006. Here it is mentioned under the heading Tropical Dry Forest Group (Mucina et al. 2006b: Table 12.1).

The occurrence of forests found under strongly seasonal climate and usually embedded within a matrix of fire-prone vegetation such as savanna grasslands and woodlands, is out of the ordinary. Unlike tropical wet forests, TDF is usually dominated by one or two species and may acquire three physiognomic faces, namely, tall forests, low (scrub) forests and dry thickets. The last-mentioned are considered part of the Tropical Dry Forest biome (see Mucina et al. 2022) due to a shared floristic composition and genesis. They are characteristically dominated by shrubs and hence of low, multi-stem habit and form dense entanglements with interlocking branches and abundant lianas. In Africa, the woody species in TDFs are deciduous and semi-deciduous (Mucina et al. 2022). These forests are characterised by poorly developed grass layer and hence are in principle fire shy. The TDFs of Africa are species- and endemic-rich, showing a high degree of floristic turnover from region to region. In Southern Africa, large portions of the tree and shrub dominants of the TDFs are of Guineo-Congolian biogeographic affinity.

In the MBSA area, TDFs are the most extreme forest type of the entire study area in terms of MAT (23 °C; the hottest), precipitation (MAP: 478 mm; the driest). These forests occur in frost-free areas.

Despite the large variability and wide global distribution of the TDF biome, only two regional biomes of TDF reach the MBSA, namely RB TDFa2 Southern African Dry Forest and RB TDFa3 Southern African Dry Thicket (the coding and naming follow Mucina et al. 2022). Both qualify as extrazonal (as they are more common in East and Central Africa; e.g., Lötter et al. 2023; Mucina 2023) and both are linked to the occurrence of deep sandy soils and hence qualify as psammobiomes.

The RB Tropical Dry Forest and RB Tropical Dry Thicket in the MBSA can sometimes lie in relatively close proximity to each other, even grading the one into the other or forming an interspersed matrix of the two regional biomes. Not surprisingly, therefore, apart from both favouring the same deep sandy substrate, they can share numerous species and overlap substantially regarding climatic conditions.

TDFa2 Southern African Dry Forest

FOz 8 Sand Forest p.p. (Mucina and Rutherford 2006)
FOz 9 Ironwood Dry Forest (Mucina and Rutherford 2006)
TDFS1 Inhamitanga Sand Forest (Lötter et al. 2023)
TDFS2 Ironwood Dry Forest (Lötter et al. 2023)
TDFS3 Madanda Sand Forest (Lötter et al. 2023)
TDFS4 Maputaland Sand Forest (Lötter et al. 2023)

Distribution The Southern African Dry Forest is fragmented but well represented in Maputaland (both in South Africa and Mozambique) but also

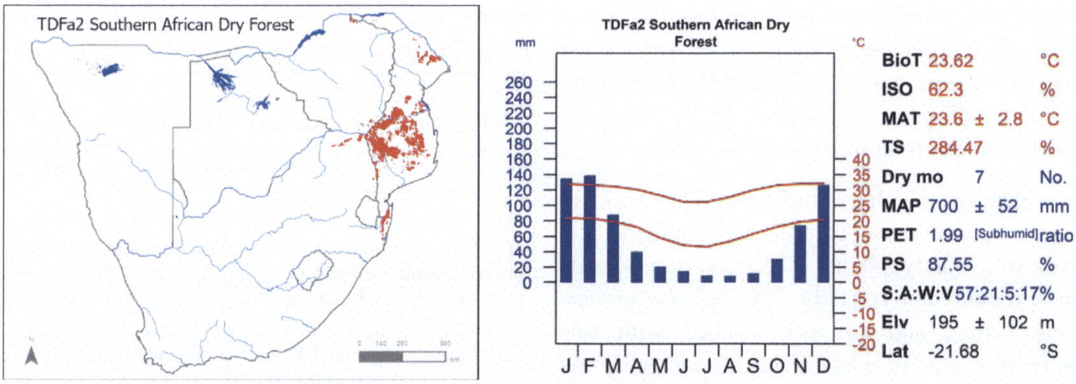

Fig. 3.37 The distribution area and the climatic diagram of the regional biome TDFa2 Southern African Dry Forest

further northwards in the northern parts of the Gaza and Inhambane Provinces and the southern parts of the Manica and Sofala Provinces of Mozambique. Another type of these tall dry forests occurs in the Zambezi River valley of Mozambique and into Zimbabwe. The most common TDF of the MBSA area is the ironwood forest which is very common, yet patchy, in the region where Zimbabwe, South Africa and Mozambique meet (see Fig. 2 in Mucina et al. 2022) and further inland in southwestern Mozambique and the Soutpansberg area of northern South Africa (Fig. 3.37).

Vegetation Physiognomy and Composition Maputaland licuati forests and the dry forests northwards towards Beira in Mozambique are medium to tall, deciduous or semi-deciduous, sometimes with broken canopy structure, as well as containing an often dense shrub layer. Dominant are *Afzelia adianthifolia, A. quanzensis, Balanites maughamii, Baphia massaiensis* subsp. *obovata, Brachystegia torrei, Cleistanthus schlechteri, Cola greenwayi, Cordyla africana, Dialium schlechteri, Diospyros natalensis, Drypetes arguta, Guibourtia conjugata, Hymenocardia ulmoides, Margaritaria discoidea, Milicia excelsa, Newtonia hildebrandtii, Rothmannia fischeri, Senegalia nigrescens, Terminalia myrtifolia, Wrightia natalensis,* and *Xeroderris stuhlmannii.* Further inland, the ironwood dry forest is typically monodominant, composed of the short to medium tall, evergreen *Androstachys johnsonii,* usually found in patches. Other frequent component species here include *Barleria matopensis, Boscia albitrunca, Cassia abbreviata, Cissus rotundifolia, Combretum paniculatum, Commiphora tenuipetiolata, Croton pseudopulchellus, Entandrophragma caudatum, Guibourtia conjugata, Leptactina delagoense, Ptaeroxylon obliquum,* and *Vitex patula. Androstachys johnsonii* has been reported to secrete secondary compounds that suppress the establishment and growth of other species under its canopy. The herbaceous layer is poorly developed and, although several grass species may be present, the lack of the common grasses of *Panicum maximum* and *Digitaria eriantha* have been highlighted in the ironwood dry forests of the Kruger National Park. It may be noted that short ironwood dry forest has sometimes been referred to as thicket. The tall dry forests in the Zambezi Valley and its associated plains are 20–25 m tall (some emergents to 30 m) with closed canopies dominated by *Adansonia digitata, Cola mossambicensis, Dalbergia boehmii, Pterocarpus lucens, Terminalia myrtifolia,,* and *Xylia torreana*, In the shrub layer *Boscia mossambicensis, Combretum elaeagnoides, C. tetrandrum, Senegalia ataxacantha,* and *Monodora junodii* are frequently found. Grasses can be virtually absent.

Climate Rainfall in this regional biome is strongly seasonal with most falling in summer and only 5% of the MAP falling in winter. The MAP of 700 mm is lower than that of Southern

Africa Dry Thicket owing in part to the inclusion of the relatively drier Ironwood Dry Forest vegetation type which has a MAP of less than 550 mm. Temperatures are relatively equitable and the area is frost free (Fig. 3.37).

Topography, Geology and Soils This regional biome is found as patches or fragments on extensive, sandy plains. The deep sandy soils are predominantly sandy clay loams, moderately acidic, and strongly leached with low nutrient status in places. In Maputaland, the soils are formed from ancient major dune cordons and are consequently more sandy with little clay. Although the ironwood dry forests also follow the pattern of patches on deep sand, they also occur on moderate to steep mountain slopes on the Lebombo Mountains and parts of the Soutpansberg, with a rock and stone surface cover up to 60% in places. Here the geology varies from sandstones of the Soutpansberg Group to rhyolite of the Lebombo Group.

Sources Myre (1964), van Wyk (1973), Moll (1978), van Rooyen (1978), van Rooyen et al. (1981a), Coetzee (1983), Gertenbach (1983b), Timberlake et al. (1988), Broderick (1990), Timberlake (1996), Kirkwood and Midgley (1999), Matthews et al. (1999, 2001), van Wyk and Smith (2001), Hoare et al. (2002), Schmidt et al. (2002), von Maltitz et al. (2003), Gaugris et al. (2004), Geldenhuys and Mucina (2006), Mucina et al. (2006b, 2022), Matthews (2007), Gandiwa et al. (2011), Cunliffe et al. (2012), Bakali et al. (2017), Nel et al. (2022), Lötter et al. (2023).

TDFa3 Southern African Dry Thicket

FOz 8 Sand Forest p.p. (Mucina and Rutherford 2006)
TDT1 Licuati Sand Thicket (Lötter et al. 2023)
TDT2 Madanda Rubber Sand Thicket (Lötter et al. 2023)
TDT4 Mazoe Gneiss Dry Thicket (Lötter et al. 2023)
TDT6 Nwambiya Sand Thicket (Lötter et al. 2023)
TDT7 Pande Sand Thicket (Lötter et al. 2023)
TDT8 Save Valley Chalk Thicket (Lötter et al. 2023)
TDT9 Zambezian Sand Thicket (Lötter et al. 2023)

Distribution Occurs in the form of patches or clusters of patches in the Licuati area in southern Mozambique, various localities in the south-central inland part of Mozambique between Villancoulos and Beira, the Zambezi Valley of northwestern Mozambique and northeastern Zimbabwe, where the sand thickets are referred to locally as 'jesse bush'. It is also found in the Nwambyia area in western Gaza Province of Mozambique extending marginally into the northern Kruger National Park of South Africa and into the Gonarezhou National Park of Zimbabwe (Fig. 3.38).

Vegetation Physiognomy and Composition The physiognomy is formed by a characteristically low, dense woody thicket generally less than 5–6 m tall, occasionally with scattered taller emergent trees protruding above the thicket, and a usually very poorly developed herbaceous layer. Woody species dominance

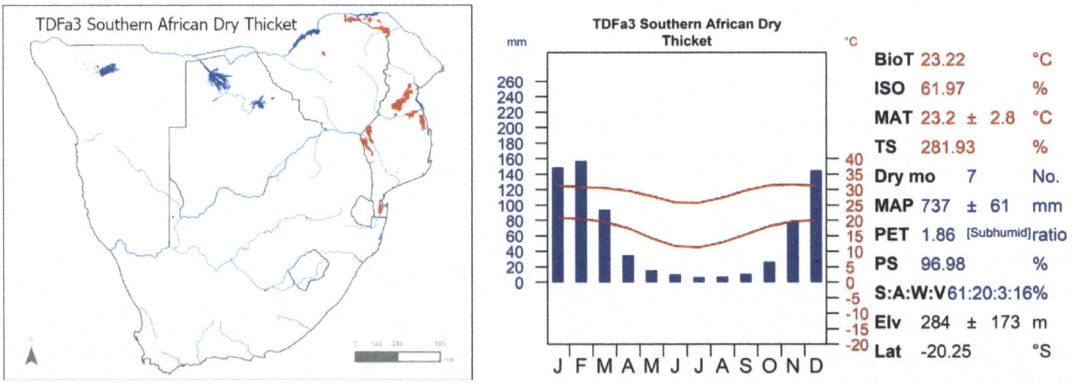

Fig. 3.38 The distribution area and the climatic diagram of the regional biome TDFa3 Southern African Dry Thicket

varies greatly over the wide geographical range of the often widely separated patches of this regional biome. The main low tree and shrub components include *Alchornea laxiflora, Baphia massaiensis, Brachylaena huillense, Cassipourea mossambicensis, Cleistochlamys kirkii, Combretum celastroides, C. collinum, C. elaeagnoides, C. hereroense, C. mkuzense, C. padoides, Croton steenkampianus, Dalbergia fischeri, Erythroxylum emarginatum, Guibourtia conjugata, Heinsia crinita, Hexalobus monopetalus, Hugonia orientalis, Hymenocardia ulmoides, Lecaniodiscus fraxinifolius, Mondora junodii, Ochna barbosae, Pavetta catophylla, Spirostachys africana, Strychnos henningsii, Tabernaemontana elegans, Terminalia brachystemma, T. myrtifolia, Tricalysia delagoensis, Turraea nilotica, Uvaria gracilipes, U. lucida, Vepris bremekampii,* and *Xylia torreana.* The occasional taller emergent trees include, in addition to some of the above listed species, *Adansonia digitata, Afzelia quanzensis, Cordyla africana, Manilkara discolor, Millettia stuhlmannii,* and *Pterocarpus lucens.* Climbers and scramblers can be common and include *Acridocarpus natalitius, Ancylobotrys petersiana, Capparis brassii, Cissampelos hirta, Landolphia kirkii, Opilia amentacea, Rhoicissus revoilii, Schlechterina mitostemmatoides, Strophanthus kombe,* and *Synaptolepis alternifolia.* Although the herbaceous component is usually poorly developed and generally devoid of grasses, there are some exceptions such as in some of the thickets in the Zambezi Valley which contain, for example, *Heteropogon melanocarpus, Panicum maximum,* and *Schmidtia pappophoroides*.

Climate Most of MAP of 737 mm falls in summer and winters are dry. Temperatures are relatively equitable and frost does not occur (Fig. 3.38).

Topography, Geology and Soils This regional biome is scattered over sandy plateaux and level upland sandy plains similar to those that support the Southern African Dry Forests described above. In the Zambezi Valley there are convex-shaped sand lenses, also described as extensive flat mounds that support jesse vegetation. The sandy soils are deep, freely drained, acidic and generally belong to the sandy clay loam textural class. The geology is very similar to that underlying the deep sands that carry Southern African Dry Forest.

Sources van der Schijff (1957), de Moor et al. (1977), van Rooyen et al. (1981a), Gertenbach (1983b), Du Toit (1993), von Maltitz et al. (2003), Stalmans et al. (2004), Geldenhuys and Mucina (2006), Mucina et al. (2006b, 2022), Lötter et al. (2023).

References

Aarrestad PA, Masunga GS, Hytteborn H et al (2011) Influence of soil, tree cover and large herbivores on field layer vegetation along a savanna landscape gradient in northern Botswana. J Arid Environ 75:290–297

Abbott A (1993) The Umtamvuna Nature Reserve. In: The Umtamvuna Nature Reserve. Umtamvuna River Trust, Port Edward, pp 2–4

Abbott ATD, van Wyk AE, Johnson DN et al (2000) Checklist of the macrofungi, lichens, bryophytes and vascular plants of the Umtamvuna Nature Reserve. Lammergeyer 46:1–74

Acocks JPH (1953) Veld types of South Africa. Mem Bot Surv S Afr 40:1–128

Acocks JPH (1988) Veld types of South Africa, 3rd edn. Mem Bot Surv S Afr 57:1–146

Adhikari S, Marcelo-Silva J, Rajakaruna N et al (2022) Influence of land use and topography on distribution and bioaccumulation of potentially toxic metals in soil and plant leaves: a case study from Sekhukhuneland, South Africa. Sci Total Environ 806:150659. https://doi.org/10.1016/j.scitotenv.2021.1506590048-9697

Akinyemi FO, Kgomo MO (2019) Vegetation dynamics in African drylands: an assessment based on the Vegetation Degradation Index in an agro-pastoral region of Botswana. Reg Environ Change 19:2027–2029

Aldworth TA, Toucher MLW, Clulow AD et al (2023) The effect of woody encroachment on evapotranspiration in a semi-arid savanna. Hydrology 10:9. https://doi.org/10.3390/hydrology10010009

Allen A (1978) A preliminary reconnaissance of the vegetation of Orapa and environs. Botsw Notes Rec 10:169–185

Anderson GD, Walker BH (1974) Vegetation composition and elephant damage in the Sengwa Wildlife Research area, Rhodesia. J S Afr Wildl Manage Assoc 4:1–14

Arraut EM, Loveridge AJ, Chamaillé-Jammes S et al (2018) The 2013–2014 vegetation structure map of Hwange National Park, Zimbabwe, produced using free satellite images and software. Koedoe 60(1):a1497. https://doi.org/10.4102/koedoe.v60i1.1497

Asner GP, Levick SR (2012) Landscape-level effects of herbivores on treefall in African savannas. Ecol Lett 15:1211–1217

Asner GP, Levick SR, Kennedy-Bowdoin T et al (2009) Large-scale impacts of herbivores on the structural diversity of African savannas. PNAS 106:4947–4952

Bakali M, Ligavha-Mbelengwa MH, Potgieter MJ et al (2017) Impact of ethnobotanical utilization on the population structure of *Androstachys johnsonii* Prain. in the Vhembe area of the Limpopo Province, South Africa. Insights For Res 1(1):50–56

Balkwill K, Williamson SD, Kidger CL et al (1997) Diversity and conservation of serpentine sites in southern Mpumalanga (Eastern Tvl.), S.A. In: Jaffré T, Reeves RD, Becquer T (eds) The ecology of ultramafic and metalliferous areas. ORSTOM, Nouméa, pp 133–138

Barclay-Smith RW (1963) A report on the ecology and vegetation of the Great Dyke within the Horseshoe Intensive Conservation area. Kirkia 4:25–34

Barrett AS, Brown LR, Barrett L et al (2010) A floristic description and utilisation of two home ranges by vervet monkeys in Loskop Dam Nature Reserve, South Africa. Koedoe 52(1):1–12

Behr CM, Bredenkamp GJ (1988) A phytosociological classification of the Witwatersrand National Botanic Garden. S Afr J Bot 54:525–533

Becker T (2000) Muster der Vegetation und ihre Determinanten in einem desertifikationsgefährdeten Raum im Nordwesten Namibias (Kaokoland). Thesis, Universität zu Köln, Köln

Bekker RP, de Wit PV (1991) Contribution to the vegetation classification of Botswana. Soil Mapping and Advisory Services Project. AG: BOT/85/011. Field Document 34. FAO/UNDP/Government of Botswana, Gaborone

Ben-Shahar R (1988) Patterns of plant species associations on a Sour Bushveld nature reserve. S Afr J Bot 54:504–506

Bezuidenhout H (2009) The classification, mapping and description of the vegetation of the Rooipoort Nature Reserve, Northern Cape, South Africa. Koedoe 51(1):Art.#695. https://doi.org/10.4102/koedoe.v51i1.695

Bezuidenhout H, Bredenkamp GJ, Elsenbroek JH (1988) Die plantegroei van die alkaligraniet en aangrensende kwartsiet in die Vredefortkoepel noordwes van Parys. S Afr Tydskr Natuurwetensk Tegnol 7:4–9

Bezuidenhout H, Bredenkamp GJ, Theron GK (1994) Syntaxonomy of the vegetation of the Fb land type in the western Transvaal grassland, South Africa. S Afr J Bot 60:72–81

Blackmore AC, Mentis MT, Scholes RJ (1990) The origin and extent of nutrient-enriched patches within a nutrient-poor savanna in South Africa. J Biogeogr 17:463–470

Bond WJ (2019) Open ecosystems. Ecology and evolution beyond the forest edge. Oxford University Press, Oxford

Boneschans RB, Coetzee MS, Siebert SJ (2015) A geobotanical investigation of the Koedoesfontein Complex, Vredefort Dome, South Africa. Austr J Bot 63:324–340

Boon R, Cockburn J, Douwes E et al (2016) Managing a threatened savanna ecosystem (KwaZulu-Natal Sandstone Sourveld) in an urban biodiversity hotspot: Durban, South Africa. Bothalia 46(2):a2112. https://doi.org/10.4102/abc.v46i2.2112

Bourliére F (ed) (1983) Ecosystems of the World. 13. Tropical savannas. Elsevier, Amsterdam

Bouwer D, le Roux PAL, van Tol J (2020) Identification of hydropedological flowpaths in Stevenson–Hamilton catena from soil morphological, chemical and hydraulic properties. Koedoe 62(2):a1584. https://doi.org/10.4102/koedoe.v62i2.1584

Bredenkamp GJ (1975) Plant communities of the Suikerbosrand Nature Reserve, Transvaal. S Afr J Sci 71:30–31

Bredenkamp GJ (1977) The grasses of the Suikerbosrand Nature Reserve: their habitat preferences and synecological significance. Proc Grassl Soc S Afr 12:135–139

Bredenkamp GJ (1982) 'n Plantekologiese studie van die Manyeleti-wildtuin. Thesis, Department of Botany, University of Pretoria, Pretoria

Bredenkamp GJ (1985) An evaluation of the relative importance of certain habitat factors that influence the distribution of the plant species in the Manyeleti Game Reserve. S Afr J Bot 51:194–196

Bredenkamp GJ (1986a) Ecological profiles of potential bush encroacher species in the Manyeleti Game Reserve. S Afr J Bot 52:53–59

Bredenkamp GJ (1986b) Vegetation studies and wildlife management in the Pietersburg Nature Reserve, South Africa. In: Environmental quality and ecosystem stability. Bar-Ilan University Press, Ramat-Gan, pp 51–61

Bredenkamp GJ (1987) An overview of the major plant communities of the Manyeleti Game Reserve, Gazankulu, South Africa. Coenoses 2:75–84

Bredenkamp GJ, Deutschländer MS (1994) The *Themedo triandrae–Setarietum incrassatae*, a new association from gabbro in the Manyeleti Game Reserve, Gazankulu, South Africa. Koedoe 37(2):43–57

Bredenkamp GJ, Deutschländer MS (1995) New azonal syntaxa from the hills and river banks of the Manyeleti Game Reserve, Northern Transvaal Province, South Africa. Koedoe 38(1):41–58

Bredenkamp GJ, Theron GK (1978) A synecological account of the Suikerbosrand Nature Reserve. I. The phytosociology of the Witwatersrand geological system. Bothalia 12:513–529

Bredenkamp GJ, Theron GK (1980) A synecological account of the Suikerbosrand Nature Reserve. II. The phytosociology of the Ventersdorp geological system. Bothalia 13:199–216

Bredenkamp GJ, Theron GK (1985) A quantitative approach to the structural analysis and classification of the vegetation of the Manyeleti Game Reserve. S Afr J Bot 51:45–54

Bredenkamp GJ, Theron GK (1988) The relationship between soil chemistry and plant species distribution in the Manyeleti Game Reserve, South Africa. S Afr J Sci 84:423–427

Bredenkamp GJ, Theron GK (1990) The vegetation of the fersiallitic soils of the Manyeleti Game Reserve. Coenoses 5:167–175

Bredenkamp GJ, Theron GK (1991) The *Eucleo divinori–Acacietum nigricentis*, a new association from the calcareous bottomland clays of the Manyeleti Game Reserve, Eastern Transvaal Lowveld, Gazankulu, South Africa. Vegetatio 93:119–130

Bredenkamp GJ, Van Vuuren DRJ (1977) The plant communities below Turfloop Dam, Lebowa. University of the North, Turfloop

Bredenkamp GJ, Theron GK, Van Vuuren DRJ (1983) Ecological interpretation of plant communities by classification and ordination of quantitative soil characteristics. Bothalia 14:691–699

Bredenkamp GJ, Deutschländer MS, Theron GK (1993) A phytosociological analysis of the *Albizio harveyi–Eucleetum divinori* from sodic bottomland clay soils of the Manyeleti Game Reserve, Gazankulu, South Africa. S Afr J Bot 59:57–64

Breebaart L, Deutschländer M (1997) The vegetation types and management units of Goedverwacht farm in the mixed bushveld of the Northern Province, South Africa. Koedoe 40:19–35

Breebaart L, O'Connor TG, Hughes JC et al (2001) The effects of topo-edaphic variables and cultivation on the woody vegetation of Weenen Nature Reserve, KwaZulu-Natal. Afr J Range Forage Sci 18:75–91

Broderick TJ (1990) An interpretation of the geology of the Cabora Bassa Basin, mid-Zambezi Valley. Ann Zimb Geol Surv 1:1–11

Brooks RR, Yang X (1984) Elemental levels and relationships in the endemic serpentine flora of the Great Dyke, Zimbabwe, and their significance as controlling factors for the flora. Taxon 33:392–399

Brown LR, Bredenkamp GJ (1994) The phytosociology of the southern section of Borakalalo Nature Reserve, South Africa. Koedoe 37(2):59–72

Brown LR, Bredenkamp GJ, van Rooyen N (1995) The phytosociology of the western section of Borakalalo Nature Reserve. Koedoe 38(2):49–64

Brown LR, Bredenkamp GJ, van Rooyen N (1996) The phytosociology of the northern section of Borakalalo Nature Reserve. Koedoe 39(1):9–24

Brown LR, Bredenkamp GJ, van Rooyen N (1997) Phytosociological synthesis of the vegetation of the Borakalalo Nature Reserve. S Afr J Bot 63:242–253

Burke A, Wittneben M (2008) A preliminary account of the vegetation of the Auas Mountains. Dinteria 30:41–91

Camp K (1999a) The bioresource groups of KwaZulu-Natal. Mistbelt. Cedara Report No. N/A/99/13. KwaZulu-Natal Department of Agriculture, Cedara

Camp K (1999b) The bioresource groups of KwaZulu-Natal. Coast & Coast Hinterland. Cedara Report No. N/A/99/12. KwaZulu-Natal Department of Agriculture, Cedara

Camp K (1999c) The bioresource groups of KwaZulu-Natal. Tall Grassveld. Cedara Report No. N/A/99/15. KwaZulu-Natal Department of Agriculture, Cedara

Camp K (1999d) The bioresource groups of KwaZulu-Natal. Thornveld. Cedara Report No. N/A/99/16. KwaZulu-Natal Department of Agriculture, Cedara

Camp K (1999e) The bioresource groups of KwaZulu-Natal. Bushveld. Cedara Report No. N/A/99/17, KwaZulu-Natal Department of Agriculture, Cedara

Camp K (2001) The classification of the units of the bioresource programme of KwaZulu-Natal (a description of the mapping process). Natural Resources Section, Department of Agriculture and Environmental Affairs, Cedara

Campbell BM, Cunliffe RN, Gambiza J (1995) Vegetation structure and small-scale pattern in Miombo Woodland, Marondera, Zimbabwe. Bothalia 25:121–126

Campbell B, Frost P, Byron N (1996) Miombo woodlands and their use: overview and key issues. In: Campbell B (ed) The miombo in transition: woodlands and welfare in Africa. CIFOR, Bognor, pp 1–10

Carbutt C (2005) The nutrient-enriching effects of urban discharges on the eThekwini Municipality's nutrient-poor grassland and forest, eThekwini Municipality. Environmental Planning and Climate Protection Department, Durban

Carbutt C, Edwards T (2001) Cape element on high-altitude and edaphic islands: historical aspects and preliminary phytogeography. Syst Geogr Plants 71:1033–1061

Child G (1968) An ecological survey of northeastern Botswana. FAO Report TA 2563. FAO, Rome

Childes SL, Walker BH (1987) Ecology and dynamics of the woody vegetation of the Kalahari sands in Hwange National Park, Zimbabwe. Vegetatio 72:111–128

Cizek A (2021) The distribution of Miombo woodland in southern Africa. Kirkia 19:141–159

Clegg BW, O'Connor TG (2012) The vegetation of Malilangwe Wildlife Reserve, south-eastern Zimbabwe. Afr J Range Forage Sci 29:109–131

Coetsee C, Botha J, Case MF et al (2023a) The hard lives of trees in African savanna—even without elephants. Austral Ecol 48:532–551

Coetsee C, February EC, Wigley BJ et al (2023b) Soil organic carbon is buffered by grass inputs regardless of woody cover or fire frequency in an African savanna. J Ecol 111:2483–2495

Coetzee JJ (1971) Die landboupotensiaal van die Noordwes-Transvaalse soetbosveld. Thesis, University of Pretoria, Pretoria

Coetzee BJ (1972) 'n Plantsosiologiese studie van die Jack Scott-natuurreservaat. Thesis, Department of Botany, University of Pretoria, Pretoria

Coetzee BJ (1974) A phytosociological classification of the vegetation of the Jack Scott Nature Reserve. Bothalia 11:329–347

Coetzee BJ (1975) A phytosociological classification of the Rustenburg Nature Reserve. Bothalia 11:561–580

Coetzee BJ (1983) Phytosociology, vegetation structure and landscapes of central district, Kruger National Park, South Africa. J. Cramer, Vaduz

Coetzee BJ, Nel PJ (1978) A phytosociological reconnaissance of Mlilwane Wildlife Sanctuary, Swaziland. Koedoe 21:1–36

Coetzee BJ, van der Meulen F, Zwanziger S et al (1976) A phytosociological classification of the Nylsvley Nature Reserve. Bothalia 12:137–160

Coetzee BJ, Gertenbach WPD, Nel PJ (1977) Korttermyn plantegroeistruktuurveranderings op basalt in die sentrale distrik, Nasionale Krugerwildtuin. Koedoe 20:53–65

Coetzee BJ, van Wyk P, Gertenbach WPD et al (1981) 'n Plantekologiese verkenning van die Waterberggebied in die noord-Transvaalse bosveld. Koedoe 24:1–23

Coetzee JP, Bredenkamp GJ, van Rooyen N (1993) The Sub-humid Warm Temperate Mountain Bushveld plant communities of the Pretoria–Witbank–Heidelberg area. S Afr J Bot 59:623–632

Coetzee JP, Bredenkamp GJ, van Rooyen N et al (1994) An overview of the physical environment and vegetation units of the Ba and Ib land types of the Pretoria–Witbank–Heidelberg area. S Afr J Bot 60:49–61

Coetzee JP, Bredenkamp GJ, van Rooyen N (1995) Plant communities of the Sub-humid Cool Temperate Mountain Bushveld in the Pretoria–Witbank–Heidelberg area, South Africa. S Afr J Bot 61:114–122

Cole MM, Brown RC (1976) The vegetation of the Ghanzi area of western Botswana. J Biogeogr 3:169–196

Combrink A (2017) Savanna woody regeneration in response to different treatments of herbivory and fire. Thesis, North-West University, Potchefstroom

Cook RM, Witkowski ETF, Henley MD (2023) Survival trends (2008–2020) of three tree species in response to elephant impact, environmental variation, and stem wire-netting protection in an African savanna. For Ecol Manage 545:121220. https://doi.org/10.1016/j.foreco.2023.121220

Cooper SM, Owen-Smith N (1985) Condensed tannins deter feeding by browsing ruminants in a South African savanna. Oecologia 67:142–146

Craig GC (1983) Vegetation survey of Sengwa. Bothalia 14:759–763

Cresswell CF, Ferrar P, Grunow JO et al (1982) Phytomass, seasonal phenology and photosynthetic studies. In: Huntley BJ, Walker BH (eds) Ecology of tropical savannas. Springer, Berlin, pp 476–497

Cronje HP, Panagos MD, Reilly BK (2008) The plant communities of the Andover Game Reserve, South Africa. Koedoe 50(1):184–201

Cunliffe R, Muller T, Mapaura A (2012) Vegetation survey of Gonarezhou National Park, Zimbabwe. Zimbabwe Parks and Wildlife Management Authority, Harare

Dahlberg AC (2000) Vegetation diversity and change in relation to land use, soil and rainfall—a case study from North-East District, Botswana. J Arid Environ 44:19–40

Davis ALV, Scholtz CH, Kryger U et al (2010) Dung beetle assemblage structure in Tswalu Kalahari Reserve: responses to a mosaic of landscape types, vegetation communities, and dung types. Environ Entomol 39:811–820

Dayaram A, Harris LR, Grobler BA et al (2019) Vegetation Map of South Africa, Lesotho and Swaziland 2018: a description of changes since 2006. Bothalia 49(1):a2452. https://doi.org/10.4102/abc.v49i1.2452

Deall GB (1985) A plant-ecological study of the Eastern Transvaal escarpment in the Sabie area. Thesis, Department of Botany, University of Pretoria, Pretoria

Deall GB, Theron GK, Westfall RH (1989) The vegetation ecology of the Eastern Transvaal escarpment in the Sabie area. 2. Floristic classification. Bothalia 19:69–89

De Cauwer V, Geldenhuys CJ, Aerts R et al (2016) Patterns of forest composition and their long term environmental drivers in the tropical dry forest transition zone of Southern Africa. For Ecosyst 3:23. https://doi.org/10.1186/s40663-016-0080-9

De Cauwer V, Becker RW, Gomes A et al (2023) The mountain top flora and vegetation of the remote Ovahimba Highlands in the Kaokoveld Centre of Endemism: a reconnaissance. Trans Roy Soc S Afr 78:109–122

De Frey WH (1999) Phytosociology of the Mpumalanga high altitude grasslands. Thesis, Centre for Wildlife Management, University of Pretoria, Pretoria

Dekker B, van Rooyen N (1995) The physical environment and plant communities of the Messina Experimental Farm. S Afr J Bot 61:158–167

de Moor PP, Pooley E, Neville G et al (1977) The vegetation of Ndumu Game Reserve, Natal: a quantitative physiognomic survey. Ann Natal Mus 23:239–272

de Wit PV, Bekker RP (1990) Explanatory note on the Land Systems map of Botswana. Soil Mapping and Advisory Services Project. AG: BOT/85/011. Field Document 31. FAO/UNDP/Government of Botswana, Gaborone

Dingaan MNV, Tsubo M (2019) Improved assessment of pasture availability in semi-arid grassland of South Africa. Environ Monit Assess 191:733. https://doi.org/10.1007/s10661-019-7918-x

Dingaan M, Walker S, Newby T et al (2013) Plant species richness, soil properties and grazing effects in semiarid grassland of South Africa. S Afr J Bot 86:174. https://doi.org/10.1016/j.sajb.2013.02.131

Dingaan MNV, Walker S, Tsubo M et al (2016) Influence of grazing on plant diversity-productivity relationship in semi-arid grassland of South Africa. Appl Ecol Environ Res 14:1–13. https://doi.org/10.15666/aeer/1404_001013
Dingaan MNV, Tsubo M, Walker S et al (2017) Soil chemical properties and plant species diversity along a rainfall gradient in semi-arid grassland of South Africa. Plant Ecol Evol 150:35–44
Dörgeloh WG (1998) A comparison of tree density and canopy cover between different plant communities in mixed bushveld, Northern Province. S Afr J Bot 64:86–87
Dörgeloh WG (1999a) Assessment of veld conditions with multivariate techniques in mixed bushveld, South Africa. Afr J Ecol 37:194–201
Dörgeloh WG (1999b) Diversity of the herbaceous layer in mixed bushveld. J Range Manage 52:519–524
Downing BH (1972) A plant ecological survey of Umfolozi Game Reserve. Thesis, Department of Botany, University of Natal, Pietermaritzburg
Downing BH (1980) Changes in the vegetation of Hluhluwe Game Reserve, Zululand, as regulated by edaphic and biotic factors over 36 years. J S Afr Bot 46:225–231
Dreber N, van Rooyen SE, Kellner K (2018) Relationship of plant diversity and bush cover in rangelands of a semi-arid Kalahari savannah, South Africa. Afr J Ecol 56:132–135
Drury CC, Ramdhani S, Naidoo S et al (2016) A lot gone but still hanging on: floristics of remnant patches of endangered KwaZulu-Natal Sandstone Sourveld. Bothalia 46(2):a2110. https://doi.org/10.4102/abc.v46i2.2110
Du Plessis F (2001) A phytosociological synthesis of Mopaniveld. Thesis, Department of Botany, University of Pretoria, Pretoria
du Preez PJ, Venter HJT (1990) The phytosociology of the woody vegetation in the southern part of the Vredefort Dome Area. Part II: Communities of the hills. S Afr J Bot 56:637–644
Du Toit R (1993) Reconnaissance vegetation survey of the Chewore–Angwa–Kanyemba area of the Zambezi Valley, Zimbabwe. Kirkia 14(1):61–77
Edwards D (1967) A plant ecological survey of the Tugela River basin. Town and Regional Planning Commission, Pietermaritzburg
Ellery WN, Scholes RJ, Scholes MC (1995) The distribution of sweetveld and sourveld in South Africa's grassland biome in relation to environmental factors. Afr J Range Forage Sci 12:38–45
Farrell JAK (1968a) Preliminary notes on the vegetation of the lower Sabi-Lundi basin, Rhodesia. Kirkia 6:223–248
Farrell JAK (1968b) Preliminary notes on the vegetation of southern Gokwe District, Rhodesia. Kirkia 6:249–257
February EC, Higgins SI (2010) The distribution of tree and grass roots in relation to soil nitrogen and water. S Afr J Bot 76:517–523
Felton IR (2002) A strategic assessment of the woody grasslands of Maputaland, South Africa, with recommendations for future conservation. Thesis, Institute of Geography and Earth Sciences, University of Wales, Aberystwyth
Fraser SW, van Rooyen TH, Verster E (1987) Soil-plant relationships in the central Kruger National Park. Koedoe 30:19–34
Frost P (1996) The ecology of miombo woodlands. In: Campbell B (ed) The miombo in transition: woodland and welfare in Africa. Center for International Forestry Research, Bogor, pp 11–57
Frost PGH (1987) The regional perspective: Nylsvley in perspective. South African National Scientific Programmes Report No. 133. Council for Scientific and Industrial Research, Pretoria
Furniss DG (2008) Can indices of Landscape Function Analysis (LFA) be derived from Ground-Based Spectroscopy? A case study from gold mines on the Highveld of South Africa, Thesis, School of Animal, Plant and Environmental Science, University of the Witwatersrand, Johannesburg
Galpin EE (1926) Botanical survey of the Springbok Flats. Mem Bot Surv S Afr 12:1–100
Gandar MV (1982) Description of a fire and its effects in the Nylsvley Nature Reserve: a synthesis report. S Afr Natl Sci Progr Rep 63:1–45
Gandiwa E, Kativu S (2009) Influence of fire frequency on *Colophospermum mopane* and *Combretum apiculatum* woodland structure and composition in northern Gonarezhou National Park, Zimbabwe. Koedoe 51(1):Art. #685. https://doi.org/10.4102/koedoe.v51i1.685
Gandiwa E, Chikorowondo G, Zisadza-Gandiwa P et al (2011) Structure and composition of *Androstachys johnsonii* woodland across various strata in Gonarezhou National Park, southeast Zimbabwe. Trop Conserv Sci 4:218–229
Gandiwa E, Zisadza-Gandiwa P, Goza D et al (2014) Diversity and structure of woody vegetation across areas with different soils in Gonarezhou National Park, Zimbabwe. Southern For 76:111–116. https://doi.org/10.2989/20702620.2014.921007
Gaugris JY, Matthews WS, van Rooyen N et al (2004) The vegetation of Tshanini Game Reserve and a comparison with equivalent units in the Tembe Elephant Park in Maputaland, South Africa. Koedoe 47:9–29
Geißler K, Blaum N, von Maltitz GP (2024) Biodiversity and ecosystem functions in southern African savanna rangelands: threats, impacts and solutions. In: von Maltitz GP, Midgley GF, Veitch J et al (eds) Sustainability of southern African ecosystems under global change. Springer, Cham, pp 407–438

Geldenhuys CJ, Mucina L (2006) Towards a new forest classification for South Africa. In: Ghazanfar S, Beentje HJ (eds) Taxonomy and ecology of African plants, their conservation and sustainable use. Royal Botanic Gardens, Kew, pp 111–129

Geppert M, Riedel F, Gummersbach VS et al (2021) Late Pleistocene hydrological settings at world heritage Tsodilo Hills (NW Kalahari, Botswana), a site of ancient human occupation. Quater Sci Adv 3:100022. https://doi.org/10.1016/j.qsa.2021.100022

Gertenbach WPD (1978) Plantgemeenskappe van die Gabbro-kompleks in die noordweste van die sentrale distrik van die Nasionale Krugerwildtuin. Thesis, Department of Plant Science, Potchefstroom University for C.H.E., Potchefstroom

Gertenbach WPD (1983a) Ekologiese studie van die suidelikste Mopaneveld in die Nasionale Krugerwildtuin. Thesis, Department of Botany, University of Pretoria, Pretoria

Gertenbach WPD (1983b) Landscapes of the Kruger National Park. Koedoe 26:9–121

Gertenbach WPD, Potgieter ALF (1978) A phytosociological classification of the Hlane Wildlife Sanctuary, Swaziland. Koedoe 21:47–65

Gertenbach WPD, Potgieter ALF (1979) Veldbrandnavorsing in die struikmopaneveld van die Nasionale Krugerwildtuin. Koedoe 22(1):a648. https://doi.org/10.4102/koedoe.v22i1.648

Giess W (1971) A preliminary vegetation map of South West Africa. Dinteria 4:1–114

Glen HF (1996) A description of the vegetation of Oribi Gorge Nature Reserve, Natal. National Botanical Institute, Pretoria

Goodier R, Phipps JB (1963) A vegetation map of the Chimanimani National Park. Kirkia 3:2–7

Goodman PS (1990) Soil, vegetation and large herbivore relations in Mkuzi Game Reserve, Natal. Thesis, Faculty of Science (Botany), University of the Witwatersrand, Johannesburg

Gordijn PJ (2010) The role of fire in bush encroachment in Ithala Game Reserve. Thesis, School of Biological and Conservation Sciences, University of KwaZulu-Natal, Pietermaritzburg

Gordon CE, Greve M, Henley M et al (2023) Elephant rewilding affects landscape openness and fauna habitat across a 92-year period. Ecol Appl 33:e2810. https://doi.org/10.1002/eap.2810

Götze AR (2002) The classification of plant communities and the evaluation of restoration technologies in different land-use areas in the Vhembe-Dongola National Park, Northern Province, South Africa. Thesis, Potchefstroom University for C.H.E., Potchefstroom

Götze AR, Cilliers SS, Bezuidenhout H et al (2008) Analysis of the vegetation of the sandstone ridges (Ib land type) of the north-eastern parts of the Mapungubwe National Park, Limpopo Province, South Africa. Koedoe 50(1):72–81

Grobler JH (1974) Aspects of the biology, population ecology and behaviour of the sable *Hippotragus niger niger* (Harris, 1938) in the Rhodes Matopos National Park, Rhodesia. Arnoldia 6:1–36

Grobler CH (2000) The vegetation ecology of urban open spaces in Gauteng. Thesis, Department of Botany, University of Pretoria, Pretoria

Grobler CH, Bredenkamp GJ, Brown LR (2002) Natural woodland vegetation and plant species richness of the urban open spaces in Gauteng, South Africa. Koedoe 45(1):19–34

Grossman D, Grunow JO, Theron GK (1981) The effect of fire with and without subsequent defoliation on the herbaceous layer of *Burkea africana* savanna. Proc Grassl Soc Southern Afr 16:117–120

Grunow JO (1965) Objective classification of plant communities: a synecological study in the sourish mixed bushveld of Transvaal. J Ecol 55:691–710

Grunow JO, Morris JW (1969) Preliminary assessment of ecological status of plant species in three South African veld types. J S Afr Bot 35:1–12

Guy PR (1977) Notes on the vegetation types of the Zambezi valley, Rhodesia, between the Kariba and Mpata Gorges. Kirkia 10:543–557

Hahn N (2002) A second species of *Brachystegia* for South Africa. SABONET News 7:110

Haimbili E (2015) Establishment of woody savannah species on various mined substrates towards rehabilitating self sustaining plant communities at Navachab gold mine. Thesis, University of Namibia, Windhoek

Haindongo PN, Kazapua VU, Shikongo PM (2021) Vegetation types and structure in the Kanovlei State Forest in Namibia. Namib J Environ Sect C 5:1–3

Harmse CJ, De Ath A, Kellner K (2023) Quantifying the effect of chemical bush control of Senegalia mellifera on vegetation production in the Eastern Kalahari Bushveld, South Africa. Afr J Range Forage Sci 40:348–358

Harris MA, Voysey MD, Jamison S-L et al (2019) Changes in bird assemblages because of vegetation homogenisation in communal livestock systems. Koedoe 61(1):a1543. https://doi.org/10.4102/koedoe.v61i1.1543

Hauwanga WN, McBenedict B, Strohbach BJ (2018) Trends of phanerophyte encroacher species along an aridity gradient on Kalahari sands, central Namibia. Eur J Ecol 4(2):41–48. https://doi.org/10.2478/eje-2018-0011

Heijnis C (2004) KwaZulu–Cape coastal forest mosaic. In: Burgess N, D'Amico Hales J, Underwood E et al (eds) Terrestrial ecoregions of Africa and Madagascar—a conservation assessment. Island Press, Washington, pp 258–259

Herbst MJ (1973) 'n Ekologiese ondersoek van die plantgemeenskappe tussen die Krokodil- en Matlabasrivier, met spesiale aandag aan die invloed van edafiese faktore op die verspreiding van houtagtige spesies. Thesis, Potchefstroom University for C.H.E., Potchefstroom

Higgins SI, Shackleton CM, Robinson ER (1999) Changes in woody community structure and composition under contrasting landuse systems in a semi-arid savanna, South Africa. J Biogeogr 26:619–627

Hoare RE, Robertson EF, Dunham KM (2002) Structure and condition of Zambezi Valley dry forests and thickets. The Zambezi Society, Harare

Hoffman MT (1983) The grasslands of the Hluleka Nature Reserve. Department of Botany, University of Cape Town, Cape Town

Holdo RM (2006) Elephant herbivory, frost damage and topkill in Kalahari sand woodland savanna trees. J Veg Sci 17:509–518

Holdo RM (2007) Elephants, fire, and frost can determine community structure and composition in Kalahari woodlands. Ecol Appl 17:558–568

Hudak AT, Wessman CA (2001) Textural analysis of high resolution imagery to quantify bush encroachment in Madikwe Game Reserve, South Africa. Int J Rem Sens 22:2731–2740

Huntley BJ (1982) Southern African savannas. In: Huntley BJ, Walker BH (eds) Ecology of tropical savannas. Springer, Berlin, pp 101–119

Huntley BJ, Morris JW (1982) Structure of the Nylsvley Savanna. In: Huntley BJ, Walker BH (eds) Ecology of tropical savannas. Springer, Berlin, pp 433–455

Huntley BJ, Walker BH (eds) (1982) Ecology of tropical savannas. Springer, Berlin

Hurter PJH, van Wyk E (2001) First distribution record for *Brachystegia spiciformis* (Caesalpinioideae) in South Africa. Bothalia 31:43–44

Hüttich C, Gessner U, Herold M et al (2009) On the suitability of MODIS time series metrics to map vegetation types in dry savanna ecosystems: a case study in the Kalahari of NE Namibia. Rem Sens 1:620–643. https://doi.org/10.3390/rs1040620

Hüttich C, Herold M, Strohbach BJ et al (2011) Integrating in-situ, Landsat, and MODIS data for mapping in southern African savannas: experiences of LCCS-based land-cover mapping in the Kalahari in Namibia. Environ Monitor Assess 176:531–547

Ingram P (1960) A preliminary account of the ecology of the vegetation of the South Matopos I.C.A. Rhodes Agric J 57:319–330

Inman EN, Hobbs RJ, Tsvuura Z et al (2020) Current vegetation structure and composition of woody species in community-derived categories of land degradation in a semiarid rangeland in Kunene region, Namibia. Land Degrad Develop 31:2996–3013. https://doi.org/10.1002/ldr.3688

Jacobsen WGB, Jacobsen WBG (1973) A checklist and discussion of the flora of a portion of the Lomagundi District, Rhodesia. Kirkia 9:139–207

Janecke BB (2020) Vegetation structure and spatial heterogeneity in the Granite Supersite, Kruger National Park. Koedoe 62(2):a1591. https://doi.org/10.4102/koedoe.v62i2.1591

Janecke BB, Van Tol J, Smit IPJ et al (2020) Biotic and abiotic connections on a granitic catena: framework for multidisciplinary research. Koedoe 62(2):a1600. https://doi.org/10.4102/koedoe.v62i2.1600

Jankowitz WJ, Venter HJT (1987) Die plantgemeenskappe van die Waterberg-platopark. Modoqua 15:97–146

Jordaan JJ, Wessels DCJ, Dannhauser CS et al (2004) Secondary succession in the mopani veld of the Limpopo valley, South Africa. Afr J Range Forage Sci 21:205–210

Joubert E (1971) The physiographic, edaphic and vegetative characteristics found in the western Etosha National Park. Modoqua 1:5–32

Joubert CJ (2013) A natural resource assessment of Chapungu Ranch, Save Valley Conservancy, Zimbabwe. Thesis, Centre for Wildlife Management, University of Pretoria, Pretoria

Jürgens N, Haarmeyer DH, Luther-Mosebach J et al (eds) (2010) Biodiversity in southern Africa. Volume 1. Patterns at local scale—the BIOTA observatories. Klaus Hess Publishers, Göttingen

Kaholongo IK, Mapaure I (2014) Vegetation composition, diversity, and structure in Uukolonkadhi Community Forest of north western Namibia. J Agric Biodivers Res 3:41–50

Käller A (2003) Growth pattern and reproduction of woody vegetation in a semi-arid savanna in southern Botswana. Minor field study 86. Committee of Tropical Ecology, Uppsala University, Uppsala

Kalwij JM, de Boer WF, Mucina L et al (2010) Tree cover and biomass increase in a southern African savanna despite growing elephant population. Ecol Appl 20:222–233

Kangombe FN (2010) The vegetation of Omusati and Oshana Regions, central-northern Namibia. Thesis, Department of Plant Science, University of Pretoria, Pretoria

Kelly RD, Walker BH (1976) The effects of different forms of land use on the ecology of a semi-arid region in south-eastern Rhodesia. J Ecol 64:553–576

Kennan TCD (1972) The effects of fire on two vegetation types in Matopos, Rhodesia. In: Proceedings annual tall timbers fire ecology conference: fire in Africa. Tall Timbers Research, Tallahassee, pp 53–98

Killick DJB (1958) An account of the plant ecology of the Table Mountain area of Pietermaritzburg, Natal. Mem Bot Surv S Afr 32:1–133

Kirkwood D, Midgley JJ (1999) The floristics of Sand Forest in northern KwaZulu-Natal, South Africa. Bothalia 29:293–304

Knoop WT, Walker BH (1985) Interactions of woody and herbaceous vegetation in two savanna communities at Nylsvley. J Ecol 73:235–253

Kyalangalilwa B, Boatwright JS, Daru BH (2013) Phylogenetic position and revised classification of *Acacia* s.l. (Fabaceae; Mimosoideae) in Africa, including new combinations in *Vachellia* and *Senegalia*. Bot J Linn Soc 172:500–523

la Grange M (2010) Vegetation classification of the proposed Heritage Park, North-West Province, South Africa. Thesis, North-West University, Potchefstroom

LeClare SK, Mduli M, Wisely SM et al (2020) Land-use diversity within an agricultural landscape promotes termite nutrient cycling services in a southern African savanna. Glob Ecol Conserv 21:e00885. https://doi.org/10.1016/j.gecco.2019.e00885

Leggett K, Fennessy J, Schneider S (2003) Seasonal vegetation changes in the Hoanib River catchment, north-western Namibia: a study of a non-equilibrium system. J Arid Environ 53:99–113. https://doi.org/10.1006/jare.2002.1027

Le Grange L (2010) The vegetation ecology of the Seringveld Conservancy, Cullinan, South Africa. Thesis, Department of Plant Science, University of Pretoria, Pretoria

Lehmann CER, Archibald SA, Hoffmann WA et al (2011) Deciphering the distribution of the savanna biome. New Phytol 191:197–209

Leistner OA (1967) The plant ecology of the southern Kalahari. Mem Bot Surv S Afr 38:1–172

Lenfers UA, Ahmady-Moghaddam N, Glake D et al (2022) Modeling the future tree distribution in a South African savanna ecosystem: an agent-based model approach. Land 11:619. https://doi.org/10.3390/land11050619

le Roux NP (1995) Grassland of Umtamvuna Nature Reserve, KwaZuluNatal: a description and recommendation for monitoring. Thesis, Department of Grassland Science, University of Natal, Pietermaritzburg

Leser H (1972) Geoökologische Verhältnisse der Pflanzengesellschaften in den Savannen des Sandveldes um den Schwarzen Nossob und um Epukiro. Dinteria 6:1–41

le Roux CJG, Grunow JO, Morris JW et al (1988) A classification of the vegetation of the Etosha National Park. S Afr J Bot 54:1–10

Lori T, Ditlhogo MK, Setshogo MP et al (2019) Classification, description and mapping of the vegetation in Khutse Game Reserve, Botswana. Botsw J Agric Appl Sci 13:8–23

Lötter M, Burrows J, Jones K et al (2023) Historical vegetation map and red list of ecosystems assessment for Mozambique—Version 2.0—Final report. USAID/SPEED+, AFD/FFEM, Maputo

Louw AJ (1970) 'n Ekologiese studie van die Mopanieveld noord van die Soutpansberg. Thesis, Department of Botany, University of Pretoria, Pretoria

Lubbe RA (1997) Vegetation and flora of the Kosi Bay Coastal Forest Reserve in Maputaland, northern KwaZulu-Natal, South Africa. Thesis, Department of Botany, University of Pretoria, Pretoria

Lubke RA, Thatcher FM (1983) Short term changes in the woody vegetation of Nylsvley. S Afr J Bot 2:85–97

Lubke RA, Morris JW, Theron GK et al (1983) Diversity, structure and pattern in Nylsvley vegetation. S Afr J Bot 2:26–41

Macdonald IAW (1980) The effect of single fires on three woody plant species in the Hluhluwe Game Reserve. Natal Parks Board, Pietermaritzburg

Macdonald IAW (1981) The management of *Euclea divinorum* woodland in Hluhluwe Game Reserve, Natal. Natal Parks Board, Pietermaritzburg

Makhabu SW, Tshireletso K, Setlalekgomo MR et al (2021) Herbaceous communities' heterogeneity in the Kalahari sandy soils environment: a case study of Kweneng, Kgatleng Districts and Mahalapye Sub-District of Botswana. J Biodivers Environ Sci 18:87–101

Makhado RA, Mapaure I, Potgieter MJ et al (2014) Factors influencing the adaptation and distribution of *Colophospermum mopane* in southern Africa's mopane savannas—a review. Bothalia 44(1):Art. #152. https://doi.org/10.4102/ABC.V44I1.152

Malan PW, van Niekerk S (2005) The extent of grass species composition in Braklaagte, Zeerust District, North-West Province, South Africa. Afr J Range Forage Sci 22:177–184

Mandinyenya B, Monks N, Mundy PJ et al (2020) Habitat choices of African buffalo (*Syncerus caffer*) and plains zebra (*Equus quagga*) in a heterogeneous protected area. Wildl Res 47:106–113. https://doi.org/10.1071/WR18201

Martin JD (1939) The Baikiaea forests of Northern Rhodesia. Emp For J 19(1):8–18

Martini F, Cunliffe R, Farcomeni A et al (2016) Classification and mapping of the woody vegetation of Gonarezhou National Park, Zimbabwe. Koedoe 58(1):a1388. https://doi.org/10.4102/koedoe.v58i1.1388

Marufu VJ (2023) A phytosociological classification of the vegetation of Etosha Heights Private Game Reserve, Namibia. Thesis, Namibia University of Science and Technology, Windhoek

Matthews WS (1991) Phytosociology of the North-eastern Mountain Sourveld. Thesis, Department of Botany, University of Pretoria, Pretoria

Matthews WS (2007) Contributions to the ecology of Maputaland, Southern Africa, with emphasis on sand forest. Thesis, Department of Botany, University of Pretoria, Pretoria

Matthews WS, van Wyk AE, van Rooyen N (1999) Vegetation of the Sileza Nature Reserve and neighbouring areas, South Africa, and its importance in conserving the woody grasslands of the Maputaland Centre of Endemism. Bothalia 29:151–167

Matthews WS, van Wyk AE, van Rooyen N et al (2001) Vegetation of the Tembe Elephant Park, Maputaland, South Africa. S Afr J Bot 67:573–594

Maurin O, Davies TJ, Burrows JE et al (2014) Savanna fire and the origins of the 'underground forests' of Africa. New Phytol 204:201–214

Mendelsohn J, Roberts C (1997) An environmental profile and atlas of Caprivi. Directorate of Environmental Affairs, Windhoek

Mendelsohn J, Obeid S, Roberts C (2000) A profile of North-central Namibia. Gamsberg. Macmillan Publishers, Windhoek

Mendelsohn J, Jarvis A, Roberts C et al (2002) Atlas of Namibia: a portrait of the land and its people. David Philip, Cape Town
Mhlanga L, Mapaure I (1999) Vegetation studies of selected islands and adjacent mainland on Lake Kariba, Zimbabwe. IES Working Paper 9. Institute of Environmental Studies, University of Zimbabwe, Harare
Midgley JJ (1986) Are 800 km disjunction in *Leucadendron* due to earlier climatic, tectonic or edaphic changes or long distance dispersal? S Afr Soc Quater Res 1:51–52
Mills AJ, Milewski AV, Sirami C et al (2012) Aerosol capture by small trees in savannas marginal to treeless grassland in South Africa. Geoderma 189–190:124–132
Milton SJ (1987) Phenology of seven *Acacia* species in South Africa. S Afr J Wildl Res 17:1–6
Mishra NB, Crews KA, Miller JA et al (2015) Mapping vegetation morphology types in southern Africa savanna using MODIS time series metrics: a case study of central Kalahari, Botswana. Land 4:197–215. https://doi.org/10.3390/land4010197
Mitchell BL (1962) Some notes on the vegetation of a portion of the Wankie National Park. Kirkia 2:200–209
Mndela M, Madakadze IC, Nherera-Chokuda FV et al (2022) Short-term responses of herbaceous vegetation to bush clearing in semi-arid rangelands of South Africa. Pastor Res Policy Pract 12:17. https://doi.org/10.1186/s13570-022-00235-7
Mndela M, Mangwane M, Ngcobo N et al (2023) Soil seed banks along a woody plant removal gradient in a semi-arid savanna of South Africa: implications for restoration. Acta Oecol 118:103891. https://doi.org/10.1016/j.actao.2023.103891
Mochesane MV (2018) Plant diversity in grasslands of selected nature reserves and adjacent grazing areas within the Gauteng province, South Africa. Thesis, University of South Africa, Pretoria
Mogonong B, van der Merwe H, Ramaswiela T et al (2023) Vegetation description around the savanna flux measurement site at Benfontein Nature Reserve, South Africa. S Afr J Bot 162:353–359
Moll EJ (1972) The distribution, abundance and utilization of the Lala palm, *Hyphaene natalensis*, in Tongaland, Natal. Bothalia 10:627–636
Moll EJ (1976) The vegetation of the Three Rivers region, Natal. Natal Town and Regional Planning Commission, Pietermaritzburg
Moll EJ (1978) The vegetation of Maputaland. Trees S Afr 29:31–59
Moll EJ (1980) Terrestrial plant ecology. In: Bruton MN, Cooper KH (eds) Studies on the ecology of Maputaland, Rhodes University, Grahamstown & Natal Branch of Wildlife Society of Southern Africa, Durban, pp 52–68
Moll EJ, White F (1978) The Indian Ocean Coastal Belt. In: Werger MJA, van Bruggen AC (eds) Biogeography and ecology of southern Africa. Dr W. Junk, The Hague, pp 561–598
Moore AE, Attwell CAM (1999) Geological controls on the distribution of woody vegetation in the central Kalahari, Botswana. S Afr J Geol 102:350–362
Morrey DR, Balkwill K, Balkwill M-J (1989) Studies on serpentine flora: preliminary analyses of soils and vegetation associated with serpentinite rock formations in the south eastern Transvaal. S Afr J Bot 55:171–177
Morrey DR, Balkwill K, Balkwill M-J et al (1992) A review of some studies of the serpentine flora of southern Africa. In: Baker AJM, Proctor J, Reeves RD (eds) The vegetation of ultramafic (serpentine) soils. Intecept Publishers, Andover, pp 253–272
Morris JW (1967) Descriptive and quantitative plant ecology of Ntshongweni, Natal. Thesis, University of Natal, Pietermaritzburg
Morris JW (1969) An ordination of the vegetation of Ntshongweni, Natal. Bothalia 10:89–120
Morris PPJ (1972) 'n Ekologiese ondersoek van die samestelling en struktuur van verteenwoordigende plantgemeentskappe van die soetbosveld, met spesiale aandag aan die invloed van edafiese faktore op die verspreiding van houtagtige species. Thesis, Potchefstroom University for C.H.E., Potchefstroom
Morris CD, Scott-Shaw R (2019) Potential grazing indicator forbs for two mesic grasslands in South Africa. Ecol Indic 107:105611. https://doi.org/10.1016/j.ecolind.2019.105611
Morris JW, Bezuidenhout JJ, Furniss PR (1982) Litter decomposition. In: Huntley BJ, Walker BH (eds) Ecology of tropical savannas. Springer, Berlin, pp 547–553
Moses M, Simbarashe M, Farai M (2017) Influence of soil drainage and elephant (*Loxodonta africana*) damage on abundance and structure of *Colophospermum mopane* in central Mana Pools National Park Zimbabwe. Poult Fish Wildl Sci 5:175. https://doi.org/10.4172/2375-446X.1000175
Mosugelo DK, Moe SR, Ringrose S et al (2002) Vegetation changes during a 36-year period in northern Chobe National Park, Botswana. Afr J Ecol 40:232–240
Moyo B, Dube S, Moyo C et al (2011) Heavily stocked 5-paddock rotational grazing effect on cross-bred Afrikaner steer performance and herbaceous vegetation dynamics in a semi-arid veld of Zimbabwe. Afr J Agric Res 6:2166–2174
Mucina L (2023) Biomes of the Southern Hemisphere. Springer Nature Switzerland, Cham
Mucina L, Hoare DB, Lötter MC et al (2006a) Grassland Biome. In: Mucina L, Rutherford MC (eds) The vegetation of South Africa, Lesotho and Swaziland. SANBI, Pretoria, pp 348–437
Mucina L, Geldenhuys CJ, Rutherford MC et al (2006b) Afrotemperate, subtropical and azonal forests. In: Mucina L, Rutherford MC (eds) The vegetation of South Africa, Lesotho and Swaziland. SANBI, Pretoria, pp 584–615
Mucina L, Scott-Shaw CR, Rutherford MC et al (2006c) Indian Ocean coastal belt. In: Mucina L, Rutherford

MC (eds) The vegetation of South Africa, Lesotho and Swaziland. SANBI, Pretoria, pp 568–583

Mucina L, Lötter MC, Rutherford MC et al (2022) Forest biomes of Southern Africa. New Zeal J Bot 60:377–428

Mucina L, Rutherford MC (eds) (2006) The vegetation of South Africa, Lesotho and Swaziland. SANBI, Pretoria

Muposhi VK, Ndlovu M, Gandiwa E et al (2014) Vegetation dynamics prior to wildlife reintroductions in southern Umfurudzi Park, Zimbabwe. J Anim Plant Sci 24:1680–1690

Muposhi VK, Chademana TC, Gandiwa E et al (2016) Edge effects: impact of anthropogenic activities on vegetation structure and diversity in western Umfurudzi Park, Zimbabwe. Afr J Ecol 54:450–459

Mwembe R, Hlatshwayo AS, Nkomboni P et al (2013) Impact of wildfires on browse availability in the semi-arid thornveld of Southeastern Zimbabwe. Sci J Pure Appl Sci 2:246–253

Myre M (1964) A vegetação do extremo sul da provincia de Moçambique. Junta de Investigaçãoes do Ultramar, Lisboa

Myre M (1971) As pastagens da região da Maputo. Mém Instit Invest Agron Moçambique 3:1–181

Nel R, Mearns KF, Jordaan M et al (2022) The role of modelling in resource management within the livelihood-conservation nexus: a socio-ecological systems approach to Sand Forest harvesting, northern KwaZulu-Natal. Ecol Inform 69:101600. https://doi.org/10.1016/j.ecoinf.2022.101600

Ngubane Z, Odindi J, Mutanga O et al (2014) Assessment of the contribution of WorldView-2 strategically positioned bands in bracken fern (*Pteridium aquilinum* (L.) Kuhn) mapping. S Afr J Geomat 3:210–223

Nicolay REE, Teder MJ, Mkhize NP et al (2023) Fire frequency influences soil organic carbon stocks and sequestration in a South African mesic grassland. ResSquare. https://doi.org/10.21203/rs.3.rs-2781556/v1

Nkosi SE, Adam E, Barrett AS et al (2022) African elephant (*Loxodonta africana*) impact on selected tree species and assessment of piosphere gradients in Venetia-Limpopo Nature Reserve. Ecol Environ Conserv 28:1182–1197

Nyazika TP, Zisadza-Gandiwa P, Chanyandura A et al (2017) Influence of fire frequency on woody vegetation structure and composition in Lake Chivero Recreational Park, northern Zimbabwe. Trop Ecol 58:583–589

Nzuza P, Ramoelo A, Odindi J et al (2021) Predicting land degradation using Sentinel-2 and environmental variables in the Lepellane catchment of the Greater Sekhukhune District, South Africa. Phys Chem Earth Parts A/B/C 124(1):102931. https://doi.org/10.1016/j.pce.2020.102931

O'Connor TG (1992) Woody vegetation-environment relations in a semi-arid savanna in the northern Transvaal. S Afr J Bot 58:268–274

O'Connor TG (2015) Long-term response of an herbaceous sward to reduced grazing pressure and rainfall variability in a semi-arid South African savanna. Afr J Range Forage Sci 32:261–270. https://doi.org/10.2989/10220119.2015.1015052

O'Connor TG, Campbell BM (1986) Classification and condition of the vegetation types of the Nyahungwe area on the Lundi River, Gonarezhou National Park, Zimbabwe. S Afr J Bot 52:117–123

O'Connor TG, Goodall VL (2017) Population size structure of trees in a semi-arid African savanna: species differ in vulnerability to a changing environment and reintroduction of elephants. Austral Ecol 42:664–676

Odindi J, Mutanga O, Rouget M et al (2016) Mapping alien and indigenous vegetation in the KwaZulu-Natal Sandstone Sourveld using remotely sensed data. Bothalia 46(2):a2103. https://doi.org/10.4102/abc.v46i2.2103

Pais AJR, Ribeiro NS, Santos RM (2021) Mapping floristic communities in Southern Africa savannas, Mozambique. Rodriguésia 72:e03422018. https://doi.org/10.1590/21757860202172010

Panagos MD (1996) The plant communities of the Onderstepoort Nature Reserve. ARC Onderstepoort Veterinary Research Institute, Pretoria

Panagos MD, Westfall RH, Scheepers JC (1986) Baseline data for the vegetation of two protected plots at the Matimba Power Station, Ellisras, NW Transvaal. Bothalia 16:89–91

Panagos MD, Westfall RH, van Staden JM et al (1998) The plant communities of the Roodeplaat Experimental Farm, Gauteng, South Africa and the importance of classification verification. S Afr J Bot 64:44–61

Pauw JC (1988) Riglyne vir die bestuur van die natuurlewe in die bosveld gemeenskappe van die Atherstone-natuurreservaat in die noordwes-Transvaal. Thesis, University of Pretoria, Pretoria

Pedro JG, Barbosa LAG (1955) A vegetação. In: Esboço do reconhecimento ecológico-agrícola de Moçambique. Volume 2. Centro de Investigação Científica Algodeira Memorias e Trabalhos Lourenço Marques [Maputo], pp 67–224

Peel MJS (1990) Determinants of veld composition on a number of ranches in the north-west Transvaal. Thesis, Centre for Wildlife Management, University of Pretoria, Pretoria

Peel MJS, Grossman D, van Rooyen N (1991) Determinants of herbaceous plant species composition on a number of ranches in the north-western Transvaal. J Grassl Soc Southern Afr 8:99–102

Peel MJS, Kruger JM, MacFayden S (2007) Woody vegetation of a mosaic of protected areas adjacent to the Kruger National Park, South Africa. J Veg Sci 18:807–814

Perkins L (1997) Aspects of the syntaxonomy and synecology of the grasslands of southern KwaZulu-Natal. Thesis, Department of Botany, University of Pretoria, Pretoria

Perkins L, Bredenkamp GJ, Granger JE (1999) The phytosociology of the incised river valleys and dry upland savanna of southern KwaZulu-Natal. S Afr J Bot 65:321–330

Pickford M, Senut B (2010) Karst geology and palaeobiology of Northern Namibia. Mem Geol Surv Namib 21:1–74

Poilecot P, Gaidet N (2011) A quantitative study of the grass and woody layers of a Mopane (*Colophospermum mopane*) savannah in the mid-Zambezi Valley, Zimbabwe. Afr J Ecol 49:150–164

Popp MR, Kalwij JM (2023) Consumer-grade UAV imagery facilitates semantic segmentation of species-rich savanna tree layers. Sci Rep 13:13892. https://doi.org/10.1038/s41598-023-40989-7

Potgieter MJ, Mashatole MM, Jordaan JJ (2017) Long-term vegetation changes in the Pietersburg Plateau Bushveld of the Limpopo Province, South Africa: a new floristic assessment of the Turfloop Nature Reserve. Insights For Res 1:57–66

Proctor J, Cole MM (1992) The ecology of ultramafic areas in Zimbabwe. In: Roberts BA, Proctor J (eds) The ecology of areas with serpentinized rocks. Kluwer Academic Publishers, Dordrecht, pp 313–331

Proctor J, Craig GC (1973) The occurrence of woodland and riverine forest on the serpentine of the Great Dyke. Kirkia 11:129–132

Rapiya M, Ramoelo A, Truter W (2023) Seasonal evaluation and mapping of aboveground biomass in natural rangelands using Sentinel-1 and Sentinel-2 data. Environ Monit Assess 195:1544. https://doi.org/10.1007/s10661-023-12133-5

Rattray JM (1962) Vegetation types of Southern Rhodesia. Kirkia 2:68–93

Ravhuhali KE (2018) Spatial variation in density, species composition and nutritive value of vegetation in selected communal areas of the North West Province. Thesis, North-West University, Potchefstroom

Revermann R (2016) Analysis of vegetation and plant diversity patterns in the Okavango Basin at different spatial scales. Integration of field based methods, remote sensing information and ecological modelling. Thesis, Department of Biology, University of Hamburg, Hamburg

Revermann R, Finckh M (2013) Okavango Basin—vegetation. Biodivers Ecol 5:29–35

Revermann R, Finckh M, Stellmes M et al (2016a) Linking land surface phenology and vegetation-plot databases to model terrestrial plant α-diversity of the Okavango Basin. Rem Sens 8:370. https://doi.org/10.3390/rs8050370

Revermann R, Gomes AL, Goncalves FM et al (2016b) Vegetation database of the Okavango Basin. Phytocoenologia 46:103–104

Ribeiro NS, Katerere Y, Chirwa PW et al (eds) (2020) Miombo woodlands in a changing environment: securing the resilience and sustainability of people and woodlands. Springer Nature Switzerland, Cham

Ringrose S, Matheson W, Wolski P et al (2003) Vegetation cover trends along the Botswana Kalahari transect. J Arid Environ 54:297–317

Roberts DC (1993) The vegetation ecology of municipal Durban, Natal. Floristic classification. Bothalia 23:271–326

Rogers CML (1993) A woody vegetation survey of Hwange National Park. A report prepared for the Department of National Parks and Wild Life Management, Zimbabwe. Causeway, Harare

Rutherford MC (1972) Notes on the flora and vegetation of the Omuverume Plateau-Mountain, Waterberg, South West Africa. Dinteria 8:3–55

Rutherford MC (1975) Aspects of ecosystem function in a woodland savanna in South West Africa. Thesis, University of Stellenbosch, Stellenbosch

Rutherford MC (1978) Primary production ecology in southern Africa. In: Werger MJA, van Bruggen AC (eds) Biogeography and ecology of southern Africa. Dr W. Junk Publishers, The Hague, pp 621–659

Rutherford MC (1980) Field identification of roots of woody plants of the savanna ecosystem study area, Nylsvley. Bothalia 13:171–184

Rutherford MC (1981) Survival, regeneration and leaf biomass changes in woody plants following spring burns in *Burkea africana-Ochna pulchra* savanna. Bothalia 13:531–552

Rutherford MC (1982) Aboveground biomass categories of woody plants in a *Burkea africana–Ochna pulchra* savanna. Bothalia 14:139–142

Rutherford MC (1983a) Herbaceous standing crop in relation to surface and subsurface rockiness. Bothalia 14:259–264

Rutherford MC (1983b) Growth rates, biomass and distribution of selected woody plant roots in *Burkea africana-Ochna pulchra* savanna. Vegetatio 52:45–63

Rutherford MC (1984) Relative allocation and seasonal phasing of growth of woody plant components in a South African savanna. Progr Biometeorol 3:200–221

Rutherford MC (1993) Empiricism and the prediction of primary production at the mesoscale: a savanna example. Ecol Model 67:129–146

Rutherford MC, Kelly RD (1978) Woody plant basal area and stem increment in *Burkea africana–Ochna pulchra* woodland. S Afr J Sci 74:307–308

Rutherford MC, Mucina L, Lötter MC et al (2006) Savanna Biome. In: Mucina L, Rutherford MC (eds) The vegetation of South Africa, Lesotho and Swaziland. SANBI, Pretoria, pp 438–539

Rutherford MC, Powrie LW (2010) Severely degraded dunes of the southern Kalahari: local extinction, persistence and natural re-establishment of plants. Afr J Ecol 48:930–938

Rutherford MC, Powrie LW, Thompson DI (2012) Impacts of high utilization pressure on biodiversity components in *Colophospermum mopane* savanna. Afr J Range Forage Sci 21:1–11

Sankaran M, Ratnam J (2013) African and Asian savannas. In: Scheiner SM (ed) Encyclopedia of biodiversity, 3rd edn, vol 1. https://doi.org/10.1016/B978-0-12-822562-2.00174-2

Sankaran M, Hanan NP, Scholes RJ et al (2005) Determinants of woody cover in African savannas. Nature 438:846–849

Scheepers JC (1977) The vegetation of Westfalia Estate on the north-eastern Transvaal escarpment. Mem Bot Surv S Afr 42:1–230

Schmidt AG, Theron GK, van Hoven W (1993) The phytosociology and structure of vegetation near Villa Nora, north-western Transvaal. S Afr J Bot 59:500–510

Schmidt E, Lötter MC, McCleland W (2002) Trees and shrubs of Mpumalanga and Kruger National Park. Jacana Education, Johannesburg

Schmiedel U, Jacke V, Hachfeld B et al (2021) Response of Kalahari vegetation to seasonal climate and herbivory: results of 15 years of vegetation monitoring. J Veg Sci 32:e12927. https://doi.org/10.1111/jvs.12927

Scholes RJ (1979) The vegetation of the Blouberg, North-Western Transvaal. B.Sc.(Hons) Project, Department of Botany, University of the Witwatersrand, Johannesburg

Scholes RJ (1990) The influence of soil fertility on the ecology of southern African dry savannas. J Biogeogr 17:415–419

Scholes RJ (1997) Savanna. In: Cowling RM, Richardson DM, Pierce SM (eds) Vegetation of southern Africa. Cambridge University Press, Cambridge, pp 258–277

Scholes RJ, Walker BH (1993) An African savanna: synthesis of the Nylsvley study. Cambridge University Press, Cambridge

Scholes RJ, Gureja N, Giannecchinni M et al (2001) The environment and vegetation of the flux measurement site near Skukuza, Kruger National Park. Koedoe 44:73–83

Scholes RJ, Dowty PR, Caylor K et al (2002) Trends in savanna structure and composition along an aridity gradient in the Kalahari. J Veg Sci 13:419–428

Schroder B, Langevelde F, Hawkins Schroder N-A et al (2024) Effect of mineral nutrition on the establishment of grazing lawns in a nutrient poor savanna. bioRxiv. https://doi.org/10.1101/2024.01.20.576489

Schulze E, Theron GK, van Hoven W (1994) The vegetation and identification of management units of the Imberbe Game Lodge in the mixed bushveld of the north-western Transvaal. S Afr J Bot 60:85–93

Seine R, Becker U, Porembski S et al (1998) Vegetation of inselbergs in Zimbabwe. Edinb J Bot 55:267–293

Setshedi KTA, Newete SW, Witkowski ETF (2022) The threat of an indigenous medicinal plant encroacher: *Artemisia afra* in Klipriviersberg Nature Reserve, Johannesburg, South Africa. Afr J Ecol 60:1133–1142

Shackleton CM (1989) The grassland dynamics of Mkambati Game Reserve. Thesis, Faculty of Science, University of The Witwatersrand, Johannesburg

Shackleton SE (1990) Socio-economic importance of *Cymbopogon validus* in Mkambati Game Reserve. S Afr J Bot 56:675–682

Shackleton CM (1992) Area and species selection by wild ungulates in coastal sour grasslands of Mkambati Game Reserve. Afr J Ecol 30:189–202

Shackleton CM, Scholes RJ (2011) Above ground woody community attributes, biomass and carbon stocks along a rainfall gradient in the savannas of the central lowveld, South Africa. S Afr J Bot 77:184–192

Shackleton SE, Shackleton CM (1994) Habitat factors influencing the distribution of *Cymbopogon validus* in Mkambati Game Reserve, Transkei. Afr J Range Forage Sci 11:1–6

Shackleton CM, Granger JE, McKenzie B et al (1991) Multivariate analysis of coastal grasslands at Mkambati Game Reserve, north-eastern Pondoland, Transkei. Bothalia 21:91–107

Shannon G, Thaker M, Vanak AT et al (2011) Relative impacts of elephant and fire on large trees in a savanna ecosystem. Ecosystems 14:1372–1381

Sianga K, Fynn R (2017) The vegetation and wildlife habitats of the Savuti-Mababe-Linyanti ecosystem, northern Botswana. Koedoe 59(2):a1406. https://doi.org/10.4102/koedoe.v59i2.1406

Siebert SJ (2001) Vegetation on the ultramafic soils of the Sekhukhuneland Centre of Endemism. Thesis, Department of Botany, University of Pretoria, Pretoria

Siebert F, Eckhardt HC (2008) The vegetation and floristics of the Nkhuhlu exclosures, Kruger National Park. Koedoe 50:126–144

Siebert SJ, van Wyk AE (2005) *Nemesia zimbabwenisis*, a new record for the FSA region with notes on its phytogeographical significance. Bothalia 35:69–71

Siebert SJ, van Wyk AE, Bredenkamp GJ (2001) Endemism in the flora of ultramafic areas of Sekhukhuneland, South Africa. S Afr J Sci 97:529–532

Siebert SJ, van Wyk AE, Bredenkamp GJ (2002a) The physical environment and major vegetation types of Sekhukhuneland, South Africa. S Afr J Bot 68:127–142

Siebert SJ, van Wyk AE, Bredenkamp GJ (2002b) Vegetation ecology of Sekhukhuneland, South Africa: *Combretum hereroense–Grewia vernicosa* Open Mountain Bushveld. S Afr J Bot 68:475–496

Siebert SJ, van Wyk AE, Bredenkamp GJ (2002c) Vegetation ecology of Sekhukhuneland, South Africa: *Kirkia wilmsii–Terminalia prunioides* Closed Mountain Bushveld. S Afr J Bot 68:497–517

Siebert SJ, Victor JE, van Wyk AE (2002d) An assessment of threatened plants and conservation in Sekhukhuneland. PlantLife 26:7–18

Siebert F, Bredenkamp GJ, Siebert SJ (2003a) A comparison of mopaneveld vegetation in South Africa, Namibia and Zimbabwe. Bothalia 33:121–134

Siebert SJ, Matthee M, van Wyk AE (2003b) Semi-arid savanna of the Potlake Nature Reserve and surrounding areas in Sekhukhuneland, South Africa. Koedoe 46(1):29–52

Siebert SJ, van Wyk AE, Bredenkamp GJ et al (2003c) Vegetation of the rock habitats of the Sekhukhuneland Centre of Plant Endemism, South Africa. Bothalia 33:207–228

Siebert F, Eckhardt HC, Siebert SJ (2010) The vegetation and floristics of the Letaba exclosures, Kruger National Park, South Africa. Koedoe 52(1):Art. #777. https://doi.org/10.4102/koedoe.v52i1.777

Siebert SJ, Siebert F, Du Toit MJ (2011) The extended occurrence of Maputaland Woody Grassland further south in KwaZulu-Natal, South Africa. Bothalia 41:341–350

Siebert F, van Staden N, Komape DM et al (2021) Effects of land-use on herbaceous vegetation in a semi-arid Mopaneveld savanna. Bothalia 51:a8. https://doi.org/10.38201/btha.abc.v51.i1.8

Simpson CD (1975) A detailed vegetation study on the Chobe river in north-east Botswana. Kirkia 10:185–227

Skarpe C (1980) Observations on two bushfires in the western Kalahari, Botswana. Acta Phytogeogr Suec 68:131–140

Skarpe C (1986) Plant community structure in relation to grazing and environmental changes along a north-south transect in the western Kalahari. Vegetatio 68:3–18

Skarpe C (1990) Shrub layer dynamics under different herbivore densities in an arid savanna, Botswana. J Appl Ecol 27:873–885

Skarpe C, Bergström R, Bråten AL et al (2000) Browsing in a heterogeneous savanna. Ecography 23:632–640

Skowno AL, Midgley JJ, Bond WJ et al (1999) Secondary succession in *Acacia nilotica* (L.) savanna in the Hluhluwe Game Reserve, South Africa. Plant Ecol 145:1–9

Smit IP, Asner GP, Govender N et al (2010) Effects of fire on woody vegetation structure in African savanna. Ecol Appl 20:1865–1875

Smit IPJ, Riddell ES, Cullum C et al (2013) Kruger National Park research supersites: establishing long-term research sites for cross-disciplinary, multi-scaled learning. Koedoe 55(1):Art. #1107. https://doi.org/10.4102/koedoe.v55i1.1107

Smith RJ (2001) Designing an integrated protected area network for Maputaland. Thesis, Durell Institute of Conservation Ecology, University of Kent at Canterbury, Canterbury

Smith P, Allen Q (2004) Field guide to the trees and shrubs of the miombo woodlands. Royal Botanic Gardens, Kew

Smith S, Balkwill K, Williamson S (2001) Compositae on serpentine in the Barberton greenstone belt, South Africa. S Afr J Sci 97:518–520

Stalmans M (1990) Vegetation survey for the scientific management of the Lekgalameetse Nature Reserve. Thesis, University of the Witwatersrand, Johannesburg

Stalmans ME (2002) Vegetation, fire and herbivore interrelationships in the Songimvelo Game Reserve: a landscape ecological approach towards analysing pattern and change. Thesis, University of the Witwatersrand, Johannesburg

Stalmans M, De Klerk G (1991) A semi-quantitative survey of the vegetation of Bewaarkloof Nature Reserve. KaNgwane Parks Corporation, Nelspruit

Stalmans M, De Klerk G (1992) Protea landscapes in the Bewaarkloof mountains, Transvaal. Protea Atlas Newsl 14:8–9

Stalmans M, Peel M (2010) Plant communities and landscapes of the Parque Nacional de Zinave, Mozambique. Koedoe 52(1):Art. #703. https://doi.org/10.4102/koedoe.v52i1.703

Stalmans M, Wishart M (2005) Plant communities, wetlands and landscapes of the Parque Nacional de Banhine, Moçambique. Koedoe 48(2):43–58

Stalmans M, Robinson ER, Balkwill K (1999) Ordination and classification of vegetation of Songimvelo Game Reserve in the Barberton Mountainland, South Africa, for the assessment of wildlife habitat distribution and quality. Bothalia 29:305–325

Stalmans M, Gertenbach WPD, Carvalho-Serfontein F (2004) Plant communities and landscapes of the Parque Nacional do Limpopo, Moçambique. Koedoe 47(2):61–81

Starke AP, Geldenhuys CJ, O'Connor TG et al (2023) Secondary vegetation provides a reservoir of non-timber forest products and agroforestry service options for forestry plantation systems, Maputaland, South Africa. Southern For 85:162–173

Staver AC (2018) Prediction and scale in savanna ecosystems. New Phytol 219:52–57

Stellmes M, Frantz D, Finckh M et al (2013a) Okavango Basin—Earth observation. Biodivers Ecol 5:23–27

Stellmes M, Frantz D, Finckh M et al (2013b) Fire frequency, fire seasonality and fire intensity within the Okavango region derived from MODIS fire products. Biodivers Ecol 5:351–362

Stevens N, Swemmer AM, Ezzy L et al (2013) Investigating potential determinants of the distribution limits of a savanna woody plant: *Colophospermum mopane*. J Veg Sci 25:363–373

Stewart T, Scogings PF, Baijnath H (2020) Dispersal of a forest liana into grasslands and post-establishment stand expansion. S Afr J Bot 131:51–55

Straub AF (2002) The vegetation of Breslau Game Farm, Northern Province, South Africa. Thesis, Department of Botany, University of Pretoria, Pretoria

Strohbach BJ (2012) Providing relevant, useful information on Namibian vegetation types. Agricola 22:7–39

Strohbach BJ (2013) Vegetation of the Eastern Communal Conservancies in Namibia: III. Annotated checklist. Dinteria 33:3–42

Strohbach BJ (2014) Vegetation of the Eastern Communal Conservancies in Namibia: I. Phytosociological descriptions. Koedoe 56(1):Art. #1116. https://doi.org/10.4102/koedoe.v56i1.1116

Strohbach BJ (2017) Vegetation of the Auas-Oanob Conservancy in the Khomas Hochland of Namibia. Namib J Environ Sect A 1:14–33

Strohbach BJ (2019) Vegetation of the thornbush savanna of central Namibia: baseline description of the present vegetation at Farm Erichsfelde, Otjozondjupa Region. Namib J Environ Sect A 3:17–36

Strohbach B (2021) Vegetation survey of the Khomas Hochland in central-western Namibia: syntaxonomical descriptions. Bothalia 51(2):a4. https://doi.org/10.38201/btha.abc.v51.i2.4

Strohbach BJ, Kutuahuripa JT (2014) Vegetation of the eastern communal conservancies in Namibia: II. Environmental drivers. Koedoe 56(1):Art. #1117. https://doi.org/10.4102/koedoe.v56i1.1117

Strohbach BJ, Strohbach MM (2023) A first syntaxonomic description of the vegetation of the Karstveld in Namibia. Veg Class Surv 4:241–284

Strohbach BJ, Adank WF, Coetzee ME et al (2019) A baseline description of the soils and vegetation of Farm Klein Boesman, Khomas Region, Namibia. Namib J Environ Sect A 3:37–55

Tafangenyasha C, Muchachavangwa S, Mvulah B et al (2016) Transient feedback in woody vegetation response in aftermath of elephant culling history at Sengwa Wildlife Area, Zimbabwe. Int J Wildl Endang Spec Conserv 1:22–36

Tedder MJ, Kirkman KP, Morris CD et al (2013) Classification and mapping of the composition and structure of dry woodland and savanna in the eastern Okavango Delta. Koedoe 55(1):Art. #1100. https://doi.org/10.4102/koedoe.v55i1.1100

Teketay D, Geeves G, Kopong I et al (2016) Diversity, stand structure and regeneration status of woody species, and spatial cover of herbaceous species in Mokolodi Nature Reserve, Southeastern Botswana. Int J Biodivers Conserv 8:180–195

Teketay D, Kashe K, Madome J et al (2018) Enhancement of diversity, stand structure and regeneration of woody species through area exclosure: the case of a mopane woodland in northern Botswana. Ecol Process 7:5. https://doi.org/10.1186/s13717-018-0116-x

Theron GK (1973) 'n Ekologiese studie van die plantegroei van die Loskopdam-natuurreservaat, Thesis, Department of Botany, University of Pretoria, Pretoria

Theron GK, Morris JW, van Rooyen N (1984) Ordination of the herbaceous stratum of savanna in the Nylsvley Nature Reserve, South Africa. S Afr J Bot 3:22–32

Theron EJ, van Aardt AC, du Preez PJ (2020) Vegetation distribution along a granite catena, southern Kruger National Park, South Africa. Koedoe 62(2):a1588. https://doi.org/10.4102/koedoe.v62i2.1588

Thomas PI, Walker BH, Wild H (1977) Relationships between vegetation and environment on an amphibolite outcrop near Nkayi, Rhodesia. Kirkia 10:503–541

Thomas RJ, Marshall CGA, Du Plessis A et al (1992) Geological studies in southern Natal and Transkei: implications for the Cape Orogeny. In: De Wit MJ, Ransome IGD (eds) Inversion tectonics of the Cape Fold Belt, Karoo and Cretaceous basins of southern Africa. A.A. Balkema, Rotterdam, pp 229–236

Tiawoun MAP, Malan PW, Comole AA (2022) Effects of soil properties on the distribution of woody plants in communally managed rangelands in Ngaka Modiri Molema District, North-West Province, South Africa. Ecologies 3:361–375. https://doi.org/10.3390/ecologies3030027

Timberlake J (1996) Sites of interest for botanical conservation in the communal lands of the Zambezi Valley, Zimbabwe. Phase I. Occasional Publications in Biodiversity No. 2. Biodiversity Foundation for Africa, Bulawayo

Timberlake JR, Childes SL (2004) Biodiversity of the four corners area: technical reviews, vol 2, chaps 5–15. Occasional Publications in Biodiversity No. 15. Biodiversity Foundation for Africa, Bulawayo & Zambezi Society, Harare

Timberlake J, Drummond RB, Maroyi A (1988) Biodiversity assessment of the Lower Guruve region, Zambezi Valley, Zimbabwe. Occasional Publications in Biodiversity No. 4. Biodiversity Foundation for Africa, Bulawayo

Timberlake JR, Nobanda N, Mapaure I (1993) Vegetation survey of the communal lands—North and West Zimbabwe. Kirkia 14:171–270

Tinley KL (1966) An ecological reconnaissance of the Moremi Wildlife Reserve, Botswana. Okavango Wildlife Society, Johannesburg

Tothill JC, Mott JJ (1985) Ecology and management of the World's savannas. Australian Academy of Science, Canberra

Trollope WSW, Trollope LA, Biggs HC et al (1998) Long-term changes in the woody vegetation of the Kruger National Park, with special reference to the effects of elephants and fire. Koedoe 41:103–112

Tshireletso K, Makhabu SW, Nsinamwa M et al (2018) Diversity patterns of woody vegetation of Kgatleng District, Botswana. Biodivers Ecol 6:416–423

Tshisikhawe MP, Baloyi O, MH Ligavha-Mbelengwa MH et al (2012) The population ecology of *Securidaca longepedunculata* Fresen. in the Nylsvley Nature Reserve, Limpopo Province, South Africa. Phyton 81:107–112

Urban M, Heckel K, Berger C et al (2020) Woody cover mapping in the savanna ecosystem of the Kruger National Park using Sentinel-1 C-Band time series data. Koedoe 62(1):a1621. https://doi.org/10.4102/koedoe.v62i1.1621

van Aardt AC, Codron D, Theron EJ et al (2020) Plant community structure and possible vegetation changes after drought on a granite catena in the Kruger National Park, South Africa. Koedoe 62(2):a1585. https://doi.org/10.4102/koedoe.v62i2.1585

van Coller H, Klem J, Siebert F (2021) Drought tolerant forb flora of a semi-arid protected savanna in the Lowveld of South Africa. Bothalia 51(1):a10. https://doi.org/10.38201/btha.abc.v51.i1.10

van der Linden JP (2004) Vegetation analysis of Owen Sitole Agricultural College campus, KwaZulu-Natal, South Africa. Thesis, University of Zululand, KwaDlangezwa

van der Linden JP, Siebert SJ, Siebert F et al (2005) Vegetation of the Owen Sitole College of Agriculture in Zululand, KwaZulu-Natal. Koedoe 48(1):33–56

van der Linden J, Siebert F, Siebert SJ et al (2008) Characterisation of the woody assemblages of Zululand coastal thornveld along the Nseleni river. Koedoe 50(1):113–125

van der Merwe H, van Rooyen N, Bezuidenhout H et al (2019) *Vachellia erioloba* dynamics over 38 years in the Kalahari Gemsbok National Park, South Africa. Koedoe 61(1):a1534. https://doi.org/10.4102/koedoe.v61i1.1534

van der Meulen F (1979) Plant sociology of the western Transvaal bushveld, South Africa: a syntaxonomic and synecological study. A. Gantner, Vaduz

van der Meulen F, Westfall RH (1979) A vegetation map of the western Transvaal bushveld. Bothalia 12:731–735

van der Meulen F, Westfall RH (1980) Structural analysis of bushveld vegetation in Transvaal, South Africa. J Biogeogr 7:337–348

van der Schijff HP (1957) Ekologiese studie van die flora van die Nasionale Krugerwildtuin Deel 1. Thesis, University of Potchefstroom for C.H.E., Potchefstroom

van der Schijff HP, Schoonraad E (1971) The flora of the Mariepskop complex. Bothalia 10:461–500

van Jaarsveld E, Hardy D (1991) The Blouberg botanical expedition, with specific reference to the succulent vegetation. Aloe 28:78–83

van Rooyen N (1978) 'n Ekologiese studie van die plantegroei van die Punda Milia-Pafuri-Wambiya-gebied in die Nasionale Krugerwildtuin. Thesis, Department of Botany, University of Pretoria, Pretoria

van Rooyen N (1983) Die plantegroei van die Roodeplaatdam-natuurreservaat. II. Die plantgemeenskappe. S Afr J Bot 2:115–125

van Rooyen N (1984) 'n Fenologiese studie van die plantegroei van die Roodeplaatdam-natuurreservaat. Thesis, Department of Botany, University of Pretoria, Pretoria

van Rooyen N (2000) Vegetation survey of the Gemsbok National Park, Botswana and mapping of the Kgalagadi Transfrontier Park. 2000 Final Project Report. Project PPF/P/23. Peace Parks Foundation, Stellenbosch

van Rooyen N, van Rooyen MW (1998) Vegetation of the south-western arid Kalahari: an overview. Trans Roy Soc S Afr 53:113–140

van Rooyen N, Theron GK, Grobbelaar N (1981a) A floristic description and structural analysis of the plant communities of the Punda Milia–Pafuri–Wambiya area in the Kruger National Park, Republic of South Africa: 2. The sandveld communities. J S Afr Bot 47:405–449

van Rooyen N, Theron GK, Grobbelaar N (1981b) A floristic description and structural analysis of the plant communities of the Punda Milia–Pafuri–Wambiya area in the Kruger National Park, Republic of South Africa. 3. The *Colophospermum mopane* communities. J S Afr Bot 47:585–626

van Rooyen N, van Rensburg DJ, Theron GK et al (1984) A preliminary report on the dynamics of the vegetation of the Kalahari Gemsbok National Park. Koedoe 1984(Suppl):83–102

van Rooyen N, Bezuidenhout D, Theron GK et al (1990) Monitoring of the vegetation around artificial watering points (windmills) in the Kalahari Gemsbok National Park. Koedoe 33:63–88

van Rooyen N, Bredenkamp GJ, Theron GK (1991a) Kalahari vegetation: veld condition trends and ecological status of species. Koedoe 34(1):61–72

van Rooyen N, Theron GK, Bredenkamp GJ (1991b) Long-term vegetation dynamics in the semi-arid Kalahari. Koedoe 34(1):91–92

van Rooyen N, Theron GK, Bezuidenhout D et al (1993) Patterns of change in the herbaceous layer of a mesic savanna, South Africa. S Afr J Bot 59:74–80

van Rooyen N, Bredenkamp GJ, Theron GK et al (1994) Vegetation gradients around artificial watering points in the Kalahari Gemsbok National Park. J Arid Environ 26:349–361

van Rooyen MW, van Rooyen N, Bothma J duP et al. (2008) Landscapes in the Kalahari Gemsbok National Park, South Africa. Koedoe 50(1):99–112

van Staden JM (1991) Die fitososiologie van die Steenbokpan omgewing in die noordwes-Transvaal. Thesis, Department of Botany, University of Pretoria, Pretoria

van Staden PJ (2002) An ecological study of the plant communities of Marakele National Park. Thesis, Centre for Wildlife Management, University of Pretoria, Pretoria

van Staden PJ, Bredenkamp GJ (2005) Major plant communities of the Marakele National Park. Koedoe 48:59–70

van Staden PJ, Bezuidenhout H, Ferreira S et al (2017) The effects of elephants and fire on vegetation at Marakele National Park, South Africa. Pachyderm 58:107–122

van Staden N, Siebert SJ, Cilliers DP et al (2020) A Floristic analysis of semi-arid mountain ecosystems of the Griqualand West centre of plant endemism, Northern Cape, South Africa. Biodiversitas J Biol Divers 21:1989–2002

van Staden PJ, Bredenkamp GJ, Bezuidenhout H et al (2021) A reclassification and description of the Waterberg Mountain vegetation of the Marakele National Park, Limpopo province, South Africa. Koedoe 63(1):a1689. https://doi.org/10.4102/koedoe.v63i1.1689

van Vegten JA (1984) Thornbush invasion in a savanna ecosystem in eastern Botswana. Vegetatio 56:3–7

van Vuuren DRJ, van der Schijff HP (1970) 'n Vergelykende ekologiese studie van die plantegroei van 'n noordelike en suidelike kloof van die Magaliesberg. Tydskr. Natuurwetensk 10:16–75
van Wyk JJP (1959) 'n Sistematies-ekologiese studie van die plantegroei van die plaas Koedoesfontein no. 746 en omgewing in die Rustenburgse distrik. Thesis, Potchefstroom University for C.H.E., Potchefstroom
van Wyk P (1973) Bome van die Nasionale Krugerwildtuin. Deel I en II. Perskor, Johannesburg
van Wyk AE (1990a) The sandstone regions of Natal and Pondoland: remarkable centres of endemism. Palaeoecol Afr 21:243–257
van Wyk AE (1990b) A new species of *Leucadendron* (Proteaceae) from Pondoland, with a discussion of its biogeography. S Afr J Bot 56:458–466
van Wyk AE (1994) Indian Ocean Coastal Belt: CPD Site AF59. Maputaland–Pondoland Region, South Africa, Swaziland and Mozambique. In: Davis SD, Heywood VH, Hamilton AC (eds) Centres of plant diversity. A guide and strategy for their conservation. Volume 1: Europe, Africa, South West Asia and the Middle East. IUCN Publications, Cambridge, pp 227–235
van Wyk AE (1996) Biodiversity of the Maputaland Centre. In: Van der Maesen LJG, Van der Burgt XM, Van Medenbach de Rooy JM (eds) The biodiversity of African plants. Kluwer, Dordrecht, pp 198–207
van Wyk E, Hurter J (2000) *Brachystegia spiciformis*—an exciting discovery. SABONET News 5:170–172
van Wyk AE, Smith GF (2001) Regions of floristic endemism in southern Africa: A review with emphasis on succulents. Umdaus Press, Hatfield
Veldhuis MP, Rozen-Rechels D, le Roux E et al (2016) Determinants of patchiness of woody vegetation in an African savanna. J Veg Sci 28:93–104
Venter HJT (1972) Die plantekologie van Richardsbaai, Natal. Thesis, Department of Botany, University of Pretoria, Pretoria
Verster E, de Villiers JM, Scheepers JC (1973) Gilgai in the Rustenburg area. Agrochemophysica 4:57–62
Visser N, van Hoeven W, Theron GK (1996) The vegetation and identification of management units of the Honnet Nature Reserve, Northern Province, South Africa. Koedoe 39(1):25–42
Vittoz P, Pellacani F, Romanens R et al (2020) Plant community diversity in the Chobe Enclave, Botswana: insights for functional habitat heterogeneity for herbivores. Koedoe 62(1):a1604. https://doi.org/10.4102/koedoe.v62i1.1604
Volder A, Briske DD, Tjoelker MG (2013) Climate warming and precipitation redistribution modify tree–grass interactions and tree species establishment in a warm-temperate savanna. Glob Change Biol 19:843–857
Volk OH, Leippert H (1971) Vegetationsverhältnisse im Windhoeker Bergland, Südwestafrika. J SWA Wiss Ges 25:5–44
von Maltitz G, Mucina L, Geldenhuys CJ et al (2003) Classification system for South African indigenous forests: an objective classification for the Department of Water Affairs and Forestry. Report ENV-P-C 2003-017. Environmentek CSIR, Pretoria
Waldram MS, Bond WJ, Stock WD (2008) Ecological engineering by a mega-grazer: white rhino impacts on a South African savanna. Ecosystems 11:101–112
Walker BH (ed) (1987) Determinants of tropical savannas. IRL Press, Oxford
Walker BH, Stone L, Henderson L et al (1986) Size structure analysis of the dominant trees in a South African savanna. S Afr J Bot 52:397–402
Wanke H, Dunkeloḧ A, Peter Udluft P et al (2013) A distributed water balance model to estimate direct groundwater recharge: a case study from the Nhoma and Khaudum catchments, Namibia. Int Sci Technol J Namib 2:12–32
Ward CJ (1962) Report on scrub control in the Hluhluwe Game Reserve. Lammergeyer 2:57–62
Ward CJ (1980) The plant ecology of the Isipingo beach area, Natal, South Africa. Mem Bot Surv S Afr 45:1–147
Watson HK, Macdonald IAW (1983) Vegetation changes in the Hluhluwe-Umfolosi Game Reserve Complex from 1937 to 1975. Bothalia 14:265–269
Weare PR, Yalala A (1971) Provisional vegetation map of Botswana. Botsw Notes Rec 3:131–147
Wedel ER (2024) Resource-use strategies of woody plants in grassy ecosystems. Thesis, Division of Biology, Kansas State University, Manhattan
Weisser PJ (1978) Conservation priorities in the dune area between Richards Bay and Mfolozi Mouth based on vegetation survey. Natal Town Reg Plan Rep 38:1–64
Weisser PJ (1987) Dune vegetation between Richards Bay and Mlalazi Lagoon and its conservation priorities in relation to dune mining. Natal Town Reg Plan Rep 19:1–71
Weisser PJ, Marques F (1979) Gross vegetation changes in the dune area between Richards Bay and the Mfolozi River. Bothalia 12:711–721
Werger MJA (1978) Vegetation structure in the southern Kalahari. J Ecol 66:933–941
Werger MJA, Coetzee BJ (1978) The Sudano-Zambezian Region. In: Werger MJA, van Bruggen AC (eds) Biogeography and ecology of southern Africa. Dr W. Junk Publishers, The Hague, pp 301–462
Werger MJA, Wild H, Drummond BR (1978a) Vegetation structure and substrate of the northern part of the Great Dyke, Rhodesia: environment and plant communities. Vegetatio 37:79–89
Werger MJA, Wild H, Drummond BR (1978b) Vegetation structure and substrate of the northern part of the Great Dyke, Rhodesia: gradient analysis and dominance-diversity relationships. Vegetatio 37:151–161
West O (1951) The vegetation of Weenen County, Natal. Mem Bot Surv S Afr 23:1–183
Westinga E, Betrana APR, de Bie CAJM et al (2020) A novel approach to optimize hierarchical vegetation

mapping from hypertemporal NDVI imagery, demonstrated at national level for Namibia. Int J Appl Earth Observ Geoinform 91:102152. https://doi.org/10.1016/j.jag.2020.102152
Westfall RH (1981) The plant ecology of the farm Groothoek, Thabazimbi District. Thesis, Department of Botany, University of Pretoria, Pretoria
Westfall RH, van Rooyen N, Theron GK (1983) The plant ecology of the farm Groothoek, Thabazimbi District. I. Ordination. Bothalia 14:785–790
Westfall RH, van Rooyen N, Theron GK (1985) The plant ecology of the farm Groothoek, Thabazimbi District. II. Classification. Bothalia 15:655–688
Whateley A, Porter RN (1983) The woody vegetation communities of the Hluhluwe-Corridor-Umfolozi Game Reserve Complex. Bothalia 14:745–758
White F (1976) The underground forests of Africa: a preliminary review. Gard Bull (Singapore) 29:55–71
White F (1983) Vegetation of Africa: a descriptive memoir to accompany the UNESCO/AETFAT/UNSO vegetation map of Africa. UNESCO, Paris
Whitecross MA, Witkowski ETF, Archibald S (2017) Savanna tree-grass interactions: a phenological investigation of green-up in relation to water availability over three seasons. S Afr J Bot 108:29–40
Whittaker RH, Morris JW, Goodman D (1984) Pattern analysis in savanna woodlands at Nylsvley, South Africa. Mem Bot Surv S Afr 49:1–51
Wigley BJ, Fritz H, Coetsee C et al (2014) Herbivores shape woody plant communities in the Kruger National Park: lessons from three long-term exclosures. Koedoe 56(1):Art. #1165. https://doi.org/10.4102/koedoe.v56i1.1165
Wild H (1965) The flora of the Great Dyke of southern Rhodesia with special reference to the serpentine soils. Kirkia 5:49–86
Wild H (1978) The vegetation of heavy metal and other toxic soils. In: Werger MJA, van Bruggen AC (eds) Biogeography and ecology of southern Africa. Dr W. Junk Publishers, The Hague, pp 1301–1332
Wild H, Barbosa LAG (1967) Vegetation map of the Flora Zambesiaca area. M.O. Collins, Salisbury [Harare]
Willis C, Smith G, Van Wyk E (2001) Further notes on South Africa's *Brachystegia spiciformis*. SABONET News 6:209–211
Wilson LJ, Hoffman MT, Ferguson AJ et al (2021) Elephant browsing impacts in a Zambezian *Baikiaea* woodland with a high density of pumped waterholes. Glob Ecol Conserv 31:e01854. https://doi.org/10.1016/j.gecco.2021.e01854
Winter PJD, Hahn N (1999) A new species of *Khadia* N.E.Br. (Mesembryanthemaceae) from the Northern Province of South Africa. Aloe 36:16–19
Winterbach R (1998) A phytosociological synthesis of *Acacia tortilis* communities in the north-western savanna of South Africa. Thesis, Department of Botany, University of Pretoria, Pretoria
Winterbach R, Bredenkamp GJ, Deutchländer MS et al (2000) Preliminary syntaxonomic scheme of vegetation classes for the Central Bushveld of South Africa. In: White PS, Mucina L, Lepš JS et al (eds) Proceedings of 41st I.A.V.S. Symposium. Opulus Press, Uppsala, pp 123–127
Wiseman R, Page BR, O'Connor TG (2004) Woody vegetation change in response to browsing in Ithala game reserve. South Africa. S Afr J Wildl Res 34:25–37
Witkowski ETF, O'Connor TG (1996) Topo-edaphic, floristic and physiognomic gradients of woody plants in a semi-arid African savanna woodland. Vegetatio 124:9–23
Yeaton RI, Frost S, Frost PGH (1988) The structure of a grass community in *Burkea africana* savanna during recovery from fire. S Afr J Bot 54:367–371
Zacharias PJK (1994) The vegetation of Madikwe—a first approximation. Bophuthatswana Parks Board, Rustenburg
Zimudzi C, Chapano C (2016) Diversity, population structure, and above ground biomass in woody species on Ngomakurira Mountain, Domboshawa, Zimbabwe. Int J Biodivers 2016:4909158. https://doi.org/10.1155/2016/4909158
Zimudzi C, Mapaura A, Chapano C et al (2013) Woody species composition, structure and diversity of Mazowe Botanical Reserve, Zimbabwe. J Biodivers Environ Sci 3(6):17–29

Subtropical Southern African Rainforests

4

Ladislav Mucina, Mervyn C. Lötter, Michael C. Rutherford and James L. Tsakalos

Abstract

The subtropical forests of southern Africa represent a remarkable vegetation and ecological anomaly in that they extend as far as latitude 34° S, well into the warm-temperate thermic zone. Their occurrence at these extreme southerly latitudes can be ascribed to the strong onshore influence of the warm Agulhas Current of the Indian Ocean. The Southern African Subtropical Rainforests in the southern half of Mozambique are a continuation of the tropical coastal forests further north along the Mozambique–Tanzania coast. Although they still can be classified as a member of the zonobiome E1 (Equatorial Zone; its subtropical global biome section), in Southern Africa, they acquire a transitional bioclimatic character through rainfall-depression becoming evident during the austral-winter season. In the study area, these forests are represented by two core regional biomes (Mozambique Tropical Lowland Forest and African Subtropical Coastal Forest) and three extrazonal (scarp) forest biomes, namely Tropical Scarp Forest, Subtropical Scarp Forest, and Albany Coastal Forest—all of the relict character.

Keywords

Albany region · Forest biomes · Coastal subtropical forest · eSwatini · Indian Ocean Coastal Belt · Maputaland · Mozambique · Scarp forest · South Africa

L. Mucina (✉) · J. L. Tsakalos
Harry Butler Institute, Murdoch University,
90 South Street, Murdoch, 6150 Perth, Australia
e-mail: Ladislav.Mucina@murdoch.edu.au

L. Mucina
Department of Geography and Environmental Studies, Stellenbosch University, Private Bag X1, Matieland 7602, Stellenbosch, South Africa

M. C. Lötter
Mpumalanga Tourism and Park Agency, P/Bag X1088, Lydenburg 1120, South Africa

School of Animal, Plants and Environmental Sciences, University of the Witwatersrand, Private Bag 3, Wits 2050, Johannesburg, South Africa

M. C. Rutherford
High Peak, Derbyshire, United Kingdom

Department of Botany and Zoology, Stellenbosch University, Private Bag X1, Matieland 7602, Stellenbosch, South Africa

J. L. Tsakalos
Plant Diversity and Ecosystems Management Unit, School of Biosciences and Veterinary Medicine, University of Camerino, Via Pontoni 5, Camerino 62032, Italy

L. Mucina and M. C. Rutherford (eds.), *Zonal Biomes of Southern Africa*, Biome Ecology 2,
https://doi.org/10.1007/978-3-031-63731-5_4

4.1 Subtropical Forests of the Southern Africa: General Patterns

The subtropical forests (Fig. 4.1) occur within the bounds of the MBSA in Mozambique, Zimbabwe, South Africa, and eSwatini. Subtropical forests of Southern Africa belong to the global biome called Subtropical Rainforest (Table 4.1), classified as a member of the Equatorial Zone E1. They exemplify a vegetation-ecological anomaly as they occur as far as 34° S along the East Coast of South Africa, extending well south into the warm-temperate thermic zone. Their distribution follows the warm seaboards of Eastern Cape and KwaZulu-Natal exposed to a strong influence of the warm offshore Agulhas Current. Towards the lower latitudes, the subtropical forests either continue on the coastal plains or ascend to higher elevations in piedmonts of east-facing flanks of the Great Escarpment. They also occupy extensive rolling plains deep inland in Mozambique (Mucina et al. 2022). In some regions, the subtropical forests co-occur (at large spatial scales) with tropical dry forests (TDF), classified as members of the Tropical Seasonal Zone E2. Mucina et al. (2022) analysed this biogeographically and biome-ecologically interesting situation and established that the MAT in the TDF units is markedly higher and the MAP markedly lower than that in the subtropical rainforests. The main difference between the latter and TDF in the study region is, however, the fact that subtropical rainforests are of zonal character while the TDF are extrazonal (hence technically, azonal) since the TDF tend to be linked to deep sandy soils.

The subtropical forests of Southern Africa occur in regions characterised by a mean annual temperature spanning 19–21 °C (Mucina et al. 2006a, b, 2022). Precipitation occurs throughout the year; hence, no pronounced dry period is detectable. However, there is a pronounced summer maximum due to cyclonic rainfall originating at the edges of the Indian Ocean anticyclone (Mucina et al. 2022). The main feature distinguishing between the subtropical and tropical rainforests in the Southern Hemisphere is the presence of distinct austral-winter precipitation

Table 4.1 A system of the forest biomes of the zonobiome E1 Equatorial Zone (its subtropical rainforest global biome part) of Southern Africa (modified after Mucina 2023)

Rank	Code	Biome	Zonality	Azonality type	Climatic diagnosis
ZB	**E1**	**Equatorial Zone**	**Zonal**		**Equally warm all-year-round; frost is rare and occurs only in marginal subtropical regions; precipitation possible all-year-round**
GB	***STF***	***Subtropical Rainforest***	***Zonal***		***Equally warm all-year-round; frost does occur, but rarely; precipitation possible all-year-round***
CB	**STFa**	***African Subtropical Forest***	**Zonal**		**Equally warm all-year-round; frost does occur, but rarely; precipitation possible all-year-round**
Core zonal regional biomes					
RB	STFa1	Mozambique Tropical Lowland Forest	Zonal		
RB	STFa2	African Subtropical Coastal Forest	Zonal		
Extrazonal regional biomes					
RB	STFa3	Tropical Scarp Forest	Extrazonal	Relict	
RB	STFa4	Subtropical Scarp Forest	Extrazonal	Relict	
RB	STFa5	Albany Coastal Forest	Extrazonal	Relict	

Fig. 4.1 Subtropical Forests (zonobiome E1). **A** Cheringoma Limestone Gorge Forest located in the Nhamfisse limestone gorge, western Cheringoma Plateau, Mozambique. The tall tree with pale stems are *Antiaris toxicaria*, accompanied by *Bombax rhodognaphalon, Breonadia salicina, Celtis philippensis, Celtis mildbraedii, Khaya anthotheca, Milletia stuhlmannii, Milicia excelsa,* and *Sterculia appendiculata*. **B** A view of the interior of the subtropical Alexandria Forest (STFa5 Albany Coastal Forest) near Alexandria, Eastern Cape Province, with prominent *Celtis africana* and *Erythrina caffra*. **C** Species-rich subtropical Pondoland Scarp Forests in the Umtamwuna Nature Reserve near Port Edward, southern KwaZulu-Natal, South Africa. **D** View of the forested river valley of the Krantzkloof Nature Reserve near Durban (eThekwini), KwaZulu-Natal Province of South Africa, hosting a patch of subtropical scarp forest. (Photo credits: A: J. Burrows; B–C: L. Mucina, D: www.zululand.co.za/ancient-forests-of-kwazulu-natal)

depression characteristic of the subtropical forests occurring clearly outside the core ITCZ.

In Moll and White (1978) and Mucina et al. (2006a), the subtropical forests were subsumed under a biogeographic concept of the Indian Ocean Coastal Belt (IOCB). Mucina et al. (2022) clarified why it had previously been done this way and rectified the biome classification accordingly.

4.2 Variability of the Subtropical Forests in Southern Africa

Mucina et al. (2022) also distinguish two groups of regional forest biomes that qualify as subtropical.

The core group (Table 4.1) contains two RBs—the STFa1 Mozambique Tropical Lowland Forest and STFa2 African Subtropical Coastal Forest. These occur on coastal plains spanning (within the borders of the MBSA) central to southern Mozambique. The latter regional biome spans Maputaland (shared between Mozambique and South Africa) and the East Coast of South Africa.

The extrazonal group of subtropical regional forest biomes contains two inland scarp forests: STFa3 Tropical Scarp Forest and STFa4 Subtropical Scarp Forest—both occurring in inland localities. The STFa5 Albany Coastal Forest, the southernmost biome unit of these forests, is found on undulating coastal plains of the Eastern Cape. All three biome units are considered relics of a warmer climatic period long gone (Mucina et al. 2022).

The link of the South African and Mozambique subtropical to the general concept of the East African Coastal Forests has been elucidated by Mucina et al. (2022).

Further summarising information on the distribution and ecology of the subtropical forests of Southern Africa are found in von Maltitz et al. (2003), Mucina et al. (2006a, b), Mucina (2018), and Lötter et al. (2023). This chapter is based on the general account of the subtropical forest biomes, as presented in Mucina et al. (2022). In the current work, we shall focus only on the characteristics of the regional forest biomes to expand on the ideas presented in the last-mentioned publication.

4.3 Descriptions of the Regional Biomes of the Subtropical Forests

The subtropical forests in the MBSA area encompass five regional biomes, namely (from south to north), Albany Subtropical Coastal Forest, African Subtropical Coastal Forest, Subtropical Scarp Forest, Tropical Scarp Forest, and Mozambique Tropical Lowland Forest (Table 4.1 and Fig. 4.1).

STFa1 Mozambique Tropical Lowland Forest

STF6 Amatonga Lowland Semideciduous Forest (Lötter et al. 2023)
STF4 Central Lowland Moist Forest (Lötter et al. 2023)
STF8 Central Mid-elevation Moist Forest (Lötter et al. 2023)
STF3 Cheringoma Limestone Gorge Forest (Lötter et al. 2023)
STF5 Northern Lowland Moist Forest (Lötter et al. 2023)
STF9 Northern Mid-elevation Moist Forest (Lötter et al. 2023)
STF7 Zambezi Delta Lowland Forest (Lötter et al. 2023)

Distribution Historically, these forests were more extensive, but due to heavy logging, they now appear much more fragmented throughout their range. They are found along the moist interior of Mozambique and the base of the Eastern (Manica) Highlands on the Mozambique–Zimbabwe border. Only a few patches extend into Zimbabwe. In Mozambique, they are found from Espungabera and Dombe in the south, northwards across the Zambezi River as far north as the foothills of the granite massifs of Nampula Province (Fig. 4.2).

Vegetation Physiognomy & Composition These are medium to tall evergreen or occasionally semi-deciduous forests. Some of the dominant or characteristic canopy species include *Blighia unijugata, Celtis gomphophylla, Chrysophyllum gorungosanum, Croton sylvaticus, Erythrophleum suaveolens, Funtumia africana, Inhambanella henriquesii, Khaya*

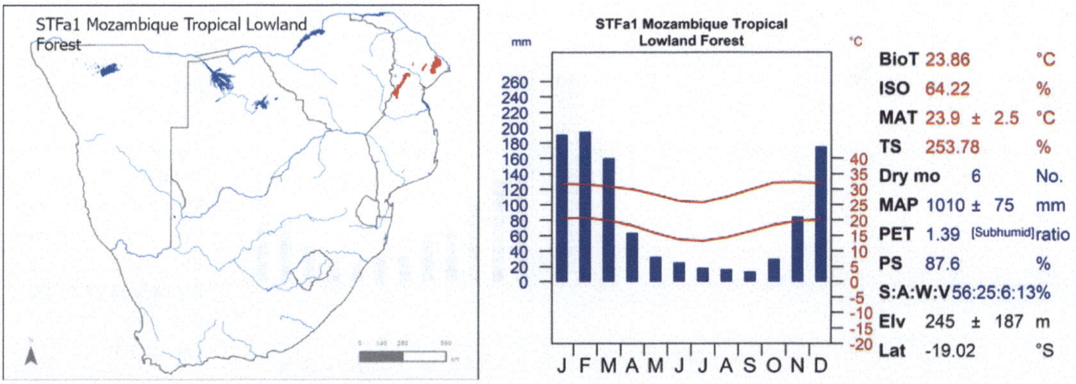

Fig. 4.2 Distribution area and the climatic diagram of the regional biome STFa1 Mozambique Tropical Lowland Forest

anthotheca, Maranthes goetzeniana, Milicia excelsa, Millettia stuhlmannii, Newtonia buchananii, Shirakiopsis elliptica, Strombosia scheffleri, Synsepalum brevipes, Treculia africana, Trichilia dregeana, Trilepisium madagascariense, and *Uapaca lissopyrena.*

Topography, Geology & Soils Even or slightly undulating terrain at the base of moist sandstone ridges and large granite massifs is the characteristic landscape of this biome. The forests are always found at elevations below 900 m. Cheringoma Limestone Gorge Forest occurs in and around limestone canyons. Soils are usually of low nutrient status and sandy.

Sources Wild and Barbosa (1967), Müller (2006), Burrows et al. (2018), Mucina et al. (2022), Lötter et al. (2023).

STFa2 African Subtropical Coastal Forest

FOz 7 Northern Coastal Forest p.p. (Mucina et al. 2006b)
STF1 Maputaland Coastal Forest (Lötter et al. 2023)
STF2 Bilene Coastal Forest (Lötter et al. 2023)

Distribution These forests are concentrated on the coastal (Indian Ocean) platforms of the southern KwaZulu-Natal, Maputaland, and northwards as far as Inhambane in southern Mozambique. The most iconic forest complexes include the Hawaan, Hlongweni (Tugela Mouth) and Dukuduku Forest (all in South Africa), as well as the Maputaland coastal forests and the Bilene forest complex in southern Mozambique (Fig. 4.3).

Vegetation Physiognomy & Composition These are usually medium-tall semi-deciduous or occasionally evergreen forests. Some of the dominant or characteristic canopy species include *Afzelia quanzensis, Albizia adianthifolia, Brachylaena discolor, Celtis africana, Chaetachme aristata, Dialium schlechteri, Diospyros inhacaensis, D. natalensis, Ficus natalensis, Filicium decipiens, Hymenocardia ulmoides, Lannea antiscorbutica, Manilkara discolor, Mimusops caffra, Morus mesozygia, Pteleopsis myrtifolia, Sclerocroton integerrimus, Sideroxylon inerme, Strychnos gerrardii,* and *Vepris lanceolata.*

Topography, Geology & Soils The forests of this regional biome are found on the flat or slightly rolling ancient coastal platform; in southern KwaZulu-Natal Province, they are found on Port Dunford clay-containing red sands and occasional Cretaceous siltstones. In Maputaland, they occur on deep dystrophic sandy soils of ancient dunes.

Sources Bews (1920), Wild and Barbosa (1967), Breen (1971), Venter (1976), Wager (1976), Weisser (1978, 1980, 1987), Ward (1980), Whateley and Porter (1983), White (1983), Cooper (1985), MacDevette (1987, 1989, 1993),

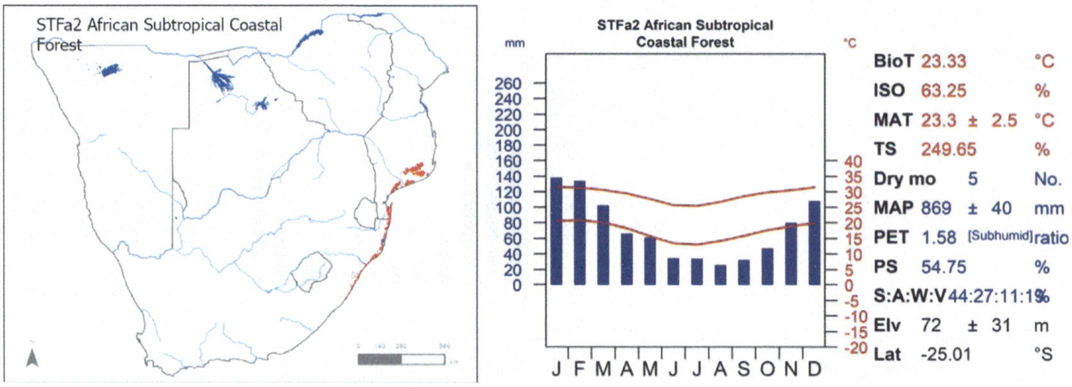

Fig. 4.3 Distribution area and the climatic diagram of the regional biome STFa2 African Subtropical Coastal Forest

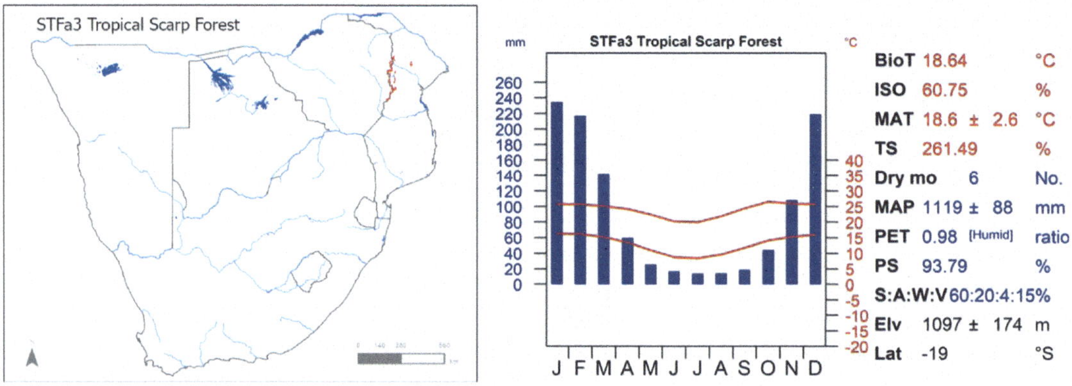

Fig. 4.4 Distribution area and the climatic diagram of the regional biome STFa3 Tropical Scarp Forest

MacDevette and Gordon (1989a, b), MacDevette and MacDevette (1989), MacDevette et al. (1989), Cooper and Swart (1992), van Wyk and Everard (1992), Weisser et al. (1992), von Maltitz et al. (1996, 2003), van Wyk et al. (1996), Lubbe (1997), Eeley et al. (1999), Geldenhuys and Mucina (2006), Mucina et al. (2006b, 2022), Burrows et al. (2018), Lötter et al. (2023).

STFa3 Tropical Scarp Forest

FOz 5 Scarp Forest (Mucina and Rutherford 2006)
FOz 6 Southern Coastal Forest p.p.

Distribution This regional forest biome is confined to central and northern Mozambique, eastern Zimbabwe, and southern Malawi (outside the MBSA area). Chirinda Forest (SE Zimbabwe) is the flagship forest of this type (Fig. 4.4).

Vegetation Physiognomy & Composition This forest type is characterised by medium-tall to tall evergreen forest with the canopy dominated by *Albizia gummifera, Chrysophyllum gorungosanum, Craibia brevicaudata, Croton sylvaticus, Diospyros abyssinica, Drypetes*

gerrardii, Ekebergia capensis, Ficus chirindensis, F. scassellatii, Garcinia kingaensis, Macaranga mellifera, Margaritaria discoidea var. *nitida, Newtonia buchananii, Ocotea kenyensis, Polyscias fulva, Strombosia scheffleri, Rawsonia lucida, Tabernaemontana ventricosa,* and *Trichilia dregeana.*

Sources Wild and Barbosa (1967), Timberlake (1993), Timberlake and Shaw (1994), Müller (1999, 2006), Mujuru and Kundhlande (2007), Mapaure (2012), Lötter (2014), Lötter et al. (2014, 2023), Burrows et al. (2018), Mucina et al. (2022)

Topography, Geology & Soils These forests occur at elevations of between 900–1300 m and are linked to steep sub-escarpment slopes and gorges. Often, they are found on and around granite or gneiss inselbergs.

STFa4 Subtropical Scarp Forest

FOz 5 Scarp Forest p.p. (Mucina et al. 2006b)
FOz 6 Southern Coastal Forest p.p. (Mucina et al. 2006b)
STF10 Lebombo-KwaZulu Natal Scarp Forest (Lötter et al. 2023)

Distribution Subtropical Scarp Forest is found in a narrow, fragmented belt from the Albany region south of Grahamstown known as the Albany Scarp Forest (Mucina et al. 2022), along the Indian Ocean seaboards of Transkei and KwaZulu-Natal and then turning inland in two nearly parallel belts—one from northern KwaZulu-Natal towards the base of Mpumalanga Escarpment, while the other along the Lebombo mountain range from Maputaland to the eSwatini and Mozambique border areas. A particular type of scarp forest, so far only poorly known, occurs on the southern flanks of the Blouberg and Soutpansberg (Fig. 4.5).

Vegetation Physiognomy & Composition Usually mid-tall, evergreen or semi-deciduous forest; some dominant or characteristic canopy taxa include *Chaetachme aristata, Chionanthus foveolatus* subsp. *foveolatus, Combretum kraussii, C. woodii, Croton sylvaticus, Englerophytum magalismontanum, Ekebergia capensis, Ficus sur, Harpephyllum caffrum, Homalium dentatum, Macaranga capensis, Manilkara concolor, M. discolor, Margaritaria discoidea, Mimusops zeyheri, Philenoptera sutherlandii, Protorhus longifolia, Rawsonia lucida, Strychnos decussata, S. mitis, Trichilia dregeana* and *Vitellariopsis marginata.*

Topography, Geology & Soils These forests occur along the lower escarpment and foothills, on a variety of geology and soil substrates, from sandstones in the south of its distribution area, rhyolites of the Lebombo Mts and ultramafics of the Barberton Greenstone Belt.

Sources Huntley (1965), van Wyk (1981, 1989), Nicholson (1982), Whateley and Porter (1983), Viljoen (1984), Cooper (1985), Acocks (1988), MacDevette et al. (1989), Cawe (1990, 1996), Cooper and Swart (1992), van Wyk and Everard (1993a, b), Everard et al. (1995), Glen (1996), Jacobs (1996), Eeley et al. (1999), van Wyk and Smith (2001), von Maltitz et al. (2003), Geldenhuys and Mucina (2006), Mucina et al. (2006b, 2022), Mostert et al. (2009), Ezemvelo KZN Wildlife (2013), Lötter (2014), Lötter et al. (2014), Burrows et al. (2018), Mucina et al. (2018b; c), Situngu et al. (2020).

STFa5 Albany Coastal Forest

FOz 6 Southern Coastal Forest p.p. (Mucina et al. (2006b)

Distribution Albany Coastal Forest is limited to the broad coastal platform of the Albany region, Eastern Cape Province, with the Alexandria Forest complex being the iconic flagship forest of this type. These are remnants of more extensive, today heavily fragmented, westernmost subtropical coastal forests in the MBSA area (Fig. 4.6).

Vegetation Physiognomy & Composition The forests characteristic of this biome are mid-tall, semi-deciduous, dominated by *Apodytes dimidiata, Celtis africana, Erythrina caffra, Maytenus undata, Mimusops obovata,* and *Sideroxylon inerme*. The shrub layer is species-rich and contains species such as *Cordia cafra, Euclea natalensis, Gymnosporia buxifolia, Strychnos decussata,* and *Teclea natalensis*. The herb layer

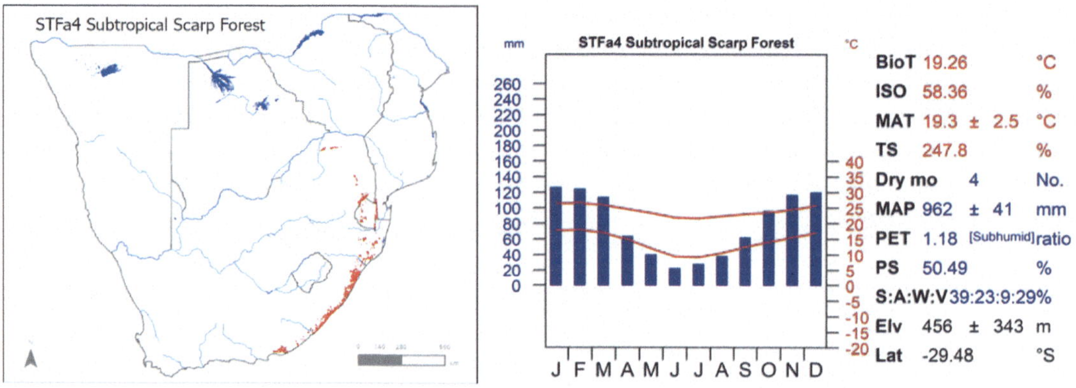

Fig. 4.5 Distribution area and the climatic diagram of the regional biome STFa4 Subtropical Scarp Forest

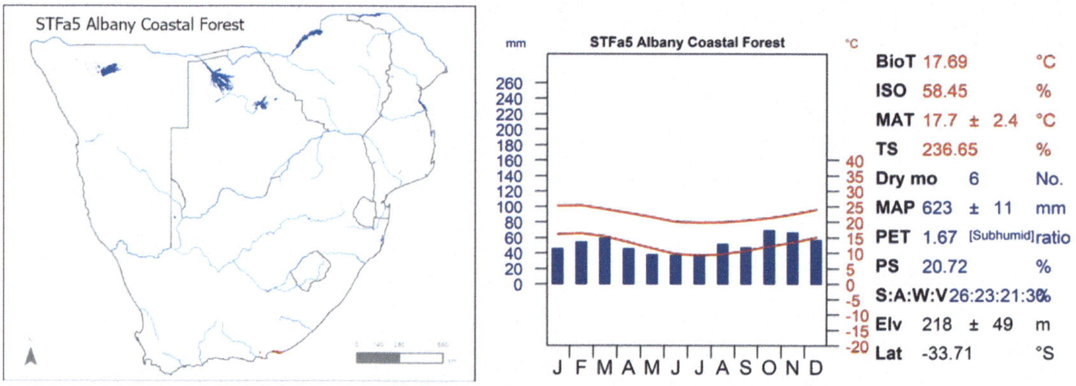

Fig. 4.6 Distribution area and the climatic diagram of the regional biome STFa5 Albany Coastal Forest

is usually dense and dominated by *Acalypha glabrata, Hypoestes aristata,* and grass *Panicum deustum.*

Topography, Geology & Soils The landscapes supporting this forest biome are broadly undulating coastal plains. Deep sands overlay the Nanaga and Alexandria Formations.

Sources Phillipson and Russell (1988), Mucina et al. (2006b, 2022), Mucina et al. (2018a).

References

Acocks JPH (1988) Veld types of South Africa, 3rd edn. Mem Bot Surv S Afr 57:1–146

Bews JW (1920) The plant ecology of the coast belt of Natal. Ann Natal Mus 4:367–467

Breen CM (1971) An account of the plant ecology of the dune forest at Lake Sibayi. Trans Roy Soc S Afr 39:223–234

Burrows JE, Burrows SM, Lötter MC et al (2018) Trees and shrubs of Mozambique. Publishing Print Matters, Noordhoek

Cawe SG (1990) Coastal Forest Survey: a classification of the Coastal Forests of Transkei and an assessment of their timber potential. Thesis, Department of Botany, University of Transkei, Umtata

Cawe SG (1996) A floristic classification of the indigenous forests of Transkei, South Africa. In: Van der Maesen LJG, Van der Burgt XM, Van Medenbach de Rooy JM (eds) The biodiversity of African plants. Kluwer, Dordrecht, pp 241–249

Cooper KH (1985) The conservation status of indigenous forests in Transvaal, Natal and O.F.S. Wildlife Society of Southern Africa, Durban, South Africa

Cooper KH, Swart W (1992) Transkei forest survey. Wildlife Society of Southern Africa, Durban

Eeley HAC, Lawes MJ, Piper SEP (1999) The influence of climate change on the distribution of indigenous forest in KwaZulu-Natal, South Africa. J Biogeogr 26:595–617
Everard DA, Midgley JJ, van Wyk GF (1995) Dynamics of some forests in KwaZulu-Natal, South Africa, based on ordinations and size-class distributions. S Afr J Bot 61:283–292
Ezemvelo KZN Wildlife (2013) Ongoye forest nature reserve KwaZulu-Natal, South Africa Integrated Management Plan: 2009–2013. Ezemvelo KZN Wildlife, Pietermaritzburg
Geldenhuys CJ, Mucina L (2006) Towards a new forest classification for South Africa. In: Ghazanfar S, Beentje HJ (eds) Taxonomy and ecology of African plants, their conservation and sustainable use. Royal Botanic Gardens, Kew, pp 111–129
Glen HF (1996) A description of the vegetation of Oribi Gorge Nature Reserve, Natal. National Botanical Institute, Pretoria
Huntley BJ (1965) A preliminary account of the Ngoye Forest Reserve, Zululand. J S Afr Bot 31:177–205
Jacobs TV (1996) Floristics of Transkei sandstone forests (South Africa). In: Van der Maesen LJG, Van der Burgt XM, Van Medebach de Rooy J (eds) The biodiversity of African plants. Kluwer, Dordrecht, pp 805–808
Lötter MC (2014) Indigenous forests of Mpumalanga Province (South Africa); Patterns and processes form inclusion in a systematic conservation plan. Thesis, University of Witwatersrand, Johannesburg
Lötter MC, Mucina L, Witkowski ETF (2014) Classification of the indigenous forests of Mpumalanga Province. South Africa S Afr J Bot 90:37–51
Lötter M, Burrows J, Jones K et al (2023) Historical vegetation map and red list of ecosystems assessment for Mozambique—Version 2.0—final report. USAID/SPEED+, AFD/FFEM, Maputo
Lubbe RA (1997) Vegetation and flora of the Kosi Bay Coastal Forest Reserve in Maputaland, northern KwaZulu-Natal, South Africa. Thesis, Department of Botany, University of Pretoria, Pretoria
MacDevette DR (1987) The dune woodlands of Zululand. In: von Gadow K (ed) South African forestry handbook. Southern African Institute of Forestry, Pretoria, pp 443–453
MacDevette DR (1989) A study of coastal dune gradients in Sodwana State Forest, Natal, South Africa. In: Gordon IG (ed) Natal indigenous forests. A preliminary collection of reports on indigenous forests in Natal. Natal Parks Board, Pietermaritzburg, pp 33–66
MacDevette DR (1993) The woody vegetation of the Zululand coastal dunes. In: van der Sijde HA (ed) South African forestry handbook/Suid-Afrikaanse Bosbouhandboek. Southern African Institute of Forestry, Pretoria, pp 633–637
MacDevette DR, Gordon IG (1989a) A report on a quantitative analysis of the Mlalazi forest. In: Gordon IG (ed) Natal indigenous forests. A preliminary collection of reports on indigenous forests in Natal. Natal Parks Board, Pietermaritzburg, pp 87–100
MacDevette DR, Gordon IG (1989b) A report on a quantitative analysis of the Yengele coastal forest. In: Gordon IG (ed) Natal indigenous forests. A preliminary collection of reports on indigenous forests in Natal. Natal Parks Board, Pietermaritzburg, pp 135–151
MacDevette DR, MacDevette DK (1989) A quantitative study of the Port Zimbali coastal forest. In: Gordon IG (ed) Natal indigenous forests. A preliminary collection of reports on indigenous forests in Natal. Natal Parks Board, Pietermaritzburg, pp 101–118
MacDevette DR, MacDevette DK, Gordon IG et al (1989) The floristics of the Natal indigenous forests. In: Gordon IG (ed) Natal indigenous forests. A preliminary collection of reports on indigenous forests in Natal. Natal Parks Board, Pietermaritzburg, pp 1–20
Mapaure I (2012) Determinants of vegetation composition and diversity of a moist forest-savanna boundary in south-eastern Zimbabwe. Int J Biodivers Conserv 4:584–591
Moll EJ, White F (1978) The Indian Ocean Coastal Belt. In: Werger MJA, van Bruggen AC (eds) Biogeography and ecology of southern Africa. Dr W. Junk, The Hague, pp 561–598
Mostert THC, Bredenkamp GJ, Klopper HL et al (2009) Major vegetation types of the Soutpansberg Conservancy and the Blouberg Nature Reserve, South Africa. Koedoe 50(1):32–48
Mucina L (ed) (2018) Vegetation survey and classification of subtropical forests of Southern Africa. Springer, Dordrecht
Mucina L, Rutherford MC (eds) (2006) The vegetation of South Africa, Lesotho and Swaziland. SANBI, Pretoria
Mucina L, Scott-Shaw CR, Rutherford MC et al (2006a) Indian Ocean Coastal Belt. In: Mucina L, Rutherford MC (eds) The vegetation of South Africa, Lesotho and Swaziland. SANBI, Pretoria, pp 568–583
Mucina L, Geldenhuys C.J, Rutherford MC et al (2006b) Afrotemperate, subtropical and azonal forests. In: Mucina L, Rutherford MC (eds) The vegetation of South Africa, Lesotho and Swaziland. SANBI, Pretoria, pp 584–615
Mucina L, Dold AP, Tichý L et al (2018a) Classification of the Albany Coastal Forests. In: Mucina L (ed) Vegetation survey and classification of subtropical forests of southern Africa. Springer, Dordrecht, pp 59–90
Mucina L, Abbott A, Tichý L (2018b) Classification of Pondoland scarp forests. In: Mucina L (ed) Vegetation survey and classification of subtropical forests of southern Africa. Springer, Dordrecht, pp 91–124
Mucina L, Lötter MC, Tichý L et al (2018c) Classification of the Eastern scarp forests. In: Mucina L (ed) Vegetation survey and classification of subtropical forests of southern Africa. Springer, Dordrecht, pp 125–226

Mucina L, Lötter MC, Rutherford MC et al (2022) Forest biomes of Southern Africa. New Zeal J Bot 60:377–428
Mujuru L, Kundhlande A (2007) Small-scale vegetation structure and composition of Chirinda Forest, south-east Zimbabwe. Afr J Ecol 45:624–632
Müller T (1999) The distribution, classification and conservation of rainforests in Zimbabwe. In: Timberlake J, Katiwu S (eds) African plants: biodiversity, taxonomy and uses. Royal Botanical Gardens, Kew, pp 221–235
Müller T (2006) The distribution, classification and conservation of rainforests in eastern Zimbabwe. Occasional Publications in Biodiversity No. 19. Biodiversity Foundation for Africa, Bulawayo
Nicholson HB (1982) The forests of the Umtamvuna River Reserve. Trees S Afr 34:2–10
Phillipson PB, Russell S (1988) Phytogeography of the Alexandria Forest (southeastern Cape Province). Monogr Syst Bot Missouri Bot Gard 25:661–670
Situngu S, Barker NP, Vetter S (2020) A snap-shot of domatial mite diversity of *Coffea arabica* in comparison to the adjacent Umtamvuna Forest in South Africa. Diversity 12:79. https://doi.org/10.3390/d12020079
Timberlake JR (1993) Chirinda forest: conservation of a rain forest. In: Pierce GD, Gumbo DJ (eds) The ecology and management of indigenous forests in Southern Africa. Zimbabwe Forestry Commission and SAREC, Harare
Timberlake J, Shaw P (1994) Chirinda Forest—a visitors' guide. Forestry Commission, Harare
van Wyk AE (1981) Vir 'n groter Umtamvuna-natuurreservaat. J Dendrol 1:106–108
van Wyk AE (1989) Floristics of the Natal/Pondoland sandstone forests. In: Geldenhuys CJ (ed) Biogeography of the mixed evergreen forests of southern Africa. FRD, Pretoria, pp 145–157
van Wyk GF, Everard DA (1992) Description and status of the communities occurring within Dukuduku forest. Report FOR DEA-564. Forestek, CSIR, Pretoria
van Wyk GF, Everard DA (1993a) The floristic composition and dynamics of Nxumeni and Hlabeni forests. Report FOR DEA-578. Forestek, CSIR, Pretoria
van Wyk GF, Everard DA (1993b) An analysis of the floristics and dynamics of Ngome forest. Report FOR DEA-586. Forestek, CSIR, Pretoria
van Wyk AE, Smith GF (2001) Regions of floristic endemism in southern Africa. A review with emphasis on succulents. Umdaus Press, Hatfield
van Wyk GF, Everard DA, Midgley JJ et al (1996) Classification and dynamics of a southern African subtropical coastal lowland forest. S Afr J Bot 62:133–142
Venter HJT (1976) An ecological study of the dune forest at Mapelana, Cape St. Lucia, Zululand. J S Afr Bot 42:211–230
Vijloen S (1984) Ngoye forest, precious possession of kwaZulu. Tree S Afr 1984:36–40
von Maltitz GP, van Wyk GF, Everard DA (1996) Successional pathways in disturbed coastal dune forest on the coastal dunes in north-east KwaZulu-Natal, South Africa. S Afr J Bot 62:188–195
von Maltitz G, Mucina L, Geldenhuys CJ et al (2003) Classification system for South African indigenous forests: an objective classification for the Department of Water Affairs and Forestry. Report ENV-P-C 2003-017. Environmentek CSIR, Pretoria
Wager VA (1976) Dwindling forests of the Natal coast. Natal Branch of the Wildlife Society of Southern Africa, Pietermaritzburg
Ward CJ (1980) The plant ecology of the Isipingo beach area, Natal, South Africa. Mem Bot Surv S Afr 45:1–147
Weisser PJ (1978) Conservation priorities in the dune area between Richards Bay and Mflozi Mouth based on vegetation survey. Natal Town Reg Plan Rep 38:1–64
Weisser PJ (1980) The dune forest of Maputaland. In: Bruton MN, Cooper KH (eds) Studies on the ecology of Maputaland. Rhodes University and Natal Branch of the Wildlife Society of southern Africa, Grahamstown, pp 78–90
Weisser PJ (1987) Dune vegetation between Richards Bay and Mlalazi Lagoon and its conservation priorities in relation to dune mining. Natal Town Reg Plan Suppl Rep 19:1–71
Weisser PJ, Smith ECA, Backer AP et al (1992) Flora and vegetation of the Mbonambi Beach Arcuate Scar on the Zululand dune barrier, Natal, South Africa. Bothalia 22:289–294
Whateley A, Porter RN (1983) The woody vegetation communities of the Hluhluwe-Corridor-Umfolozi Game Reserve Complex. Bothalia 14:745–758
White F (1983) Vegetation of Africa: a descriptive memoir to accompany the UNESCO/AETFAT/UNSO vegetation map of Africa. UNESCO, Paris
Wild H, Barbosa LAG (1967) Vegetation map of the Flora Zambesiaca area. M.O. Collins, Salisbury [Harare]

Fynbos: A Jewel of the Ethesial Zone

5

Ladislav Mucina, Johan Wasserman, Michael C. Rutherford and James L. Tsakalos

Abstract

The Fynbos biome (at the continental biome level) is a member of the global warm-temperate Ethesial Zone (zonobiome S1). The dominating feature of the vegetation supported by this biome are species- and endemic-rich shrublands. The occurrence of bioclimatically and pedologically marginal coastal thickets as well as inland fire-shy Cape thickets is briefly discussed. The relationship between pockets of fire-shy afrotemperate forests (belonging to another zonobiome) and the fynbos shrublands are also analysed. The CB Fynbos is here divided into 19 regional biomes, of which most are intrazonal pedobiomes, and two are extra-zonal; only four regional biomes (all characterised by renosterveld vegetation) are of true zonal nature. This chapter presents descriptions of all 19 regional biomes.

Keywords

Azonal regional biomes · Cape region · Ethesial biome · Fynbos · Palaeoclimatology · Peinobiome · Renosterveld · South Africa · Strandveld

L. Mucina (✉) · J. Wasserman · J. L. Tsakalos
Harry Butler Institute, Murdoch University,
90 South Street, Murdoch, Perth 6150, Australia
e-mail: Ladislav.Mucina@murdoch.edu.au

L. Mucina
Department of Geography and Environmental Studies, Stellenbosch University, Private Bag X1, Matieland, Stellenbosch 7602, South Africa

M. C. Rutherford
High Peak, Derbyshire, United Kingdom

M. C. Rutherford
Department of Botany and Zoology, Stellenbosch University, Private Bag X1, Matieland, Stellenbosch 7602, South Africa

J. L. Tsakalos
Plant Diversity and Ecosystems Management Unit, School of Biosciences and Veterinary Medicine, University of Camerino, Via Pontoni 5, Camerino 62032, Italy

5.1 Continental Biome Fynbos

The Fynbos Biome (defined most precisely in terms of vegetation typology by Rebelo et al. 2006) is an African instalment of the five-member warm-temperate Ethesial zonobiome (Mucina 2023). It is a complex region characterised by diverse and endemic-rich zonal and azonal shrublands, thickets, and rare open woodlands (Fig. 5.1). As such, it has been a subject of intensive taxonomic, biogeographic, vegetation-ecological, and bioclimatic research (for crucial references, see Linder 2003; Rebelo et al. 2006; Allsopp et al. 2014; Mucina 2023).

L. Mucina and M. C. Rutherford (eds.), *Zonal Biomes of Southern Africa*, Biome Ecology 2,
https://doi.org/10.1007/978-3-031-63731-5_5

Fig. 5.1 South African ethesial (mediterranean-type) woodlands and scrub. **A** Sandstone fynbos with *Leucadendron gandogeri* (Proteaceae; yellow-flowering shrub in the foreground) and pink-flowering *Phaenocoma prolifera* (Asteraceae) in the background; slopes of the Potberg Mt, De Hoop Nature Reserve, Overberg. Western Cape, South Africa. **B** Sand fynbos dominated by Restionaceae (*Willdenowia, Elegia*) and aromatic shrubs of the Rutaceae (white flowering in the foreground); above Kommetjie, Cape Peninsula, Cape Town, Western Cape, South Africa. **C** A mosaic of tussocks grasslands and scrub of the shale renosterveld dominated by Asteraceae shrubs; Potberg section of the De Hoop Nature Reserve. **D** Southwestern Strandveld in De Hoop Nature Reserve near Struisbaai, dominated by *Acmadenia mundiana, Agathosma collina, Helichrysum crispum,* and *Searsia glauca.* Photo credits: All L. Mucina

The delimitation of the Fynbos (as a biome entity) since this term was introduced (Kruger 1978; Rutherford and Westfall 1986) has never been a matter of serious controversy (but see Bergh et al. 2014). The notable examples of headaches surrounding the contents of the Fynbos Biome are the identity and position of the (1) 'strandveld' vegetation, (2) thickets embedded within the iconic Cape shrublands, and (3) the relationship of afrotemperate forests to fynbos within which they are embedded. The notions of zonality and azonality were used by some authors (e.g., Campbell 1985), yet they never used an apparent reference to Walter's zonal-azonal conceptual framework. We suggest that this conceptual flaw has been the primary source of doubts about whether renosterveld belongs to Fynbos Biome or not (see Bergh et al. 2014), despite clear definitions and delimitations of the Fynbos Biome constituents offered by Rebelo et al. (2006).

5.1.1 The Problem of the Strandveld

Strandveld (as a concept and term) was introduced (see Acocks 1988: 85 for references to two works of the previous century) to classify those Cape shrublands which occur in a habitat close to the sea yet not directly influenced by the sea breeze and salt deposition, preferring nutrient-rich soils of sedimentary origin, and characterised by fire-shy, low-grown and often succulent scrub (Rebelo et al. 2006: 75–76). Because of their physiognomic, floristic, and ecological peculiarities (compared to other traditionally recognised vegetation types of the Cape), their position in a South African biome system has always been controversial and has become food for recurring and often fruitless discussions. Acocks (1988) recognised Strandveld as one of his 'veld types' (no. 34). Low and Rebelo (1996a, b) classified the Cape strandveld (all strandveld vegetation along the West Coast and South Coast as far as the Tsitsikamma region), leaning on problematic physiognomic similarities, as part of the 'Thicket' biome with its core distribution in the Albany region of the Eastern Cape Province. Today, these thickets are called 'Albany Thicket' (Hoare et al. 2006) or Subtropical Albany Thicket (Dayaram et al. 2019). Rebelo et al. (2006: 197) reconsidered the classification of the Cape strandveld and erected an informal unit called 'Western Strandveld' to be placed within the brackets of the Fynbos Biome.

The controversial classification of the strandveld vegetation as 'Thicket Biome' and its inclusion into the Albany Thicket Biome by Low and Rebelo (1996a, b) made an ill-fated comeback in Bergh et al. (2014). These authors have attempted an interesting 'floristic' analysis involving the VegMap2006 (Mucina and Rutherford 2006) vegetation units of the Fynbos, Succulent Karoo, Albany Thicket, and a few azonal and forest units. They have scored the number of species per genus (genus being the variable of their data matrix) in these vegetation units using correspondence analysis to show that the Strandveld units (code FS in Rebelo et al. 2006) are 'Thicket'. We reanalysed the data and found that inferences from the Bergh et al. (2014) analysis are highly questionable since the first two CA axes accounted only for 5.7 and 3.2% variability when the two forest samples were removed. This points to the naïve use of genera as an informative variable in biogeographic or ecological analyses of vegetation. The horseshoe effect, as seen in Fig. 1.1 in Bergh et al. (2014), is witness to an extremely high value of genus turnover yet does not provide any tangible proof for the floristic (albeit only at the genus level) similarity between the Albany Thicket and Strandveld. Even if the horseshoes are removed, the FS (strandveld) and AT (Albany Thicket) units remain separated along axis 1 despite the poor informative content of this axis.

How to view Strandveld in the light of the criteria adopted here while aiming to find the most plausible scientific framework that reflects its ecology and evolutionary assembly? Firstly, it is clear that Strandveld, as defined and understood in the sources listed above (most notably Rebelo et al. 2006; Bergh et al. 2014), is not a bioclimatically homogenous unit. The Strandveld units of the West Coast are bioclimatically well-defined and recognised here (see Chap. 9, Sect. 9.4) as an ecotone between the Fynbos and Succulent Karoo. The remaining units of what was called 'Western Strandveld' by Rebelo et al. (2006) are azonal (more precisely intrazonal) and treated as such in our biome classification (Table 5.1).

5.1.2 The Problem of the 'Fynbos Thicket'

There are several types of thicket formations found within the Capensis. Some form narrow strips along the mountain rivers; those were classified, albeit poorly mapped, as Fynbos Riparian Thicket by Mucina et al. (2006b). Some cover broad alluvia of the Cape mountain-range piedmonts and lowlands and have been classified either as part of the Cape Alluvial Vegetation (Rebelo et al. 2006) or Cape Riparian Woodland (Mucina et al. 2022). The third group, that of the non-strandveld thickets, was recognised by Rebelo et al. (2006: 78) and called 'Fynbos Thicket'. Due to their very fine-scale occurrence (as small-sized patches are mostly less than 1 ha large), they have not been mapped by Mucina and Rutherford (2006). An attempt was made by Fourie (2011; see also Fourie et al. 2012) to develop a GIS routine which would

Table 5.1 A system of regional biomes within the CB Fynbos (GB Southern Ethesial Woodland; ZB S1 Ethesial Zone)

Rank	Code	Name of biome	Zonality	Azonality type	Climatic diagnosis
ZB	**S1**	**Ethesial Zone**	**Zonal**		**Warm-temperate ethesial; subarid, seasonal winter rainfall**
GB	***MEDs***	***Southern Ethesial Woodland***	***Zonal***		***Warm-temperate ethesial; subarid, seasonal winter rainfall***
CB	**FYN**	**Fynbos**	**Zonal**		**warm-temperate ethesial; subarid, seasonal winter rainfall**
RB	F01	Northwestern Arid Fynbos	Intrazonal	Peinobiome	
RB	F02	Southwestern Rocky Fynbos	Intrazonal	Peinobiome	
RB	F03	Western Sand Fynbos	Intrazonal	Peinobiome	
RB	F04	Western Alluvium Fynbos	Intrazonal	Peino-amphibiome	
RB	F05	Langeberg Rocky Fynbos	Intrazonal	Peinobiome	
RB	F06	Overberg Rocky Fynbos	Intrazonal	Peinobiome	
RB	F07	Southern Limestone Fynbos	Intrazonal	Peinobiome	
RB	F08	Southern Sand Fynbos	Intrazonal	Peinobiome	
RB	F09	Western Inland Renosterveld	Zonal		
RB	F10	Eastern Fynbos	Intrazonal	Peinobiome	
RB	F11	West Coast Renosterveld	Zonal		
RB	F12	South Coast Renosterveld	Zonal		
RB	F13	Karoo Renosterveld	Zonal		
RB	F14	Namaqualand Montane Cape Shrubland	Extrazonal/relict	Lithobiome	
RB	F15	Namaqualand Dune Fynbos	Extrazonal/relict	Psammobiome	
RB	F16	Inland Arid Fynbos	Intrazonal	Peinobiome	
RB	F17	Eastern Renosterveld	Zonal		
RB	F18	Southwestern Cape Strandveld	Intrazonal	Psammobiome	
RB	F19	Southeastern Cape Strandveld	Intrazonal	Psammobiome	

assist in the automatic recognition of such thickets; however, the success was only limited. Since the patches of those thickets are recognisable using the topographic imagery, many have been digitised and classified under the RB Cape Talus Forest (see Mucina et al. 2022). The separation of the talus forest, on the one hand, and fynbos thickets as Vegetation Units is still outstanding.

A tentative solution to map and characterise the Fynbos Thicket was offered by Mucina et al. (2022), who have incorporated the tall-shrub vegetation of the Cape in several RBs, namely the zonal AF1 Cape Afrotemperate Forest and zonal RBs – ATF1 Cape Talus Forest and ACF1 Cape Milkwood Woodland. Each of these RBs contains vegetation defined in physiognomic terms as either 'forest' or 'thicket' (= tall scrub). The distinction between these physiognomic structures as Vegetation Units (as defined and used by Mucina et al. 2006a) is challenging using remotely sensed data but should be attempted.

5.1.3 Fynbos Biome *Versus* Forest

Numerous afrotemperate (and riparian) forest patches are embedded within the Fynbos Biome. In the Knysna-Tsitsikamma region, the afrotemperate forest dominates (Mucina et al. 2022) and occurs in an intricate mosaic with the fynbos shrublands. It is not surprising that this spatial constellation gave birth to a series of hypotheses explaining the temporal and spatial cohabitation of vegetation structures underpinning two disparate zonobiomes—zonobiome IV (now zonobiome S1) and zonobiome 'X' (read: ten) now considered a member of the zonobiome T3 by Mucina (2023). Rebelo et al. (2006) paid close attention to the Fynbos/Forest delimitation. A new alternative approach that attempts to explain this relationship has emerged under the banner of the 'alternative stable states' (ASS) theory, summarised by Pausas and Bond 2020). The spatial and temporal cohabitation of the afrotemperate forests and fynbos shrublands has been addressed in detail by Mucina (2023). Perhaps the most important message here is that the forest–fynbos spatial relationship results from complex climatic changes characterising the glacial–interglacial climatic dynamics, including the trade-off between the contribution of winter versus summer rainfall. Quick et al. (2016) postulate that during the MIS 5d stage, increased summer rainfall under warmer conditions may have offset the increased potential evapotranspiration, allowing for the development of more extensive forests. MacPherson et al. (2019), in their study of the dynamics/stability of the forest–fynbos ecotone, also implied underlying ASS and postulated fynbos expansion during drier periods characterised by more frequent fire, while forest expansion was suggested to occur during wetter and less fire-prone periods. Contrary to this expectation, they revealed 'remarkable stability of the fynbos–forest ecotone over the past 3000 years, despite significant changes in climate, fire and land-use'. These authors proposed why their expectations were not fulfilled yet failed to address perhaps the most apparent reason—the very short time span (last 3000 years) of their observation window. Precessional forcing was found to underpin the fynbos–forest dynamics by Dupont et al. (2022), using a more convincing, 300,000-year broad time window.

In summary, fynbos shrublands and afrotemperate forests are biomes belonging to different zonobiomes and, as such, cannot be seen as elements of an alternative stable state system. Depending on the time scale, either can be seen as a relict biome. Fynbos is in a relict state in times when the climate is favourable to the expansion of forest, while forest is in a relict state when it has retreated into refugial habitats because of the expansion of fynbos. Indeed, the two biomes are involved in an intricate dynamic tango.

5.1.4 The Global Position of the CB Fynbos

The global position of the Fynbos Biome is perhaps the least problematic issue surrounding the concept of Fynbos. Its core is a representative of one of five continental biomes known as 'mediterranean-type ecosystems' (MTEs), of which the Chilean, South African, and Australian MTEs belong to the GB Southern Ethesial Woodland (a member of the zonobiome S1 Ethesial Zone). For more details, see Mucina (2023).

5.1.5 New Subdivisions of the CB Fynbos

Rutherford et al. (2006) and Rebelo et al. (2006), in two chapters of the same book, presented two alternative subdivisions of the albeit small-sized, still very complex Fynbos Biome.

Rebelo et al. (2006) classified Fynbos vegetation units into traditionally accepted informal groups of vegetation types, including Fynbos, Renosterveld, and (Western) Strandveld. The system recognises Fynbos Thicket as a unit at the same level, yet the Fynbos Thicket has not been mapped (see the section devoted to Fynbos Thicket above). In principle, the subdivision of the VUs within these informal groups follows geology.

Rutherford et al. (2006) subdivided the Fynbos into 12 bioregions by grouping geographically close (neighbouring) and ecologically similar vegetation units. However, this spatial consolidation principle was also followed in the case of two bioregions containing fynbos and renosterveld. This was done to avoid fine-scale local fragmentation of the bioregions concerned. These bioregions in Fynbos (and possibly also in some areas outside Fynbos) are more of a geo-regional or orographic nature rather than purely of an ecological nature.

This chapter presents a new classification of the CB Fynbos into 19 regional biomes (Table 5.1). Of these, most are intrazonal pedobiomes, and two are extrazonal; only four regional biomes (all characterised by renosterveld vegetation) are of true zonal nature. Most of the recognised regional biomes are identical to the bioregions, with some notable exceptions:

(1) The 'mixed' (fynbos/renosterveld) bioregions were abandoned, and the respective parts were assigned to either fynbos or renosterveld regional biomes.
(2) The broadly conceived bioregion Namaqualand Cape Shrublands, uniting fynbos-like shrublands in extrazonal positions such as the Kamiesberg (Namaqualand) and deep sandy dune systems along the Namaqualand coast, were subdivided into two regional biomes to reflect the different ecology as well as biogeography of those two disparate regions.
(3) Five of the original nine VUs representing the so-called 'Western Strandveld' were extracted and recognised as an ecotone called 'WCS West Coast Arid Strandveld' (see Sect. 4.6 for more detailed reasoning below; also see Chap. 9 in this book). The remaining strandveld units served as the basis for describing two regional biomes—RB Southwestern Cape Strandveld and RB Southeastern Cape Strandveld differing in precipitation seasonality (winter-rainfall versus year-round rainfall, respectively).
(4) The altimontane fynbos VUs have been extracted to form an ecotonal biome called 'CAP Cape Altimontane Fynbos' (see Chap. 9 in this book).
(5) Lastly, the Southern Cape Valley Thicket (Hoare et al. 2006) was extracted from the Albany Thicket Biome and used as the basis for the description of a new ecotonal biome called 'CVT Southern Cape Valley Thicket' (see Chap. 9 in this book).

5.1.6 Descriptions of the Regional Biomes of the CB Fynbos

F01 Northwestern Arid Fynbos

FFb 1 Northern Inland Shale Band (Dayaram et al. 2019)
FFh 1 Kouebokkeveld Shale Fynbos (Dayaram et al. 2019)
FFq 2 Swartruggens Quartzite Fynbos (Dayaram et al. 2019)
FFs 1 Bokkeveld Sandstone Fynbos (Dayaram et al. 2019)
FFs 2 Graafwater Sandstone Fynbos (Dayaram et al. 2019)
FFs 3 Olifants Sandstone Fynbos (Dayaram et al. 2019)
FFs 4 Cederberg Sandstone Fynbos (Dayaram et al. 2019)
FFs 5 Winterhoek Sandstone Fynbos (Dayaram et al. 2019)
FFs 6 Piketberg Sandstone Fynbos (Dayaram et al. 2019)
FFs 7 North Hex Sandstone Fynbos (Dayaram et al. 2019)
FFs 8 South Hex Sandstone Fynbos (Dayaram et al. 2019)
FFs 32 Nardouw Sandstone Fynbos (Dayaram et al. 2019)

Distribution This regional biome is a core unit of the CB Fynbos, involving several major mountain ranges trending south–north of the Cape Fold Mountains. The most prominent orographic unit is the Cederberg and the adjacent Swartruggens, Olifants and Bokkeveld Mountains. Outliers of this unit (to the west of the central mountain axis) are found around Graafwater, Winterhoek, and Piketberg. The southern extension of this regional biome is represented by the Hex River Mts. Kouebokkeveld fynbos communities on shale and alluvium are also considered part of the regional biome F01 Northwestern Arid Fynbos (Fig. 5.2).

Vegetation Physiognomy and Composition This regional azonal biome (peinobiome) contains vegetation characterised by typical fynbos physiognomy, involving fynbos scrub dominated by restios, ericas (mainly in wetter habitats), and proteaceous shrubs (*Leucadendron, Leucospermum, Protea, Serruria*). Also included are shale bands supporting fynbos scrub of different floristic composition compared to the sandstone substrate, and are embedded especially within the Cederberg Sandstone Fynbos. They support vegetation functionally, very similar to that of their sandstone counterparts.

Climate This is the driest of the core Fynbos regional biomes. Rain falls prevalently in autumn and winter (73%), yet the overall MAP is relatively low – 319 mm, although the range is considerable: 287 mm for FFs 1 Bokkeveld Sandstone Fynbos and 953 mm for FFs 8 South Hex Sandstone Fynbos (Rebelo et al. 2006: 94). Accordingly, the number of dry months for the entire area of the unit is high (11) (Fig. 5.2).

Topography, Geology and Soils Topography is mountainous, characterised by steep rugged mountain ranges and deep valleys. The dominating geology is Nardouw Sandstone, which, in places, is metamorphosed into quartzite (e.g., in Swartruggens). There are also some sandstone inselbergs, such as Piketberg and Heerenlogement. Only Kouebokkeveld—a broad valley exposing the Bokkeveld Shale—has a smooth, flat landscape appearance. The Glenrosa and Mispah soils are acidic lithosols, invariably nutrient-poor and shallow.

Sources Linder (1976), Milton (1978), Kruger (1979), van Jaarsveld (1982), Mustart et al. (1993), Taylor (1996), Lechmere-Oertel and Cowling (2000, 2001), Rebelo et al. (2006).

F02 Southwestern Rocky Fynbos

FFb 2 Western Coastal Shale Band Vegetation (Dayaram et al. 2019)
FFg 2 Boland Granite Fynbos (Dayaram et al. 2019)
FFg 3 Peninsula Granite Fynbos (Dayaram et al. 2019)
FFh 4 Breede Shale Fynbos (Dayaram et al. 2019)
FFh 5 Cape Winelands Shale Fynbos (Dayaram et al. 2019)
FFh 6 Elgin Shale Fynbos (Dayaram et al. 2019)
FFh 7 Greyton Shale Fynbos (Dayaram et al. 2019)
FFh 11 Peninsula Shale Fynbos (Dayaram et al. 2019)
FFq 4 Breede Quartzite Fynbos (Dayaram et al. 2019)
FFs 9 Peninsula Sandstone Fynbos (Dayaram et al. 2019)
FFs 10 Hawequas Sandstone Fynbos (Dayaram et al. 2019).
FFs 11 Kogelberg Sandstone Fynbos (Dayaram et al. 2019).
FFs 13 North Sonderend Sandstone Fynbos (Dayaram et al. 2019)
FFs 14 South Sonderend Sandstone Fynbos (Dayaram et al. 2019)

Distribution Most of this regional biome occurs in the southwestern corner of the Western Cape Province in the Boland and Overberg regions. It covers the roughly triangular area from near Tulbagh across the lower inland slopes of the Cape Fold Mountains to near Swellendam and converging on the eastern side of False Bay

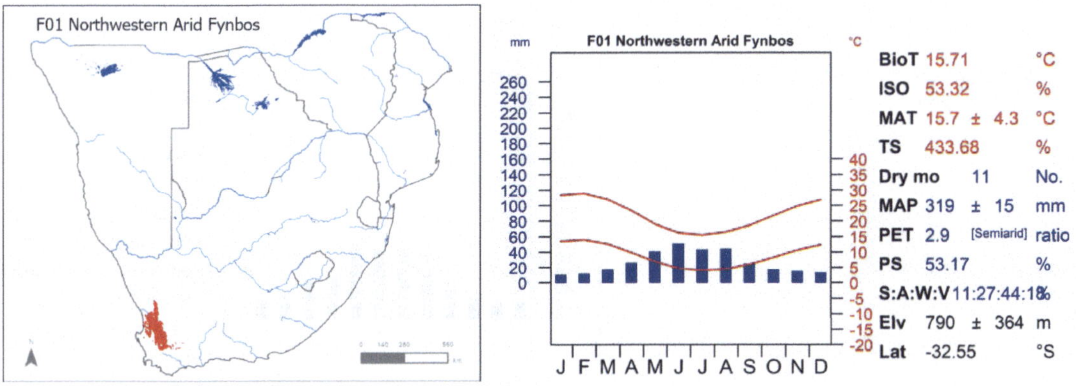

Fig. 5.2 The distribution area and the climatic diagram of the regional biome F01 Northwestern Arid Fynbos. The construction of the climatic diagrams and the distribution maps of the biomes is described in Chap. 1

at Pringle Bay on the coast. This area incorporates many prominent mountain ranges, including Groot Winterhoek, Hex River, Langeberg, Riviersonderend, Hawequas, Hottentots Holland, and Kogelberg, and intermontane valleys like the Breede River, Slanghoek, Jonkershoek, and Franschhoek Valleys, among many others. A separate fragment lies to the west on the Cape Peninsula, covering the area from Lion's Head and Table Mountain down to Cape Point. The elevation ranges from 50 to 1 800 m (Fig. 5.3).

Vegetation Physiognomy and Composition The topographical and geological heterogeneity of the region supports a diversity of fynbos structural communities. Extensive sandstone plains and slopes host generally dense and low to fairly tall ericoid shrubland with a tall proteoid element as an open overstorey or as scattered emergent shrubs. Ericaceous fynbos is often dominant at the highest parts of slopes, while proteoid and restioid fynbos are generally prevail mid-slope. Asteraceous shrublands, including waboomveld (asteraceous fynbos on talus slopes with the emergent waboom *Protea nitida*), are common at lower elevations in some areas but rare in other areas like the Kogelberg region. The sandstone fynbos in the Kogelberg boasts an astonishing level of endemism, with many endemic species occurring only in vulnerable wetland habitats or sheltered rocky areas like cliff faces. Vegetation over shale is relatively tall, dense, and generally dominated by proteoid and asteraceous elements. Ericaceous and restioid shrublands can occur in wetter areas. Renosterveld and waboomveld (at lower elevations) can be found on shale bands of this regional biome. Graminoid elements are also well represented in some areas like around Greyton and parts of the Cape Peninsula, where tall silver trees (*Leucadendron argenteum*) form a canopy over a grassy understorey in some areas. Moderately tall and dense asteraceous and proteoid shrublands (with some scattered trees) also occur over granite in parts of the Boland and Cape Peninsula regions. This granite fynbos is diverse, with various Proteaceae species being dominant on dry and moist slopes, and patches of ericaceous and restioid fynbos can be found in wetter areas. On the Cape Peninsula, dense groves of silver trees can be found on wetter slopes. Waboomveld is also widespread in granite fynbos, and granite domes sometimes host distinct patches of succulents and geophytes. The easterly patches of this vegetation that lie over quartzite ridges and patches consist of tall open shrublands within a matrix of somewhat tall and dense asteraceous, restioid and proteoid shrublands.

Climate Rainfall is strongly seasonal, peaking in winter (May to August) with MAP ranging widely (190–3000 mm; mean: 669 mm). In summer, dense clouds brought in by south-easterly trade winds deliver precipitation (in

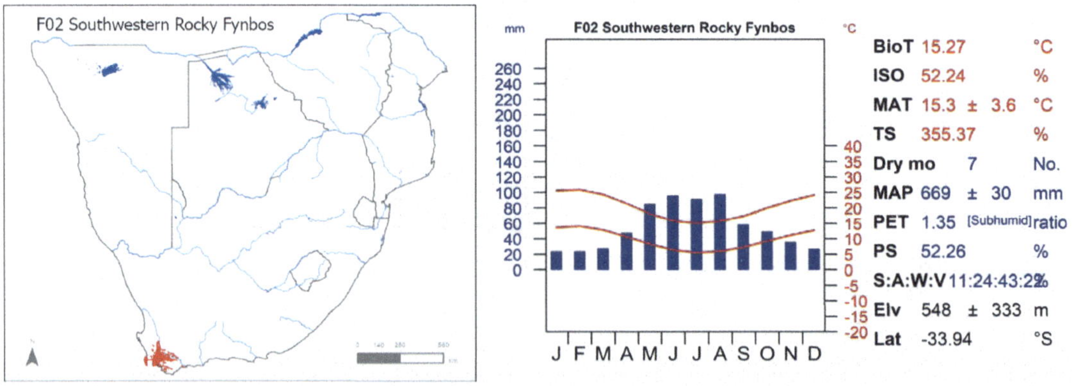

Fig. 5.3 The distribution area and the climatic diagram of the regional biome F02 Southwestern Rocky Fynbos

the form of mist) to the summits and south- and east-facing slopes of various mountains, including the Hawequas, Kogelberg, and Table Mountain. Mean daily minimum and maximum temperatures are around 4.5 °C in July and 26 °C in February, respectively. Frost generally occurs 2–3 days a year but more frequently (up to 10 days a year) on the inland mountain ranges (Fig. 5.3).

Topography, Geology and Soils The landscape is topographically and geologically complex, occurring across a variety of gentle hills, high mountains with steep slopes and river valleys. The largest area lies on acidic lithosol soils (usually apedal or Glenrosa and Mispah forms) over Table Mountain sandstones (e.g., the Riviersonderend, Kogelberg, and Hawequas mountains and the Cape Peninsula). Much of this vegetation also occurs on moist, acidic clay-loamy soils over shale derived from the Malmesbury, Bokkeveld, and Table Mountain Groups, including several shale bands at high elevations. Deep, well-drained, loamy to sandy soils over granite (with sizeable granite domes and sheets in some areas) of the Cape Granite Suite occur in the Boland and Cape Peninsula regions. To the east of the main body of this regional biome, this fynbos also occurs on a series of parallel ridges, hills and low mountains in the southern Breede River valley characterised by skeletal sandy soils over quartzites derived from the Witteberg Group.

Sources Adamson (1925, 1927, 1935), Duthie (1929), Acocks (1935), Adamson and Salter (1950), Rycroft (1953), van der Merwe (1962, 1966), Taylor (1969, 1981, 1983, 1984a, b), Andrag (1970), Boucher (1972, 1977, 1978, 1983, 1987, 1988, 1994a, 1997a, 1999a, e), Werger (1972), Werger et al. (1972a, b), Kruger (1974, 1979), Cowling (1976), McKenzie (1976), McKenzie et al. (1977), Glyphis et al. (1978), Laidler et al. (1978), Parkman (1978), Lamprecht (1979), Rutherford (1978), Durand (1981), Kathan (1981), van Wilgen and Kruger (1981, 1985), Swart (1983), Burman and Bean (1985), Chesselet (1985), McDonald (1985, 1987, 1988), Neethling (1986), Jeffery and Wilson (1987), Le Maitre (1987), Davis (1988), Lötter and van Wageningen (1988), Privett (1998), Boschoff (1989), Barnes (1992), Boucher and Stindt (1992), Greenfield (1992), Kirkwood (1992), Irwing (1992), Rule (1992), Simmons (1992, 1996), Tritton (1992), van Wilgen et al. (1992), Lotz (1993), McDonald (1993b), Rode (1994), Cowling et al. (1996), Schwilk et al. (1997), Raitt (1999), Privett et al. (2001), Sieben (2003), Agenbag and Esler (2004a, b), Agenbag et al. (2004), Holmes and Newton (2004), Sieben et al. (2004), Bean and Johns (2005), Rebelo et al. (2006), Rutherford et al. (2011), Marais (2012), Dzikiti et al. (2014), Galloway et al. (2017), Verboom et al. (2024).

F03 Western Sand Fynbos

FFd 2 Leipoldtville Sand Fynbos (Dayaram et al. 2019)
FFd 3 Hopefield Sand Fynbos (Dayaram et al. 2019)
FFd 4 Atlantis Sand Fynbos (Dayaram et al. 2019)
FFd 5 Cape Flats Sand Fynbos (Dayaram et al. 2019)
FFd 6 Hangklip Sand Fynbos (Dayaram et al. 2019)
FFd 8 Breede Sand Fynbos (Dayaram et al. 2019)

Distribution This regional biome has a fragmented distribution along the West Coast of the Western Cape Province spanning from roughly between Doringbaai and Klawer in the north to the Cape Peninsula and Hermanus in the south. The largest, most contiguous areas lie roughly between Strandfontein and Yzerfontein. They can extend far inland in some places, such as the foot of the Graafwater Mountains in the north and Piketberg further south. The largest patches south of Yzerfontein stretch along the coast to the Cape Point Nature Reserve in the south and inland across parts of the Cape Flats. To the east, small outlying areas hug the coast between Rooiels and Hermanus. Other small outlying areas can also be found inland around Eendekuil, Citrusdal, Riebeek-Kasteel, and in the Breede River Valley between Brandvlei and Robertson. The elevation range is 20–350 m (Fig. 5.4).

Vegetation Physiognomy and Composition This fynbos consists of tall ericoid shrublands with scattered tall emergent shrubs. Proteoid and restioid fynbos are most common, but asteraceous fynbos occurs in drier areas and ericaceous fynbos can be found in wetter areas around seeps near watercourses. In the driest areas on the west coast, Proteaceae are rare and Ericaceae are absent, and annuals and geophytes (often with affinities to renosterveld vegetation) can be abundant in the understorey in wet years. In the southern coastal region, east of False Bay, thicket and forest patches can be found along dune edges and other areas protected from fires.

Climate Rainfall is strongly seasonal, peaking from May to August. MAP generally ranges between 210 and 660 mm (mean: 384 mm) but can be far more significant in the southernmost coastal areas. Mist is common in winter throughout most of the region. Mean daily minimum and maximum temperatures are generally 6.6 °C in July and 30.2 °C in February, respectively. Frost occurs about 3 days a year but can be more frequent in the patches in the Breede River region (Fig. 5.4).

Topography, Geology and Soils The topography varies from flat to undulating coastal sandplains, dunes, and some inland dune fields. Soils are generally deep and comprised of acidic Tertiary sands and can be white, grey, pale yellow or reddish brown. The inland patches around the Breede River occur on dune plumes and patches with aeolian sands deposited by the river.

Sources Adamson (1927), Acocks (1935), Boucher (1972, 1974, 1977, 1978, 1983, 1986, 1987, 1989a, 1991, 1992, 1997a, 1998d, f, 1999a), Taylor (1972, 1983, 1984a, b), Boshoff (1985), Hoffman et al. (1987), Milton (1978), Lane (1980), Boucher and Shepherd (1988), Jarman (1988), Witkowski and Mitchell (1989), Musil and de Witt (1990), Cowling (1991), Boix (1992), Wood et al. (1994), Simmons (1996), McKenzie and Rebelo (1997), Privett (1998), Boucher and Rode (1999), Maze and Rebelo (1999a, b), Stofberg (2000), Helme (2002), Rebelo et al. (2006), Knauer (2011), Keith et al. (2013), Bergh et al. (2014), Dzikiti et al. (2014), Hall (2018), Hall et al. (2021), Ngwenya et al. (2023), Retief (2022).

F04 Western Alluvial Fynbos

FFa 1 Kouebokkeveld Alluvium Fynbos (Dayaram et al. 2019)
FFa 2 Breede Alluvium Fynbos (Dayaram et al. 2019)
FFa 3 Swartland Alluvium Fynbos (Dayaram et al. 2019)
FFa 4 Lourensford Alluvium Fynbos (Dayaram et al. 2019)

Distribution This regional biome has a patchy distribution on alluvial lowlands and valley bottoms in the Western Cape Province, spanning from the Kouebokkeveld and Groot Winterhoek Mountains near Porterville and across the Swartland and Drakenstein regions to the coast at Somerset West and Gordon's Bay. Large areas occur in the Breede River, Hex River, Brandwag, Paarl, Franshoek, and Slanghoek Valleys, among others. The elevation mostly ranges between 20 and 350 m, but this fynbos

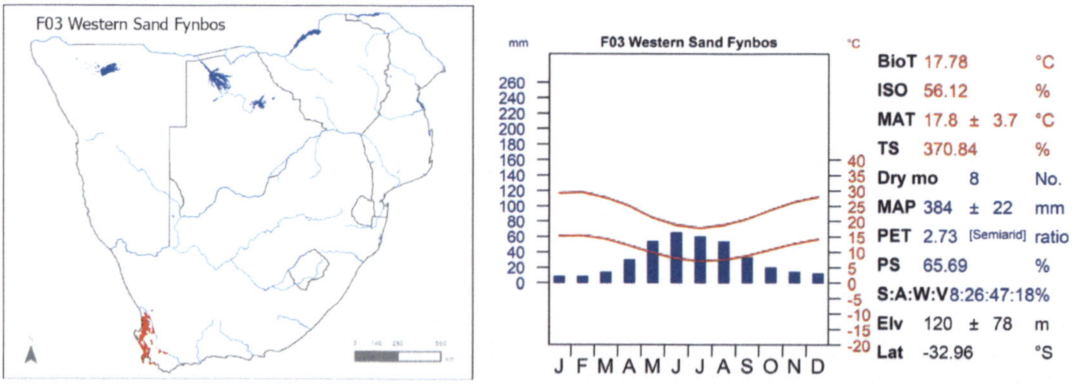

Fig. 5.4 The distribution area and the climatic diagram of the regional biome F03 Western Sand Fynbos

also occurs at the elevation span of 850–1000 m in the Kouebokkeveld region (Fig. 5.5).

Vegetation Physiognomy and Composition The vegetation is characterised by evergreen fynbos shrubland over a short but distinct graminoid layer. The shrubland structure mainly comprises tall, open emergent fynbos but can be low and moderately dense (e.g., in riparian areas). Asteraceous and proteoid fynbos are typically dominant, but ericaceous and restioid fynbos is locally dominant in some areas (particularly seeps). These shrublands transition into shale fynbos in the Kouebokkeveld and renosterveld in the Swartland and Breede River Valley.

Climate Experiences seasonal winter rainfall with MAP ranging widely (240–980 mm; mean: 558 mm) but even more variable in the Breede River and Hex River valleys, where it can be much lower. Mean daily minimum and maximum temperatures are around 4 °C in July and 30 °C in February, respectively. Rainfall and temperature maxima of the inland areas are influenced by elevation and rain shadows, while the southernmost coastal area experiences a maritime influence. Frost is generally infrequent but can occur between 10 and 30 days a year in the colder valleys of the Kouebokkeveld in the north (Fig. 5.5).

Topography, Geology and Soils Occurs on alluvium in low-lying undulating plains in mountain valleys and river basins, often on alluvial fans and near rivers and streams. Soils may be silty, sandy or loamy and contain alluvial cobbles, pebbles and gravel. The underlying formations generally consist of metasediments of the Malmesbury Group and shales of the Bokkeveld Group but also include Cape Granite and Malmesbury Group sandstones to the south.

Sources Boucher (1983, 1987, 1994a), Jones (1986), Diemer (2000), Rebelo et al. (2006), Walton (2006), Forbes et al. (2010), Dzikiti et al. (2014).

F05 Langeberg Rocky Fynbos

FFb 4 Central Coastal Shale Band Vegetation (Dayaram et al. 2019)
FFg 4 Robertson Granite Fynbos (Dayaram et al. 2019)
FFh 8 Montagu Shale Fynbos (Dayaram et al. 2019)
FFs 15 North Langeberg Sandstone Fynbos (Dayaram et al. 2019)
FFs 16 South Langeberg Sandstone Fynbos (Dayaram et al. 2019)

Distribution This regional biome forms a continuous longitudinal band across the southern slopes of the Langeberg Mountains in the Western Cape Province, spanning from Worcester in the west to Herbertsdale in the east. A relatively large area also occurs on the Langeberg's northern slopes roughly between north of Worcester and Montagu, with some patches lying nearby to the east. A separate smaller band occurs on the Aasvoëlberg hills

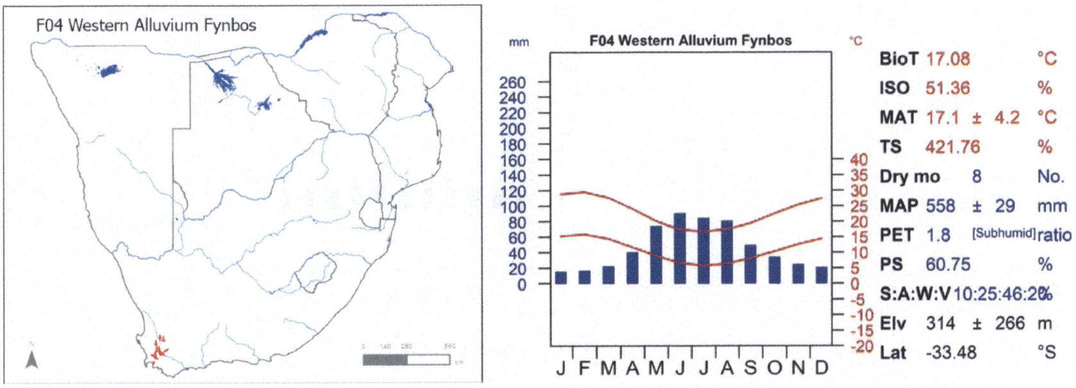

Fig. 5.5 The distribution area and the climatic diagram of the regional biome F04 Western Alluvium Fynbos

between Albertinia and Mossel Bay. The elevation ranges broadly from 50 to 1800 m, with the lowest points generally occurring in the south-eastern coastal areas (Fig. 5.6).

Vegetation Physiognomy and Composition North-facing sandstone slopes are dominated by proteoid and restioid fynbos, but ericaceous fynbos can be found at higher elevations and asteraceous fynbos on lower slopes. South-facing sandstone slopes are dominated by ericaceous and restioid fynbos at higher elevations and tall-grown proteoid fynbos scrub on the middle and lower slopes. Also, afrotemperate forests can be found in ravines protected from fires. Scrub fynbos and restioid shrublands occur where groundwater is abundant. Dense proteoid fynbos and graminoid fynbos can be found in some confined areas overlying granite on south-facing foothills. Fragmented areas overlying shale are generally characterised by tall, dense proteoid and asteraceous fynbos with scattered emergent proteoids and localised areas of waboomveld. In contrast, shale bands support a mixture of fynbos communities.

Climate The westernmost parts of the regional biome experience a winter-rainfall regime (peaking from May to August), while relatively even seasonality with a slight low over summer months is predominant to the east. MAP ranges from 240 to 1560 mm (mean: 392 mm) with north-facing slopes generally being notably drier. At high elevations in some areas, rainfall is supplemented by mist precipitation brought in by southeasterly clouds. Mean daily minimum and maximum temperatures are approximately 4 °C in July and 28 °C in January, respectively. Frost generally occurs between 3 and 20 days a year but can occur up to 30 days a year in some areas lying inland of the Langeberg (Fig. 5.6).

Topography, Geology and Soils Most of this fynbos occurs on gentle to steep slopes of the Langeberg Mountains with acidic lithosol soils over sandstones of the Table Mountain Group. On some steep southern foothills, there are also patches of deep loamy sands over Cape Granite. Shale patches with acidic clay-loamy soils (Glenrosa and Mispah forms) derived from the Bokkeveld Group are scattered across uplands and steep foothills on the northern side of the Langeberg, and a narrow shale band with clay derived from the Cedarberg Formation spans the length of the Langeberg. Similar shale bands occur on Grootberg and Amandelbosberg north of Heidelberg and Riversdale and hills west of Mossel Bay.

Sources Muir (1929), Boucher (1972, 1977, 1978, 1988), Kruger (1979), Rebelo et al. (1991, 2006), McDonald (1993a, b, c, 1995, 1999), McDonald et al. (1996), van Blerk et al. (2021).

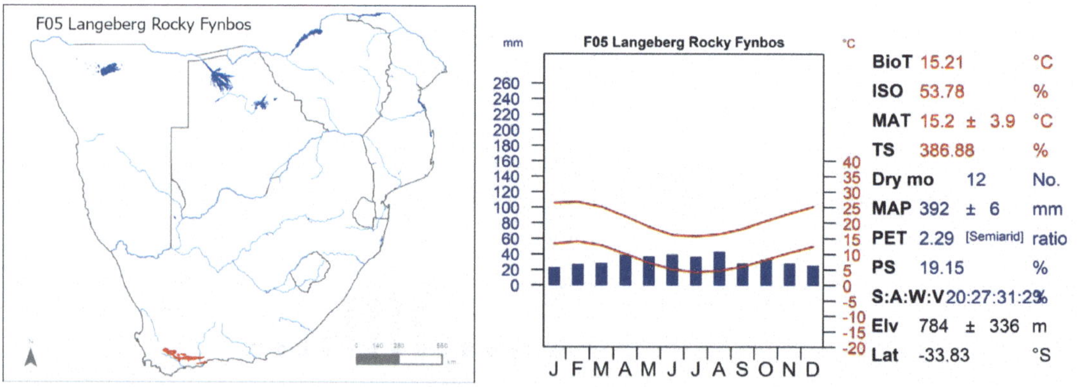

Fig. 5.6 The distribution area and the climatic diagram of the regional biome F05 Langeberg Rocky Fynbos

F06 Overberg Rocky Fynbos

FFc 1 Swellendam Silcrete Fynbos (Dayaram et al. 2019)
FFf 1 Elim Ferricrete Fynbos (Dayaram et al. 2019)
FFf 2 Potberg Ferricrete Fynbos (Dayaram et al. 2019)
FFs 12 Overberg Sandstone Fynbos (Dayaram et al. 2019)
FFs 17 Potberg Sandstone Fynbos (Dayaram et al. 2019)

Distribution This regional biome lies fragmented on the coastal forelands of the Overberg and Garden Route regions of the Western Cape Province. The largest area lies between Bot River and Hawston in the west and Bredasdorp and Arniston in the east. A small outlying patch occurs on the Swartberg Mountain north of Caledon. To the northeast, an interrupted longitudinal band occurs on the southern foothills of the Langeberg Mountains from near Swellendam to the north of Klein Brakrivier. Another small patch also occurs on Potberg Mountain near Cape Infanta. The elevation typically ranges from sea level to 400 m, although there are some higher peaks like Potberg (611 m) and Babilonstoringberg (1167 m) (Fig. 5.7).

Vegetation Physiognomy and Composition The vegetation is dense and relatively tall, or low with scattered tall shrubs. A mixture of asteraceous, ericaceous, restioid and proteoid fynbos occur, with dominant elements varying across edaphic gradients (particularly pH, soil depth and rock cover). The Agulhas Plain boasts abundant endemic species, especially Proteaceae. There are both fynbos and renosterveld floristic elements on the Langeberg, and graminoid fynbos can be found on summits and disturbed northern slopes, which can be transformed into species-poor renosterveld by overgrazing. Large areas have been transformed for cultivation, particularly in the Agulhas Plain, Potberg, and Langeberg regions, leaving just remnant patches of this fynbos vegetation.

Climate MAP ranges from 300 to 860 mm (mean: 517 mm), peaking between May and August in the large area in the Overberg but occurring relatively evenly throughout the year (with a slight summer low) around the Langeberg and Potberg to the east. Mist from southeasterly clouds provide additional precipitation to some higher-elevation slopes in summer. Mean daily minimum and maximum temperatures are about 5.5 °C in July and 26 °C in January, respectively. Frost occurs about 2–3 days a year (Fig. 5.7).

Topography, Geology and Soils Most of this regional biome occurs across undulating plains, hills and low mountains. The geology is comprised of a mixture of lithosol and colluvial soils derived from sandstones of the Table Mountain Group, shales of the Bokkeveld Group, Cape Granite, ferricrete, and silcrete (iron- and silica-cemented hardpans, respectively). Shallow and dry loamy sand over silcrete and conglomerate occurs on the southern foothills of the Langeberg Mountains.

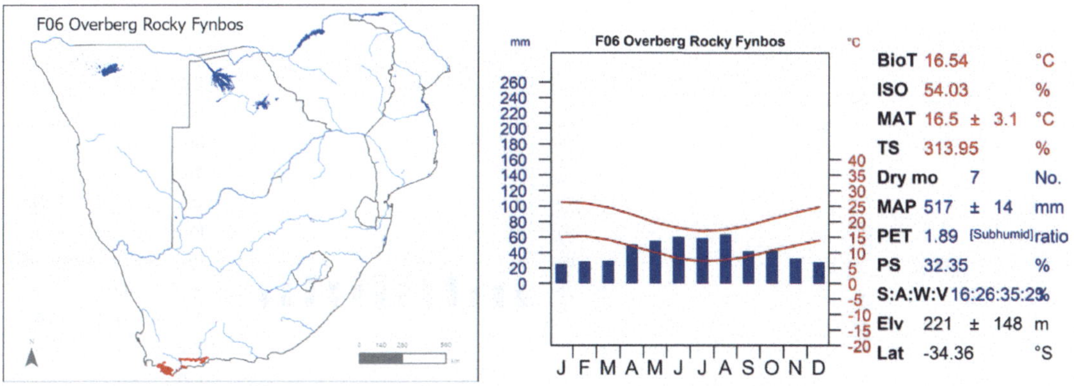

Fig. 5.7 The distribution area and the climatic diagram of the regional biome F06 Overberg Rocky Fynbos

Sources Grobler and Marais (1967), Taylor (1972), Cowling et al. (1988), Thwaites and Cowling (1988), Rebelo et al. (1991), de Lange (1992), McDonald (1993b), Richards (1994), Richards et al. (1995, 1997a, b), Mustart et al. (1997), Rebelo et al. (2006), Mergili and Privett (2008), Joubert et al. (2009), Mills et al. (2012).

F07 Southern Limestone Fynbos

FFl 1 Agulhas Limestone Fynbos (Dayaram et al. 2019)
FFl 2 De Hoop Limestone Fynbos (Dayaram et al. 2019)
FFl 3 Canca Limestone Fynbos (Dayaram et al. 2019)

Distribution This regional biome lies on limestone coastal forelands of the Western Cape Province, spanning from Hermanus in the west to Mossel Bay in the east. The western portion (between Hermanus and Arniston) is fragmented, while larger, relatively uninterrupted portions occur to the east. Some small outlying patches occur towards Groot Brak east of Mossel Bay. The elevation ranges between 20 to 400 m, although some patches occur up to 500 m (Fig. 5.8).

Vegetation Physiognomy and Composition The vegetation mainly comprises low and somewhat dense proteoid and asteraceous fynbos with some tall emergent proteoids. However, restioid fynbos occurs on sandy areas, skeletal soils, limestone pavements, and low-lying waterlogged areas. The easternmost corner (around Vleesbaai) is less species-rich, with no proteoid overstorey but a predominance of Rutaceae and notably more geophytes and succulents, marking the transition of limestone fynbos to succulent-rich thicket.

Climate Rainfall generally peaks between April and October, although the eastern parts of this fynbos extend into the seasonally-even rainfall region, with MAP ranging from 250 to 660 mm (mean: 458 mm). Temperatures are mild, with mean daily minimum and maximum temperatures of about 6 °C in July and 25.5 °C in February, respectively. Frost is rare, occurring approximately 3 days a year (Fig. 5.8).

Topography, Geology and Soils The topography is characterised by low undulating hills on inland plains and undulating plains on coastal forelands with shallow regic grey sands (neutral to alkaline) and Glenrosa and Mispah soils over limestone of the Bredasdorp Group. Distinct and sometimes extensive (> 2 km long), karstic valleys and sinkhole depressions have formed in the limestone in some areas between Struisbaai and Cape Infanta.

Sources Muir (1929), Boucher (1997b, 1998c1972, 1977, 1978, 1994b, 1995a), van der Merwe (1977a), Kruger (1979), Uys (1983), Cowling et al. (1988), Thwaites and Cowling (1988), Cowling (1990a, b), Rebelo et al. (1991, 2006), Mustart and Cowling (1993a, b), Heydenrych (1994), Richards (1994), Boucher

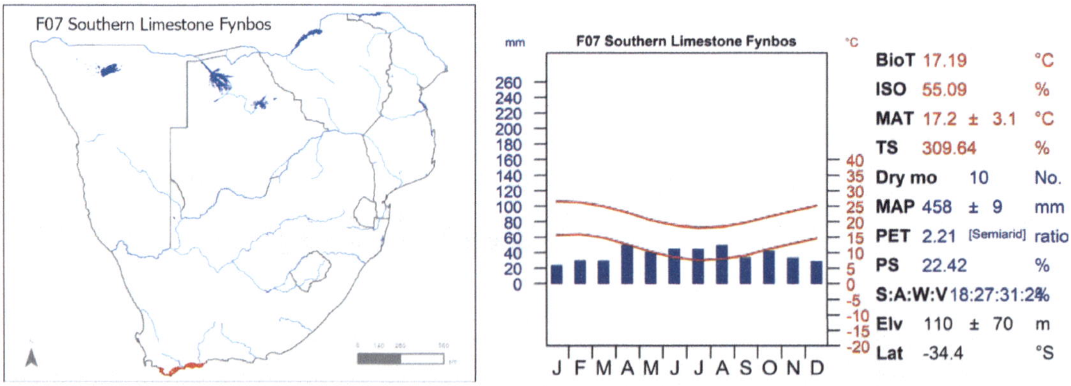

Fig. 5.8 The distribution area and the climatic diagram of the regional biome F07 Southern Limestone Fynbos

and Rode (1995a, b), Richards et al. (1995, 1997a, b), Willis and Cowling (1996), Willis et al. (1996), Mustart et al. (1997), Mergili and Privett (2008), Mills et al. (2012).

F08 Southern Sand Fynbos

FFd 7 Agulhas Sand Fynbos (Dayaram et al. 2019)
FFd 9 Albertinia Sand Fynbos (Dayaram et al. 2019)
FFd 10 Knysna Sand Fynbos (Dayaram et al. 2019)

Distribution This regional biome occurs as a fragmented longitudinal band on low coastal plains in the Western Cape Province between the east of Gansbaai and the Robberg Peninsula at Plettenberg Bay. Isolated patches are concentrated in areas between Gansbaai and Arniston, Potberg and the Gouritz River as well as between Wilderness and Robberg. The elevation range is 20–300 m (Fig. 5.9).

Vegetation Physiognomy and Composition The vegetation typically consists of a dense, medium to fairly tall layer of ericoid shrubs and hemicryptophytes, often with a sparse to open layer of emergent shrubs. Structurally, the communities are proteoid and restioid (especially along watercourses and coastal fringes) fynbos. However, some asteraceous fynbos can also be found (e.g., areas in the Agulhas Plain where sand fynbos grades into ferricrete fynbos). The dominant fynbos components are determined by edaphic characteristics like pH, soil depth and rock cover. Thicket can occur in fire-protected or perennially wet areas, particularly embedded in sand and limestone fynbos between Potberg and the Gouritz River.

Climate Rainfall occurs relatively evenly throughout the year with a slight low over summer. However, there is a slight peak in winter in the westernmost areas, no distinct peak in the central areas, and a slight peak in autumn and spring in the easternmost areas. Overall rainfall peaks between April and October with an approximate MAP of 475 mm, but this can range from 230–660 mm between Gansbaai and the Gouritz River and 670–1090 mm to the east of Wilderness. Mean daily minimum and maximum temperatures are about 6.4 °C in July and 27.3 °C in January–February, respectively. Frost can occur up to 3 days a year (Fig. 5.9).

Topography, Geology and Soils Patches mainly occur on low coastal plains and undulating hills. Between Potberg and the Gouritz River, these patches are always bordered by limestone fynbos, often occurring in extensive depressions, dune slacks, and at lower slopes of valleys. Soils mainly comprise deep neutral to acid sands, generally over limestone of the Bredasdorp Formation.

Sources Muir (1929), Taylor (1970b), Van der Merwe (1977a), Drews (1980a), Cowling et al. (1988), Thwaites and Cowling (1988), Rebelo et al. (1991), Boucher and Rode (1995a, b), Boucher (1997b, 1998c, 1995a, b), Richards

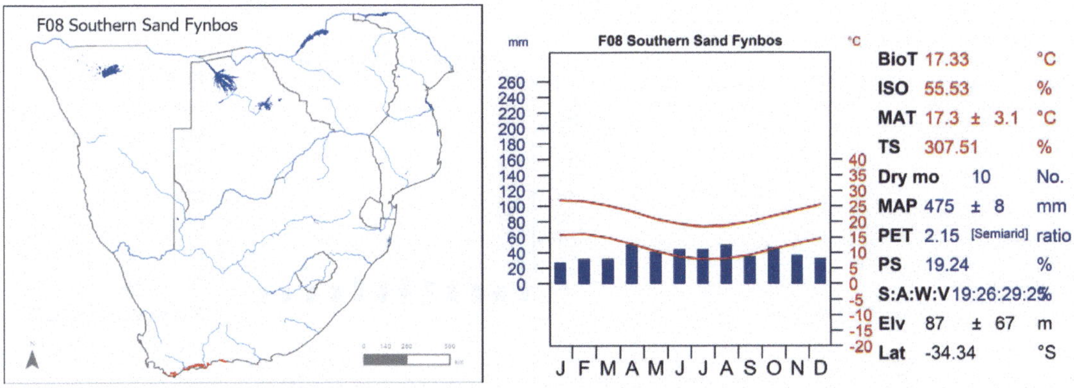

Fig. 5.9 The distribution area and the climatic diagram of the regional biome F08 Southern Sand Fynbos

et al. (1995, 1997a, b), Mustart et al. (1997), Rebelo et al. (2006), Mergili and Privett (2008), February et al. (2021).

F09 Western Inland Renosterveld

FRl 1 Kango Limestone Renosterveld (Dayaram et al. 2019)
FRs 4 Ceres Shale Renosterveld (Dayaram et al. 2019)
FRs 6 Matjiesfontein Shale Renosterveld (Dayaram et al. 2019)
FRs 7 Montagu Shale Renosterveld (Dayaram et al. 2019)
FRs 15 Swartberg Shale Renosterveld (Dayaram et al. 2019)

Distribution This renosterveld biome has a patchy distribution across the Klein Karoo (Western Cape Province) from the Warm Bokkeveld Valley near Ceres in the west to the Vanwyksdorp and Calitzdorp region in the east. This includes parts of the Kouebokkeveld, Langeberg, Swartberg, Gamkaberg, Anysberg, Rooiberg, Langeberg Mts, and associated ridges, slopes, foothills, and intermontane valleys and plains. Many of the northernmost patches border the Succulent Karoo biome. The elevation range spans 200–1300 m (Fig. 5.10).

Vegetation Physiognomy and Composition Vegetation mainly consists of open to somewhat dense, generally low (but occasionally tall) cupressoid and leptophyllous shrublands dominated by renosterbos (*Dicerothamnos rhinocerotis*). Dense succulent thickets can be found in some ravines and steep rocky slopes protected from fires. Heuweltjies occur in some places but generally at low densities.

Climate Rainfall occurs relatively evenly throughout the year, with a slight peak in March and from May to August and a minor low in January. MAP ranges from 140 to 720 mm (mean: 311 mm). Mean daily minimum and maximum temperatures are about 2.3 °C in July and 31 °C in January, respectively. Frost occurs for 10–40 days a year (Fig. 5.10).

Topography, Geology and Soils The topography is highly variable, covering undulating plains, hills, slopes, low mountains, mid-elevation plateaus and broad intermontane valleys. The soils (Glenrosa and Mispah forms dominant) consist of clays and loams derived from shale (and sandstone in some areas) from various groups of the Cape Supergroup (Table Mountain, Witteberg, and Bokkeveld Groups) and Karoo Supergroup (Dwyka Group). In the Swartberg region north of Oudtshoorn, the clays and loams are derived from the shale of the Cango Caves Group and limestone of the Namibian Erathem, with some areas overlying clastic sediments from the Uitenhage Group. Extensive areas with quartzite pebble fields can be found in some parts.

Sources Muir (1929), Moffett and Deacon (1977), Acocks (1988), Taylor (1996), Vlok and Euston-Brown (2002), Rebelo et al. (2006), van Blerk et al. (2021).

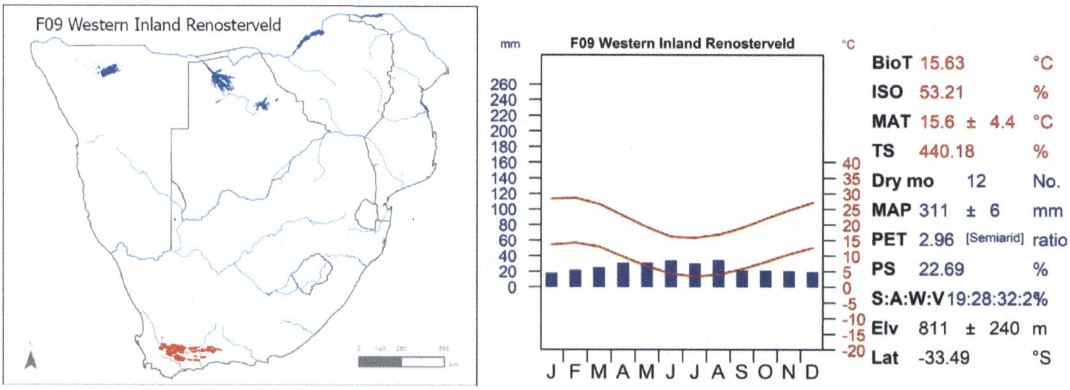

Fig. 5.10 The distribution area and the climatic diagram of the regional biome F09 Western Inland Renosterveld

F10 Eastern Fynbos

FFb 5 Eastern Inland Shale Band Vegetation (Dayaram et al. 2019)
FFb 6 Eastern Coastal Shale Band Vegetation (Dayaram et al. 2019)
FFg 5 Garden Route Granite Fynbos (Dayaram et al. 2019)
FFh 9 Garden Route Shale Fynbos (Dayaram et al. 2019)
FFs 18 North Outeniqua Sandstone Fynbos (Dayaram et al. 2019)
FFs 19 South Outeniqua Sandstone Fynbos (Dayaram et al. 2019)
FFs 20 Tsitsikamma Sandstone Fynbos (Dayaram et al. 2019)
FFs 27 Kouga Sandstone Fynbos (Dayaram et al. 2019)
FFs 28 Kouga Grassy Sandstone Fynbos (Dayaram et al. 2019)
FFs 29 Algoa Sandstone Fynbos (Dayaram et al. 2019)
FFt 2 Loerie Conglomerate Fynbos (Dayaram et al. 2019)

Distribution This regional biome spans across the mountainous coastal Garden Route region in the Western Cape Province toward Algoa Bay in the Eastern Cape Province. It stretches from the Outeniqua, Baviaanskloof, and Groot Winterhoek Mts inland toward the coast at Klein Brak in the west and Gqeberha (formerly Port Elizabeth) in the east. It covers the Kouga, Langkloof, Tsitsikamma, Suuranysberge, Kareedouw, Elandsberge and Van Stadensberg Mts, among others. To the west of the main body, an outlying patch occurs on the coastal foothills of Langeberg Mountains to the north of Heidelberg and Riversdale region. The elevation range is extensive: 0–1758 m (Fig. 5.11).

Vegetation Physiognomy and Composition The sandstone mountain slopes, valleys and plains in the Garden Route region support open to somewhat dense fynbos with a moderately tall and dense understorey of shrubs, often with tall emergent Proteaceae shrubs scattered throughout. Proteoid and restioid elements are mostly dominant, but ericaceous fynbos can be found at higher elevations and asteraceous (and sometimes graminoid) fynbos at lower latitudes. Scrub fynbos and thicket can be found in riparian (or otherwise wet) areas. The sandstone mountain slopes to the north and northeast of the Tsitsikamma Mountains (e.g., Kouga, Baviaanskloof, and Groot Winterhoek) host low fynbos at high elevations, typical sandstone fynbos with three layers at intermediate elevations, and low grassy shrublands or grasslands with scattered shrubs at lower elevations. Shale bands throughout the region support a variety of renosterveld, thicket and fynbos (mostly graminoid fynbos, but other structural types are also present). Shales on the coastal forelands overlying shale support tall, dense shrublands with a predominance of proteoid and ericaceous fynbos in wetter areas and graminoid fynbos in drier areas. Most shales in this region support afrotemperate forests, but fynbos occurs on extensive flat areas that are more fire-prone. Coastal areas overlying granite support dense shrubby grasslands with a prevalence of proteoid and graminoid fynbos elements and ericaceous fynbos elements occurring in seeps. The shales and conglomerates in the Gamtoos Valley host low shrubland or grassland with scattered tall shrubs with asteraceous, graminoid and proteoid elements with a diversity of geophytes and succulents, and have broad transitional communities grading into neighbouring regional biome F17 Eastern Renosterveld. Pine plantations occur across large areas in the Garden Route, and relatively frequent prescribed burns in the surrounding fynbos have altered community structure and likely recruitment dynamics.

Climate Rainfall occurs year-round, with a slight bimodal peak in autumn and spring and a slight low in winter. MAP ranges from 240–1170 mm (mean: 563 mm), with the southwestern- and easternmost areas generally receiving the least rainfall. Daily minimum and maximum temperatures are about 2.9 °C in July and 27.8 °C in January and February, respectively, with the areas furthest inland tending to have slightly hotter summers and colder winters. Frost generally occurs for just 2–3 days a year in the coastal forelands and for 2–10 days throughout most of the inland mountainous areas but can occur for up to 40 days a year at the highest elevations. Unlike the rest of the Fynbos Biome, fires in this region are aseasonal (Fig. 5.11).

Topography, Geology and Soils Most of this regional biome occurs on gentle to steep mountain slopes, intramontane valleys, and flat to

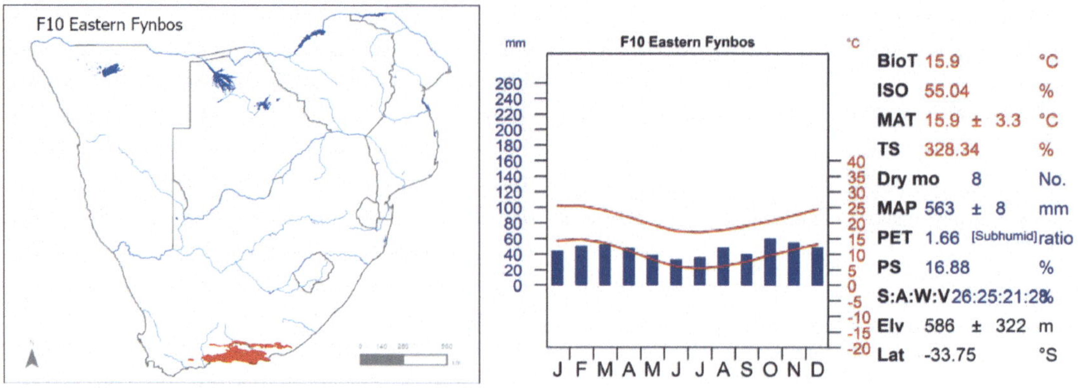

Fig. 5.11 The distribution area and the climatic diagram of the regional biome F10 Eastern Fynbos

undulating plains with acidic lithosols soils over sandstones of the Table Mountain Group as well as quartzite sandstones of the Witteberg Group in the eastern regions. Long and narrow (80–200 m wide) longitudinal shale bands belonging to the Cedarberg Formation with clay soils are common across the region's various mountain ranges and coastal plains. There are areas of acidic clay-loam over shale from the Kaaimans Group (in the west) and the Ecca Group (in the east) spanning the undulating coastal forelands, as well as areas with deep duplex soils over Cape Granite (between Klein Brak and Karatara). To the east, the undulating plains in the Gamtoos Valley have moist acidic clay-loam and lithosol soils derived from shales and conglomerates of the Uitenhage Group.

Sources Taylor (1970b), Bond (1978a, b, 1981), Cameron (1980), Drews (1980a, b), Cowling and Campbell (1983a, b, 1984), Cowling (1984), van Daalen (1984), Campbell (1985), Bond et al. (1988), Hanekom et al. (1989), Rebelo et al. (1991, 2006), Euston-Brown (1995), Vermeulen (1995), Boshoff et al. (2000), Hoare et al. (2000), Vlok and Euston-Brown (2002), Kraaij et al. (2011, 2012a, b, 2013, 2017, 2018), Bergh et al. (2014), Aiello-Lammens et al. (2017), Baard et al. (2023), Smit et al. (2024).

F11 West Coast Renosterveld

FRa 2 Swartland Alluvium Renosterveld (Dayaram et al. 2019)
FRc 1 Swartland Silcrete Renosterveld (Dayaram et al. 2019)
FRg 2 Swartland Granite Renosterveld (Dayaram et al. 2019)
FRs 9 Swartland Shale Renosterveld (Dayaram et al. 2019)
FRs 10 Peninsula Shale Renosterveld (Dayaram et al. 2019)
FRs 20 Citrusdal Shale Renosterveld (Dayaram et al. 2019)

Distribution This regional biome covers a large area of the low-lying Swartland and Boland regions in the Western Cape Province from Redelinghuys and Citrusdal in the north to Strand in the south. Some outlying patches occur near Graafwater in the north, between Citrusdal and Clanwilliam in the northeast, and on the northern slopes of Table Mountain, Devil's Peak and Signal Hill to the southwest. The elevations range between the sea level and 350 m (Fig. 5.12).

Vegetation Physiognomy and Composition These are open, low to reasonably tall leptophyllous shrublands, microphyllous renosterbos-dominated shrublands, and mosaics of grasslands and herblands. Succulents can be abundant over silcrete patches, and geophytes are often abundant around granite outcrops. Heuweltjies are

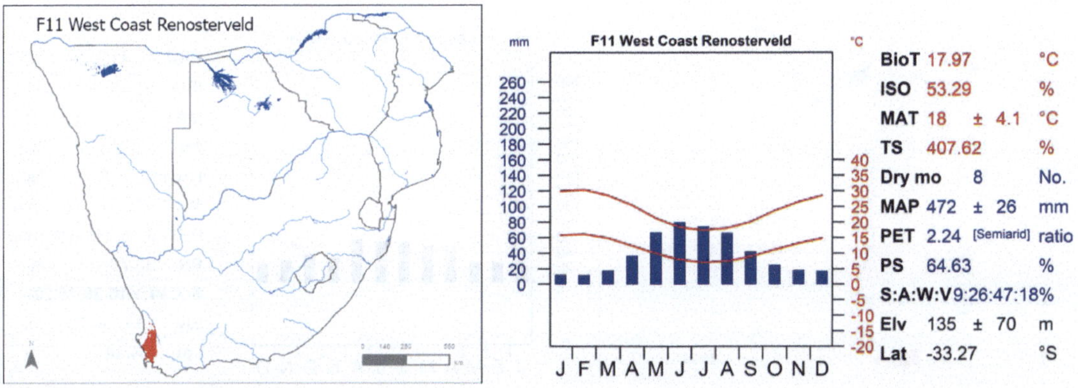

Fig. 5.12 The distribution area and the climatic diagram of the regional biome F11 West Coast Renosterveld

common throughout much of the region, especially over shale, and often support stunted trees and thickets. Large areas of this region have been transformed for agriculture and development, leaving the native vegetation highly fragmented and often disturbed in some areas.

Climate This regional biome experiences strong seasonal winter rainfall peaking from May to August, with a MAP of 250–870 mm (mean: 472 mm). Mist is common in winter. The mountain slopes around the Cape Town city bowl are notably wetter than the rest of the region. Mean daily minimum and maximum temperatures are about 6.3 °C in July and 29.6 °C in February, respectively. Frost occurs 2–4 days a year (Fig. 5.12).

Topography, Geology and Soils This renosterveld occurs on undulating plains, lower mountain slopes and lowlands. Most of the region lies over shales of the Malmesbury Group with shallow clay soils. A substantial portion also occurs on coarse sandy to loamy soils over Cape Granite. Narrow riverine plains in southern Swartland have fine silty and sandy alluvial soils derived from Cape Granite. Remnant patches of silcrete overlying shale of the Malmesbury Group and Cape Granite with shallow soils are scattered throughout the region.

Sources Adamson (1927), Acocks (1935), Linder (1976), Boucher (1977, 1978, 1980, 1983, 1987, 1989a, 1999b), Boucher and Moll (1981), Paterson (1982), Jacobs (1984), Jones (1986), Lötter and van Wageningen (1988), Landman and Nel (1989), Joubert (1991), Joubert and Moll (1992), Duvenhage (1993), Boucher and Rode (1994, 1999), Boucher (1972), Britton and Jackelman (1995, 1996), Nel (1995), Cowling et al. (1996), Simmons (1996), von Hase et al. (2003), Musil et al. (2005), Walton (2006), Marais (2012), Cousins et al. (2017), Verboom et al. (2024).

F12 South Coast Renosterveld

FRa 1 Breede Alluvium Renosterveld (Dayaram et al. 2019)
FRc 2 Ruens Silcrete Renosterveld (Dayaram et al. 2019)
FRg 3 Robertson Granite Renosterveld (Dayaram et al. 2019)
FRs 8 Breede Shale Renosterveld (Dayaram et al. 2019)
FRs 11 Western Ruens Shale Renosterveld (Dayaram et al. 2019)
FRs 12 Central Ruens Shale Renosterveld (Dayaram et al. 2019)
FRs 13 Eastern Ruens Shale Renosterveld (Dayaram et al. 2019)
FRs 14 Mossel Bay Shale Renosterveld (Dayaram et al. 2019)

Distribution This regional biome covers a large area on the coastal forelands of the Langeberg and Riviersonderend Mts in the Western Cape Province. It spans from between Tulbagh and Bot River in the west to the Mossel Bay region in the east. It is bounded in the south by the

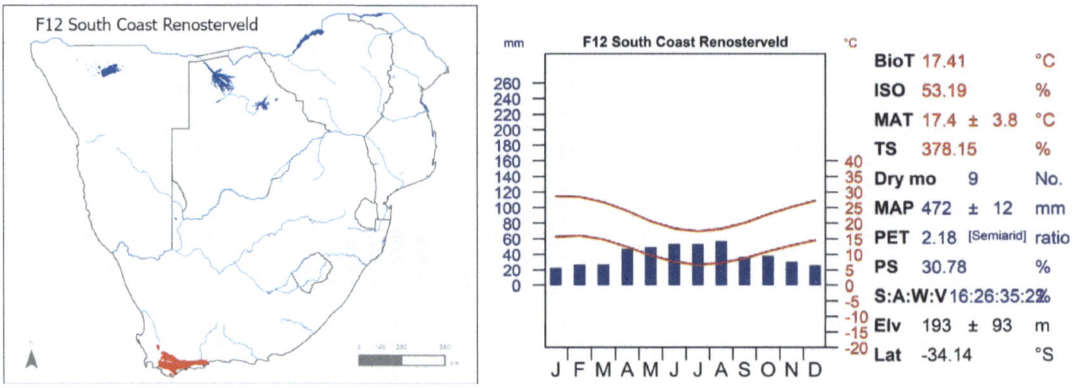

Fig. 5.13 The distribution area and the climatic diagram of the regional biome F12 South Coast Renosterveld

regional biomes F06 Overberg Rocky Fynbos and F07 Southern Limestone Fynbos. Elevation ranges from 20 to 450 m but can reach up to 650–850 m on mountain slopes in the Breede River Valley (Fig. 5.13).

Vegetation Physiognomy and Composition The vegetation comprises open to fairly dense and low to somewhat tall grassy shrublands, usually dominated by renosterbos and with various cupressoid and small-leaved shrubs. Succulents are often abundant on the fragmented silcrete patches. The granite dome slopes of Tierberg also support many succulents and some small trees. Heuweltjies are present on shales throughout many areas. However, they are generally inconspicuous, except in the Breede River Valley, where they support succulent shrubs or bush clumps in dry and wet habitats, respectively, and in the easternmost region, where they support thicket clumps. Thicket patches are common in the easternmost region, likely due to the prevalence of fire-safe habitats across the rugged landscape. Large areas of this region have been transformed by agriculture, with some remnants of the vegetation remaining only on steep slopes and in national parks and nature reserves.

Climate Rainfall generally peaks between May and August throughout most of the region but becomes more even to the east. MAP ranges from 170 to 770 mm (mean: 472 mm), with the least occurring in the Breede River Valley (between Worcester and Ashton). Mean daily minimum and maximum temperatures are, respectively, about 4 °C in July and 29.8 °C in February, with the extremes being more prominent in the Breede River Valley. Frost occurrence is roughly 3 days a year throughout most of the region but more frequent in the Breede River Valley (up to 10 days a year) (Fig. 5.13).

Topography, Geology and Soils Most of the region occurs on flat to moderately undulating plains, hills, pans, and some lower mountain slopes with shallow clays and loams (often Glenrosa and Mispah forms) over shale of the Bokkeveld Group and some localised patches over Uitenhage, Witteberg and Malmesbury Group shales. Highly fragmented patches of shallow soils covering silcrete caps over shales of the Bokkeveld Group are scattered throughout the region. They are especially concentrated in the area south of Swellendam and Riversdale. Large patches are generally associated with drainage systems, with substantial areas occurring around the Bot, Breede, Duiwenhoks, and Goukou Rivers. Small patches dot the uplands and summits of hills and plains. On the bottomlands of the Breede River Valley, there are river terraces and alluvial fans with gravel- and cobble-rich fine loamy sands over various types of rock from the Cape and Karoo Supergroups and Uitenhage Group. A small area also occurs over shallow loamy soils on the slopes of a Cape Granite dome at Tierberg near Robertson.

Sources Muir (1929), Olivier (1966), Grobler and Marais (1967), Taylor (1970a), Mustart et al. (1997), Norton (1977), van der Merwe (1977b), Boucher and Moll (1981), Chesselet (1985), Cowling et al. (1988), Boschoff (1989), Smitheman and Perry (1990), Wood (1990), Rebelo et al. (1991, 2006), Boucher (1999d), Kemper et al. (1999), Schmiedel (2002), Raitt (2005), Schmiedel and Mucina (2006), Mills et al. (2013), Kuhel (2023).

F13 Karoo Renosterveld

FRd 1 Nieuwoudtville-Roggeveld Dolerite Renosterveld (Dayaram et al. 2019)
FRd 2 Hantam Plateau Dolerite Renosterveld (Dayaram et al. 2019)
FRs 1 Vanrhynsdorp Shale Renosterveld (Dayaram et al. 2019)
FRs 2 Nieuwoudtville Shale Renosterveld (Dayaram et al. 2019)
FRs 3 Roggeveld Shale Renosterveld (Dayaram et al. 2019)
FRs 5 Central Mountain Shale Renosterveld (Dayaram et al. 2019)

Distribution This renosterveld lies in the southwestern corner of the Great Karoo in the Northern and Western Cape Provinces and constitutes the furthest inland distribution of the Fynbos Biome. Most of this regional biome occurs on the Great Escarpment, separate from the main body of the biome and embedded in the continental biomes Nama-Karoo and Succulent Karoo, where it is restricted to higher elevations in the Hantam and Roggeveld regions. This portion covers stretches from the Hantam Plateau near Calvinia in the northwest to the Roggeveld near Fraserburg and Laingsburg to the southeast. Smaller areas of this renosterveld also occur to the west of Calvinia on the northern and western slopes and valleys of the Bokkeveld Escarpment and Matsikamma Mountains in the region of Nieuwoudtville and Klawer (Fig. 5.14).

Vegetation Physiognomy and Composition The Hantam and Roggeveld region has long been recognised as a centre of endemism, particularly for geophytes (but also notably for Asteraceae and Scrophulariaceae), and hosts some of the richest geophyte communities in the world. Most of the region is characterised by low to medium tall cupressoid shrublands dominated by renosterbos with an abundance of annuals and geophytes in the wet season. Geophytes are particularly prevalent in species-rich herblands associated with dolerite ridges, plains and plateaus and in wet and rocky habitats such as temporary wetlands on the Hantamberg Plateau and koppies on the Roggeveld Plateau.

Climate This is one of the more arid regions of the Fynbos Biome with a MAP ranging between 140–530 mm (mean: 248 mm), peaking slightly from March to August. Mean daily minimum and maximum temperatures throughout most of the region are about 1.5 °C in July and 31.2 °C in February, respectively. Frost mostly occurs 10–50 days a year but for just 3–10 days in the eastern patches around Nieuwoudtville and Klawer. The Roggeveld region, however, is notably colder in winter with a mean daily minimum temperature of 0.2 °C in July and frost occurring for 30–70 days a year (Fig. 5.14).

Topography, Geology and Soils This renosterveld occurs on broad ridges and slopes of low mountains, hills, plateaus, escarpments, and in valleys. The largest area in the Roggeveld region occurs on clayey soils derived from mudstones and sandstones from the Beaufort Group. Clays and clayey loams are derived from mudstone and siltstone of the Vanrhynsdorp Group and schist and phyllite of the Gariep Supergroup. Closely associated with the Table Mountain Group are quartzites and shales on slopes and in valleys below the sandstone cliffs of the Bokkeveld Escarpment and Matsikamma Mts. Clays associated with diamictites of the Dwyka Group and some skeletal clays over the Nardouw Subgroup rocks occur on the Bokkeveld Plateau. There are also dolerite plains and ridges comprised of koppies in the Nieuwoudtville region and parts of the Roggeveld and dolerite tafelberg plateaus on the Hantam Plateau with clay soils associated with the Karoo Dolerite Suite. Soils are mostly of Glenrosa and Mispah forms.

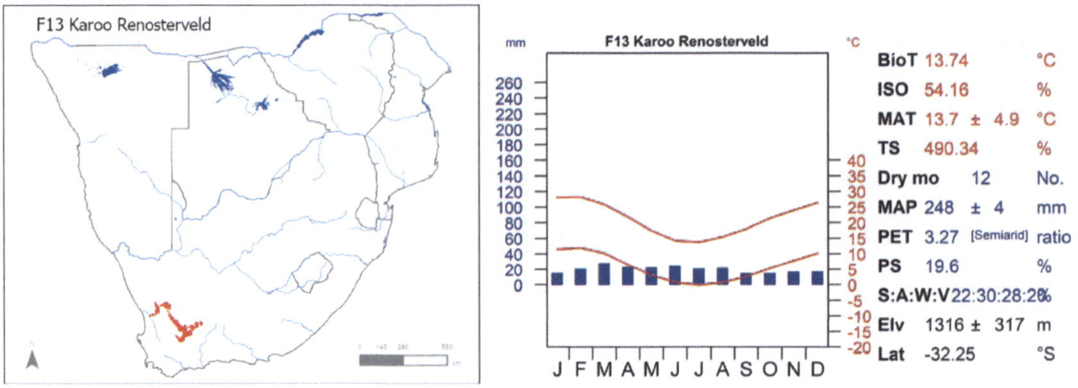

Fig. 5.14 The distribution area and the climatic diagram of the regional biome F13 Karoo Renosterveld

Sources Acocks (1988), van Jaarsveld (1982), Snijman and Perry (1987), Manning and Goldblatt (1997), van Wyk and Smith (2001), Rebelo et al. (2006).

F14 Namaqualand Montane Cape Shrubland

FFg 1 Kamiesberg Granite Fynbos (Dayaram et al. 2019)
FFq 1 Stinkfonteinberge Quartzite Fynbos (Dayaram et al. 2019)
FRg 1 Namaqualand Granite Renosterveld (Dayaram et al. 2019)

Distribution This is an extrazonal regional biome scattered on upper slopes and summits of various mountains in the Namaqualand and Richtersveld regions in the Northern Cape Province. The patches in the Richtersveld and Namaqualand are restricted to higher elevations embedded in the regional biomes SK1 Richtersveld-Aus Upland Karoo and SK3 Namaqualand Hardeveld Karoo, respectively. The largest area is concentrated around Leliefontein in Namaqualand, including the peaks of Kamiesberg, Rooiberg, Sneeukop, Sittenberg, Eselkop, Grasberg and Stalberg (among others), and on some higher peaks lying further north between Komaggas and Springbok. Further north, in the Richtersveld, two narrow belts occur on the ridges of the Vandersterrberg and Stinkfonteinberg Mts directly north of Eksteenfontein, marking the northernmost distribution of fynbos. Small outlying areas can also be found on remote peaks in the northwestern corner of the Western Cape Province. The elevation range is 1100–1705 m (Fig. 5.15).

Vegetation Physiognomy and Composition The vegetation in the Namaqualand region consists of fynbos embedded in renosterveld with boundaries based on rainfall and the composition of granite. Transitions between these communities can be distinct or gradual. Fynbos is generally restricted to higher elevations (> 1450 m) but can be found in lower areas (as low as 1200 m) such as flats between mountains and amongst renosterveld in alluvial washes, seepages, and in rocky and moist facies. Renosterveld patches can also occur at higher elevations within fynbos where soils are sufficiently deep or on south-facing slopes with clay-rich soils. Structurally, the fynbos is sparse and moderately tall (1–2 m) asteraceous fynbos with a predominance of malacophyllous shrubs. The renosterveld is dense and shorter than the fynbos (up to 1.5 m tall) and is characterised by renosterbos and various other asteraceous shrubs. The lower boundary of renosterveld gradually transitions into Namaqualand Hardeveld Karoo vegetation. A variety of communities are present in the Richtersveld, reflecting the complex geomorphology of the region. Areas with sufficient water availability throughout most of the year (due to soil depth and rock structure) support dense shrublands. In contrast, flat plateaus with leached quartzite-rich soils support open

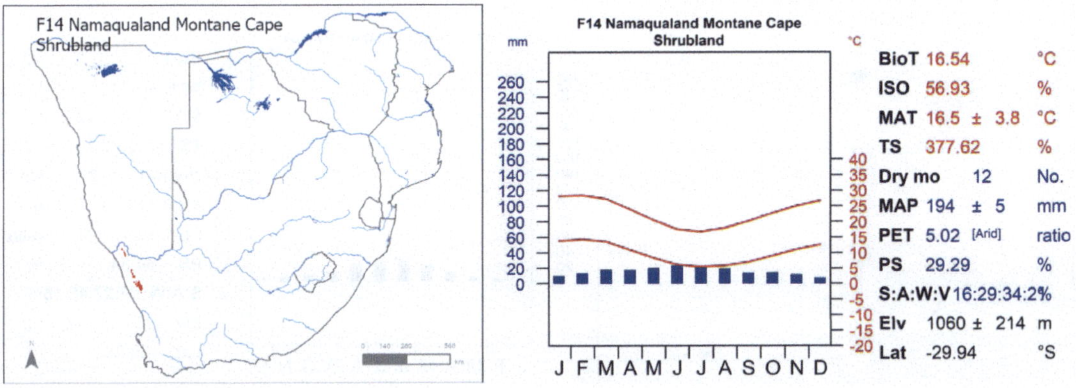

Fig. 5.15 The distribution area and the climatic diagram of the regional biome F14 Namaqualand Montane Cape Shrubland

grasslands dominated by *Tenaxia dura*, while the areas with shallow soils and exposed rock support shrublands dominated by dwarf leaf succulents.

Climate Rainfall is low with MAP ranging from 130 to 450 mm (mean: 194 mm) peaking from May to August. In the Richtersveld, fog supplements precipitation, especially on western slopes. Mean daily minimum and maximum temperatures are about 2.1 °C in July and 28.8 °C in January, respectively. Frost can occur for 10–30 days a year in Namaqualand and 10–40 days in the Richtersveld (Fig. 5.15).

Topography, Geology and Soils The region in Namaqualand is characterised by round-top mountains, plateaus and broad ridges with prominent granite domes and slabs. The sandy to loamy soils are shallow and skeletal and lie over granites and gneisses of the Stalhoek Complex, as well as Kamieskroon Gneiss and Nababeep Gneiss, which is partially overlain by quartzite and metasediments of the Bushmanland Group. The Richtersveld region is topographically complex and comprises long mountain ridges, plateaus, and embedded mountain tops among saddles and valleys. The soils range in depth and comprise loam to loamy sand, mostly lying over quartzite but also over minor volcanic rocks and metamorphosed clastic sediments of the Gariep Supergroup.

Sources Adamson (1958), Boucher and Moll (1981), van Jaarsveld (1981), van Wyk and Smith (2001), Schüttler (2002), Jürgens (2004), Rebelo et al. (2006), Samuels et al. (2022), Simons et al. (2024).

F15 Namaqualand Dune Fynbos

FFd 1 Namaqualand Sand Fynbos (Dayaram et al. 2019)

Distribution This is an extrazonal regional biome occurring on a series of isolated inland dune fields embedded on the coastal plains of Namaqualand and the West Coast, where it is mostly surrounded by the regional biome SK4 Namaqualand Sandveld but also borders on the SK3 Namaqualand Hardeveld Karoo to the interior. These patches span from near Komaggas in the north to the northeast of Lambert's Bay in the south. The elevation ranges from 60 to 300 m (Fig. 5.16).

Vegetation Physiognomy and Composition This fynbos is characterised by moderately tall shrubs dispersed among a short and somewhat sparse restioid layer. Asteraceous and restioid elements are dominant but patches of proteoid fynbos also occur, especially on dune ridges. Dune slacks are dominated by restios and often support succulents and annuals (often genera shared with renosterveld vegetation). No ericoid elements are present. Unlike the rest of the

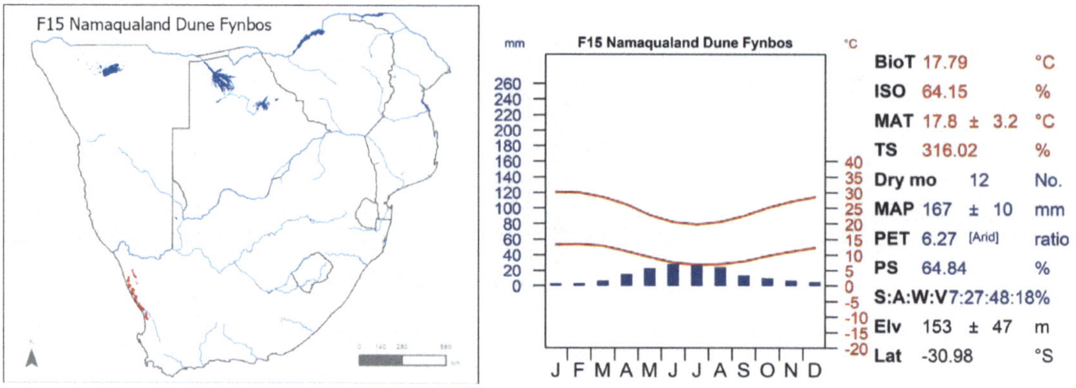

Fig. 5.16 The distribution area and the climatic diagram of the regional biome F15 Namaqualand Dune Fynbos

Fynbos Biome, fire does not play a significant (if any) role in maintaining this vegetation – it appears that dune field dynamics are of primary importance.

Climate Almost all rainfall occurs during winter. This is the most arid region of the CB Fynbos, with MAP ranging as low as 70 mm (mean: 167 mm). Thick sea fog is common in winter, supplementing precipitation. Mean daily minimum and maximum temperatures are about 7 °C in July and 28.4 °C in February, respectively. Frost occurs for 1–2 days a year (Fig. 5.16).

Topography, Geology, and Soils This fynbos occurs on fragmented dune fields with deep, loose red and acidic aeolian sand over marine or other sediments. These thick sand deposits can store water, resulting in localised aquifers that supply water to the fynbos.

Sources Boucher and le Roux (1989, 1994), Rebelo et al. (2006), Bergh et al. (2014).

F16 Inland Arid Fynbos

FFb 3 Central Inland Shale Band Vegetation (Dayaram et al. 2019)
FFh 2 Matjiesfontein Shale Fynbos (Dayaram et al. 2019)
FFh 3 Swartberg Shale Fynbos (Dayaram et al. 2019)
FFh 10 Suurberg Shale Fynbos (Dayaram et al. 2019)
FFq 3 Matjiesfontein Quartzite Fynbos (Dayaram et al. 2019)
FFq 5 Grootrivier Quartzite Fynbos (Dayaram et al. 2019)
FFq 6 Suurberg Quartzite Fynbos (Dayaram et al. 2019)
FFs 21 North Rooiberg Sandstone Fynbos (Dayaram et al. 2019)
FFs 22 South Rooiberg Sandstone Fynbos (Dayaram et al. 2019)
FFs 23 North Swartberg Sandstone Fynbos (Dayaram et al. 2019)
FFs 24 South Swartberg Sandstone Fynbos (Dayaram et al. 2019)
FFs 25 North Kammanassie Sandstone Fynbos (Dayaram et al. 2019)
FFs 26 South Kammanassie Sandstone Fynbos (Dayaram et al. 2019)
FFt 1 Kango Conglomerate Fynbos (Dayaram et al. 2019)

Distribution This fynbos forms a long semi-continuous band across the longitudinal inland mountain ranges from Saalberg near Karoopoort in the Western Cape Province, stretching across the Klein Karoo and all the way to the Kaprivierberge east of Makhanda (Grahamstown) in the Eastern Cape Province. This distribution includes the Bontberg, Voetpadsberg, Koegaberge, Witteberg, Anysberg, Elandsberg, Klein and Groot Swartberge, Touwsberg, Sandberg, Rooiberg, Gamkaberg, Kammanassie, Slypsteenberg, Kleinwinterberg, Suurberg, and Baviaanskloof mountains. Elevation ranges from 400 m to 1800 m (Fig. 5.17).

Vegetation Physiognomy and Composition Moderately dense fynbos of all structural types are present throughout the extensive quartzite hills, ridges, and scarps. Generally, asteraceous and proteoid fynbos are most common in the west, and restioid and ericoid fynbos are most common in the central regions. Grasses become increasingly common eastward, with grassy fynbos being the dominant vegetation in the easternmost region. Shale slopes and hills in the Klein Karoo region support fairly tall and dense proteoid and asteraceous fynbos. Those in the easternmost areas of the regional biome support closed ericoid shrubland or grassland with a grassy and restioid understorey, as well as graminoid fynbos with concentrated patches of proteoid fynbos. Sandstone and conglomerate slopes and dolerite intrusions throughout the Klein Karoo also support all structural types of fynbos. Typically, asteraceous fynbos (and some waboomveld) dominates the lower slopes, proteoid and restioid fynbos are prevalent mid-slope, and ericaceous fynbos occur at higher elevations. Graminoid fynbos is present but generally rarer over sandstone. Deep fire-protected ravines on north-facing or south-facing slopes can support thickets or afromontane forest remnants, respectively. High-altitude shale bands in the Klein Karoo host a diversity of communities, including fynbos of all structural types, but with generally asteraceous shrublands and waboomveld at lower elevations and renosterveld and karoo shrublands at lowest elevations.

Climate Rainfall is generally fairly even throughout the year with a slight low in June–July, although there are localised peaks in winter months in the westernmost areas and a bimodal peak in October–November and February–March in the easternmost areas. MAP ranges 140–1150 mm (mean: 334 mm). Mean daily minimum and maximum temperatures are about 1.6 °C in July and 30 °C in January, respectively. Frost can occur on 10–40 days a year but is generally less frequent in the easternmost areas (Fig. 5.17).

Topography, Geology and Soils Quartzite hills, low mountains and parallel ridges and scarps associated with the Witteberg Group with skeletal sandy soils span across the entire regional biome, also marking its northern boundary. Shale mountain slopes and hills associated with the Cape Supergroup (Cango, Bokkeveld and Witteberg Groups, the latter two being associated with quartzite) with moist and shallow acidic clay-loam soils occur in the Klein Karoo and the eastern extreme around Kleinpoort in Makhanda. Large areas of the Klein Karoo comprise steep to gentle, often deeply dissected, sandstone slopes and some high plateaus associated with the Table Mountain Group with acidic lithosol soils. A small area in the Klein Karoo is also covered with high, dissected foothills with flat summits comprising Cango Group sandstones, the conglomerates of the Uitenhage Group, and dolerite intrusions with soils variable in depth. Lastly, narrow shale bands (80–200 m) with clay soils (associated with the Cedarberg Formation) occur on high slopes of many of the Klein Karoo's mountains.

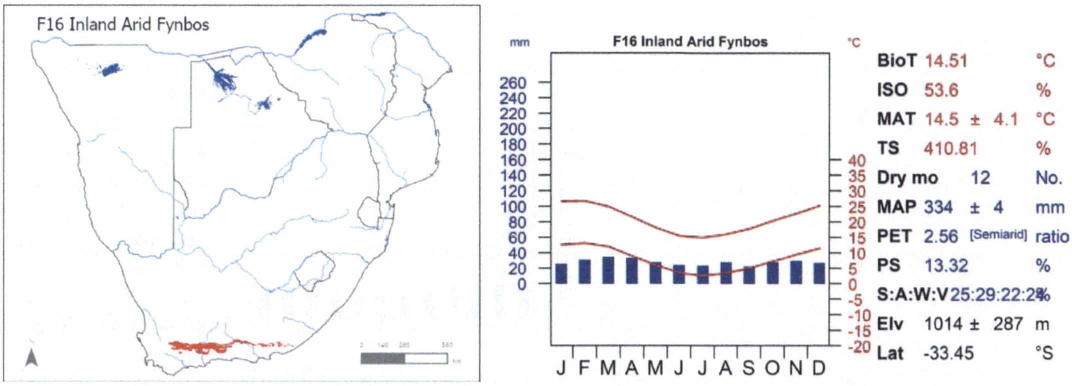

Fig. 5.17 The distribution area and the climatic diagram of the regional biome F16 Inland Arid Fynbos

Sources Dyer (1937), Martin and Noel (1960), Martin (1965, 1966), Jessop and Jacot Guillarmod (1969), Moffett and Deacon (1977), Kruger (1979), Taylor (1979), Bond (1981), Taylor and van der Meulen (1981), Cowling (1983b), Lubke (1983), Richardson et al. (1984), Campbell (1985), Lubke et al. (1986), van Wyk et al. (1988), Euston-Brown (1995), Kemper et al. (1999), Vlok and Euston-Brown (2002), Cleaver et al. (2005), Rebelo et al. (2006).

F17 Eastern Renosterveld

FRs 16 Uniondale Shale Renosterveld (Dayaram et al. 2019)
FRs 17 Langkloof Shale Renosterveld (Dayaram et al. 2019)
FRs 18 Baviaanskloof Shale Renosterveld (Dayaram et al. 2019)
FRs 19 Humansdorp Shale Renosterveld (Dayaram et al. 2019)

Distribution This regional biome has a disjointed distribution in the roughly triangular area from the Outeniqua Mountains between Oudtshoorn and George in the southwest to St Francis Bay in the southeast and the Grootrivierberge northeast of Willowmore. The largest areas occur in the Klein Karoo on the northern slopes of the Outeniqua Mountains, the nearby slopes of the Kammanassie and Kouga Mountains, and on the mountainous surrounds of Willowmore. Other notable areas include a narrow band in the Langkloof Valley, two patches on the southern hillslopes of the Baviaanskloof Valley, and patches that span inland from the coast between St Francis Bay and the Gamtoos River mouth, and the easternmost patch near Rocklands. The elevation ranges between 20 m and 1150 m (Fig. 5.18).

Vegetation Physiognomy and Composition The vegetation comprises low, somewhat dense cupressoid and graminoid shrublands dominated by renosterbos. Clumps of thicket can be found on some north-facing slopes, fire refugia, and termitaria in the eastern areas. Most of this renosterveld transitions into F10 Eastern Fynbos at higher elevations. Large areas of vegetation have been transformed by agriculture in the southern and eastern regions.

Climate Rainfall is relatively even throughout the year with a slight bimodal peak in March and October. MAP ranges from 170–850 mm (mean: 484 mm) and mean daily minimum and maximum temperatures are about 2.4 °C in July and 29.6 °C in January, respectively. The areas inland of the Outeniqua and Kouga Mountains are notably drier and experience more pronounced temperature extremes than the lower lying southern and eastern areas. Frost can occur for 2–10 days in the southern and eastern areas but generally occurs for 10–40 days in the upland areas (Fig. 5.18).

Topography, Geology and Soils This renosterveld can be found across various lower mountain slopes, flat mountain bases, intermontane

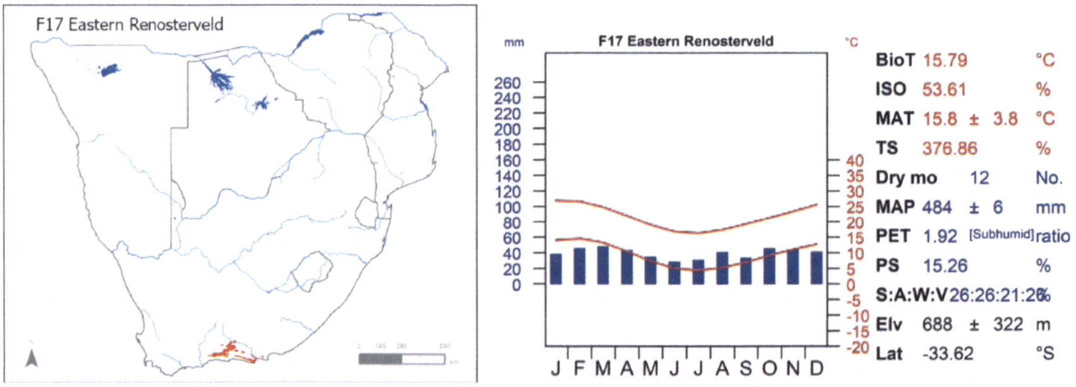

Fig. 5.18 The distribution area and the climatic diagram of the regional biome F17 Eastern Renosterveld

valleys, and moderately undulating plains and hills. It occurs on clays and loams associated with the Bokkeveld, Witteberg, and Table Mountain Groups.

Sources Boucher and Moll (1981), Cowling (1984), Euston-Brown (1995), Kemper et al. (1999), Vlok and Euston-Brown (2002), Rebelo et al. (2006), Aiello-Lammens et al. (2017).

F18 Southwestern Cape Strandveld

FS 6 Cape Flats Dune Strandveld (Dayaram et al. 2019)
FS 7 Overberg Dune Strandveld (Dayaram et al. 2019)

Distribution This regional biome forms a series of interrupted narrow patches along the southwestern coastline of the Western Cape Province. The most substantial area can be found along the False Bay coast, extending through the Cape Flats as far inland as Belville. Another notable area spans across the coast of Table Bay to Melkbosstrand, including the large Atlantis dune plume. It can also be found offshore on Robben Island. To the east, narrow stretches occur along the coast all the way to Cape Infanta, almost always adjacent to limestone formations. The elevation generally ranges between the sea level and 100 m but reaches up to 200 m in some areas close to Cape Town (Fig. 5.19).

Vegetation Physiognomy and Composition This strandveld is characterised by tall, evergreen, hard-leaved shrublands. Grasses and annual herbs are abundant in openings between the shrub canopy. Near the coast, shrublands are more developed in moist dune slacks and sheltered valleys, while shorter thicket occurs in exposed, windy areas.

Climate The largest areas of this regional biome (around the Cape Flats and Melkbosstrand) experience strongly seasonal winter rainfall, which transitions towards even seasonal rainfall towards the easternmost areas. Winter rains prevail, however, and occur alongside cold weather and strong north-westerly winds. MAP ranges between 350–600 mm (mean: 547 mm). Mean daily minimum and maximum temperatures are about 7.5 °C in July and 26 °C in February, respectively. Dense mist can also be common throughout the Overberg region in colder months. Frost is a very scarce occurrence (Fig. 5.19).

Topography, Geology and Soils The landscape is characterised by deep calcareous marine sands on flat to moderately undulating dune fields. Limestone outcrops (hardpans) occur around the False Bay area. Shallower sandstones and limestones (Bredasdorp Group) overlying Table Mountain sandstone occur in parts of the eastern stretches of this regional biome.

Sources Taylor (1972, 1984a), Boucher (1974, 1977, 1978, 1982, 1983, 1987, 1994b, 1995, 1997a, 1998a, 1998b, 1999a, 1999c, g), Milton (1976), van der Merwe (1977a), Cowling et al. (1988), O'Callaghan (1990), Berry (1991),

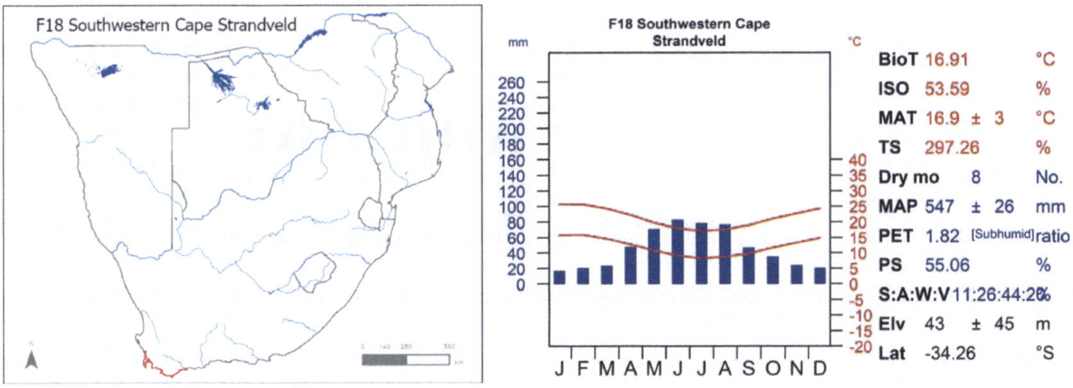

Fig. 5.19 The distribution area and the climatic diagram of the regional biome F18 Southwestern Cape Strandveld

Taylor and Boucher (1993), Privett (1998), De Hoop Nature Reserve Planning Committee (2001), Rebelo et al. (2006), Zietsman and Bredenkamp (2006, 2007).

F19 Southeastern Cape Strandveld

FS 8 Blombos Strandveld (Dayaram et al. 2019)
FS 9 Groot Brak Strandveld (Dayaram et al. 2019)

Distribution This regional biome occurs in intermittent narrow strips along the Southern Cape coast in the Western Cape Province. These strips can be found at Witsand, Klein Brak, Groot Brak, and between Herolds Bay and Victoria Bay. Elevation ranges from 0 m to 180 m (Fig. 5.20).

Vegetation Physiognomy and Composition The vegetation consists of dense evergreen shrublands and thickets dominated by sclerophyllous spiny scrub, sometimes up to 3 m in height. The undergrowth is generally sparse and poorly developed. Dune slacks that are protected from persistent salt spray from the sea and sporadic fires from inland generally host the most well-developed thickets.

Climate Rainfall has slight peaks in spring and autumn, and MAP ranges approximately between 300 and 750 mm (mean: 449 mm). Mean daily minimum and maximum temperatures are about 6.4 °C in July and 26.8 °C in February, respectively. Coastal fogs are common on winter mornings. Easterly and westerly sea breezes are prevalent and influence the coastal vegetation. Around the end of autumn, strong and dry berg winds are common. Frost is infrequent (Fig. 5.20).

Topography, Geology and Soils The landscape ranges from coastal flats to undulating stabilised dunes, although some steep slopes are present. The western area lies on limestones and sandstones of the Bredasdorp Group. The eastern portions lie mostly over clastic sedimentary rocks of the Uitenhage Group and some metamorphic rocks of the Kaaimans Group towards the east. The edges of coastal cliffs here are made up of Cape Granite. Some coastal areas are covered by unconsolidated white dune sands, sometimes shelly or mixed with calcareous or sandstone cobblestones.

Sources Muir (1929), Rebelo et al. (1991, 2006), Boucher and Rode (1995a, b), Boucher (1997b), Vlok and Euston-Brown (2002), Vlok et al. (2003).

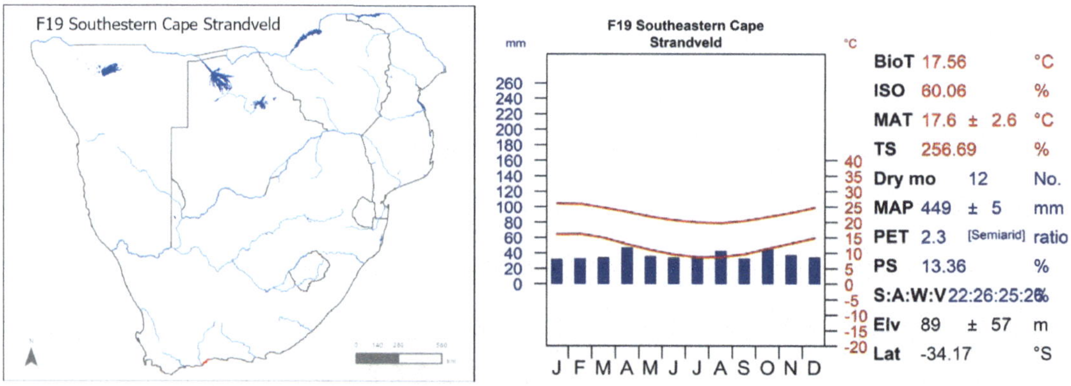

Fig. 5.20 The distribution area and the climatic diagram of the regional biome F19 Southeastern Cape Strandveld

References

Acocks JPH (1935) Vegetation of a portion of the Cape Flats and list of plants so far found there. Thesis, Department of Botany, University of Cape Town, Cape Town

Acocks JPH (1988) Veld types of South Africa, 3rd edn. Mem Bot Surv S Afr 57:1–146

Adamson RS (1925) The native vegetation of Kirstenbosch. J Bot Soc S Afr 11:19–23

Adamson RS (1927) Plant communities of Table Mountain: preliminary account. J Ecol 15:278–284

Adamson RS (1935) The plant communities of Table Mountain. III. A six years' study of regeneration after burning. J Ecol 23:44–55

Adamson RS (1958) The Cape as an ancient African flora. Adv Sci 58:1–10

Adamson RS, Salter TM (1950) Flora of the Cape Peninsula. Juta & Co, Cape Town

Agenbag L, Esler KJ (2004a) An environmental gradient reveals how climate influences phenophases in fynbos growth forms. In: Arianoutsou M, Papanastasis V (eds) 10th MEDECOS conference April 25–May 1, 2004, Rhodes. Millpress, Rotterdam

Agenbag L, Esler KJ (2004b) Studies along an environmental gradient reveal how biome boundaries may respond to climate change. In: Arianoutsou M, Papanastasis V (eds) 10th MEDECOS conference April 25–May 1, 2004, Rhodes. Millpress, Rotterdam

Agenbag L, Rutherford M, Midgley G et al (2004) Fundamental and realized niches of fynbos and karoo seedlings as revealed by reciprocal transplants. In: Arianoutsou M, Papanastasis V (eds) Proceedings 10th MEDECOS conference April 25–May 1, 2004, Rhodes. Millpress, Rotterdam

Aiello-Lammens ME, Slingsby JA, Merow C et al (2017) Processes of community assembly in an environmentally heterogeneous, high biodiversity region. Ecography 40:561–576

Allsopp N, Colville JF, Verboom GA (eds) (2014) Fynbos. Ecology, evolution and conservation of a megadiverse region. Oxford University Press, Oxford

Andrag RH (1970) Plantopname van Helderberg-Natuurreservaat. BSc(Hons) Project, Forestry Faculty, University of Stellenbosch, Stellenbosch

Baard JA, Grobler BA, Kraaij T (2023) Passive restoration of fynbos after afforestation with exotic pines, South Africa. Restor Ecol 32:e14037. https://doi.org/10.1111/rec.14037

Barnes K (1992) The community structure of the Bokkop Steenberg Region of Silvermine Nature Reserve. 3rd-Year Project, Department of Botany, University of Cape Town, Cape Town

Bean A, Johns A (2005) South African wild flower guide 5. Stellenbosch to Hermanus, including Kogelberg and Hottentots Holland. Botanical Society of South Africa, Cape Town

Bergh NG, Verboom GA, Rouget M et al (2014) Vegetation types of the greater Cape Floristic Region. In: Allsopp N, Colville JF, Verboom GA et al (eds) Fynbos: ecology, evolution and conservation of a megadiverse region. Oxford University Press, Oxford, pp 1–25

Berry MG (1991) Die flora en plantegroei van die Swart-Produkte eiendom, te Kaapse Vlakte. BSc(Hons) Project, Department of Botany, University of Stellenbosch, Stellenbosch

Boix C (1992) A vegetation survey of the Wildevoëlvlei flats and adjacent areas. 3rd-Year Project, Department of Botany, University of Cape Town, Cape Town

Bond W (1978a) A management plan of De Vasselot Nature Reserve. Department of Forestry, Saasveld Forestry Research Station, George

Bond W (1978b) A management plan of Ysternek Nature Reserve. Department of Forestry, Saasveld Forestry Research Station, George

Bond W (1981) Vegetation gradients in southern Cape mountains. Thesis, Department of Botany, University of Cape Town, Cape Town

Bond W, Midgley J, Vlok J (1988) When is an island not an island? Insular effects and their causes in fynbos shrublands. Oecologia 77:515–521

Boschoff CR (1989) Structure, distribution and phenology of perennial plant species in the Worcester Veld Reserve. Thesis, Department of Botany, University of Cape Town, Cape Town

Boshoff A, Cowling RM, Kerley G (2000) The Baviaanskloof conservation area. TERU Report 27. Terrestrial Ecology Research Unit, University of Port Elizabeth, Port Elizabeth

Boshoff CR (1985) A vegetation survey at Kenilworth Race Course—implications for conservation. BSc(Hons) Project, Department of Botany, University of Cape Town, Cape Town

Boucher C (1972) The vegetation of the Cape Hangklip area. Thesis, Department of Botany, University of Cape Town, Cape Town

Boucher C (1974) Dune vegetation in the South Western Cape. Veld Flora 4(4):67–69

Boucher C (1977) Cape Hangklip area. I. The application of association-analysis, homogeneity functions and Braun-Blanquet techniques in the description of south-western Cape vegetation. Bothalia 12:293–300

Boucher C (1978) Cape Hangklip area. II. The vegetation. Bothalia 12:455–497

Boucher C (1980) Notes on the use of the term 'Renosterveld.' Bothalia 13:237

Boucher C (1982) Some thoughts about the West Coast Strandveld vegetation. Veld Flora 68(4):108–109

Boucher C (1983) Floristic and structural features of the coastal foreland vegetation south of the Berg River, Western Cape Province, South Africa. Bothalia 14:669–674

Boucher C (1986) The vegetation of the Greater Silverstroomstrand Development Area. Ecological Research Report 1. Department of Botany, University of Stellenbosch, Stellenbosch

Boucher C (1987) A phytosociological study of transects through the Western Cape coastal foreland, South

Africa. Thesis, Department of Botany, University of Stellenbosch, Stellenbosch
Boucher C (1988) A report on the flora and vegetation of the Buffelstalberg Flats. Ecological Research Report 2. Department of Botany, University of Stellenbosch, Stellenbosch
Boucher C (1989a) A brief report on some aspects of the flora and vegetation observed on the Farm Modderrivier 721, Malmesbury. Ecological Research Report 6. Department of Botany, University of Stellenbosch, Stellenbosch
Boucher C (1989b) Comments on the restoration of disturbed areas on Portion 1 of Farm 1014 near Paternoster (Groot Paternoster), Western Cape. Ecological Research Report 5b. Department of Botany, University of Stellenbosch, Stellenbosch
Boucher C (1991) Vegetation survey of Rondegat farm, Clanwilliam. Ecological Research Report 8. Department of Botany, University of Stellenbosch, Stellenbosch
Boucher C (1992) A vegetation survey of Granzekraal Recreational Area. Ecological Research Report 9. Department of Botany, University of Stellenbosch, Stellenbosch
Boucher C (1994a) Botanical evaluation of two potential dam sites in the Molenaars River, Du Toitskloof. Ecological Research Report 11. Department of Botany, University of Stellenbosch, Stellenbosch
Boucher C (1994b) The vegetation of Erf 1092, Gansbaai. Ecological Research Report 12. Department of Botany, University of Stellenbosch, Stellenbosch
Boucher C (1995) Hoek-van-die-Berg: brief comments about the vegetation. Ecological Research Report 18. Department of Botany, University of Stellenbosch, Stellenbosch
Boucher C (1997a) Conservation worthy vegetation on the Capricorn Park site: being an evaluation of a report submitted by A.B. Low (1966). Ecological Research Report 48. Department of Botany, University of Stellenbosch, Stellenbosch
Boucher C (1997b) Vegetation description and management guidelines for the Blombos Private Nature Reserve, Riversdale District. Ecological Research Report 26. Department of Botany, University of Stellenbosch, Stellenbosch
Boucher C (1998a) Kleinmond Harbour Project: botanical scoping report. Ecological Research Report 34. Department of Botany, University of Stellenbosch, Stellenbosch
Boucher C (1998b) Report on preliminary scoping of the proposed Pelican Park Extension: botany. Ecological Research Report 33. Department of Botany, University of Stellenbosch, Stellenbosch
Boucher C (1998c) The importance of the flora and vegetation on Rein's Nature Reserve. Ecological Research Report 32. Department of Botany, University of Stellenbosch, Stellenbosch
Boucher C (1998d) The vegetation of portions of the farms Modderfontein A & E, Aurora. Ecological Research Report 38. Department of Botany, University of Stellenbosch, Stellenbosch
Boucher C (1998e) The vegetation of the farm Stofbergsfontein 365, West Coast National Park. Ecological Research Report 37. Department of Botany, University of Stellenbosch, Stellenbosch
Boucher C (1998f) Vegetation and management recommendations for 'Die Duine'. Ecological Research Report 35. Department of Botany, University of Stellenbosch, Stellenbosch
Boucher C (1999a) Botanical aspects relevant to the lower Silvermine River development. Ecological Research Report 41. Department of Botany, University of Stellenbosch, Stellenbosch
Boucher C (1999b) Die plantegroei van die parkie te Simonswyk, Stellenbosch. Ecological Research Report 53. Department of Botany, University of Stellenbosch, Stellenbosch
Boucher C (1999c) Kleinmond Harbour Project: evaluation of impacts of proposed development on botanical features. Ecological Research Report 41. Department of Botany, University of Stellenbosch, Stellenbosch
Boucher C (1999d) The riparian vegetation of the lower Hopies River, Hawston. Ecological Research Report 39. Department of Botany, University of Stellenbosch, Stellenbosch
Boucher C (1999e) The vegetation of the western portion adjacent to the lower Silvermine River, Clovelly, Cape Peninsula. Ecological Research Report 51. Department of Botany, University of Stellenbosch, Stellenbosch
Boucher C, le Roux A (1989) Heavy metal sand-mining project assessment—vegetation report. Ecological Research Report 4a. Department of Botany, University of Stellenbosch, Stellenbosch
Boucher C, le Roux A (1994) Vegetation of the proposed national park between the Spoeg and Groen Rivers on the Namaqualand coast. Ecological Research Report 4b. Department of Botany, University of Stellenbosch, Stellenbosch
Boucher C, Moll EJ (1981) South African mediterranean shrublands. In: di Castri F, Goodall DW, Specht RL (eds) Ecosystems of the world 11. Mediterranean-type shrublands. Elsevier, Amsterdam, pp 233–248
Boucher C, Rode E (1994) Plankenburg River: explanation of present condition including a rehabilitation plan. Ecological Research Report 15. Department of Botany, University of Stellenbosch, Stellenbosch
Boucher C, Rode E (1995a) The flora and vegetation of Rein's Nature Reserve, Riversdale District. Ecological Research Report 16. Department of Botany, University of Stellenbosch, Stellenbosch
Boucher C, Rode E (1995b) The flora and vegetation of Rein's Nature Reserve, Riversdale District. Supplementary report. Ecological Research Report 17. Department of Botany, University of Stellenbosch, Stellenbosch
Boucher C, Rode E (1999) Vegetation conservation importance ratings for the Saldanha-Vredenburg regional structure plan. Ecological Research Report 50. Department of Botany, University of Stellenbosch, Stellenbosch

Boucher C, Shepherd PA (1988) Plant communities of the Pella site. In: Jarman ML (ed) A description of the Fynbos Biome Project intensive study site at Pella. FRD, Pretoria, pp 38–76
Boucher C, Stindt M (1992) The vegetation of the Haasvlakte, Houhoek, South Africa. S Afr J Bot 58:510–524
Britton P (1972) A report on the proposed nature reserve at Swartklip. BSc(Hons) Project, Department Forestry Conservation, University of Stellenbosch, Stellenbosch
Britton P, Jackelman J (1995) Wolfgat Nature Reserve. A desolate wasteland destined for housing or a dynamic community conservation project? Veld Flora 81(2):46–48
Britton P, Jackelman J (1996) Biological management of Renosterveld: some questions and lessons from Signal Hill and Devil's Peak. In: Low AB, Jones FE (eds) The sustainable use and management of renosterveld remnants in the Cape Floristic Region. FCC Report 1995/4. Botanical Society of South Africa, Kirstenbosch, pp 58–62
Burman L, Bean A (1985) South African wild flower guide 5. Stellenbosch to Hermanus. Botanical Society of South Africa, Kirstenbosch
Cameron MJ (1980) Fynbos islands in the Knysna forests. S Afr J 112:27–29
Campbell BM (1985) A classification of the mountain vegetation of the Fynbos Biome. Mem Bot Surv S Afr 50:1–121
Chesselet P (1985) Vegetation analyses in the Worcester ecotone. 3rd-Year Project, Department of Botany, University of Cape Town, Cape Town
Cleaver G, Brown LR, Bredenkamp GJ (2005) The phytosociology of the Vermaaks, Marnewicks and Buffelsklip valleys of the Kammanassie Nature Reserve, Western Cape. Koedoe 48:1–16
Cousins SR, Witkowski ETF, Esler KJ (2017) Influence of fire on critically endangered Swartland Shale Renosterveld in the Cape Floristic Region. Appl Veg Sci 21:561–576
Cowling R (1976) The ecological status of the understorey communities of pine forests on Table Mountain. In: Moll EJ, Campbell BM (eds) The ecological status of Table Mountain. Appendix 1. Department of Botany, University of Cape Town, Cape Town
Cowling RM (1983) The occurrence of C3 and C4 grasses in fynbos and allied shrublands in the southeastern Cape, South Africa. Oecologia 58:121–127
Cowling RM (1984) A syntaxonomic and synecological study in the Humansdorp region of the Fynbos Biome. Bothalia 15:175–227
Cowling RM (1990a) Diversity components in a species-rich area of the Cape Floristic Region. J Veg Sci 1:699–710
Cowling R (1990b) Who will save Hagelkraal? Veld Flora 76(2):52–53
Cowling RM (1991) A preliminary report on the flora and vegetation of the Wildevoëlvlei Flats in the Noordhoek-Kommetjie basin. FCC Report 91/2. Botanical Society of South Africa, Kirstenbosch
Cowling RM, Campbell BM (1983a) A comparison of fynbos and non-fynbos coenoclines in the lower Gamtoos River Valley, southeastern Cape, South Africa. Vegetatio 53:161–178
Cowling RM, Campbell BM (1983b) The definition of leaf consistence categories in the Fynbos Biome and their distribution along an altitudinal gradient in the South Eastern Cape. J S Afr Bot 49:87–101
Cowling RM, Campbell BM (1984) Beta diversity along fynbos and non-fynbos coenoclines in the lower Gamtoos River Valley, south-eastern Cape. S Afr J Bot 3:187–189
Cowling RM, Campbell BM, Mustart P et al (1988) Vegetation classification in a floristically complex area: the Agulhas Plain. S Afr J Bot 54:290–300
Cowling RM, MacDonald IAW, Simmons MT (1996) The Cape Peninsula, South Africa: physiographical, biological and historical background to an extraordinary hot-spot of biodiversity. Biodivers Conserv 5:527–550
Davis G (1988) Description of a proteoid-restioid stand in Mesic Mountain Fynbos of the south-western Cape and some aspects of its ecology. Bothalia 18:279–287
Dayaram A, Harris LR, Grobler BA et al (2019) Vegetation Map of South Africa, Lesotho and Swaziland 2018: a description of changes since 2006. Bothalia 49(1):a2452. https://doi.org/10.4102/abc.v49i1.2452
De Hoop Nature Reserve Planning Committee (2001) Management plan for De Hoop Nature Reserve. Cape Nature Conservation Board Report. Cape Nature Conservation Board, Cape Town
de Lange C (1992) Floristic analysis of the Vogelgat Nature Reserve, Cape Province, South Africa. Thesis, Department of Botany, University of Cape Town, Cape Town
Diemer N (2000) Fire management at Elandsberg Nature Reserve. BSc(Hons) Project, Department of Conservation, University of Stellenbosch, Stellenbosch
Drews BK (1980a) Report on conservation worthy areas of fynbos in the Knysna–Wilderness–Plettenberg Bay guide-plan area. Botanical Research Institute, Pretoria
Drews BK (1980b) Report on conservation worthy areas of fynbos in the George guide-plan area. Botanical Research Institute, Pretoria
Dupont LM, Zhao X, Charles C et al (2022) Continuous vegetation record of the Greater Cape Floristic Region (South Africa) covering the past 300 000 years (IODP U1479). Clim Past 18:1–21. https://doi.org/10.5194/cp-18-1-2022
Durand BJ (1981) A study of the short-term responses of fynbos to fire in the Kogelberg State Forest, South Africa. Thesis, Department of Botany, University of Cape Town, Cape Town
Duthie AV (1929) Vegetation and flora of the Stellenbosch Flats. Ann Univ Stellenbosch Sect A7:1–22

Duvenhage AJ (1993) Die voorkoms van uitheemse kruidgewasse in die natuurlike plategroei van die Stellenbosch omgewing. Thesis, Department of Botany, University of Stellenbosch, Stellenbosch

Dyer RA (1937) The vegetation of the divisions of Albany and Bathurst. Mem Bot Surv S Afr 17:1–138

Dzikiti S, Jovanovic NZ, Bugan R et al (2014) Measurement and modelling of evapotranspiration in three fynbos vegetation types. Water SA 40:189–198

Euston-Brown D (1995) Environmental and dynamic determinants of vegetation distribution in the Kouga and Baviaanskloof Mountains, Eastern Cape. Thesis, Department of Botany, University of Cape Town, Cape Town

February EC, Munyai NM, Tucker CP et al (2021) The causes and effects of indigenous C4 grass expansion into a hyper-diverse fynbos shrubland. Oecologia 195:421–433

Forbes CJ, Gillson L, Hoffman MT (2010) Shifting baselines in a changing world: identifying management targets in endangered heathlands of the Cape Floristic Region, South Africa. Anthropocene 22:81–93

Fourie C (2011) One class object based system for sparse geographic feature identification. Thesis, Stellenbosch University, Stellenbosch

Fourie C, van Niekerk A, Mucina L (2012) Semi-automated segment generation for geographic novelty detection using edge and area metrics. S Afr J Geomat 1:133–148. http://www.ajol.info/index.php/sajg/article/viewFile/107006/96913

Galloway AD, Holmes PM, Gaertner M et al (2017) The impact of pine plantations on fynbos above-ground vegetation and soil seed bank composition. S Afr J Bot 113:300–307

Glyphis J, Moll EJ, Campbell BM (1978) Phytosociological studies on Table Mountain, South Africa: 1. The back table. J S Afr Bot 44:281–289

Greenfield A (1992) A vegetation survey of the Swartkopberge at an altitude greater than 500 m above sea level. 3rd-Year Project, Department of Botany, University of Cape Town, Cape Town

Grobler PJ, Marais J (1967) Die plantegroei van die Nasionale Bontebokpark, Swellendam. Koedoe 10:132–148

Hall ASW (2018) Restoration potential of alien-invaded Lowland Fynbos, Thesis, Department of Conservation Ecology and Entomology, Faculty of AgriSciences, Stellenbosch University, Stellenbosch

Hall SA, Holmes PM, Gaertner M et al (2021) Active seed sowing can overcome constraints to passive restoration of a critically endangered vegetation type. S Afr J Bot 138:249–261

Hanekom N, Southwood A, Ferguson M (1989) A vegetation survey of the Tsitsikamma Coastal National Park. Koedoe 32:47–66

Helme N (2002) The vegetation. In: Maze K (ed) Biodiversity of the Edith Stephens Wetland Park. Botanical Society of South Africa, Kirstenbosch

Heydenrych B (1994) Limestone Fynbos—unique but threatened: an investigation into the conservation status of Limestone Fynbos in the southern Cape. FCC Report 94/1. Botanical Society of South Africa, Kirstenbosch

Hoare DB, Victor JE, Lubke RA et al (2000) Vegetation of the coastal fynbos and rocky headlands south of George, South Africa. Bothalia 30:87–96

Hoare DB, Mucina L, Rutherford MC et al (2006) Albany Thicket Biome. In: Mucina L, Rutherford MC (eds) The vegetation of South Africa, Lesotho and Swaziland. SANBI, Pretoria, pp 540–567

Hoffman MT, Moll EJ, Boucher C (1987) Post fire succession at Pella, a South African lowland fynbos site. S Afr J Bot 53:370–374

Holmes PM, Newton RJ (2004) Patterns of seed persistence in South African fynbos. Plant Ecol 172:143–158

Irving N (1992) A study of the fynbos communities of Else's Peak. 3rd-Year Project, Department of Botany, University of Cape Town, Cape Town

Jacobs LD (1984) Die plantegroei en flora van die Duthie-natuurreservaat. BSc(Hons) Project, Department of Botany, University of Stellenbosch, Stellenbosch

Jarman ML (ed) (1988) A description of the Fynbos Biome Project. Intensive study site at Pella. FRD, Pretoria

Jeffery DJ, Wilson N (1987) A survey of the plant communities in the Greater Kirstenbosch region. Kirstenbosch National Botanic Gardens, Cape Town

Jessop JP, Jacot Guillarmod A (1969) The vegetation of Thomas Baines Nature Reserve. J S Afr Bot 35:367–392

Jones C (1986) Vegetation analysis of the Eensaamheid Reserve and extension. 3rd-Year Project, Department of Botany, University of Cape Town, Cape Town

Joubert C (1991) History and description of contemporary vegetation, Signal Hill, Cape Town. Thesis, Department of Botany, University of Cape Town, Cape Town

Joubert C, Moll EJ (1992) A phytosociological study of Signal Hill, Cape Town, utilizing both perennial and ephemeral species. Bothalia 22:255–282

Joubert L, Esler KJ, Privett SDJ (2009) The effect of ploughing and augmenting natural vegetation with commercial fynbos species on the biodiversity of Overberg Sandstone fynbos on the Agulhas Plain, South Africa. S Afr J Bot 75:526–531

Jürgens N (2004) A first classification of the vegetation of the Richtersveld (RSA) and directly adjacent regions in Namibia and South Africa. Schumannia 4:149–180

Kathan L (1981) A study of certain autoecological aspects pertaining to a *Leucadendron laureolum* community at the Silvermine Nature Reserve, South Africa. Thesis, Department of Botany, University of Cape Town, Cape Town

Keith DA, Rodríguez JP, Rodríguez-Clark KM et al (2013) Scientific Foundations for an IUCN Red List of Ecosystems. PLoS ONE 8(5):e62111. https://doi.org/10.1371/journal.pone.0062111

Kemper J, Cowling RM, Richardson DM (1999) Fragmentation of South African renosterveld shrublands: effects on plant community structure and conservation implications. Biol Conserv 90:103–111

Kirkwood D (1992) A vegetation survey of Constantiaberg's eastern slopes. 3rd-Year Project, Department of Botany, University of Cape Town, Cape Town

Knauer K (2011) Monitoring ecosystem health of Fynbos remnant vegetation in the City of Cape Town using remote sensing. Thesis, Lehrstuhl für Fernerkundung, Institut für Geographie, Julius-Maximilians-Universität Würzburg, Würzburg

Kraaij T, Cowling RM, van Wilgen BW (2011) Past approaches and future challenges to the management of fire and invasive alien plants in the new Garden Route National Park. S Afr J Sci 107:1–11

Kraaij T, Baard JA, Cowling RM et al (2012a) Historical fire regimes in a poorly understood, fire-prone ecosystem: eastern coastal fynbos. Int J Wildl Fire 22:277–287

Kraaij T, Cowling RM, van Wilgen BW (2012b) Lightning and fire weather in eastern coastal fynbos shrublands: seasonality and long-term trends. Int J Wildl Fire 22:288–295

Kraaij T, Cowling RM, van Wilgen BW (2013) Fire regimes in eastern coastal fynbos: imperatives and thresholds in managing for diversity. Koedoe 55(1) (Art. #1104). https://doi.org/10.4102/koedoe.v55i1.1104

Kraaij T, Cowling RM, van Wilgen BW et al (2017) Vegetation responses to season of fire in an aseasonal, fire-prone fynbos shrubland. PeerJ. https://doi.org/10.7717/peerj.3591

Kraaij T, Baard JA, Arndt J et al (2018) An assessment of climate, weather, and fuel factors influencing a large, destructive wildfire in the Knysna region, South Africa. Fire Ecol 14:4. https://doi.org/10.1186/s42408-018-0001-0

Kruger FJ (1974) The physiography and plant communities of Jakkalsrivier catchment. Thesis, Faculty of Forestry, University of Stellenbosch, Stellenbosch

Kruger FJ (1978) A description of the Fynbos Biome Project. South African National Scientific Programmes Report 28. CSIR, Pretoria

Kruger FJ (1979) South African heathlands. In: Specht RL (ed) Ecosystems of the world 9A. Heathlands and related shrublands of the world. A. Descriptive studies. Elsevier, Amsterdam, pp 19–80

Kuhel I (2023) Effects of wild grazers on carbon sequestration in the renosterveld of De Hoop Nature Reserve. Thesis, University of Utrecht, Utrecht

Laidler D, Moll EJ, Campbell BM et al (1978) Phytosociological studies on Table Mountain, South Africa: 2. The Front Table. J S Afr Bot 44:291–295

Lamprecht E (1979) 'n Floristiese opname van 'n deel van die Jan Maraisnatuurreservaat. BSc(Hons) Project, Department of Botany, University of Stellenbosch, Stellenbosch

Landman A, Nel E (1989) Opname van die inheemse en uitheemse plantegroei te Pappegaaiberg, Stellenbosch. BSc(Hons) Project, Department of Botany, University of Stellenbosch, Stellenbosch

Lane SB (1980) Interpretation of Digital Landsat-1 Imagery from Verlorenvlei, South Western Cape. Thesis, University of Cape Town, Cape Town

Le Maitre DC (1987) Effects of season of burn on species populations and composition of fynbos in the Jonkershoek Valley. S Afr J Bot 53:284–292

Lechmere-Oertel RG, Cowling RM (2000) Combining floristic and growth form composition in a gradient-directed vegetation survey of Matjiesrivier Nature Reserve, Western Cape, South Africa. Bothalia 30:161–173

Lechmere-Oertel RG, Cowling RM (2001) Abiotic determinants of the fynbos/succulent karoo boundary, South Africa. J Veg Sci 12:75–80

Linder HP (1976) A preliminary study of the vegetation of Piketberg Mountain, Cape Province. BSc(Hons) Project, Department of Botany, University of Cape Town, Cape Town

Linder HP (2003) The radiation of the Cape flora, Southern Africa. Biol Rev 78:597–638

Lötter M, van Wageningen M (1988) Die plantegroei van die berggrond bokant Coetzenburg. BSc(Hons) Project, Department of Botany, University of Stellenbosch, Stellenbosch

Lotz CN (1993) Ordination of plant communities between Simonstown and Miller's Point, Cape Peninsula. 3rd-Year Project, Department of Botany, University of Cape Town, Cape Town

Low AB, Rebelo A(T)G (eds) (1996a) Vegetation of South Africa, Lesotho and Swaziland. Department of Environmental Affairs and Tourism, Pretoria

Low AB, Rebelo A[T]G (eds) (1996b) Vegetation of South Africa, Lesotho and Swaziland. A companion to the vegetation map of South Africa, Lesotho and Swaziland. Department of Environmental Affairs and Tourism, Pretoria

Lubke RA (1983) A survey of the coastal vegetation near Port Alfred, Eastern Cape. Bothalia 14:725–738

Lubke RA, Everard DA, Jackson SY (1986) The biomes of the Eastern Cape with emphasis on their conservation. Bothalia 16:251–261

MacPherson AJ, Gillson L, Hoffman MT (2019) Between- and within-biome resistance and resilience at the fynbos-forest ecotone, South Africa. Holocene 29:1801–1816

Manning J, Goldblatt P (1997) South African Wild Flower Guide 9. Nieuwoudtville, Bokkeveld Plateau & Hantam. Botanical Society of South Africa, Kirstenbosch

Marais KE (2012) Postfire regeneration of mountain fynbos by resprouting: a comparison of species with different life history type. Thesis, Department of Conservation Ecology, Stellenbosch University, Stellenbosch

Martin ARH (1965) Plant ecology of the Grahamstown Nature Reserve. I. Primary communities and plant succession. J S Afr Bot 31:1–54

Martin ARH (1966) Plant ecology of the Grahamstown Nature Reserve. II. Some effects of burning. J S Afr Bot 32:1–39

Martin ARH, Noel ARA (1960) The flora of Albany and Bathurst. Rhodes University, Grahamstown

Maze KE, Rebelo AG (1999a) Core flora conservation areas on the Cape Flats. FCC Report 99/2. Botanical Society of South Africa, Kirstenbosch

Maze KE, Rebelo AG (1999b) Core flora conservation areas on the lowlands of the Cape Metropolitan Area. FCC Report 99/1. Botanical Society of South Africa, Kirstenbosch

McDonald DJ (1985) The plant communities of Swartboschkloof, Jonkershoek. S Afr Natl Sci Progr Rep 104:1–54

McDonald DJ (1987) Ordination by detrended correspondence analysis (DCA) of the vegetation of Swartboschkloof, Jonkershoek, Cape Province. Bothalia 17:121–129

McDonald DJ (1988) A synopsis of the plant communities of Swartboschkloof, Jonkershoek, Cape Province. Bothalia 18:233–260

McDonald DJ (1993a) The vegetation of the southern Langeberg, Cape Province. 1. The plant communities of the Boosmansbos Wilderness Area. Bothalia 23:129–151

McDonald DJ (1993b) The vegetation of the southern Langeberg, Cape Province. 2. The plant communities of the Marloth Nature Reserve. Bothalia 23:153–174

McDonald DJ (1993c) The vegetation of the southern Langeberg, Cape Province. 3. The plant communities of the Bergfontein, Rooiwaterspruit and Phesantefontein Areas. Bothalia 23:239–263

McDonald DJ (1995) Phytogeography, endemism and diversity of the fynbos of the southern Langeberg. Thesis, Department of Botany, University of Cape Town, Cape Town

McDonald DJ (1999) Montane flora of the southern Langeberg, South Africa: a checklist of the flowering plants and ferns. Bothalia 29:119–137

McDonald DJ, Cowling RM, Boucher C (1996) Vegetation environment relationships on a species-rich coastal mountain range in the Fynbos Biome (South Africa). Vegetatio 123:165–182

McKenzie B (1976) A phytosociological study of Orange Kloof, Table Mountain. In: Moll EJ, Campbell BM (eds) The ecological status of Table Mountain. Department of Botany, University of Cape Town, Cape Town

McKenzie B, Rebelo T (1997) Core flora conservation areas on the Cape Flats. FCC Report 97/1. Botanical Society of South Africa, Kirstenbosch

McKenzie B, Moll EJ, Campbell BM (1977) A phytosociological study of Orange Kloof, Table Mountain, South Africa. Vegetatio 34:41–53

Mergili M, Privett S (2008) Vegetation and vegetation-environment relationships at Grootbos Nature Reserve, Western Cape, South Africa. Bothalia 38:89–102

Mills J, Birch SC, Stephenson JD et al (2012) Carbon stocks in fynbos, pastures and vineyards on the Agulhas Plain, South Africa: a preliminary assessment. S Afr J Plant Soil 29:191–193

Mills AJ, Birch S-JC, Stanway R et al (2013) Sequestering carbon and restoring renosterveld through fallowing: a practical conservation approach for the Overberg, Cape Floristic Region, South Africa. Conserv Lett 6(4 July/August):255–263

Milton SJ (1976) A phytosociological survey of the dune vegetation of part of the northern False Bay coast. BSc(Hons) Project, University of Cape Town, Cape Town

Milton SJ (1978) Plant communities of Andriesgrond. Clanwilliam District. Department of Botany, University of Cape Town, Cape Town

Moffett RO, Deacon HJ (1977) The flora and vegetation in the surrounds of Boomplaas Cave: Cango Valley. S Afr Archaeol Bull 32:127–145

Mucina L (2023) Biomes of the southern hemisphere. Springer Nature Switzerland, Cham

Mucina L, Rutherford MC (eds) (2006) The vegetation of South Africa, Lesotho and Swaziland. SANBI, Pretoria

Mucina L, Rutherford MC, Powrie LW (2006a) Logic of the map: Approaches and procedures. In: Mucina L, Rutherford MC (eds) The vegetation of South Africa, Lesotho and Swaziland. SANBI, Pretoria, pp 12–29

Mucina L, Rutherford MC, Powrie LW et al (2006b) Inland azonal vegetation. In: Mucina L, Rutherford MC (eds) The vegetation of South Africa, Lesotho and Swaziland. SANBI, Pretoria, pp 616–657

Mucina L, Lötter MC, Rutherford MC et al (2022) Forest biomes of Southern Africa. New Zeal J Bot 60:377–428

Muir J (1929) The vegetation of the Riversdale area Cape Province. Mem Bot Surv S Afr 13:1–82

Musil CF, de Witt DM (1990) Post-fire regeneration in a sand plain lowland fynbos community. S Afr J Bot 56:167–184

Musil CF, Milton SJ, Davis GW (2005) The threat of alien invasive grasses to lowland Cape floral diversity: an empirical appraisal of the effectiveness of practical control strategies. S Afr J Sci 101:337–344

Mustart P, Cowling RM (1993a) Effects of soil and seed characteristics on seed germination and their possible roles in determining field emergence patterns of four Agulhas Plain (South Africa) Proteaceae. Can J Bot 71:1363–1368

Mustart P, Cowling RM (1993b) The role of regeneration stages in the distribution of edaphically restricted fynbos Proteaceae. Ecology 74:1490–1499

Mustart P, Cowling RM, Albertyn J (1997) South African wild flower guide 8. Southern Overberg. Botanical Society of South Africa, Kirstenbosch

Mustart PJ, Moll EJ, Taylor HC (1993) The efficient use of small plots in a fynbos phytosociological study in the northern Cederberg: a quick way to collect plant-environment data. Bothalia 23:265–269

Neethling D-M (1986) 'n Strukturele en floristiese opname van die plantegroei in die kleinsoogdier-navorsingsterrein te Swartboschkloof, Jonkershoek naby Stellenbosch. BSc(Hons) Project, Department of Botany, University of Stellenbosch, Stellenbosch

Nel PL (1995) The vegetation and flora of the Veldwachters River, Stellenbosch. BSc(Hons) Project, Department of Botany, University of Stellenbosch, Stellenbosch

Ngwenya DK, Holmes PM, Geerts S et al (2023) Scaling up restoration efforts by simulating the effects of fire to circumvent prescribed burns when preparing restoration sites in South African fynbos ecosystems. Austral Ecol 48:158–181

Norton J (1977) Vegetation survey of the Worcester Botanic Gardens. BSc(Hons) Project, Department of Botany, University of Cape Town, Cape Town

O'Callaghan M (1990) The ecology of the False Bay estuarine environments, Cape, South Africa. 1. The coastal vegetation. Bothalia 20:105–111

Olivier MC (1966) Die plantegroei en flora van die Worcesterse Veldreservaat. Thesis, Department of Botany, University of Stellenbosch, Stellenbosch

Parkman NC (1978) A structural-functional classification of the fynbos vegetation of the Bain's Kloof area. BSc(Hons) Project, Department of Botany, University of Cape Town, Cape Town

Paterson CC (1982) A phytosociological classification of the vegetation of the Tygerberg Nature Reserve and De Grendel Farm, Bellville. 3rd-Year Project, Department of Botany, University of Cape Town, Cape Town

Pausas JG, Bond WJ (2020) Alternative biome states in terrestrial ecosystems. Trends Plant Sci 25:250–265

Privett SDJ (1998) Determinants of pattern in fynbos vegetation. Physical site factors, disturbance regime, species attributes and temporal change. Thesis, Department of Botany, University of Cape Town, Cape Town

Privett SDJ, Cowling RM, Taylor HC (2001) Thirty years of change in the fynbos vegetation of the Cape of Good Hope Nature Reserve, South Africa. Bothalia 31:99–115

Quick LJ, Meadows ME, Bateman MD et al (2016) Vegetation and climate dynamics during the last glacial period in the fynbos-afrotemperate forest ecotone, southern Cape, South Africa. Quater Int 404:136–149

Raitt GR (1999) Management plan for the Silwerboom Kloof Natural Heritage Site. BSc(Hons) Project, Forestry Conservation Department, University of Stellenbosch, Stellenbosch

Raitt GR (2005) *Themeda triandra* renosterveld in the Heidelberg District. Thesis, Department of Botany & Zoology, Stellenbosch University, Stellenbosch

Rebelo AG, Cowling RM, Campbell BM et al (1991) Plant communities of the Riversdale Plain. S Afr J Bot 57:10–28

Rebelo AG, Boucher C, Helme N et al (2006) Fynbos Biome. In: Mucina L, Rutherford MC (eds) The vegetation of South Africa, Lesotho and Swaziland. SANBI, Pretoria, pp 52–219

Retief L (2022) A determination of Cape Flats Sand Fynbos restoration success in the Western Cape, South Africa, based on vegetation establishment and pollinator activity. Thesis, Faculty of Applied Sciences, Cape Peninsula University of Technology, Cape Town

Richards MB (1994) Soil factors and competition as determinants of fynbos plant species distributions in the south-western Cape, South Africa. Thesis, Department of Botany, University of Cape Town, Cape Town

Richards MB, Cowling RM, Stock WD (1995) Fynbos plant communities and vegetation-environment relationships in the Soetanysberg hills, Western Cape. S Afr J Bot 61:298–305

Richards MB, Cowling RM, Stock WD (1997a) Soil factors and competition as determinants of the distribution of six fynbos Proteaceae species. Oikos 79:394–406

Richards MB, Stock WD, Cowling RM (1997b) Soil nutrient dynamics and community boundaries in the fynbos vegetation of South Africa. Plant Ecol 130:143–153

Richardson GR, Lubke RA, Jacod Guillarmod A (1984) Regeneration of grassy fynbos near Grahamstown (Eastern Cape) after fire. S Afr J Bot 3:153–162

Rode E (1994) Die plantegroei van die plaas Solva in die Elgin-kom, Suidwes-Kaap. BSc(Hons) Project, Department of Botany, University of Stellenbosch, Stellenbosch

Rule S (1992) A vegetation survey of Elsepeak in Fishhoek, Cape Province. 3rd-Year Project, Department of Botany, University of Cape Town, Cape Town

Rutherford MC (1978) Karoo-Fynbos biomass along an elevational gradient in the western Cape. Bothalia 12:555–560

Rutherford MC, Westfall RH (1986) Biomes of southern Africa: an objective categorization. Mem Bot Surv S Afr 54:1–98

Rutherford MC, Mucina L, Powrie LW (2006) Biomes and bioregions of Southern Africa. In: Mucina L, Rutherford MC (eds) The vegetation of South Africa, Lesotho and Swaziland. SANBI, Pretoria, pp 30–51

Rutherford MC, Powrie LW, Husted LB et al (2011) Early post-fire plant succession in Peninsula Sandstone Fynbos: the first three years after disturbance. S Afr J Bot 77:665–674

Rycroft HB (1953) A quantitative ecological study of the vegetation of Biesievlei, Jonkershoek. Thesis, University of Cape Town, Cape Town

Samuels MI, Allsopp N, Hoffman MT (2022) Stocking density and distribution in relation to vegetation in a biodiverse semi-arid pastoral system in South Africa. Rangel J 44:1–12. https://doi.org/10.1071/RJ21046

Schmiedel U (2002) The quartz fields of southern African flora, phytogeography, vegetation, and habitat ecology. Thesis, Botanisches Institut, Universität zu Köln, Köln

Schmiedel U, Mucina L (2006) Vegetation of quartz fields in the Little Karoo, Tanqua Karoo and eastern Overberg (Western Cape Province, South Africa). Phytocoenologia 36:1–44

Schüttler G (2002) Floristische und vegetationsökologische Analyse eines Höhengradienten im südlichen Richtersveld, Republik Südafrika. Thesis, Botanisches Institut, Universität zu Köln, Köln

Schwilk DW, Keeley JE, Bond WJ (1997) The intermediate disturbance hypothesis does not explain fire and diversity pattern in fynbos. Plant Ecol 132:77–84

Sieben EJJ (2003) The riparian vegetation of the Hottentots Holland Mountains, Western Cape, South Africa. Thesis, Department of Botany, University of Stellenbosch, Stellenbosch

Sieben EJJ, Boucher C, Mucina L (2004) Vegetation of high-altitude fens and restio marshlands of the Hottentots Holland Mountains, Western Cape, South Africa. Bothalia 34:141–153

Simmons M (1992) Ordination using two methods of plant communities on Chapman's Peak, Cape Province. 3rd-Year Project, Department of Botany, University of Cape Town, Cape Town

Simmons MT (1996) Vegetation, diversity and rarity on the Cape Peninsula, South Africa. Thesis, Department of Botany, University of Cape Town, Cape Town

Simons M, Samuels I, Morris C et al (2024) The effects of fire and stocking density on the unpalatable species *Elytropappus rhinocerotis* in Renosterveld vegetation in South Africa. Afr J Ecol 62:e13190. https://doi.org/10.1111/aje.13190

Smit IPJ, Baard JA, van Wilgen BW (2024) Fire regimes and management options in mixed grassland-fynbos vegetation, South Africa. Fire Ecol 20:29. https://doi.org/10.1186/s42408-024-00262-2

Smitheman J, Perry P (1990) A vegetation survey of the Karoo National Botanic Garden Reserve, Worcester. S Afr J Bot 56:525–541

Snijman D, Perry P (1987) A floristic analysis of the Nieuwoudtville Wild Flower Reserve, north western Cape. S Afr J Bot 53:445–454

Stofberg T (2000) Management plan for the greater lakes of Betty's Bay: Groot Wit Vlei and Rondevlei. BSc(Hons) Project, Department of Forestry Conservation, University of Stellenbosch, Stellenbosch

Swart EC (1983) Die plantegroei en flora van die Jan Marais-natuurreservaat. BSc(Hons) Project, Department of Botany, University of Stellenbosch, Stellenbosch

Taylor HC (1969) A vegetation survey of the Cape of Good Hope Nature Reserve. Thesis, Department of Botany, University of Cape Town, Cape Town

Taylor HC (1970a) I.B.P. checksheet for the Riversdale Nature Reserve. Botanical Research Institute, Stellenbosch

Taylor HC (1970b) I.B.P. checksheet for the Robberg Nature Reserve. Botanical Research Institute, Stellenbosch

Taylor HC (1972) Notes on the vegetation of the Cape Flats. Bothalia 10:637–646

Taylor HC (1979) Observations on the flora and phytogeography of Rooiberg, a dry fynbos mountain in the southern Cape Province, South Africa. Phytocoenologia 6:524–531

Taylor HC (1981) Red Hill, Simons Town flora and vegetational survey. Preliminary hydrological survey, Informal Settlement 1991. FCC Report. Botanical Society of South Africa, Kirstenbosch

Taylor HC (1983) The vegetation of the Cape of Good Hope Nature Reserve. Bothalia 14:779–784

Taylor HC (1984a) A vegetation survey of the Cape of Good Hope Nature Reserve. I. The use of association-analysis and Braun-Blanquet methods. Bothalia 15:245–258

Taylor HC (1984b) A vegetation survey of the Cape of Good Hope Nature Reserve. II. Descriptive account. Bothalia 15:259–291

Taylor HC (1996) Cederberg vegetation and flora. National Botanical Institute, Pretoria

Taylor HC, Boucher C (1993) Dry coastal ecosystems of the South African South Coast. In: van der Maarel E (ed) Ecosystems of the world 2B. Dry coastal ecosystems. Africa, America and Oceania. Elsevier, Amsterdam, pp 89–107

Taylor HC, van der Meulen F (1981) Structural and floristic classifications of Cape Mountain Fynbos on Rooiberg, southern Cape. Bothalia 13:557–567

Thwaites RN, Cowling RM (1988) Soil-vegetation relationships on the Agulhas Plain, South Africa. CATENA 15:333–345

Tritton R (1992) Sandy Bay vegetation survey. 3rd-Year Project, Department of Botany, University of Cape Town, Cape Town

Uys CJ (1983) Portrait of a fynbos habitat extraordinary—De Hoop Vlei. Afr Wildl 37:5–8

van Blerk JJ, West AG, Altwegg R et al (2021) Does a trade-off between growth plasticity and resource conservatism mediate post-fire shrubland responses to rainfall seasonality? New Phytol 230:1407–1420

van Daalen JC (1984) Distinguishing features of forest species on nutrient-poor soils in the southern Cape. Bothalia 15:229–239

van der Merwe CV (1977a) 'n Plantegroei-opname van die De Hoop-natuurreservaat. Bontebok 1:1–29

van der Merwe CV (1977b) 'n Plantegroeibeskrywing van die Vrolijkheidnatuurbewaringstasie. Departament van Natuur- en Omgewingsbewaring, Cape Town

van der Merwe P (1962) Floristiese opname van Swartboskloof, Stellenbosch en die herstel van die flora na 'n brand. Thesis, University of Stellenbosch, Stellenbosch
van der Merwe P (1966) Die flora van Swartboskloof, Stellenbosch en die herstel van die soorte na 'n brand. Ann Univ Stellenbosch Ser A 41:687–736
van Jaarsveld E (1981) A preliminary report on the vegetation of the Richtersveld with specific reference to the trees and shrubs of the area. Trees S Afr 33:58–84
van Jaarsveld E (1982) Plantegroei van die Gifberg met spesifieke verwysing na bome, struike en die gifboom. Trees S Afr 34:10–27
van Wilgen BW, Kruger FJ (1981) Observations on the effects of fire in mountain fynbos at Zachariashoek, Paarl. J S Afr Bot 47:195–212
van Wilgen BW, Kruger FJ (1985) The physiography and fynbos vegetation communities of the Zachariashoek catchments, south western Cape Province. S Afr J Bot 51:379–398
van Wilgen BW, Richardson DM, Kruger FJ et al (eds) (1992) Fire in South African mountain fynbos: ecosystem, Community and species response at Swartboskloof. Springer-Verlag, Berlin
van Wyk AE, Smith GF (2001) Regions of floristic endemism in southern Africa. A review with emphasis on succulents. Umdaus Press, Hatfield
van Wyk B-E, Novellie PA, van Wyk CM (1988) Flora of the Zuurberg National Park. 1. Characterization of major vegetation units. Bothalia 18:211–220
Verboom AG, Slingsby JA, Cramer MD (2024) Fire-modulated fluctuations in nutrient availability stimulate biome-scale floristic turnover in time, and elevated species richness, in low-nutrient fynbos heathland. Ann Bot 133:819–832
Vermeulen WJ (1995) 'n Fitososiologiese studie van die Ysternek-natuurreservaat in Suid-Kaapland. Thesis, Department of Botany, University of Stellenbosch, Stellenbosch
Vlok JHJ, Euston-Brown DIW (2002) Subtropical Thicket Ecosystem Planning Project (STEP). Biological Survey Report (Plants and Birds). Terrestrial Ecology Research Unit, University of Port Elizabeth, Port Elizabeth
Vlok JHJ, Euston-Brown DIW, Cowling RM (2003) Acocks's Valley Bushveld 50 years on: new perspectives on the delimitation, characterisation and origin of subtropical thicket vegetation. S Afr J Bot 69:27–51
von Hase A, Rouget M, Maze K et al (2003) A fine-scale conservation plan for Cape Lowlands Renosterveld: Technical Report. Report CCU2/03. Botanical Society of South Africa, Kirstenbosch
Walton BA (2006) Vegetation patterns and dynamics of renosterveld at the Agter-Groenberg Conservancy, Western Cape, South Africa. Thesis, Department of Conservation Ecology, Stellenbosch University, Stellenbosch
Werger MJA (1972) Species-area relationship and plot size: with some examples from South African vegetation. Bothalia 10:583–594
Werger MJA, Kruger FJ, Taylor HC (1972a) A phytosociological study of the Cape Fynbos and other vegetation at Jonkershoek, Stellenbosch. Bothalia 10:599–614
Werger MJA, Kruger FJ, Taylor HC (1972b) Pflanzensociologische Studie der Fynbosvegetation am Kap der Guten Hoffnung. Vegetatio 24:71–89
Willis CK, Cowling AT (1996) Patterns of endemism in the limestone flora of South African lowland fynbos. Biodivers Conserv 5:55–73
Willis CK, Lombard AT, Cowling RM et al (1996) Reserve systems for limestone endemic flora of the Cape lowland fynbos: iterative versus linear programming. Biol Conserv 77:53–62
Witkowski ETF, Mitchell DT (1989) The effects of nutrient additions on above ground phytomass and its phosphorus and nitrogen contents of sand plain lowland fynbos. S Afr J Bot 55:243–249
Wood JC (1990) The natural vegetation of the Robertson Karoo: an evaluation of its conservation status. Thesis, Department of Botany, University of Cape Town, Cape Town
Wood JC, Low AB, Donaldson JS et al 1994. Threats to plant species diversity through urbanization and habitat fragmentation in the Cape Metropolitan Area, South Africa. In: Huntley BJ (ed) Botanical diversity in southern Africa. National Botanical Institute, Pretoria, pp 259–274
Zietsman MM, Bredenkamp GJ (2006) Dune vegetation and coastal thicket plant communities in threatened limestone fynbos of Andrew's Field and Tsaba-Tsaba Reserve, Struisbaai, Western Cape. Koedoe 49:33–47
Zietsman MM, Bredenkamp GJ (2007) Threatened Limestone Fynbos plant communities of Andrew's Field and Tsaba-Tsaba Nature Reserve, Western Cape. Bothalia 37:89–102

Warm-Temperate Zonobiome: Afrotemperate Forests

6

Ladislav Mucina, Mervyn C. Lötter, Michael C. Rutherford and James L. Tsakalos

Abstract

Afrotemperate forests are a representative of the zonobiome T3 Warm-Temperate Zone. The flagship regional biome of these evergreen forests dominated by species of such genera as *Afrocarpus, Curtisia, Diospyros, Ilex, Nuxia, Ocotea, Olea, Olinia, Podocarpus, Pterocelastrus, Rapanea, Widdringtonia* and many more, are the coastal Knysna-Tsitsikamma forests straddling the borders of the Western and Eastern Cape Provinces of South Africa. Starting on the east-facing slopes of Table Mountain, surrounded by Cape Town, these forests are found further eastwards and north-eastwards as well as further inland towards lower latitudes. These inland forests, which in the MBSA reach as far as the Eastern Highlands of Zimbabwe and Mozambique, are of a mistbelt nature—they occur in climatic belts under the influence of precipitating sea-borne cloud/mist masses. The floristically relatively uniform afrotemperate forests also occur as extrazonal regional biomes in deep valleys—often associated with main mountain streams. Here, they occur on steep talus slopes. The extrazonality of these forests is underlined by them being embedded within the Fynbos and Highveld biomes or straddling the borders between the Highveld grasslands and Mesic Savanna. In total, there are four zonal (Cape Afrotemperate Forest, Southern Mistbelt Forest, Northern Mistbelt Forest and Manica-Mulanje Mistbelt Forest) and four extrazonal (Cape Talus Forest, Bushveld Talus Forest, Drakensberg Afrotemperate Forest and Eastern Highveld Afrotemperate Forest) regional biomes.

L. Mucina (✉) · J. L. Tsakalos
Harry Butler Institute, Murdoch University,
90 South Street, Murdoch, Perth 6150, Australia
e-mail: Ladislav.Mucina@murdoch.edu.au

L. Mucina
Department of Geography and Environmental Studies, Stellenbosch University, Private Bag X1, Matieland, Stellenbosch 7602, South Africa

M. C. Lötter
Mpumalanga Tourism and Park Agency, P/Bag X1088, Lydenburg 1120, South Africa

School of Animal, Plants and Environmental Sciences, University of the Witwatersrand, Private Bag 3, Wits, 2050, Johannesburg, South Africa

M. C. Rutherford
High Peak, Derbyshire, United Kingdom

Department of Botany and Zoology, Stellenbosch University, Private Bag X1, Matieland, Stellenbosch 7602, South Africa

J. L. Tsakalos
Plant Diversity and Ecosystems Management Unit, School of Biosciences and Veterinary Medicine, University of Camerino, Via Pontoni 5, Camerino 62032, Italy

L. Mucina and M. C. Rutherford (eds.), *Zonal Biomes of Southern Africa*, Biome Ecology 2,
https://doi.org/10.1007/978-3-031-63731-5_6

Keywords

Cape forests · Temperate forest biomes · Knysna-Tsitsikamma region · South African mistbelts · Talus forests

6.1 Afrotemperate Forests: General Patterns

Southern Africa is home to a wide variety of zonal and azonal forests. The zonal section is represented by afrotemperate and subtropical forest. This chapter is about the zonal forests.

The term 'afrotemperate' originates in two features characteristic of these forests. Firstly, they occur in the warm-temperate section of the bioclimatic envelope, and (2) hardly surprisingly, they are geographically confined to Africa. Especially in South Africa they were formerly called 'afromontane' (see White 1978) although many of these forests are found close to sea level, hence not exclusively in the mountains. For a more detailed discussion of the current terminology, consult Mucina et al. (2006a).

The afrotemperate forests, in South Africa, have been the subject of intensive vegetation-ecological and silvicultural research as summarised by von Maltitz et al. (2003), Mucina et al. (2006a, 2022), and Lötter et al. (2014). Müller (1999, 2006) and Lötter et al. (2023) studied and mapped these forests in Zimbabwe and Mozambique.

As described by Mucina et al. (2022: Figs. 2 and 3), the afrotemperate forests show an interesting distribution pattern, demonstrating the latitude vs elevation bioclimatic trade-off as observed by Ohsawa (1993) for the vegetation belt system of the Asian mountain ranges. As one moves from the high latitudes of the W and S Cape inland, the forested vegetation belts are found at increasingly higher elevations (see Mucina et al. 2022 for a more detailed discussion and graphical analysis of this phenomenon). Four zonal regional biomes have been recognised along the increasing elevation axis. These are the Cape Afrotemperate Forest, Southern Mistbelt Forest, Northern Mistbelt Forest and Manica-Mulanje Mistbelt Forest (Table 6.1; Fig. 6.1). The largest continuous patch in the Southern Cape is known as the Knysna-Tsitsikamma complex, while the spatially largest, albeit very fragmented, regional biome is the Southern Mistbelt Forest.

The bioclimate of the Cape Afrotemperate Forest is characterised by high precipitation and, perhaps more importantly, year-round rainfall distribution. As established by Mucina (2023), the region receives year-round precipitation—a combination of the winter rainfall brought in by the Westerlies and the trade wind onshore moisture flow in summer. The mistbelt forest units (see von Maltitz et al. 2003) are also considered 'afrotemperate'. These are mistbelt (hence zonal) forests as they occupy habitats under a direct influence of regional rainfall and regional fog-formation due to ascending moisture-bearing onshore air flow systems that originate above the Indian Ocean (Mucina et al. 2022; Mucina 2023).

Besides the core zonal, the afrotemperate forests also occur in the MBSA as extrazonal units, occurring in refugial habitats, protected from excessive desiccation and frequent fires (Mucina et al. 2022). Although not under its current name (coined by L. Mucina in von Maltitz et al. 2003), these forests have been known under the Afrikaans terms 'dasbos' (van der Merwe 1966) or 'kloof forest' and 'ravine forest'. Since it is found in topographically complex situations (steep slopes, deep ravines), the substrate supporting these forests is usually skeletal, taking the form of boulder fields and coarse-gravel screes. The talus forests (Table 6.1; Fig. 6.2) occur in the MBSA in both winter-rainfall Cape and summer-rainfall Bushveld regions (see von Maltitz et al. 2003; Mucina et al. 2006a).

The Cape Talus Forests share the same meso-climate (regional climate at the spatial scale of up to 100 km) as the surrounding fynbos shrublands and show a typical warm-temperate bioclimatic envelope (Mucina et al. 2022). They are possibly a relict character of the typical zonal Cape Afrotemperate of the Knysna-Tsitsikamma forests (zonobiome T3; see below) yet embedded meso-climatically within the CB Fynbos belonging to the zonobiome S1.

Table 6.1 A system of biomes of the zonobiome T3 warm-temperate zone of Southern Africa (following Mucina 2023)

Rank	Code	Biome	Zonality	Azonality type	Climatic diagnosis
ZB	**T3**	**Warm-Temperate Zone**	**Zonal**		**Warm to hot summers, mild winters; weak dry season; precipitation maximum in austral summer**
GB	**WTZs**	**Southern Warm-Temperate Forest**	**Zonal**		**Warm to hot summers, mild winters; weak dry season; precipitation maximum in austral summer**
CB	**ATFs**	**Southern African Afrotemperate Forest**	**Zonal**		**Warm to hot summers, mild winters; weak dry season; precipitation maximum in austral summer**
		Core zonal regional biomes	Zonal		
RB	AF1	Cape Afrotemperate Forest	Zonal		
RB	AF2	Southern Mistbelt Forest	Zonal		
RB	AF3	Northern Mistbelt Forest	Zonal		
RB	AF4	Manica-Mulanje Mistbelt Forest	Zonal		
		Extrazonal regional biomes			
RB	AF5	Cape Talus Forest	Extrazonal	Lithobiome/relict	
RB	AF6	Drakensberg Afrotemperate Forest	Extrazonal	Lithobiome/relict	
RB	AF7	Bushveld Talus Forest	Extrazonal	Lithobiome/relict	
RB	AF8	Eastern Highveld Afrotemperate Forest	Extrazonal	Lithobiome/relict	

The coding of the regional biomes follows Mucina et al. (2022)

The Drakensberg Afrotemperate Forests, Eastern Highveld Afrotemperate Forests and Bushveld Talus, all confined to summer-rainfall regions of Southern Africa, are dominated by the floristic elements of the afrotemperate forest. They also occur in refugial habitats (e.g., Adie et al. 2017) such as mesic scarp and deep ravine habitats, where they appear to be embedded either within the CB Mesic Savanna (zonobiome E2; see Chap. 3 in this book) or within the zonoecotone AMG African Montane Grassland (the mesic section of the Highveld (see Chap. 9 in this book). The microclimatic conditions underpin the relict nature of their occurrence, surrounded by fire-prone savanna and other landscapes dominated by C4-grasslands. Thus, this climatically decouples them from the meso-climatic scale. This decoupling defines their extrazonal character in relation to the zonal afrotemperate forests (Mucina et al. 2022).

For more information on the general physiognomy, floristic composition, and the palaeobotanical roots of the formation of these forests, we refer to von Maltitz et al. (2003), Mucina et al. (2006a, 2022) and Mucina (2023).

6.2 Descriptions of the Afrotemperate Regional Forest Biomes

AF1 Cape Afrotemperate Forest

FOz 1 Southern Afrotemperate Forest p.p. (Mucina and Rutherford 2006)

Fig. 6.1 Afrotemperate Mistbelt Forests (zonobiome T3). **a** A stand of mistbelt forest at Almeida near Barberton, Barberton Mountainland, Mpumalanga Province, *with Brachylaena transvaalensis, Cassipourea malosana, Chionanthus foveolatus* subsp. *major, Combretum kraussii, Cussonia spicata,* and *Ochna gamostigmata.* **b** Afrotemperate mistbelt forest in the Calodendrum Gorge in Buffelskloof Private Nature Reserve near Lydenburg, Mpumalanga Province of South Africa. The most prominent species in this forest are *Celtis africana, Combretum kraussii, Cryptocarya transvaalensis, Ekebergia capensis, Gymnosporia harveyana. Harpephyllum caffrum, Heteropyxis canescens, Scolopia mundii*, and *Zanthoxylum davyi.* **c** Ngome Forest, 60 km east of Vryheid, KwaZulu-Natal, South Africa, with dominating trees *Allophylus dregeanus, Cassipourea malosana, Chionanthus peglerae, Cryptocarya woodii, Dalbergia obovata, Rinorea angustifolia, Searsia chirindensis, Strophanthus speciosus, Syzygium gerrardii,* and *Xymalos monospora.* Photo credits: All photos M. C. Lötter

Fig. 6.2 Selected extrazonal talus forests. **a** Ravine forest on Table Mountain Sandstone in a deep kloof of Vogelgat Private Nature Reserve near Hermanus, Western Cape Province. **b** A forest patch of boulder scree under a steep Clarens Sandstone cliff, Golden Gate Highlands National Park near Clarens, Free State Province. The surrounding vegetation is a mixed, C3-dominated and C4-dominated high-elevation grassland. **c** Talus forests (dark-green patches), surrounded by wet *Leucadendron* and *Erica* dominated fynbos shrublands on steep sandstone slopes of a deep kloof in Tradouw Pass near Suurbraak, Western Cape Province. **d** Kloof forest flanking a stream and a waterfall on dolomite cliffs in the Strydom Pass, Northern Drakensberg Escarpment, Limpopo Province. Photo credits: All photos L. Mucina

Distribution This forest biome has the largest continuous, albeit fragmented due to historical logging, forest complex in South Africa, known as the Knysna-Tsitsikamma forest complex. In the Southern Cape, it occupies the coastal platform (incl. coastal scarp) approximately between George and Humansdorp. Outlier forest patches are found on low-elevation slopes of the Outeniqua and Langeberg ranges. The westernmost outlying patch is located on the east-facing slopes of Table Mountain (Cape Town) (Fig. 6.3).

Vegetation Physiognomy and Composition These are tall, multilayered forests dominated by *Apodytes dimidiata, Cassine peragua, Curtisia dentata, Diospyros whyteana,*

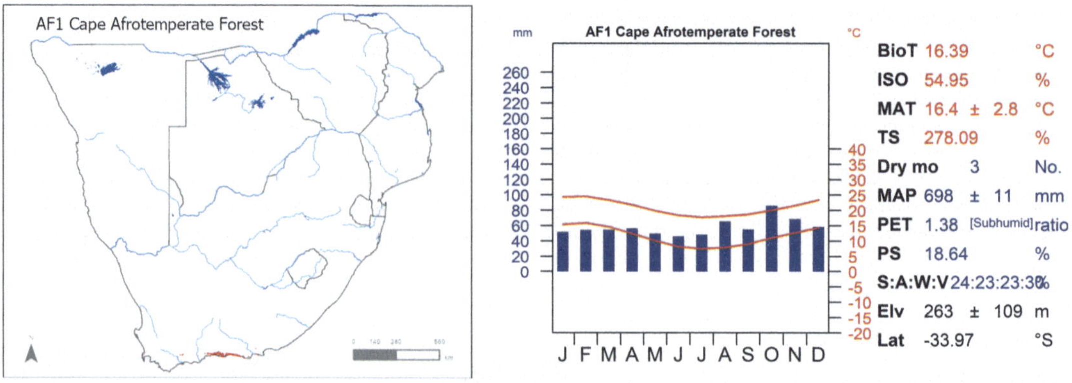

Fig. 6.3 Distribution area and the climatic diagram of the regional biome AF1 Cape Afrotemperate Forest

Nuxia floribunda, Ocotea bullata, Olea capensis subsp. *macrocarpa, Podocarpus latifolius, Pterocelastrus tricuspidatus* and *Rapanea melanophloeos*. The shrub layer hosts *Alsophila capensis* (tree fern), *Burchellia bubalina, Trichocladus crinitus,* and *Sparmannia africana* on forest edges. Lianas are common, such as those of the genera *Capparis, Rhoicissus, Scutia,* and *Secamone*. The herbaceous layer is rich in ferns; *Oplismenus hirtellus* and some *Carex* species (formerly *Schoenoxiphium*) are the most important understory graminoids.

Climate Most of the biome experiences year-round precipitation. Winter rainfall is prevalent in the western regions, while the rainfall in the core Knysna-Tsitsikamma region is mainly a result of eastern onflow (see Mucina 2023: 106–107). Thermically, this is a warm-temperate biome (Fig. 6.3).

Topography, Geology, and Soils The dominant topography of the region supporting this forest biome is flat (Tsitsikamma region) or convex (Knysna region, Outeniqua and Langeberg piedmonts). In the latter area, the forests occurring on semi-steep slopes, i.e. the coastal scarp forests of the Knysna-Tsitsikamma region, are found on steep sea-facing slopes. The underlying geology is predominantly Table Mountain Sandstone, but Cape shale, Cape granites, and conglomerates also occur.

Sources Adamson (1927), Phillips (1931), Laughton (1937), Taylor (1955, 1984), von Breitenbach (1972), van Laar and Lewark (1973), van Laar and Geldenhuys (1975), Campbell and Moll (1977), McKenzie et al. (1977), Glyphis et al. (1978), McKenzie (1978), Geldenhuys (1982, 1986, 1987, 1989a, b, 1992a, b, 1993a, b, 1994a, b, 1996, 1997a), Cowling (1984), van Daalen (1984), Koen and Crowe (1987), Jeffery and Wilson (1987), Hanekom et al. (1989), Masson and McKenzie (1989), Masson (1990), Balfour and Bond (1993), McDonald (1993a, b), Vermeulen (1993, 1994, 1995), von Maltitz et al. (2003), Wardell-Johnson and Calver (2005), Geldenhuys and Mucina (2006), Mucina et al. (2006a, 2022), Coetsee and Wigley (2013), Seifert et al. (2014), South African National Parks (2014), Coetsee et al. (2015), Poulsen and Hoffman (2015), Quick et al. (2016), Moir et al. (2020), Naudé et al. (2021), Swart et al. (2021), Giddey et al. (2022a, b), Kraaij et al. (2023), Mucina (2023), Townsend et al. (2024).

AF2 Southern Mistbelt Forest

FOz 3 Southern Mistbelt Forest (Mucina et al. 2006a)

Distribution This regional forest biome has a fragmented distribution at mid-elevations of the Kwa-Zulu Natal Midlands near Ulundi in the north and southwards to the Kokstad vicinity, into the Eastern Cape through areas of Mount Ayliff and Engcobo to the southern Amathole

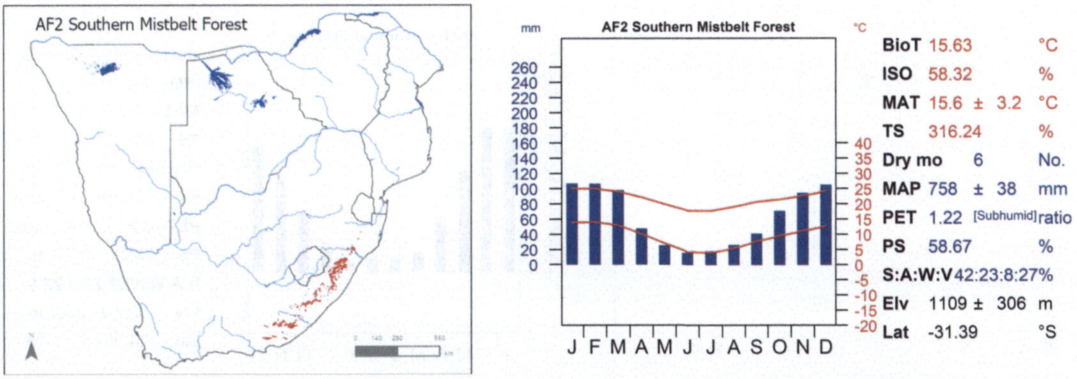

Fig. 6.4 Distribution area and the climatic diagram of the regional biome AF2 Southern Mistbelt Forest

and Winterberg Mts between Stutterheim and Fort Beaufort, with outliers as far as Somerset East and the Zuurberg further west.

Vegetation Physiognomy and Composition The forests characterising this biome are 15–20 m tall and multilayered in Amathole, Transkei Escarpment and the KwaZulu-Natal Midlands. Those found on low-elevation scarps are low-grown and, although less structured into different tree layers, are still species-rich (Mucina et al. 2006a). The prominent tree species include *Afrocarpus falcatus, Calodendrum capense, Celtis africana, Kiggelaria africana, Podocarpus latifolius, Rapanea melanophloeos, Xymalos monospora,* and *Zanthoxylum davyi. Podocarpus henkelii* is an essential element in the KwaZulu-Natal Midlands.

Climate A warm-temperate climate is characteristic of this biome, which experiences mainly summer rainfall and frequent heavy mist/fog in spring and summer as an effect of the onshore flow of air masses from the Indian Ocean (Fig. 6.4).

Topography, Geology, and Soils These forests are found along the escarpments, mountain slopes, and scarps, including broad intermountain valleys, major river valleys, and gullies. They occur mainly on mudstone, shale, sandstone, and dolerite of the Ecca and Beaufort Groups.

Sources Fourcade (1889), Story (1952), Edwards (1967), Moll (1972, 1976), Moll and Woods (1972), Cooper (1985), Cawe (1986, 1996), Phillipson (1987), Cawe and McKenzie (1989a, b, c, d), Everard (1992), Geldenhuys (1992a), Everard and Hardy (1993), Everard et al. (1995), Geldenhuys and Rathogwa (1997), Bester (1998), Lawes et al. (2000, 2005, 2006a, b, 2007a, b), Kruger and Midgley (2001), Wirminghaus et al. (2001), von Maltitz et al. (2003), Downs and Symes (2004), Robertson and Lawes (2005), Mucina et al. (2006a, 2022), Geldenhuys (2007), Kotze and Lawes (2007), Abbott (2010), Adie et al (2013), Hart et al. (2013), Lawes et al. (2004), Grieve and Downs (2015), Leaver et al. (2019), Moir et al. (2020), Hall et al. (2022), Mulvaney et al. (2022), Ngcobo et al. (2022), Parker et al. (2022), Sibanda et al. (2022), Bitani et al. (2023), Rea et al. (2023), Sosibo et al. (2023, 2024).

AF3 Northern Mistbelt Forest

FOz 4 Northern Mistbelt Forest p.p. (Mucina et al. 2006a)

Distribution The distribution area spans the Mpumalanga Escarpment (Barberton, Long Tom Pass, Blyde and Marieskop areas), extending northwards along the North Eastern Escarpment as far north as Soutpansberg in the west (Fig. 6.5).

Vegetation Physiognomy and Composition These are tall, moist, evergreen forests rich in epiphytes and comprised of *Xymalos monospora, Brachylaena transvaalensis, Bersama tysoniana, Cassipourea malosana, Chionanthus foveolatus* subsp. *major, Cryptocarya*

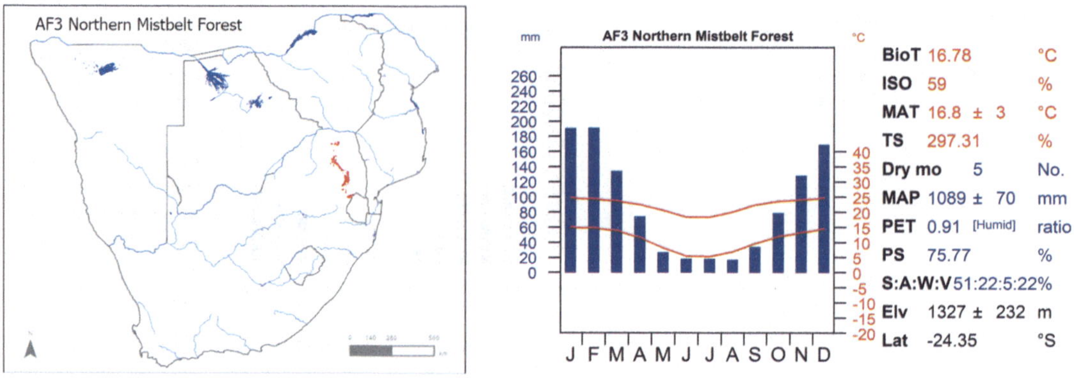

Fig. 6.5 Distribution area and the climatic diagram of the regional biome AF3 Northern Mistbelt Forest

transvaalensis, Curtisia dentata, Olea capensis subsp. *macrocarpa, Oxyanthus speciosus* subsp. *gerrardii, Maytenus acuminata* var. *acuminata, Podocarpus latifolius, Psydrax obovata* subsp. *elliptica, Rapanea melanophloeos, Rothmannia capensis, Schefflera umbellifera, Scolopia mundii, Searsia chirindensis, Syzygium gerrardii,* and *Zanthoxylum davyi.*

Topography, Geology, and Soils These forests are found primarily along the cooler east-facing fire-protected slopes of the high-lying and wet escarpment areas or in moist, sheltered kloofs. Bedrock varies and comprises shale, diabase, dolomite, quartzite, and granite.

Sources van der Schijff and Schoonraad (1971), Scheepers (1978), Deall et al. (1989), von Breitenbach (1990), van Jaarsveld and Hardy (1991), Matthews et al. (1992), Geldenhuys and Murray (1993), Geldenhuys and Pieterse (1993), Masson (1994), Geldenhuys (1997b, 2007), Morgenthal and Cilliers (1999, 2000), Stalmans et al. (1999), Geldenhuys and Venter (2002), von Maltitz et al. (2003), Lötter and Beck (2004), Geldenhuys and Mucina (2006), Mucina et al. (2006a, 2022), Mostert et al. (2008), Lötter et al. (2014), Mensah et al. (2016a, b, 2017a, b, 2018a, b, c, 2020a, b), Moir et al. (2020), Sinthumule and Mashau (2020), Fisher (2024).

AF4 Manica-Mulanje Mistbelt Forest

Afrocrania volkensii montane forest (Müller 1999)
Ilex mitis-Schefflera umbellifera-Maesa lanceolata montane forest (Müller 1999)
Regenerating montane forest (Müller 1999)
Syzygium guineense subsp. *afromontanum* montane forest (Müller 1999)
Syzygium masukuense montane forest (Müller 1999)
Widdringtonia nodiflora forest (Müller 1999)
Gorongosa Montane Forest (Stalmans and Beilfuss 2008)
Afromontane forest (Timberlake et al. 2016)
ATF1 Central Montane Forest (Lötter et al. 2023)
ATF2 Central Submontane Forest (Lötter et al. 2023)

Distribution This multi-patch regional biome is confined to central Mozambique and Zimbabwe but extends northwards into northern Mozambique and adjacent Malawi outside the MBSA area. It is a characteristic feature of the upland mountainous regions of Chimanimani, Nyanga, Tsetserra, Serra Choa, and Serra Gorongosa (Fig. 6.6).

Vegetation Physiognomy and Composition These are tall moist evergreen forests, rich in epiphytes and ferns. The canopy is dominated by *Albizia gummifera, Aphloia theiformis, Astropanax goetzenii, Cornus volkensii, Cassipourea malosana, Chionanthus foveolatus* subsp. *major, Cryptocarya liebertiana, Curtisia dentata, Faurea rubriflora, F. wentzeliana, Macaranga mellifera, Maytenus*

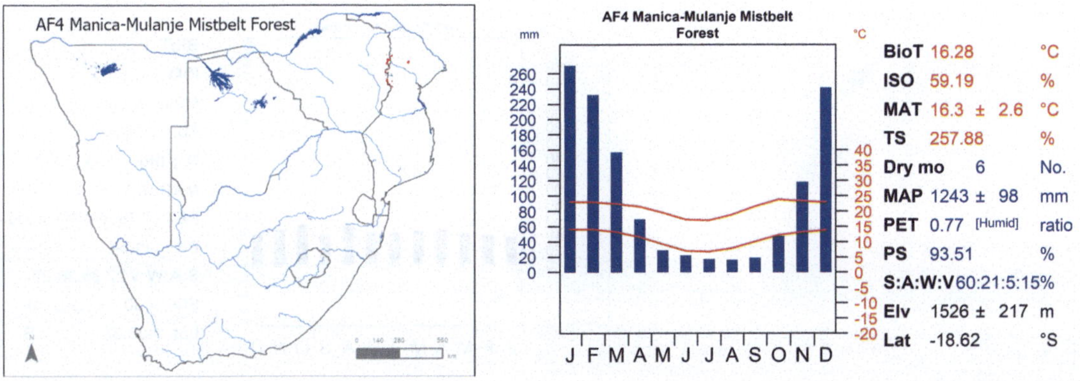

Fig. 6.6 Distribution area and the climatic diagram of the regional biome AF4 Manica-Mulanje Mistbelt Forest

acuminata, Neoboutonia melleri, Olea capensis subsp. *hochstetteriana, Pittosporum viridiflorum, Podocarpus milanjianus, Polyscias fulva, Rapanea melanophloeos, Rothmannia urcelliformis, Strombosia scheffleri, Syzygium afromontanum,* and *Xymalos monospora.*

Topography, Geology, and Soils The Manica-Mulanje forests are found along the moist and cool slopes or summits of the Eastern (Manica) Highlands escarpment and on large inselbergs at elevations above 1300 m. The geology predominantly comprises quartzite, granite, gneiss, and rarely gabbro.

Sources Müller (1999, 2006), Müller et al. (2005, 2012), Timberlake et al. (2007), Burrows et al. (2018), Mucina et al. (2022), Lötter et al. (2023).

AF5 Cape Talus Forest

Western Cape Scree Forest (von Maltitz et al. 2003)
FOz 1 Southern Afrotemperate Forest p.p. (Mucina et al. 2006a)
AZa 1 Fynbos Riparian Vegetation p.p. (Mucina et al. 2006b)

Distribution The Cape Talus Forest occurs in isolated patches along the north-south trending mountain chains from Bokkeveld (Hantam) through Cederberg to the mountain ranges of the southwestern Cape. They also extend eastwards along the Cape Fold Mountains, mainly in the Riversonderend, Langeberg, Outeniqua and Tsitsikamma Mts as far east as to the Baviaanskloof and Groot Winterhoek Mts areas in the Eastern Cape. This forest biome is also found in the Swartberg, Rooiberg, and Kammanassie Mts (Fig. 6.7).

Vegetation Physiognomy and Composition These are medium-tall and especially low (scrub) forests with dominants including *Apodytes dimidiata, Cassine schinoides, Cunonia capensis, Ilex mitis, Heeria argentea, Kiggelaria africana, Podocarpus elongatus* (in the Bokkeveld area), *Rapanea melanophloeos. Brabejum stellatifolium, Platylophus trifoliatus,* and *Virgilia* spp. The fynbos thickets, also classified within this biome, are composed of *Cassine peragua, C. schinoides, Diospyros glabra, Heeria argentea, Hyphaene globosa, Maurocenia frangula, Maytenus acuminata, M. oleoides* and *Olea europaea* subsp. *africana.*

Climate Winters with dry summers are typical for the far western occurrence of these forests. The precipitation becomes non-seasonal in the eastern Western Cape and western Eastern Cape. The total rainfall spans an extensive range in MAP, from 400 to 1600 mm. The deep valleys create microclimatic (and fire) refugia, supporting the persistence of these fragmented forest patches (Fig. 6.7).

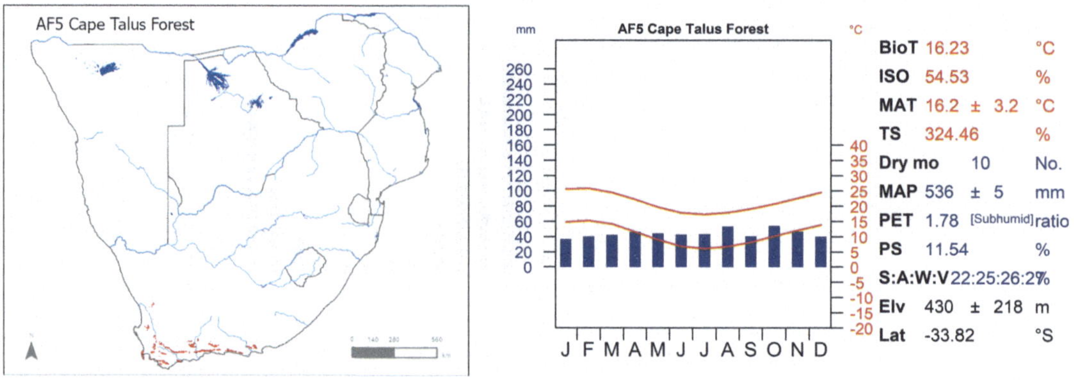

Fig. 6.7 Distribution area and the climatic diagram of the regional biome AF5 Cape Talus Forest

Topography, Geology, and Soils The Cape Talus Forests occur in narrow gullies and on scree slopes, mainly on the sandstone of the Table Mountain Group. The skeletal soils typically contain large boulders, stones, and gravel.

Sources Adamson (1927), van der Merwe (1966), Campbell and Moll (1977), McKenzie et al. (1977), Boucher (1978), Laidler et al. (1978), McDonald (1987, 1988), Masson and McKenzie (1989), Masson (1990), McKenzie et al. (1990), van Wilgen et al. (1990), Manders and Richardson (1992), Kruger (1994), Taylor (1996), Watson and Cameron (2002), von Maltitz et al. (2003), Geldenhuys and Mucina (2006), Mucina et al. (2006a, 2022), Hylander and Hedderson (2007), Poulsen and Hoffman (2015), Naudé et al. (2021).

AF6 Drakensberg Afrotemperate Forest

FOz 2 Northern Afrotemperate Forest p.p. (Mucina et al. 2006a)

Distribution This forest biome occurs in patches along the uKhahlamba Escarpment between KwaZulu-Natal and Lesotho (and marginally northwards to Van Reenen's Pass). It extends southwards along the Great Escarpment and the northeastern Eastern Cape (Fig. 6.8).

Vegetation Physiognomy and Composition These are low- to medium-tall forests, with dominants such as *Halleria lucida, Ilex mitis, Kiggelaria africana, Maytenus acuminata, M. undata, Olinia emarginata, Podocarpus latifolius, Rapanea melanophloeos, Scolopia mundii, Pittosporum viridiflorum,* and *Pterocelastrus rostratus.*

Climate The region supporting this forest biome experiences summer rainfall. Thermically, it is transitional between a cool-temperate and warm-temperate climate. Occasional winter snowfall also occurs (Fig. 6.8).

Topography, Geology, and Soils These forests occur in narrow gorges, lower scarps and slopes, commonly below cliffs formed by sandstone of the Clarens Formation. Soils are sandy, acidic, and shallow. Boulder fields and screes (talus) are typical habitats supporting these forests.

Sources Roberts (1961), Killick (1963), Van Zinderen Bakker (1971, 1973), Cooper (1985), Everard (1986), du Preez and Bredenkamp (1991a, b), du Preez et al. (1991), Hill (1996), von Maltitz et al. (2003), Geldenhuys and Mucina (2006), Mucina et al. (2006a, 2022), Lawes et al. (2007a), Adie and Lawes (2009a, b), Mthimkhulu (2009), Uys et al. (2010), Adie et al. (2017), Daemane et al. (2021).

AF7 Bushveld Talus Forest

Marekele Afromontane Forest (von Maltitz et al. 2003)
Northern Highveld Forest (von Maltitz et al. 2003)
FOz 2 Northern Afrotemperate Forest p.p. (Mucina et al. 2006a)

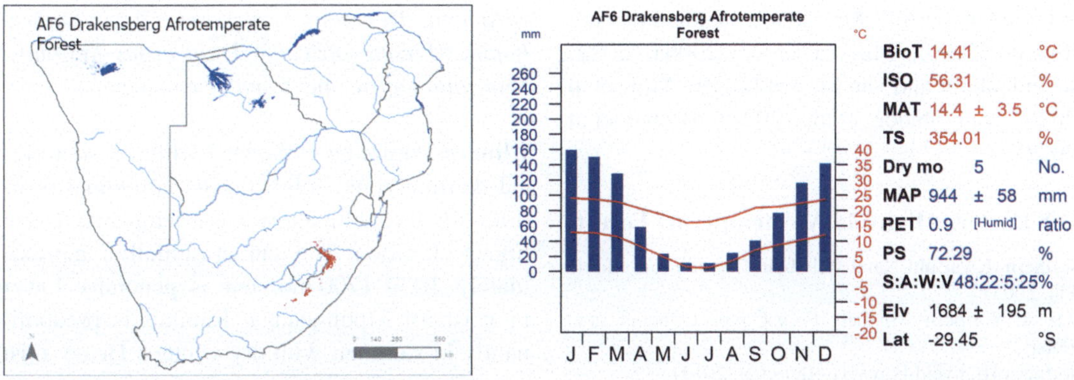

Fig. 6.8 Distribution area and the climatic diagram of the regional biome AF6 Drakensberg Afrotemperate Forest

Distribution This regional forest biome occurs in isolated patches spanning the Pilanesberg and Waterberg in the Limpopo Province in the north, to the western Magaliesberg in the North West Province and Gauteng in the south (Fig. 6.9).

Vegetation Physiognomy and Composition These are generally low, and rarely also medium-tall, forests dominated by *Buxus macowanii* (only locally), *Celtis africana, Diospyros lycioides, D. whyteana, Empogona lanceolata, Erythrina lysistemon, Kirkia wilmsii, Myrsine africana, Olea europaea* subsp. *africana, Osyris lanceolata, Pittosporum viridiflorum, Podocarpus latifolius, Searsia lancea, S. leptodictya,* and *Widdringtonia nodiflora.*

Topography, Geology, and Soils Deep kloofs that follow ancient crevices and steep scarps are the typical topographic features of this biome. The geological substrate is formed mainly by sandstones and quartzites of the Soutpansberg and Waterberg Groups. They support shallow soils predominantly of the Mispah Form.

Sources van Vuuren (1961), van Vuuren and van der Schijff (1970), Coetzee (1974, 1975), van der Meulen (1978, 1979), Bredenkamp and Theron (1978, 1980), Westfall (1981), Cooper (1985), Westfall et al. (1985), Behr and Bredenkamp (1988), Ellery et al. (2001), Siebert (2001), van Staden (2002), Siebert et al. (2003), von Maltitz et al. (2003), van Staden

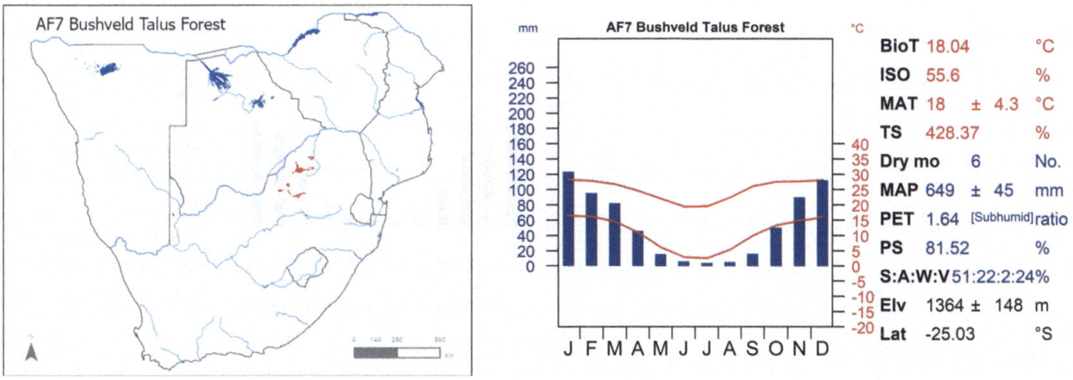

Fig. 6.9 Distribution area and the climatic diagram of the regional biome AF7 Bushveld Talus Forest

and Bredenkamp (2005, 2006), Geldenhuys and Mucina (2006), Mucina et al. (2006a, 2022), Bezuidenhout and van Staden (2009), Moir et al. (2020), van Staden et al. (2021), Brown et al. (2022).

AF8 Eastern Highveld Afrotemperate Forest

Northern KwaZulu-Natal Mistbelt Forest (von Maltitz et al. 2003)
FOz 4 Northern Mistbelt Forest p.p. (Mucina et al. 2006a)
Northern Highveld Forest (Lötter et al. 2014)
ATF5 Northern Highveld Afrotemperate Forest (Mucina et al. 2022)

Distribution These forests occur in patches along the eastern side of the low escarpment separating KwaZulu-Natal and the Free State. They extend further on to the Highveld around Vryheid and Wakkerstroom in southern Mpumalanga and occur further north, mainly on the western (rain-shadow) flank of the Great Escarpment in northern Mpumalanga (e.g., in the Lydenburg area) (Fig. 6.10).

Vegetation Physiognomy and Composition These are medium-tall, multilayered forests with well-developed understory and herb layers. The dominant species include *Afrocarpus falcatus, Allophylus africanus, Celtis africana, Diospyros whyteana, Dombeya burgessiae, Greyia sutherlandii, Myrsine africana, Podocarpus latifolius, Senegalia caffra,* and *Xymalos monospora.*

Climate Although embedded within a subtropical thermic zone (following its latitudinal position), the local climate is warm-temperate due to higher elevation. The annual rainfall is approximately 1000–1700 mm but is generally lower in northern Mpumalanga. Rainfall is predominantly in summer, with dry winters. Heavy mist occurs in the southern region of these forests in spring and summer (Fig. 6.10).

Topography, Geology, and Soils In KwaZulu-Natal and southern Mpumalanga Provinces, the substrates are mainly sandstone and mudstones of the Vryheid and Volksrust Formations, while in the northern Mpumalanga, the geological formations of the Transvaal Supergroup predominate. The soils are shallow, classified as Glenrosa and Mispah Forms, with boulders being common.

Sources Cooper (1982, 1985), Smit et al. (1993), Eckhardt et al. (1993, 1997), von Maltitz et al. (2003), Geldenhuys and Mucina (2006), Mucina et al. (2006a, 2022), Lötter (2014), Lötter et al. (2014).

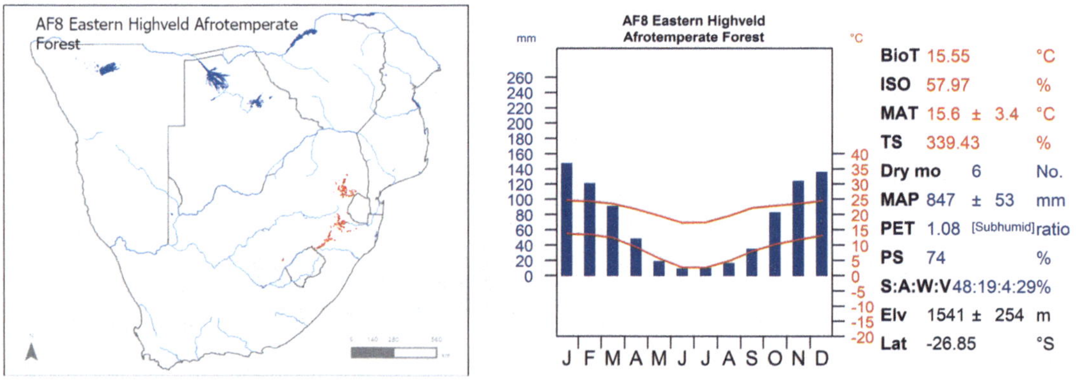

Fig. 6.10 Distribution area and the climatic diagram of the regional biome AF8 Eastern Highveld Afrotemperate Forest

References

Abbott ATD (2010) Checklist of Ngele Mountain and environs Weza State Forest 3029DA collectors. Nicholson/Abbott Herbarium, Port Edward

Adamson RS (1927) The plant communities of Table Mountain: preliminary account. J Ecol 15:278–309

Adie H, Lawes MJ (2009a) Explaining conifer dominance in Afrotemperate forests: shade tolerance favours *Podocarpus latifolius* over angiosperm species. For Ecol Manage 259:176–186

Adie H, Lawes MJ (2009b) Role reversal in the stand dynamics of an angiosperm-conifer forest: colonising angiosperms precede a shade-tolerant conifer in Afrotemperate forest. For Ecol Manage 258:159–168

Adie H, Rushworth I, Lawes MJ (2013) Pervasive, long-lasting impact of historical logging on composition, diversity and above ground carbon stocks in Afrotemperate forest. For Ecol Manage 310:887–895. https://doi.org/10.1016/j.foreco.2013.09.037

Adie H, Kotze DJ, Lawes MJ (2017) Small fire refugia in the grassy matrix and the persistence of Afrotemperate forest in the Drakensberg mountains. Sci Rep 7:6549. https://doi.org/10.1038/s41598-017-06747-2

Balfour DA, Bond WJ (1993) Factors limiting climber distribution and abundance in a southern African forest. J Ecol 81:93–99

Behr CM, Bredenkamp GJ (1988) A phytosociological classification of the Witwatersrand National Botanic Garden. S Afr J Bot 54:525–533

Bester SP (1998) Vegetation and flora of the Southern Drakensberg Escarpment and adjacent areas. Thesis, Department of Botany, University of Pretoria, Pretoria

Bezuidenhout H, van Staden P (2009) Botanical assessment of the Apiesrivierpoort forest. Department of Water Affairs and Forestry, Pretoria

Bitani N, Cordier CP, Ehlers Smith DA et al (2023) Microhabitat requirements and occupancy of understorey bird forest specialists in Southern Mistbelt Forests of KwaZulu-Natal, South Africa. For Ecol Manage 549:121484. https://doi.org/10.1016/j.foreco.2023.121484

Boucher C (1978) Cape Hangklip area II. The vegetation. Bothalia 12:455–497

Bredenkamp GJ, Theron GK (1978) A synecological account of the Suikerbosrand nature reserve. I. The phytosociology of the Witwatersrand geological system. Bothalia 12:513–529

Bredenkamp GJ, Theron GK (1980) A synecological account of the Suikerbosrand nature reserve. II. The phytosociology of the Ventersdorp geological system. Bothalia 13:199–216

Brown LR, Magagula IP, Barrett AS (2022) A vegetation classification and description of Telperion Nature Reserve, Mpumalanga, South Africa. Veg Classif Surv 3:199–219. https://doi.org/10.3897/VCS.85209

Burrows JE, Burrows SM, Lötter MC et al (2018) Trees and shrubs of Mozambique. Publishing Print Matters, Noordhoek, Cape Town

Campbell BM, Moll EJ (1977) The forest communities of Table Mountain, South Africa. Vegetatio 34:105–115

Cawe SG (1986) A quantitative and qualitative survey of the inland forests of Transkei. Thesis, University of Transkei, Umtata

Cawe SG (1996) A floristic classification of the indigenous forests of Transkei, South Africa. In: Van der Maesen LJG, Van der Burgt XM, Van Medenbach de Rooy JM (eds) The biodiversity of African plants. Kluwer, Dordrecht, pp 241–249

Cawe SG, McKenzie B (1989a) The Afromontane forests of Transkei, southern Africa. I: The importance of phytogeography and past utilization to the study of forest patches and a description of a sampling strategy. S Afr J Bot 55:22–30

Cawe SG, McKenzie B (1989b) The afromontane forests of Transkei, southern Africa. II: A floristic classification. S Afr J Bot 55:31–39

Cawe SG, McKenzie B (1989c) The afromontane forests of Transkei, southern Africa. III: A structural classification. S Afr J Bot 55:40–44

Cawe SG, McKenzie B (1989d) The Afromontane forests of Transkei, southern Africa. IV: Aspects of their exploitation potential. S Afr J Bot 55:45–55

Coetsee C, Wigley BJ (2013) *Virgilia divaricata* may facilitate forest expansion in the Afrotemperate forests of the southern Cape, South Africa. Koedoe 55(1):Art.#1128. https://doi.org/10.4102/koedoe.v55i1.1128

Coetsee C, Bond WJ, Wigley BJ (2015) Forest and fynbos are alternative states on the same nutrient poor geological substrate. S Afr J Bot 101:57–65

Coetzee BJ (1974) A phytosociological classification of the vegetation of the Jack Scott Nature Reserve. Bothalia 11:329–347

Coetzee BJ (1975) A phytosociological classification of the Rustenburg Nature Reserve. Bothalia 11:561–580

Cooper K (1982) Nelson's Kope forest. Highest altitude forest in South Africa. Afr Wildl 36:101–103

Cooper KH (1985) The conservation status of indigenous forests in Transvaal, Natal and O.F.S., South Africa. Wildlife Society of Southern Africa, Durban

Cowling RM (1984) A syntaxonomic and synecological study in the Humansdorp region of the Fynbos Biome. Bothalia 15:175–227

Daemane ME, Ramoelo A, Adelabu S (2021) The spatial distribution of the woodland communities and their associated environmental drivers in the Golden Gate Highlands National Park, South Africa. Koedoe 63(1):a1672. https://doi.org/10.4102/koedoe.v63i1.1672

Deall GB, Theron GK, Westfall RH (1989) The vegetation ecology of the Eastern Transvaal Escarpment in the Sabie area. 2. Floristic classification. Bothalia 19:69–89

Downs CT, Symes CT (2004) Snag dynamics and forest structure in Afromontane forests in KwaZulu-Natal, South Africa: implications for the conservation of cavity-nesting avifauna. S Afr J Bot 70:265–276

du Preez PJ, Bredenkamp GJ (1991a) The syntaxonomy and synecology of the forests in the eastern Orange Free State, South Africa. II. The *Pittosporetalea viridiflorum*. S Afr J Bot 57:207–212

du Preez PJ, Bredenkamp GJ (1991b) Vegetation classes of the southern and eastern Orange Free State (Republic of South Africa) and highlands of Lesotho. Navors Nasion Mus Bloemfontein 7:477–526

du Preez PJ, Bredenkamp GJ, Venter HJT (1991) The syntaxonomy and synecology of the forests in the eastern Orange Free State, South Africa. I. The *Podocarpetalia latifolii*. S Afr J Bot 57:198–206

Eckhardt HC, van Rooyen N, Bredenkamp GJ (1993) The phytosociology of the thicket and woodland vegetation of the north-eastern Orange Free State. S Afr J Bot 59:401–409

Eckhardt HC, van Rooyen N, Bredenkamp GJ (1997) Plant communities of the forests, woodlands and thickets in northern KwaZulu-Natal. Koedoe 40:91–112

Edwards D (1967) A plant ecological survey of the Tugela River basin. Town and Regional Planning Commission, Pietermaritzburg

Ellery WN, Balkwill K, Ellery K et al (2001) Conservation of the vegetation on the Melville Ridge, Johannesburg. S Afr J Bot 67:261–273

Everard DA (1986) The effects of fire on the *Podocarpus latifolius* forests of the Royal Natal National Park, Natal Drakensberg. S Afr J Bot 52:60–66

Everard DA (1992) On the feasibility of developing a functional classification as a decision support system for the management of the indigenous forests in Natal. Report FOR-DEA 458. Division of Forest Science and Technology, CSIR, Pretoria

Everard DA, Hardy SD (1993) Composition, structure and dynamics of the Amathole forests. Report FOR-DEA 810. Division of Forest Science and Technology, CSIR, Pretoria

Everard DA, Midgley JJ, van Wyk GF (1995) Dynamics of some forests in KwaZulu-Natal, South Africa, based on ordinations and size-class distributions. S Afr J Bot 61:283–292

Fisher JT (2024) Optimizing aboveground carbon mapping in Afrotemperate forests to fulfil IPCC carbon reporting standards. For Ecol Manage 552:121583

Fourcade HG (1889) Forestry pamphlets. Africa. Vol. II. Report on the Natal forests. W.M. Watson, Pietermaritzburg. https://doi.org/10.5962/bhl.title.23002

Geldenhuys CJ (1982) The management of the southern Cape indigenous forests. S Afr For J 121:4–10

Geldenhuys CJ (1986) Nature's Valley: a refugium for rare Southern Cape forest species. Palaeoecol Afr 17:173–181

Geldenhuys CJ (1987) Distribution and classification of the indigenous evergreen forest and deciduous woodland in South Africa. In: von Gadow K (ed) South African Forestry Handbook. Suid-Afrikaanse Bousbouhandboek. Southern African Institute of Forestry, Pretoria, pp 443–453

Geldenhuys CJ (1989a) Environmental and biogeographic influences on the distribution and composition of the Southern Cape forests (veld type 4). Thesis, Department of Botany, University of Cape Town, Cape Town

Geldenhuys CJ (1989b) Phytogeography of the Southern Cape forest flora. In: Geldenhuys CJ (ed) Biogeography of the mixed evergreen forests of southern Africa. Occasional Report No 45. FRD, Pretoria, pp 158–181

Geldenhuys CJ (1992a) Disjunctions and distribution limits of forest species in the Southern Cape. S Afr For J 161:1–13

Geldenhuys CJ (1992b) Richness, composition and relationships of the floras of selected forests in southern Africa. Bothalia 22:205–233

Geldenhuys CJ (1993a) Floristic composition of the southern Cape forest flora with an annotated checklist. S Afr J Bot 59:26–44

Geldenhuys CJ (1993b) Composition and dynamics of plant communities in the southern Cape forests. Report FOR-DEA 612. Division of Forest Science and Technology, CSIR, Pretoria

Geldenhuys CJ (1994a) Bergwind fires and the location pattern of forest patches in the southern Cape landscape, South Africa. J Biogeogr 21:49–62

Geldenhuys CJ (1994b) Phytogeography of the Southern Cape forest flora. In: Seyani JH, Chikuni AC (eds) Proceedings of the XIIIth plenary meeting of AETFAT. National Herbarium and Botanic Gardens, Zomba, pp 969–992

Geldenhuys CJ (1996) Fruit/seed characteristics and germination requirements of tree and shrub species of the southern Cape forests. Report FOR-DEA 954. Division of Forest Science and Technology, CSIR, Pretoria

Geldenhuys CJ (1997a) Composition and biogeography of forest patches on the inland mountains of the Southern Cape. Bothalia 27:57–74

Geldenhuys CJ (1997b) Native forest regeneration in pine and eucalypt plantations in Northern Province, South Africa. For Ecol Manage 99:101–115

Geldenhuys CJ (2007) Growth, ingrowth and mortality patterns over stands and species in the forest growth sites at Weza and Ngome indigenous forests, KwaZulu-Natal: Report on first remeasurement, 2003/2004'. Project no. 2008-450. Department of Water Affairs and Forestry, Pretoria

Geldenhuys CJ, Mucina L (2006) Towards a new forest classification for South Africa. In: Ghazanfar S, Beentje HJ (eds) Taxonomy and ecology of African plants, their conservation and sustainable use. Royal Botanic Gardens, Kew, pp 111–129

Geldenhuys CJ, Murray B (1993) Floristic and structural composition of Hanglip forest in the Soutpansberg, Northern Transvaal. S Afr for J 165:9–20

Geldenhuys CJ, Pieterse FJ (1993) Floristic and structural composition of Wonderwoud forest in the Wolkberg, north-eastern Transvaal. S Afr for J 164:9–19

Geldenhuys CJ, Rathogwa NR (1997) Growth and mortality patterns over stands and species in four Amathole Forest Growth Study Sites: Report on 1996 measurements. Report ENV/P/C/97006. Division of Water, Environment and Forestry Technology, CSIR, Pretoria

Geldenhuys CJ, Venter S (2002) Plant communities and biodiversity of the Limpopo Province forests: relevance and management options. In: Seydack AHW, Vorster T, Vermeulen WJ et al (eds) Multiple use management of natural forests and woodlands: policy refinements and scientific progress. DWAF, Pretoria, pp 23–37

Giddey BL, Baard JA, Kraaij T (2022a) Fire severity and tree size affect post-fire survival of Afrotemperate forest trees. Fire Ecol 18:5. https://doi.org/10.1186/s42408-022-00128-5

Giddey BL, Baard JA, Kraaij T (2022b) Verification of the differenced Normalised Burn Ratio (dNBR) as an index of fire severity in Afrotemperate Forest. S Afr J Bot 146:348–353

Glyphis J, Moll EJ, Campbell BM (1978) Phytosociological studies on Table Mountain, South Africa: 1. The back table. J S Afr Bot 44:281–289

Grieve GRH, Downs CT (2015) A checklist of the plants of the forests and grasslands in the Weza district, southern KwaZulu-Natal and a review of their status in the Red Data List. Koedoe 57(1):Art.#1237. https://doi.org/10.4102/koedoe.v57i1.1237

Hall TF, Scholes MC, Scholes RJ (2022) The Pongola bush: tree diversity assessment in a KwaZulu-Natal forest patch. Koedoe 64(1):a1724. https://doi.org/10.4102/koedoe.v64i1.1724

Hanekom N, Southwood A, Ferguson M (1989) A vegetation survey of the Tsitsikamma Coastal National Park. Koedoe 32:47–66

Hart L, Grieve G, Downs CT (2013) Fruiting phenology and implications of fruit availability in the fragmented Ngeli Forest Complex, KwaZulu-Natal, South Africa. S Afr J Bot 88:296–305. https://doi.org/10.1016/j.sajb.2013.08.003

Hill TR (1996) Description, classification and ordination of the dominant vegetation communities, Cathedral Peak, KwaZulu-Natal Drakensberg. S Afr J Bot 62:263–269

Hylander K, Hedderson TAJ (2007) Does the width of isolated ravine forests influence moss and liverwort diversity and composition?—A study of temperate forests in South Africa. Biodivers Conserv 16:1441–1458

Jeffery DJ, Wilson N (1987) A survey of the plant communities in the Greater Kirstenbosch region. Kirstenbosch National Botanical Garden, Cape Town

Killick DJB (1963) An account of the plant ecology of the Cathedral Peak Area of the Natal Drakensberg. Mem Bot Surv S Afr 34:1–178

Koen JB, Crowe TM (1987) Animal-habitat relationships in the Knysna Forest, South Africa: discrimination between forest types by birds and invertebrates. Oecologia 72:414–422

Kotze DJ, Lawes MJ (2007) Viability of ecological processes in small Afromontane forest patches in South Africa. Austral Ecol 32:294–304

Kraaij T, Baard JA, Grobler BA et al (2023) Effects of *Acacia melanoxylon*, an alien tree species to South Africa, on Afrotemperate forest tree sapling composition. Southern Forests 85:74–83. https://doi.org/10.2989/20702620.2023.2198093

Kruger LM (1994) Dynamics and conservation status of the Swartkransberg Forests. BSc (Hons) Project, Department of Botany, University of Cape Town, Cape Town

Kruger LM, Midgley JJ (2001) The influence of resprouting forest canopy species on richness in Southern Cape forests, South Africa. Glob Ecol Biogeogr 10:567–572

Laidler D, Moll EL, Campbell BM et al (1978) Phytosociological studies on Table Mountain, South Africa. J S Afr Bot 44:291–295

Laughton FS (1937) The sylviculture of the indigenous forests of the Union of South Africa. Government Printer, Pretoria

Lawes MJ, Mealin PE, Piper SE (2000) Patch occupancy and potential metapopulation dynamics of three forest mammals in fragmented Afromontane forest in South Africa. Conserv Biol 14:1088–1098

Lawes MJ, Macfarlane DM, Eeley HAC (2004) Forest landscape pattern in the KwaZulu-Natal midlands, South Africa: 50 years of change or stasis? Austral Ecol 29:613–623. https://doi.org/10.1111/j.1442-9993.2004.01396.x

Lawes MJ, Kotze DJ, Bourquin SL (2005) Epigaeic invertebrates as potential ecological indicators of Afromontane forest condition in South Africa. Biotropica 37:109–118

Lawes ML Fly S, Piper SE (2006a) Gamebird vulnerability to forest fragmentation: patch occupancy of the crested guineafowl (*Guttera edouardi*) in Afromontane forests. Anim Conserv 9:67–74

Lawes ML, Midgley JJ, Bordreau S et al (2006b) Lack of disturbance as an explanation for the additive basal area phenomenon in a stratified Afrotemperate forest. Austral Ecol 31:471–477

Lawes MJ, Adie H, Eeley HAC et al. (2007a) An assessment of the forests of the Maloti-Drakensberg Transfrontier bioregion, with reference to important ecosystem processes. Final report. Forest Biodiversity Research Unit, School of Biological and Conservation Sciences, University of KwaZulu-Natal, Scottsville

Lawes MJ, Griffiths ME, Boudreau S (2007b) Colonial logging and recent subsistence harvesting affect the composition and physiognomy of a podocarp dominated Afrotemperate forest. For Ecol Manage 247:48–60

Leaver J, Carstens JC, Cherry MI (2019) Harvesting of forest products and implications for Afrotemperate bird communities in a montane forest of the Eastern Cape, South Africa. For Ecosyst 6:48. https://doi.org/10.1186/s40663-019-0207-x

Lötter MC (2014) Indigenous forests of Mpumalanga Province (South Africa); patterns and processes form inclusion in a systematic conservation plan. Thesis, Faculty of Science, University of Witwatersrand, Johannesburg

Lötter MC, Beck HT (2004) Preliminary inventory and classification of indigenous Afromontane forests on the Blyde River Canyon Nature Reserve, Mpumalanga, South Africa. BMC Ecol 4:9.https://doi.org/10.1186/1472-6785-4-9

Lötter MC, Mucina L, Witkowski ETF (2014) Classification of the indigenous forests of Mpumalanga Province, South Africa. S Afr J Bot 90:37–51

Lötter M, Burrows J, Jones K et al (2023) Historical vegetation map and red list of ecosystems assessment for Mozambique—version 2.0—final report. USAID/SPEED+, AFD/FFEM, Maputo

Manders PT, Richardson DM (1992) Colonization of Cape fynbos communities by forest species. For Ecol Manage 48:277–293

Masson P (1990) The dynamics of the Afromontane forest remnants in the southwestern Cape. Thesis, Department of Botany, University of Cape Town, Cape Town

Masson P (1994) Forest composition and conservation status in the Swaziland highveld. In: Seyani JH, Chikuni AC (eds) Proceedings of the XIIIth Plenary Meeting of AETFAT. National Herbarium and Botanic Gardens, Zomba, pp 993–1005

Masson P, McKenzie B (1989) Floristics of southwestern Cape forests. In: Geldenhuys CJ (ed) Biogeography of the mixed evergreen forests of southern Africa. FRD, Pretoria, pp 182–193

Matthews WS, Bredenkamp GJ, van Rooyen N (1992) The vegetation of the dry dolomitic regions of the northeastern mountain sourveld of the Transvaal Escarpment, South Africa. Phytocoenologia 20:467–488

McDonald DJ (1987) Ordination by detrended correspondence analysis (DCA) of the vegetation of Swartboschkloof, Jonkershoek, Cape Province. Bothalia 17:121–129

McDonald DJ (1988) A synopsis of the plant communities of Swartboschkloof, Jonkershoek, Cape Province. Bothalia 18:233–260

McDonald DJ (1993a) The vegetation of the southern Langeberg, Cape Province. 1. The plant communities of the Boosmansbos Wilderness Area. Bothalia 23:129–151

McDonald DJ (1993b) The vegetation of the southern Langeberg, Cape Province. 2. The plant communities of the Marloth nature reserve. Bothalia 23:153–174

McKenzie B (1978) A quantitative and qualitative study of the indigenous forests of the southwestern Cape. Thesis, Department of Botany, University of Cape Town, Cape Town

McKenzie B, Moll EJ, Campbell BM (1977) A phytosociological study of Orange Kloof, Table Mountain, South Africa. Vegetatio 34:41–53

McKenzie B, Moll E, Denman S (1990) The forests of the Swartkransberg, South Africa. Mitt Inst Bot Hamburg 23:119–128

Mensah S, Glèlè Kakaï R, Seifert T (2016a) Patterns of biomass allocation between foliage and woody structure: the effects of tree size and specific functional traits. Ann For Res 59:49–60

Mensah S, Veldtman R, du Toit B et al (2016b) Aboveground biomass and carbon in a South African mistbelt forest and the relationships with tree species diversity and forest structures. Forests 79:1–17

Mensah S, Veldtman R, Seifert T (2017a) Potential supply of floral resources to managed honey bees in natural mistbelt forests. J Environ Manage 189:160–167

Mensah S, Veldtman R, Seifert T (2017b) Allometric models for height and above ground biomass of dominant tree species in South African mistbelt forests. Southern Forests 79:19–30

Mensah S, Salako VK, Assogbadjo AE (2018a) Differential responses of taxonomic, structural, and functional diversity to local-scale environmental variation in afromontane Forest in South Africa. Trop Conserv Sci 11:1–13

Mensah S, du Toit B, Seifert T (2018b) Diversity–biomass relationship across forest layers: implications for niche complementarity and selection effects. Oecologia 187:783–795

Mensah S, Pienaar OL, Kunneke A et al (2018c) Height–diameter allometry in South Africa's indigenous high forests: assessing generic models performance and function forms. For Ecol Manage 410:1–11

Mensah S, Salako KV, Assogbadjo A et al (2020a) Functional trait diversity is a stronger predictor of multifunctionality than dominance: evidence from an Afromontane forest in South Africa. Ecol Indic 115:106415. https://doi.org/10.1016/j.ecolind.2020.106415

Mensah S, Egeru A, Assogbadjo AE et al (2020b) Vegetation structure, dominance patterns and height growth in an Afromontane forest. Southern Africa. J For Res 31:453–462

Moir M, Richard LR, Cherry MI, Rambau RV (2020) Demographic responses of forest-utilizing bats to past climate change in South Africa. Biol J Linn Soc 20:1–19

Moll EJ (1972) The current status of mistbelt mixed *Podocarpus* forest in Natal. Bothalia 10:595–598

Moll EJ (1976) The vegetation of the Three Rivers region, Natal. Natal Town and Regional Planning Commission, Pietermaritzburg

Moll EJ, Woods DFB (1972) The rate of forest tree growth and a forest ordination at Xumeni, Natal. Bothalia 10:451–460

Morgenthal TL, Cilliers SS (1999) Vegetation analysis of Pedlar's Bush, Mpumalanga, and its conservation. S Afr J Bot 65:270–280
Morgenthal TL, Cilliers SS (2000) Species composition and phytogeographical significance of an Afromontane forest fragment in the Mpumalanga province, South Africa. S Afr For J 189:81–102
Mostert THC, Bredenkamp GJ, Klopper HL et al (2008) Major vegetation types of the Soutpansberg Conservancy and the Blouberg nature reserve. South Africa. Koedoe 50(1):32–48
Mthimkhulu O (2009) Understanding the present and historic forest resource use of the Ntabamhlophe indigenous state forest by rural communities. Thesis, School of Environmental Science, University of KwaZulu-Natal, Pietermaritzburg
Mucina L (2023) Biomes of the Southern Hemisphere. Springer Nature Switzerland, Cham
Mucina L, Rutherford MC (eds) (2006) The vegetation of South Africa, Lesotho and Swaziland. SANBI, Pretoria
Mucina L, Geldenhuys CJ, Rutherford MC et al (2006a) Afrotemperate, subtropical and azonal forests. In: Mucina L, Rutherford MC (eds) The vegetation of South Africa, Lesotho and Swaziland. SANBI, Pretoria, pp 584–615
Mucina L, Rutherford MC, Powrie LW et al. (2006b) Inland azonal vegetation. In: Mucina L, Rutherford MC (eds) The vegetation of South Africa, Lesotho and Swaziland. SANBI, Pretoria, pp 616–657
Mucina L, Lötter MC, Rutherford MC et al (2022) Forest biomes of Southern Africa. New Zeal J Bot 60:377–428
Müller T (1999) The distribution, classification and conservation of rainforests in Zimbabwe. In: Timberlake J, Katiwu S (eds) African plants: biodiversity, taxonomy and uses. Royal Botanical Gardens, Kew, pp 221–235
Müller T (2006) The distribution, classification and conservation of rainforests in eastern Zimbabwe. Occasional Publications in Biodiversity No. 19. Biodiversity Foundation for Africa, Bulawayo
Müller T, Mapaure A, Wursten B et al (2012) Vegetation survey of Mount Gorongosa. Occasional Publications in Biodiversity No. 23. Biodiversity Foundation for Africa, Bulawayo
Müller T, Sitoe A, Mbunda R (2005) Assessment of the forest reserve network in Mozambique. WWF Mozambique, Maputo
Mulvaney JM, Cherry MI, Matthee CA (2022) Climate refugia for three Afromontane forest-dependent bird species in south-eastern South Africa. J Biogeogr 49:1352–1366
Naudé A-L, Snyman RG, Odendaal JP (2021) Aluminium and iron contamination of soil, leaf litter and bioindicators in selected South African forest pockets. Water Air Soil Pollut 232:304. https://doi.org/10.1007/s11270-021-05246-5
Ngcobo SP, Ehlers Smith DA, Gumede ST et al (2022) Using a multifaceted approach to reveal avian community responses to natural and anthropogenic effects in a fragmented Southern Mistbelt Forest system, South Africa. Landsc Ecol 37:1885–1903
Ohsawa M (1993) Latitudinal pattern of mountain vegetation zonation in southern and eastern Asia. J Veg Sci 4:13–18
Parker DM, Junkuhn K, Barker NP (2022) Note on the small mammals of small, isolated forest patches in the Eastern Cape, South Africa. Afr J Ecol 60:784–787
Phillips JFV (1931) Forest-succession and ecology in the Knysna region. Government Printer, Pretoria
Phillipson PB (1987) A checklist of vascular plants of the Amathole Mountains, eastern Cape Province/Ciskei. Bothalia 17:237–256
Poulsen ZC, Hoffman MT (2015) Changes in the distribution of indigenous forest in Table Mountain National Park during the 20th Century. S Afr J Bot 101:49–56
Quick LJ, Meadows ME, Bateman MD et al (2016) Vegetation and climate dynamics during the last glacial period in the fynbos-afrotemperate forest ecotone, southern Cape, South Africa. Quater Int 404:136–149
Rea M, Elliot J, Carstens JC et al (2023) Using acoustic recording units to investigate the effects of logging of indigenous trees in the Amathole forests, South Africa on Cape Parrot *Poicephalus robustus* breeding and the presence of three primary cavity-excavating bird species. Bird Conserv Int 33:e59. https://doi.org/10.1017/S0959270923000084
Roberts BR (1961) Preliminary notes on the vegetation of Thaba 'Nchu. J S Afr Bot 27:241–251
Robertson J, Lawes MJ (2005) User perceptions of conservation and participatory management of iGxalingenwa forest, South Africa. Environ Conserv 32:64–75. https://doi.org/10.1017/S0376892905001979
Scheepers JC (1978) The vegetation of Westfalia Estate on the north-eastern Transvaal escarpment. Mem Bot Surv S Afr 42:1–230
Seifert T, Seifertl S, Seydack A (2014) Competition effects in an afrotemperate forest. For Ecosyst 1:13. http://www.forestecosyst.com/content/1/1/13
Sibanda M, Gumede N, Mutanga O (2022) Estimating leaf area index of the Yellowwood tree (*Podocarpus* spp.) in an indigenous Southern African Forest, using Sentinel 2 Multispectral Instrument data and the Random Forest regression ensemble. Geocarto Int 37:6953–6974. https://doi.org/10.1080/10106049.2021.1959654
Siebert SJ (2001) Vegetation of the ultramafic soils of the Sekhukhuneland Centre of Endemism. Thesis, Department of Botany, University of Pretoria, Pretoria
Siebert SJ, van Wyk AE, Bredenkamp GJ et al (2003) Vegetation of the rock habitats of the Sekhukhuneland Centre of Plant Endemism, South Africa. Bothalia 33:207–228
Sinthumule NI, Mashau ML (2020) Traditional ecological knowledge and practices for forest conservation in Thathe Vondo in Limpopo Province, South

Africa. Glob Ecol Conserv 22: e00910. https://doi.org/10.1016/j.gecco.2020.e00910
Smit CM, Bredenkamp GJ, van Rooyen N (1993) Woodland plant communities of the Fa land type in the Newcastle–Memel–Chelmsford Dam area. S Afr J Bot 59:14–20
Sosibo MT, Ehlers Smith DA, Ehlers Smith YC et al (2023) Influence of microhabitat structure on large- and medium-sized mammals in South African forests. Afr J Ecol 61:617–627
Sosibo MT, Ehlers Smith DA, Ehlers Smith YC et al (2024) Patterns and predictors of mammalian taxonomic and functional species diversity in naturally fragmented Southern Mistbelt Forests in South Africa. For Ecol Manage 559:121820. https://doi.org/10.1016/j.foreco.2024.121820
South African National Parks (2014) Monitoring for sustainable indigenous forest management in the Garden Route National Park. South African National Parks, Knysna
Stalmans M, Beilfuss R (2008) Landscapes of the Gorongosa National Park. Condensed Version. Gorongosa National Park, Mozambique
Stalmans M, Robinson ER, Balkwill K (1999) Ordination and classification of the vegetation of Songimvelo Game Reserve in the Barberton Mountainland, South Africa for the assessment of wildlife habitat distribution and quality. Bothalia 29:305–325
Story R (1952) A botanical survey of the Kieskammahoek District. Government Printer, Pretoria
Swart RC, Samways MJ, Roets F (2021) Latitude, paleo-history and forest size matter for Afromontane canopy beetle diversity in a world context. Biodivers Conserv 30:659–672
Taylor HC (1955) Forest types and floral composition of Grootvadersbosch. J S Afr For Assoc 26:33–46
Taylor HC (1984) A vegetation survey of the Cape of Good Hope Nature Reserve. I. The use of association-analysis and Braun-Blanquet methods. Bothalia 15:245–258
Taylor HC (1996) Cederberg vegetation and flora. National Botanical Institute, Pretoria
Timberlake J, Bayliss J, Alves, T et al (2007) The biodiversity and conservation of Mount Chiperone, Mozambique. Royal Botanic Gardens, Kew
Timberlake J, Darbyshire I, Wursten B et al (2016) Chimanimani Mountains: botany and conservation. Royal Botanic Gardens, Kew
Townsend G, van Rooyen E, Hill M et al (2024) Invasion of an Afrotemperate forest complex by the polyphagous shot hole borer beetle. Entomol Exper Appl 172:354–369
Uys C, Hamer M, Slotow R (2010) Step process for selecting and testing surrogates and indicators of afrotemperate forest invertebrate diversity. PLoS ONE 5(2):e9100. https://doi.org/10.1371/journal.pone.0009100
van Daalen JC (1984) Distinguishing features of forest species on nutrient-poor soils in the Southern Cape. Bothalia 15:229–239
van der Merwe P (1966) Die flora van Swartboskloof, Stellenbosch en die herstel van die soorte na 'n brand. Ann Univ Stellenbosch Ser A 41:687–736
van der Meulen F (1978) Progress with vegetation studies in the Sourish Mixed Bushveld of the western Transvaal. Bothalia 12:531–536
van der Meulen F (1979) Plant sociology of the western Transvaal Bushveld, South Africa: a syntaxonomic and synecological study. A. Gantner, Vaduz
van der Schijff HP, Schoonraad E (1971) The flora of the Mariepskop complex. Bothalia 10:461–500
van Jaarsveld E, Hardy D (1991) The Blouberg botanical expedition, with specific reference to the succulent vegetation. Aloe 28:72–83
van Laar A, Geldenhuys CJ (1975) Distribution and correlation patterns in an indigenous forest. S Afr For J 94:20–28
van Laar A, Lewark S (1973) Sampling for forest inventories in the indigenous forests in the Southern Cape Province. Forests S Afr 14:35–43
van Staden PJ (2002) An ecological study of the plant communities of Marekele National Park. Thesis, Centre for Wildlife Management, University of Pretoria, Pretoria
van Staden PJ, Bredenkamp GJ (2005) Major plant communities of the Marekele National Park. Koedoe 48(2):59–70
van Staden PJ, Bredenkamp GJ (2006) A floristic analysis of forest and thicket vegetation of the Marakele National Park. Koedoe 49(1):15–32
van Staden PJ, Bredenkamp GJ, Bezuidenhout H et al (2021) A reclassification and description of the Waterberg Mountain vegetation of the Marakele National Park, Limpopo province. South Africa. Koedoe 63(1):a1689. https://doi.org/10.4102/koedoe.v63i1.1689
van Vuuren DRJ (1961) 'n Ekologiese studie van 'n noordelike en suidelike kloof van die Magaliesberge. Thesis, Department of Botany, University of Pretoria, Pretoria
van Vuuren DRJ, van der Schijff HP (1970) 'n Vergelykende ekologiese studie van die plantegroei van 'n noordelike en suidelike kloof van die Magaliesberg. Tydskr Natuurwetensch 10:16–75
van Wilgen BW, Higgins KB, Bellstedt DU (1990) The role of vegetation structure and fuel chemistry in excluding fire from forest patches in the fire-prone fynbos shrublands of South Africa. J Ecol 78:210–222
Van Zinderen Bakker Jr EM (1971) Ecological investigations on ravine forests of the Eastern Orange Free State (South Africa). Thesis, Department of Botany, University of the Orange Free State, Bloemfontein
Van Zinderen Bakker Jr EM (1973) Ecological investigations of forest communities in the eastern Orange

Free State and the adjacent Natal Drakensberg. Vegetatio 28:299–334
Vermeulen WJ (1993) Management of indigenous evergreen high forests with specific reference to the Southern Cape. In: von Gadow K (ed) South African Forestry Handbook. Suid-Afrikaanse Bousbouhandboek. Southern African Institute of Forestry, Pretoria, pp 617–632
Vermeulen WJ (1994) The multiple-use management of the indigenous evergreen high forests of the Southern Cape and Tsitsikamma. Department of Water Affairs and Forestry, Knysna
Vermeulen WJ (1995) 'n Fitososiologiese studie van die Ysterneknatuurreservaat in Suid Kaapland. Thesis, Department of Botany, University of Stellenbosch, Stellenbosch
von Breitenbach F (1972) Indigenous forests of the Southern Cape. Veld Flora 58:18–47
von Breitenbach F (1990) Reports on indigenous forests Part 1. Introduction and methods, Southeastern Transvaal forests, Kaapsehoop forests, Uitsoek forests. Department of Environment Affairs, Forestry Branch, Pretoria
von Maltitz G, Mucina L, Geldenhuys CJ et al (2003) Classification system for South African indigenous forests: an objective classification for the Department of Water Affairs and Forestry. Report ENV-P-C 2003-017. Environmentek CSIR, Pretoria
Wardell-Johnson G, Calver M (2005) Toward sustainable management: Southern Africa's Afromontane, and Western Australia's jarrah forests. In: Calver M, Bigler-Cole H, Bolton G et al (eds) Proceedings 6th national conference of the Australian Forest History Society Inc. Millpress, Rotterdam, pp 729–739
Watson LH, Cameron MJ (2002) Forest tree and fern species as indicators of an unnatural fire event in a southern Cape mountain forest. S Afr J Bot 68:357–361
Westfall RH (1981) The plant ecology of the farm Groothoek, Thabazimbi District. Thesis, Department of Botany, University of Pretoria, Pretoria
Westfall RH, van Rooyen N, Theron GK (1985) The plant ecology of the farm Groothoek, Thabazimbi District. II. Classification. Bothalia 15:655–688
White F (1978) The Afromontane region. In: Werger MJA, van Bruggen AC (eds) Biogeography and ecology of southern Africa. Dr W. Junk, The Hague, pp 463–513
Wirminghaus JO, Downs CT, Symes CT et al (2001) Fruiting in two afromontane forests in KwaZulu-Natal, South Africa: the habitat type of the endangered Cape Parrot *Poicephalus robustus*. S Afr J Bot 67:329–336

Deserts and Semideserts of Southern Africa

7

Michael C. Rutherford, Ladislav Mucina, Johan Wasserman, James L. Tsakalos and Leslie W. Powrie

Abstract

Southern Africa is ecologically a region of steep climatic gradients and vegetation contrasts. The dry corner of the broad spatial heterogeneity is formed by extensive deserts and semideserts that occupy the western and southwestern portions of the subcontinent. The region is home to three continental biomes of deserts, of which the iconic coastal-fog Namib Desert is the largest and best-researched. The smaller Gariep Desert continental biome was recognised only in 2006, whereas the smallest Tankwa Desert continental biome is recognised for the first time in this book. The semideserts are more extensive and varied (in bioclimatic and floro-genetic terms), although only two continental biomes have been recognised, namely Nama-Karoo and Succulent Karoo. All these deserts and semideserts, except the exclusively South African Tankwa Desert, are shared between South Africa and Namibia. Only the Namib Desert and Nama-Karoo also extend beyond the MBSA study area into Angola. The five continental biomes and their 18 constituent regional biomes are described. The continental biome Succulent Karoo is the richest in regional biomes within the zonobiome S2 Hot Arid Zone in the MBSA area.

Keywords

Coastal fog deserts · Gariep Desert · Nama-Karoo · Namaqualand · Namib Desert · Namibia · Rain shadow effect · South Africa · Succulent Karoo · Tankwa Desert

M. C. Rutherford (✉)
High Peak, Derbyshire, United Kingdom

Department of Botany and Zoology, Stellenbosch University, Private Bag X1, Matieland, Stellenbosch 7602, South Africa

L. Mucina · J. Wasserman · J. L. Tsakalos
Harry Butler Institute, Murdoch University, 90 South Street, Murdoch, Perth 6150, Australia

L. Mucina · L. W. Powrie
Department of Geography and Environmental Studies, Stellenbosch University, Private Bag X1, Matieland, Stellenbosch 7602, South Africa

J. L. Tsakalos
Plant Diversity and Ecosystems Management Unit, School of Biosciences and Veterinary Medicine, University of Camerino, Via Pontoni 5, Camerino 62032, Italy

L. W. Powrie
Kirstenbosch Research Centre, South African National Biodiversity Institute, Private Bag X7 Claremont, Cape Town, South Africa

L. Mucina and M. C. Rutherford (eds.), *Zonal Biomes of Southern Africa*, Biome Ecology 2,
https://doi.org/10.1007/978-3-031-63731-5_7

7.1 Diverse Faces of the Southern African Deserts and Semideserts

The biomes (and the constituent vegetation units) making up the zonobiome S2 in Southern Africa have been described in three chapters of the Vegetation of South Africa, Lesotho and Swaziland (Mucina and Rutherford 2006), namely in Jürgens et al. (2006) and Mucina et al. (2006a; b). In this sequel, we expand substantially to cover Namibia and shall focus on the biome identity of various types of Southern African deserts and semideserts, placing these into a global perspective for the first time. The newly re-defined continental biomes will also be characterised by accounts of their component regional biomes.

Jürgens et al. (2006) suggested that in South Africa, the term 'Desert' is limited to the hyper-arid climatic zone of the northernmost western seaboards of Namaqualand and Richtersveld. The South African section is tiny compared to the 'Desert' that continues northwards along the Namibia coast, commonly called the Namib Desert. The term Namib means 'vast place' in the Khoekhoegowab language. Its relatively mesic eastern fringes are sometimes (especially in vernacular sources) considered to be a part of the Namib Desert. This region is commonly referred to as the Pro-Namib (Afrikaans: Voornamib; German: Vornamib) at the foot of the Escarpment. In our bioclimatic classification, we recognise this relatively mesic region as complying with the definition of Nama-Karoo. The Namib Desert does extend outside our study region for about 350 km from Namibia into Angola, narrowly tapering to terminate around Lucira (M.C. Rutherford, personal observation) on the coast between Moçâmedes and Benguela.

SW Southern Africa is home to a broad range of arid ecosystems, including the harsh deserts (continental biomes Namib, Gariep, and Tankwa) and semideserts (continental biomes Succulent Karoo and Nama-Karoo), that represent four global biomes, namely DESc Subtropical Coastal Hot Deserts, DESi Subtropical Inland Hot Deserts, SDESi Subtropical Inland Hot Semideserts, and SDESc Subtropical Coastal Hot Semideserts (Fig. 7.1; Table 7.1; Mucina 2023). All the arid biomes found in Southern Africa find their analogues on other continents (see Table 1.1 in Mucina 2023). Geographically, all arid biomes in southern Africa form a contiguous, compact unit, today well separated from the next-closest large-scale desert/semidesert grouping of the Saharo-Sindian region (incl. Sahara, Horn of Africa, Arabian Peninsula, Iran and the broader Indus region). Several biogeographic analyses and molecular-phylogenetic studies have suggested the existence of a temporary arid corridor, which appeared regularly during the Pleistocene and assisted the two-way migration of flora. This is the well-known Arid Corridor (Balinsky 1962; Verdcourt 1969; see also Jürgens 1997; Pokorny et al. 2015; Mucina 2017 for additional literature sources and discussion). Simple modelling (Fig. 7.2) demonstrates how this corridor could have been formed.

The term 'desert' in the so-called 'Kalahari Desert' (in Botswana, Namibia and South Africa) is an unfortunate popular historical misnomer that arose possibly through sole considerations of ubiquitous sand and lack of surface water. The Kalahari forms a part of the relatively well-watered Savanna Biome despite extending across semiarid and arid climatic zones (according to Middleton and Thomas 1997).

There are three desert continental biomes (Namib Desert, Gariep Desert, Tankwa Deserts) and two semidesert continental biomes (Succulent Karoo, Nama-Karoo) in Southern Africa (Fig. 7.3; Table 7.1).

7.2 Continental Biome Namib Desert (DN)

The Namib Desert forms a coastal belt, up to about 100 km broad, stretching from slightly south of the border between the Northern Cape Province of South Africa and Namibia in the vicinity of the lower Gariep (Orange) River Valley, northwards, through its major area of

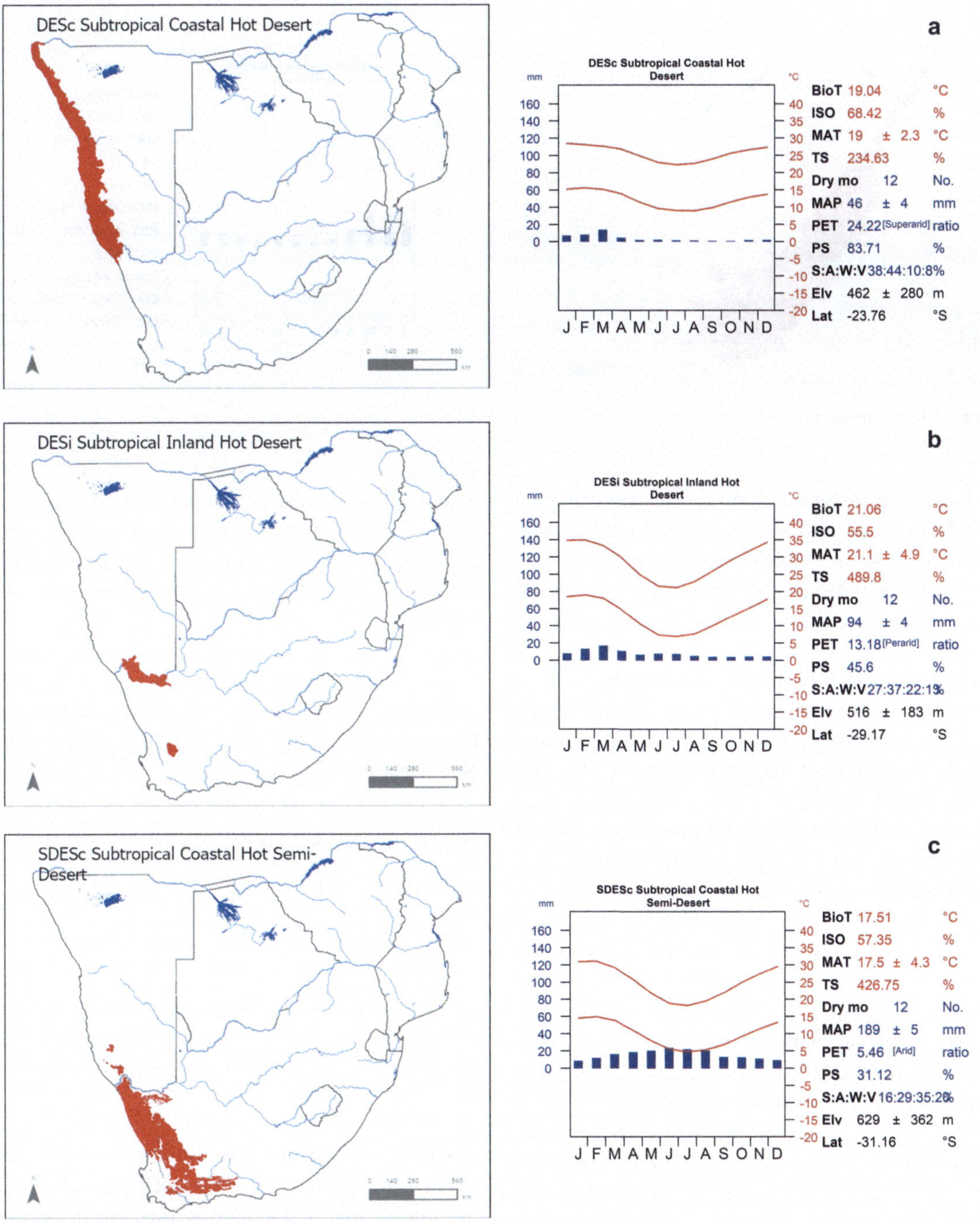

Fig. 7.1 Distribution areas and the climatic diagrams of the global biomes within the zonobiome S2 Hot Arid Zone. The construction of the climatic diagrams and the distribution maps of the biomes is described in Chap. 1

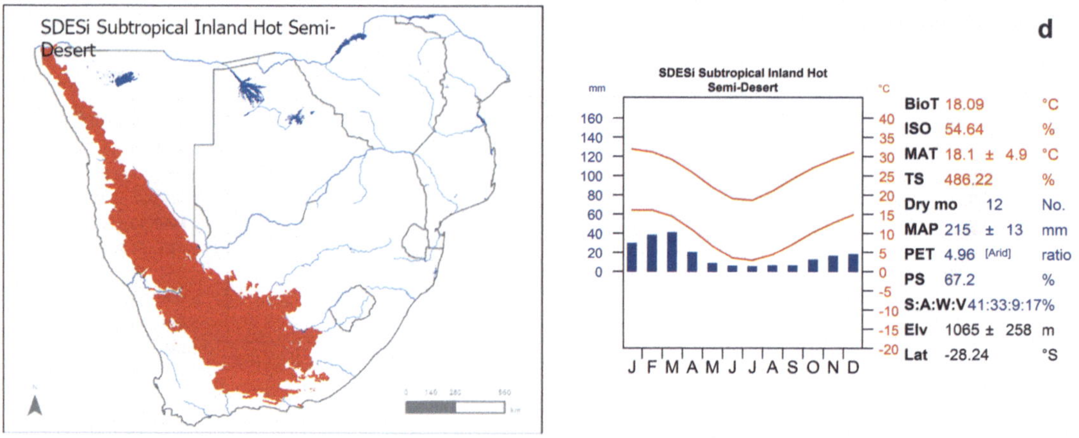

Fig. 7.1 (continued)

occurrence in Namibia, up to the Kunene River and extending into southwestern Angola, where it is also known as the Mossamedes Desert.

The Namib Desert is, after the South American Sechura-Atacama Desert, probably the most well-known fog desert in the World. The hyperarid Namib Desert Biome has a mean annual rainfall that ranges from around 20 mm to only about 80 mm, but the temperature ranges widely from cool near the coast to hot farther inland. Much of the desert biome vegetation is directly rain-fed and hence lies dormant most of the time. Only when the relatively rare rain events occur, often in thunderstorms in the north and hinterland areas, but with the greater predictability of cyclonic winter rains in the southwest, does this vegetation actively respond through growth and reproduction. The typically sparse vegetation cover can also temporarily transform into a high cover of annual plants in some areas following downpours of sufficient magnitude. However, with increasing distance toward the coast in the Namib Desert, fog becomes very frequent, and several studies have shown that many plant species in or near the coastal belt can survive through fog utilisation (Van Damme 1991). Prominent lichen fields are also closely associated with high fog frequency in areas with stable substrate and away from wind corridors of sand blasting, particularly in the coastal south (Fig. 7.4a–c). Interestingly, it has been shown in coastal parts of the Southern Namib Desert regional biome that, except for rare rain showers, atmospheric water vapour can be an even more critical water source than fog for certain lichen species (Maphangwa et al. 2012).

The Namib Desert continental biome contains three regional biomes—two zonal (Southern Namib Desert and Northern Namib Desert) and one intrazonal regional biome—the Namib Sand Sea Desert.

DN1 Southern Namib Desert

Dn 1 Alexander Bay Coastal Duneveld (Dayaram et al. 2019)
Dn 2 Namib Lichen Fields (Dayaram et al. 2019)
Dn 3 Western Gariep Plains Desert (Dayaram et al. 2019)
Dn 4 Western Gariep Lowland Desert (Dayaram et al. 2019)
Dn 5 Western Gariep Hills Desert (Dayaram et al. 2019)
Dn 6 Holgat Coastal Duneveld (newly proposed VU)
SKs 1 Richtersveld Coastal Duneveld p.p. (Dayaram et al. 2019)

Distribution This regional biome is found on plains and some smaller inselbergs extending northwards to Hottentots Bay, the Koichab Depression, and small unmapped coastal enclaves such as Spencer Bay (Namibia) further north (Fig. 7.5). It borders on Succulent Karoo in the south between the vicinity of the Holgat River and Khubus in South Africa. It is interrupted by the larger inselbergs of Succulent Karoo (e.g., Aurusberg, Rooiberg,

Table 7.1 A system of global biomes (GB), continental biomes (CB), and regional biomes (RB) of the Zonobiome S2 Hot Arid Zone of Southern Africa

Rank	Code	Biome	Zonality	Azonality type	Climatic diagnosis
ZB	**S2**	**Hot Arid Zone**	**Zonal**		**Subtropical-warm-temperate summer-hot subarid to arid; low rainfall**
GB	***DESc***	***Subtropical Coastal Hot Desert***	***Zonal***		***Subtropical arid poorly predictable mainly low winter rainfall; fog effect***
CB	**DN**	**Namib Desert**	**Zonal**		**Subtropical arid poorly predicta-ble mainly low winter rainfall; fog effect**
RB	DN1	Southern Namib Desert	Zonal		
RB	DN2	Namib Sand Sea Desert	Intrazonal	Psammobiome	
RB	DN3	Northern Namib Desert	Zonal		
GB	***DESi***	***Subtropical Inland Hot Desert***	***Zonal***		***Subtropical-warm-temperate; arid poorly predictable***
CB	**DG**	**Gariep Desert**	**Zonal**		**Subtropical-warm-temperate; arid poorly predictable mainly summer rainfall**
RB	DG1	Gariep Desert	Zonal		
CB	**DT**	**Tankwa Desert**	Zonal		**Warm-temperate; arid poorly predictable mainly winter rainfall; rainshadow effect**
RB	DT1	Tankwa Desert	Zonal		
GB	***SDESi***	***Subtropical Inland Hot Semidesert***	***Zonal***		***Subtropical summer-hot; subarid***
CB	**NK**	**Nama-Karoo**	**Zonal**		**Subtropical summer-hot; subarid predictable summer or bimodal rainfall**
RB	NK1	Kaokoland Pediment Karoo	Zonal		
RB	NK2	Namaland-Bushmanland Karoo	Zonal		
RB	NK3	Upper Karoo	Zonal		
RB	NK4	Lower Karoo	Zonal		
RB	NK5	Escarpment Karroid Thicket	Zonal		
GB	***SDESc***	***Subtropical Coastal Hot Semidesert***	***Zonal***		***Subtropical; subarid predictable winter rainfall; partial fog effect***
CB	**SK**	**Succulent Karoo**	**Zonal**		**Subtropical; subarid predictable winter rainfall; partial fog effect**
RB	SK1	Richtersveld-Aus Upland Karoo	Zonal		
RB	SK2	Bushmanland Inselberg Karoo	Extrazo-nal	Lithobiome	
RB	SK3	Namaqualand Hardeveld Karoo	Zonal		
RB	SK4	Namaqualand Sandveld	Intrazonal	Psammobiome	

(continued)

Table 7.1 (continued)

Rank	Code	Biome	Zonality	Azonality type	Climatic diagnosis
RB	SK5	Knersvlakte Karoo	Intrazonal	Lithobiome	
RB	SK6	Western Escarpment Karoo	Zonal		
RB	SK7	Transmontane Rainshadow Karoo	Zonal		
RB	SK8	Little Karoo	Zonal		

The codes of the zonobiomes follow Mucina (2023)

and Boegoeberg). The area of the unit remains west of the uplands of Nama-Karoo and of the main outliers of Succulent Karoo in the vicinities of Aus and Rosh Pinah. Most of the area in Namibia lies within the Tsau ǁKhaeb Sperrgebiet National Park, with a relatively small part conserved within the Richtersveld National Park in South Africa.

Vegetation Physiognomy and Composition The vegetation physiognomy of this regional biome is formed by sparse, low, leaf-succulent scrub in most areas, with generally higher cover in the south. However, it appears reduced and patchy in the north. It also includes some desert grasslands (e.g., with *Stipagrostis geminifolia*) on sand and gravel plains in the eastern sections of the unit. In the coastal zone and towards the central sand plains *Cladoraphis cyperoides*, *Conophytum saxetanum*, *Dregeochloa pumila*, *Mesembryanthemum arenosum*, *Othonna lasiocarpa*, and *Pelargonium ceratophyllum* feature more prominently. Very sparse vegetation occurs in sandblasted habitats in the coastal belt. Some lichen fields are found outside the sandblasted areas. Here, the widespread species include *Augea capensis*, *Euphorbia gummifera*, *Monsonia patersonii*, *Psammophora modesta*, and *Tetraena prismatocarpa*. Important plants on some rocky outcrops and inselbergs include *Bulbine rhopalophylla*, *Crassula ausensis* subsp. *ausensis*, *Hoodia alstonii*, *Pelargonium crassicaule*, and *Tetraena clavata*.

Climate Rain can fall at any time of the year in this hyper-arid regional biome, but the cyclonic winter rainfall is the most important to the growth of much of its flora. Fog is an essential additional source of precipitation. MAT is the lowest of all the desert regional biomes. Frequent strong southern winds occur in the coastal belt (Fig. 7.5).

Topography, Geology and Soils Habitats are mainly coastal dunes, sand plains, scattered mobile dune systems, gravel plains, rocky outcrops, and smaller arid inselbergs. The predominant geological substrates are gneiss, quartzite, sandstone, shale, dolerite, schist, granite, granodiorite, phyllite, breccia, arkose grit, limestone and dolomite. In earlier works (e.g. Rutherford and Westfall 1986), this regional biome was designated a hyper-arid part of the Succulent Karoo biome.

Sources Giess (1971), Jürgens (1991), Williamson (1997), Burke (2004, 2005a, 2020), Burke and Mannheimer (2004), Burke et al. (2004), Jürgens et al. (2006), Maphangwa et al. (2012), Burke and Loots (2020).

DN2 Namib Sand Sea Desert

Distribution This regional biome encompasses the main area of the continuous sea of mobile dunes of the Namib. It stretches east of the Atlantic Ocean from the Kuiseb River in the north to the region of the Koichab Depression in the south (Fig. 7.6). The sand sea desert unit is bounded on the east by the Pronamib (Vornamib) area of the Nama-Karoo biome. Almost all of the area lies within the Namib-Naukluft Park.

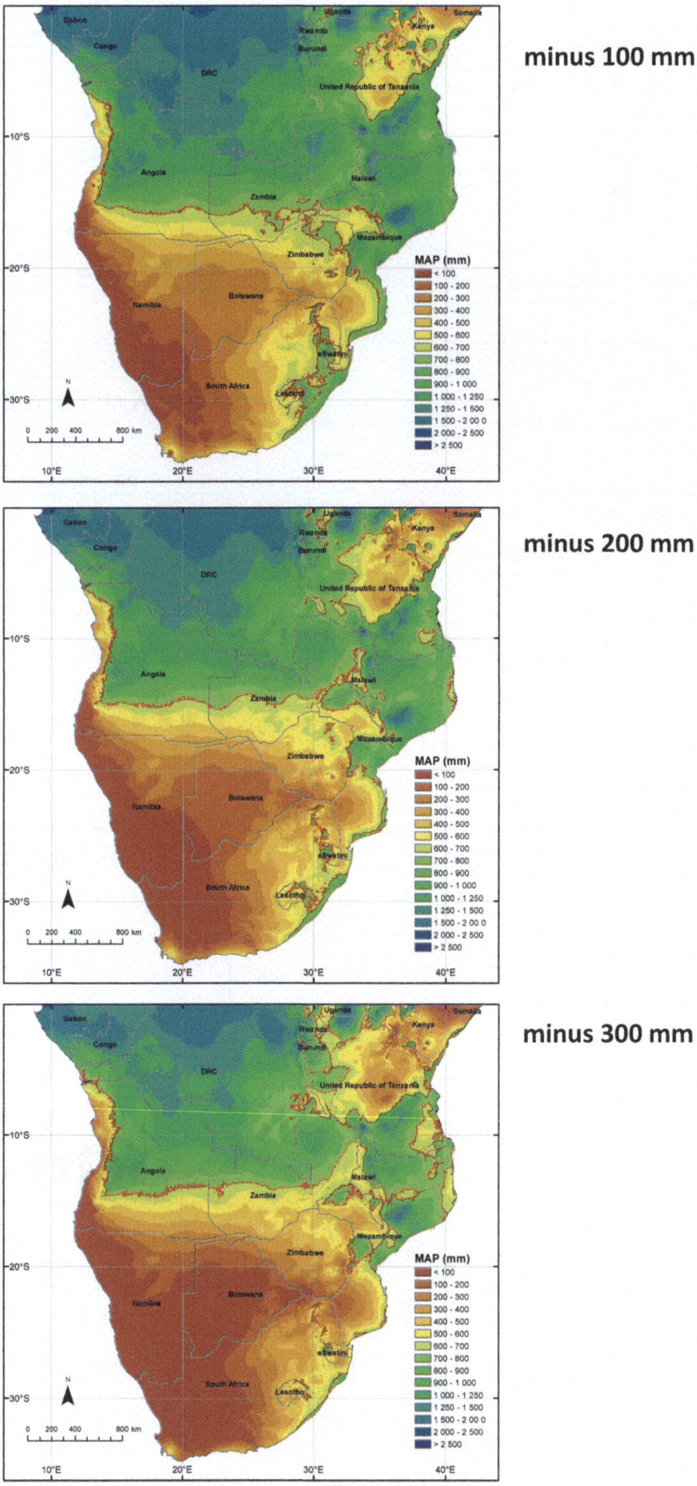

Fig. 7.2 Simulation of the opening of the Arid Corridor opening a putative migration route for flora linking the Horn of Africa and SW Southern African arid region. The current pattern of precipitation (MAP) is based on Average Monthly precipitation data for the period 1970–2000 and was sourced from WorldClim 2.1 (http://worldclim.org). The MAP was calculated from the average monthly precipitation raster layers using raster algebra in ArcMap 10.8.1. Then scenarios for MAP minus 100, 200, and 300 mm (simulating projected progressive decrease of precipitation) were performed using the raster calculator in ArcMap 10.8.1. All raster maps were reclassified to the given bin sizes for mapping purposes

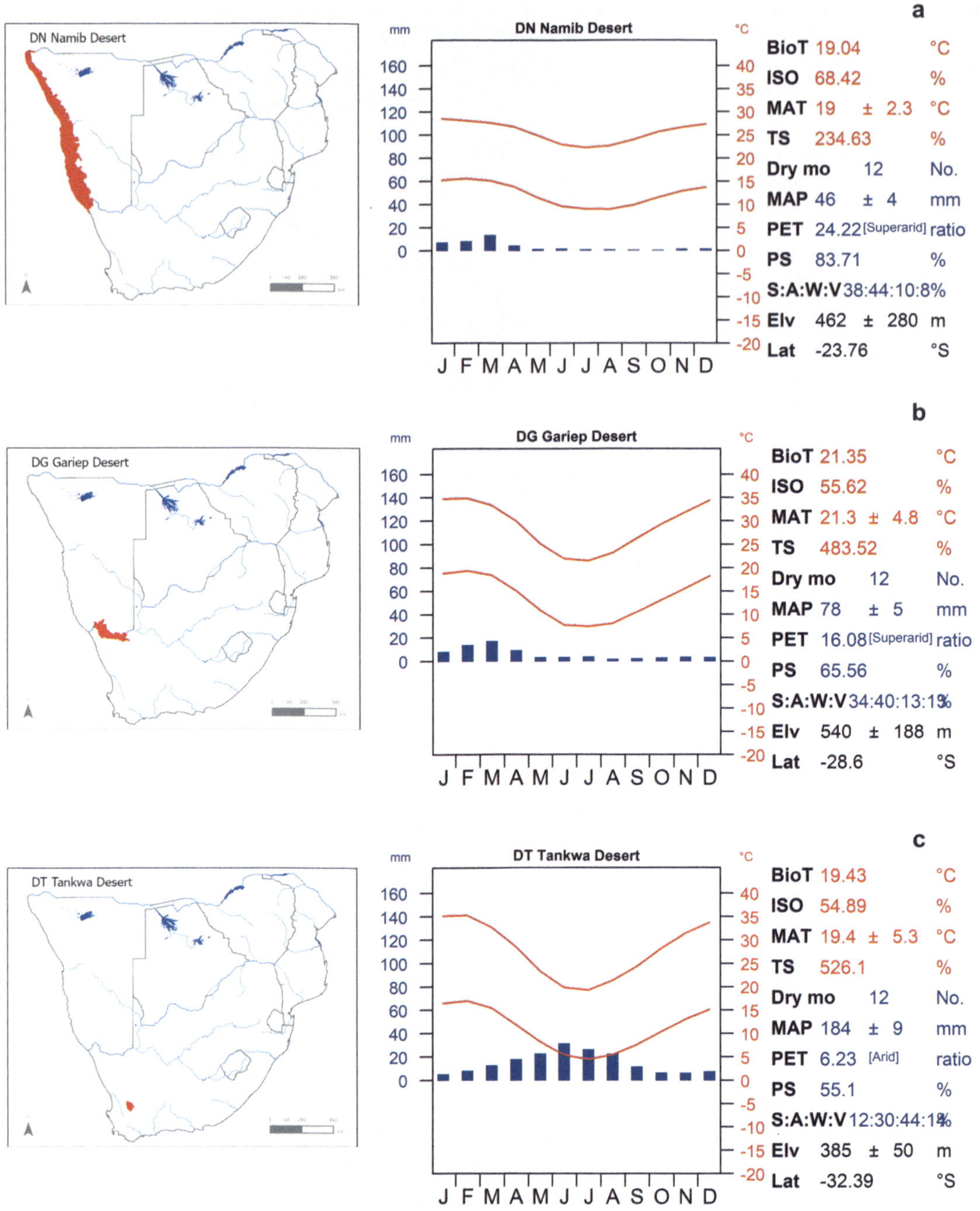

Fig. 7.3 Distribution areas and the climatic diagrams of the continental biomes within the zonobiome S2 Hot Arid Zone

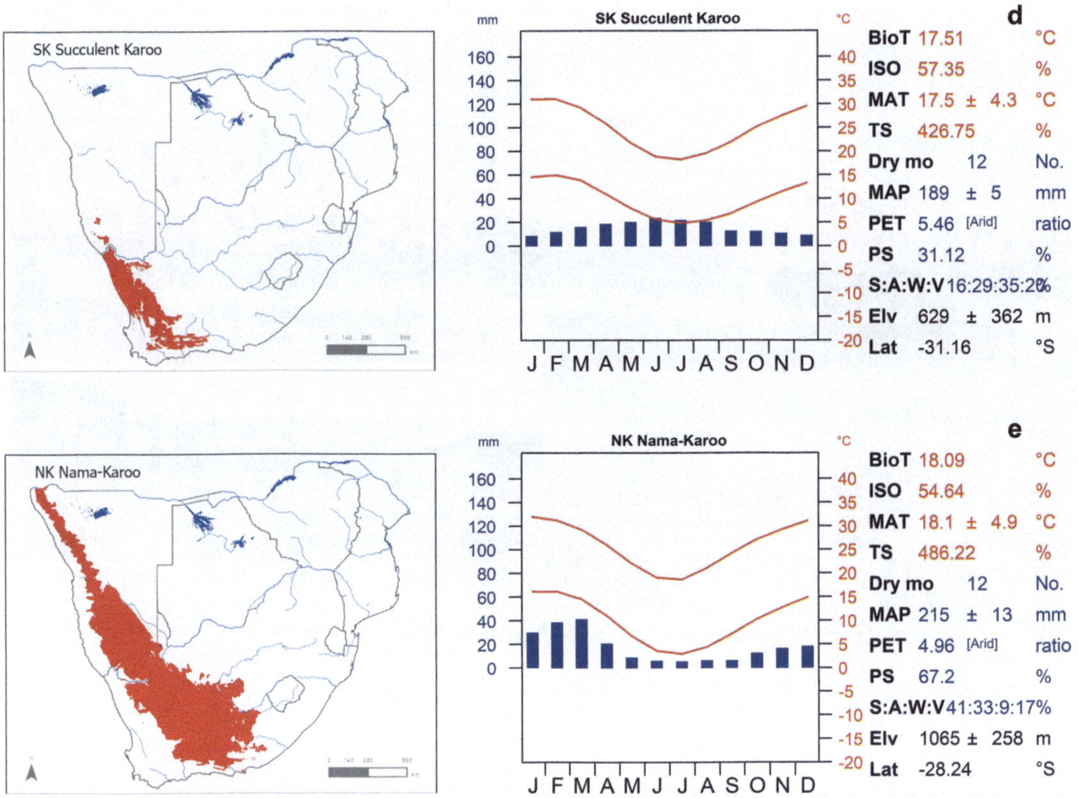

Fig. 7.3 (continued)

Vegetation Physiognomy and Composition The vegetation physiognomy is determined by the very low cover of mainly graminoids. The lowest vegetation cover is found on the upper parts of the mobile dunes, built almost exclusively of the mound-building perennial graminoid *Stipagrostis sabulicola*, accompanied by a leaf-succulent *Trianthema hereroensis* (also characteristic on coastal dunes). Endemic and near-endemic species include *Hermannia minimifolia* and *Stipagrostis seelyae.* Species richness increases towards the inland margin of this unit and can include species such as *Cladoraphis spinosa*, *Stipagrostis ciliata*, and *S. lutescens.* After extensive rain, annuals can become temporarily dominant. Inselbergs (e.g. the Hauchab Mts) have stronger floristic relationships to the tropical north than to the south.

Climate The low annual rainfall is concentrated in summer and early autumn and is lowest near the coast, including the area of the Hauchab Mts located in the southern sand sea. In the south, this regional biome does experience some influence of winter rainfall. Temperatures are cool near the coast and hot in the interior. The predominant strong south-westerly winds produce the transverse dunes near the coast. The areas further from the coast are subject to warm, desiccating berg winds originating in the eastern interior (Fig. 7.6).

Topography, Geology and Soils The North–South trending complex linear dunes (spaced 1–2 km apart and up to about 200 m high in places) dominate the interior. In contrast, there is a belt of lighter-coloured transverse crescentic dunes (with barchanoid ridges) nearer the coast. The interdune valleys usually comprise irregular mobile sand sheets, but these become more stable towards the eastern interior margin of the unit. A few prominent inselbergs are embedded

a
b
c
d
e
f
g
h

Fig. 7.4 Physiognomy of selected southern African deserts and semideserts. **a** Northern Namib Desert: gravel hammadas with gymnosperm *Welwitschia mirabilis* (Welwitschiaceae) on Welwitschia Plains east of Swakopmund. **b** Northern Namib Desert, near Hentiesbaai, north of Swakopmund, dominated by epilithic and epiphytic lichen *Teloschistes capensis* (Teloschistaceae). **c** Namib Sand Sea Desert, Namibia: Vornamib dunes near Wolwedans, NamibRand Nature Reserve. **d** Gariep Desert, South Africa, a view of the Helskloof Canyon and the Nababeepberge in Richtersveld. **e** Tankwa Desert, NE of Ceres, Western Cape, South Africa—rare spring flower display with dominating *Euryops annuus* (Asteraceae). **f** Succulent Karoo, South Africa: succulent scrub with *Eriocephalus* spp., *Tripteris sinuata* (all Asteraceae), *Euphorbia decussata* (Euphorbiaceae), *Scholtzia leipoldtii* and many other Aizoaceae, overlooking granodiorite koppies alternating with flats in Goegap Nature Reserve near Springbok, northern Namaqualand. **g** Nama-Karoo, South Africa: typical winter view of the Eastern Upper Karoo shrublands with species of *Pentzia*, *Eriocephalus*, *Rosenia* (all Asteraceae) and *Lycium* (Solanaceae) and scattered 'white grasses' (species of *Aristida*, *Eragrostis*, *Stipagrostis*) south of Richmond. **h** Kaokoland Pediment Karoo (a member of the CB Nama-Karoo) with low-grown *Commiphora* spp. (Burseraceae) and *Welwitschia mirabilis*, in Petrified Forest, near Khorixas, northern Namibia. The prolonged, broken cylindric rock in the foreground is a petrified log of *Dadoxylon arberi* (a conifer-like cordaitales) from Lower Permian. Photo credits: All images by L. Mucina

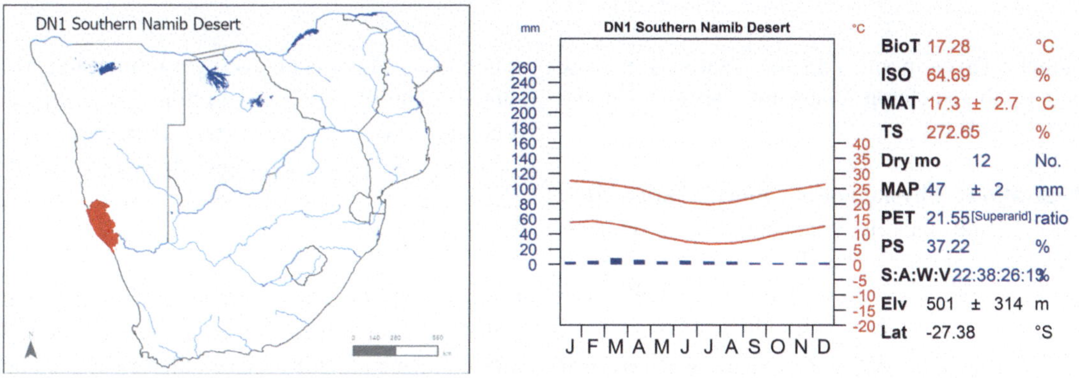

Fig. 7.5 The distribution area and the climatic diagram of the regional biome DN1 Southern Namib Desert

within this regional biome, most prominently—the Hauchab Mts. The substrate for plant growth is, unsurprisingly, made up of mobile dunes, inter-dispersed with some areas of more consolidated sand. This is the only coastal desert regional biome in which the aboveground water flow of all ephemeral rivers from the interior is blocked from reaching the ocean by the sand sea.

Sources Yeaton (1988), Boyer (1989), Burke et al. (1998), Giess (1971), Craven (2009), Jürgens et al. (2013), Livingstone (2013), Aushiku et al. (2015), Kaseke et al. (2016), Qiao and Wang (2022), Chase et al. (2023).

DN3 Northern Namib Desert

Distribution The Northern Namib Desert forms a belt of mainly flat to gently sloping and undulating gravel plains 60–90 km wide, parallel to the Atlantic Ocean coast, from the Kuiseb River in the south to the Kunene River in the north, on the Angolan border (Fig. 7.7). To the east, it extends up the broad, low-lying valley of the Kunene River to the vicinity of the northernmost end of the Marienfluss Plain. The eastern boundary remains west of the Hartmann Berge, Endendeka Hills, Grootberg, and the line of inselbergs that runs southwards from the Brandberg and Spitzkoppe to the vicinity of Heinrichberg and Barrowberg. Most of the southernmost section (commonly called the Central Namib) is conserved within the Namib-Naukluft Park, while much of the northern half lies within the Skeleton Coast Park. The plains are often dissected by rocky azonal corridors, with usually dry beds of rivers originating further inland. In the north are two relatively restricted sand ergs: the Kunene Erg south of the

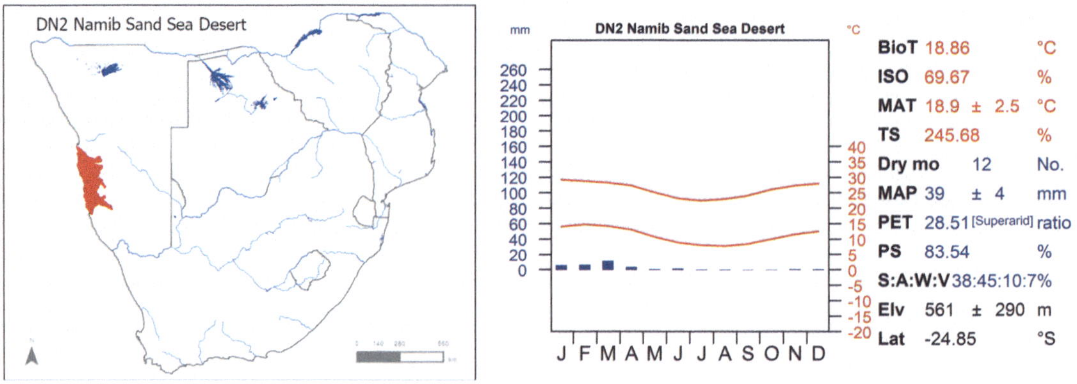

Fig. 7.6 The distribution area and the climatic diagram of the regional biome DN2 Namib Sand Sea Desert

Kunene River mouth and the relatively narrow belt of the Skeleton Coast Erg between Möwe and Torra Bays.

Vegetation Physiognomy and Composition This biome unit is characterised by ephemeral desert grassland and localised patches of dwarf shrub; vegetation cover is usually very low with extremely patchy distribution. Closer to the coast, *Arthraerua leubnitziae* and *Tetraena stapfii* are common, together with species such as *Chloris flabellata*, *Cladoraphis cyperoides*, and *Stipagrostis hermannii* in the north. At least in the southern half of this regional biome, the proportion of *Tetraena stapfii* increases further away from the coast. However, it becomes replaced towards the inland margin by perennial *Stipagrostis ciliata* and *S. obtusa* and annual *S. hirtigluma* subsp. *hirtigluma*, and *S. uniplumis* var. *intermedia*. *S. giessii* is more important in the north. On the dune fields in the north, *Brachiaria psammophila*, *Centropodia glauca*, *Indigofera cunenensis*, *Distimake multisectus*, and *Stipagrostis subacaulis* are more common. Widely scattered populations of *Welwitschia mirabilis* are endemic to this regional biome and extend to the Mossamedes Desert of Angola. The coastal fog belt of this regional biome supports the most extensive Namib Desert lichen fields, extending from near the coast to as much as 30 km inland.

Climate The climate is hyper-arid, with MAP ranging from about 85 mm to less than 20 mm at the coast. There is an average summer to early autumn rainfall peak, and rainfall variability is very high. Fog is an important source of water for many plants, for example, *Arthraerua leubnitziae*, close to the coast. MAT is the highest of the three Namib regional biomes (Fig. 7.7).

Topography, Geology and Soils This regional biome forms a belt of flat to gently sloping and undulating plains with rocky outcrops and some localised linear dune fields. Gravel plains and gypsum flats are the most common. The vegetation of the belt closer to the coast is often closely associated with gypsum crusts in contrast to the calcrete-associated vegetation further inland. Rock types vary considerably and include basalt, chert, conglomerate, diorite, dolerite, gneiss, granite, granodiorite, graywacke, limestone, marble, marl, quartzite, sandstone, shale, schist and tillite. Extensive shallow washes include the typical habitat of the iconic, often highly isolated populations of the deep-rooted *Welwitschia mirabilis*. On the southern boundary of this unit lies Gobabeb Research Station, which for over 50 years has served as a major research base for ecological investigations in and around the Namib Desert.

Sources Kers (1967), Giess (1971, 1981), Robinson (1976), Moisel and Moll (1981), Jacobsen and Moss (1987), Loris and Schieferstein (1992), Günster (1993), Jürgens et al. (1997, 2010, 2013, 2021), Becker (2000),

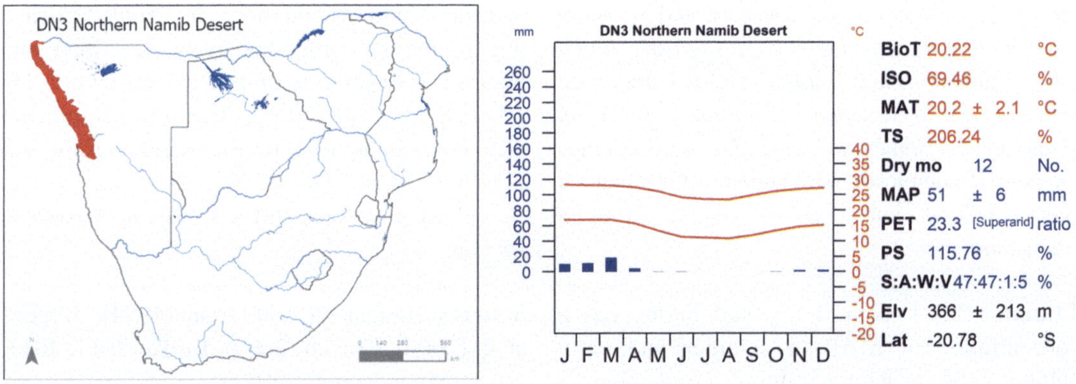

Fig. 7.7 The distribution area and the climatic diagram of the regional biome DN3 Northern Namib Desert

Henschel and Seely (2000), Burke (2002c, 2007), Lalley and Viles (2005), Schultz (2006), Eckardt et al. (2013), Henschel and Lancaster (2013), Jacobson and Jacobson (2013), Kaseke et al. (2016), Gottlieb et al. (2019), Rohde et al. (2019).

7.3 Continental Biome Gariep Desert (DG)

The Gariep Desert (Fig. 7.3b) is a member of the global biome DESi Subtropical Inland Hot Deserts (Fig. 7.1b). It is distinguished from the Namib Desert continental biome by climatic factors associated with its continental, inland position in contrast to the coastal-influenced climate of the Namib Desert on the seaboard of the Atlantic Ocean. Fog plays a crucial role in much of the Namib Desert but is generally insignificant in the Gariep Desert and Tankwa Desert. The Gariep Desert exhibits greater temperature extremes than the Namib Desert. It has a more uniform hot-season rainfall regime than the far more variable rainfall seasonality in the Namib Desert, which ranges from winter rainfall in the south to summer rainfall in the north. The Gariep Desert is hotter than the Tankwa Desert (especially regarding winter daytime temperatures), and the influence of winter rainfall is markedly less important than in the Tankwa Desert, where winter rainfall is critical to biota.

DG1 Gariep Desert

Dg 1 Noms Mountain Desert (Dayaram et al. 2019)
Dg 2 Richtersberg Mountain Desert (Dayaram et al. 2019)
Dg 3 Richtersveld Sheet Wash Desert (Dayaram et al. 2019)
Dg 4 Kwaggarug Mountain Desert (Dayaram et al. 2019)
Dg 5 Kahams Mountain Desert (Dayaram et al. 2019)
Dg 6 Helskloof Canyon Desert (Dayaram et al. 2019)
Dg 7 Northern Nababiepsberge Mountain Desert (Dayaram et al. 2019)
Dg 8 Southern Nababiepsberge Mountain Desert (Dayaram et al. 2019)
Dg 9 Eastern Gariep Plains Desert (Dayaram et al. 2019)
Dg 10 Eastern Gariep Rocky Desert (Dayaram et al. 2019)

Distribution This biome comprises the inland lower Gariep River valley and covers the lower altitude region between the Lorelei area in the west and around Onseepkans in the east (Fig. 7.8). It also includes the valley of the lower Fish River tributary (and much of the Fish River Canyon itself) in Namibia and the eastern parts of the Richtersveld in South Africa (Fig. 7.4d). The landscapes are characterised by a very rugged terrain overall, with deeply incised valleys and plains such as the Springbokvlakte in South Africa and the Gamkab Plains in Namibia. Some westernmost parts are conserved within the Richtersveld National Park in South Africa and the |Ai-|Ais National Park in Namibia.

Vegetation Physiognomy and Composition Vegetation in this biome is characterised by a sparse cover of low shrubs such as *Calicorema capitata*, *Commiphora namaensis*, *Petalidium*

setosum, *Portulacaria namaquensis*, and *Sisyndite spartea.* Graminoids include *Stipagrostis brevifolia, S. anomala* and *S. obtusa.* Other important taxa include *Myxopappus acutilobus*, *Sesuvium sesuvioides*, *Trianthema parvifolia*, and *Tribulus cristatus*. Endemic or near-endemic taxa include *Bowiea volubilis* subsp. *gariepensis* and *Jatropha orangeana*.

Climate This hyperarid regional biome has a concentration of MAP in summer and early autumn, and a winter rainfall component is unimportant, except possibly in the west. MAP is higher than in the neighbouring Southern Namib Desert, but this is offset by having a higher MAT, which reduces water availability. Fog is generally an insignificant water source for plants of this regional biome. A high-temperature seasonality characterises the Gariep Desert compared to the regional biomes of the Namib Desert (Fig. 7.8).

Topography, Geology and Soils Often rocky with very little or extremely shallow soils are characteristic of this biome. The substrate includes deep sandy alluvial material on valley bottoms, sheet washes, and colluvial material. Soils are freely drained soils with high base status on the upland marginal plateaus. The main rock types include granodiorite, adamellite, leucogranite, syenite, gneiss, quartzite, schist, tonalite, diorite, limestone, shale, granite, andesite, and diamictite. In South Africa, this biome is comprehensively described and mapped as vegetation units Dg 1 through to Dg 10 (Jürgens et al. 2006). The most significant difference compared to the neighbouring RB Southern Namib Desert flora is an increase in Acanthaceae species and a decline in Aizoaceae species.

Sources Jürgens (1991), Irish (1994), Jürgens et al. (2006), Craven (2009), Burke (2011, 2013, 2015), Mvondo et al. (2011).

7.4 Continental Biome Tankwa Desert (DT)

The CB Tankwa Desert (Fig. 7.3c) is another member of the global biome DESi Subtropical Inland Hot Deserts Inland Desert (Fig. 7.1b) and shares a relatively inland position with the Gariep Desert continental biome but is distinguished from the latter by cooler temperatures and a significant influence of winter rainfall. It lacks the strong winds associated with the more southerly regions of the CB Namib Desert and, notably, mostly lacks the incidence of fog, which is crucial in much of the Namib Desert. The Tankwa Desert does not experience the very low extremes of MAP (less than 70 mm) attained in many parts of the other two desert continental biomes.

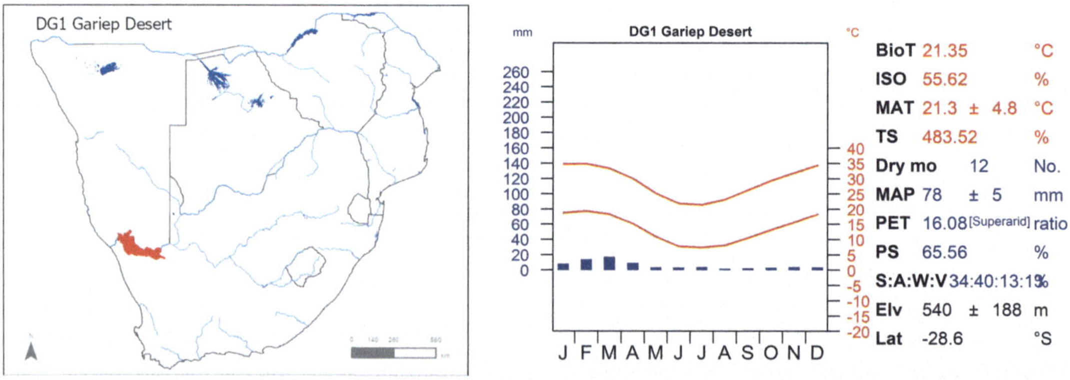

Fig. 7.8 The distribution area and the climatic diagram of the regional biome DG1 Gariep Desert

This desert unit follows van der Merwe et al.'s (2008) unit '*Stipagrostis obtusa* Central Tanqua Grassy Plains', except for a few areas at higher elevations.

DT1 Tankwa Desert

SKv 5 Tanqua Karoo p.p. (Dayaram et al. 2019)

Distribution This small, localised rain shadow desert regional biome comprises the lowest-lying (elevations of approx. 280–450 m) portion of the gently sloping plains of the Tankwa River basin, situated between the Cederberg Mountain Ranges and the Roggeveld Escarpment and straddling the Northern and Western Cape Provinces of South Africa (Fig. 7.9). It is by far the smallest desert unit within our study area and covers only about 2375 km^2. It is surrounded by the CB Succulent Karoo, which also occurs on some small inselbergs within the northeastern section of the unit. Conspicuous broad washes (AZi 7 Tanqua Wash Riviere; Mucina et al. 2006c) associated with the Tankwa and Ongeluks Rivers traverse this desert (Fig. 7.10). About one-third of the area of this desert is conserved within the current boundary of the Tankwa Karoo National Park. In the past, the term 'Tanqua' was used more often (see Mucina et al. 2006a).

Vegetation Physiognomy and Composition Shrub cover is low, but grass cover ranges from 10 to 90% in extraordinarily high rainfall years. Important perennial species are *Stipagrostis obtusa*, *S. brevifolia*, *S. ciliata*, and *Augea capensis*, with species such as *Limeum aethiopicum*, *Tetraena macrocarpa*, and *Cladoraphis spinosa* also frequent. Potentially common annuals include *Amellus tridactylus*, *Enneapogon desvauxii*, *Leyssera tenella*, *Tribulus terrestris*, and *Tetragonia microptera*. Ephemeral mass flowering of *Euryops annuus* and other annual (and geophytic) species occurs under favourable conditions (Fig. 7.4e), which can be less specific and sometimes asynchronous relative to the growth of other vegetation. The seed density of annuals in the soil seed bank exceeds that of the perennial shrubs (Saaed et al. 2018). Near endemic species to the biome (excluding candidates associated with intervening washes) include *Cyphia comptonii*, *Mesembryanthemum eurystigmatum*, *Moraea tanquana*, and *Nemesia suaveolens* (Bester et al. 2012; Steyn et al. 2013).

Climate Although this desert biome experiences a hyper-arid climate, it does not extend into areas of extremely low average rainfall, such as those found in the other regional desert biomes of Southern Africa. We estimate that most of the area has a MAP of between 70 and 90 mm. The MAP of 72.3 mm recorded at the sole long-term rainfall station (Elandsvlei No. 86079; alternatively calculated as 69 mm due to the discontinuous record; Fig. 7.11) in the biome is probably, based on its position in the landscape, very close to the lowest MAP limit for this biome (see Saaed et al. 2020 for analysis of rainfall patterns at Elandsvlei). Current global climatic models probably vastly overestimate the MAP of this biome and are likely to have been unduly influenced by data from other stations in the higher rainfall surrounding areas. For example, the modelled MAP of 184 mm indicated in the climate diagram for this biome is almost certainly a gross overestimation, but the basic mean seasonal rainfall pattern given should be realistic (Fig. 7.9). The rainfall is highly erratic but includes a peak of cyclonic showers in winter and some convective storms in the summer. Fog is generally unimportant. Summers are hot, and winters are cool with occasional frost.

Topography, Geology and Soils Geology is predominantly Ecca shale, siltstone, and mudstone, as well as Dwyka tillite, with a limited extent of Bokkeveld sandstone and siltstone in the northwest. More than 300 mm deep soils occur in the western lowest part of the basin (below elev. of 400 m). Extensive tracts of these and other soils in the biome are covered in a monolayer of course gravels and pebbles (clasts), which, with associated accretionary soil material, forms a stone or desert pavement (McAuliffe et al. 2018). These pavements

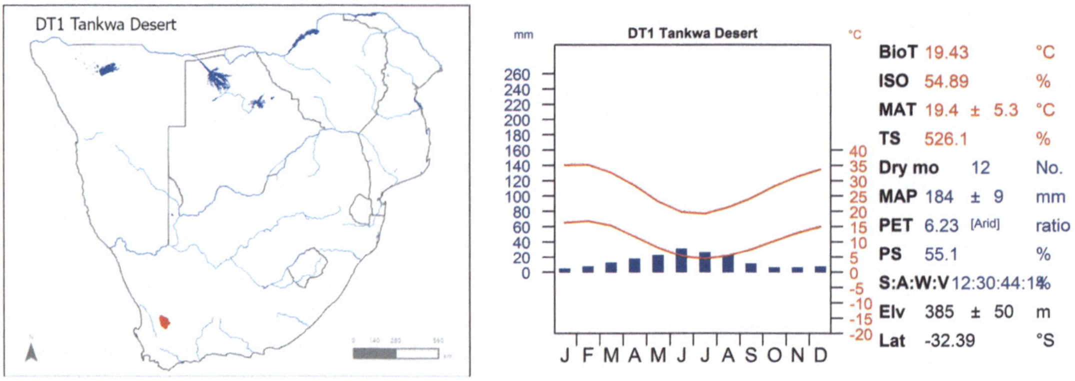

Fig. 7.9 The distribution area and the climatic diagram of the regional biome DT1 Tankwa Desert

greatly restrict water infiltration and raise salinity, resulting in a lack or absence of perennial vegetation in such areas. Soils are generally poorly developed, with loamy sands, sandy loams, lime commonly present, and infrequent rock outcrops, including some dolerite intrusions (Rubin 1998; Osman et al. 2019). It is commonly believed that much of this biome has been subjected to long-term intense overgrazing by domestic stock since the early nineteenth century (Acocks 1988; van der Merwe 2010). Even current fence lines that show stark contrasts between moderately vegetated and virtually bare ground in some of this biome suggest that herbivore pressure, in addition to low water availability, can exacerbate desert conditions. Large tracts of land within the Tankwa Karoo National Park are highly degraded.

Sources Acocks (1988), Rubin (1998), Mucina et al. (2006a), van der Merwe et al. (2008, 2015), van der Merwe (2009, 2010), van der Merwe and van Rooyen (2011a, 2011b), Bester et al. (2012), Steyn et al. (2013), Osman et al. (2019), Saaed et al. (2020, 2022).

7.5 Continental Biome Nama-Karoo (NK)

The term 'Nama-Karoo Biome' was coined by Rutherford and Westfall (1986) to establish the common identity of the extensive arid grassy shrublands of the central-west plateau of South Africa and those of the southern interior of neighbouring Namibia (Figs. 7.1d and 7.3e). The term stands in contrast with the CB Succulent Karoo, which has a generally much-reduced grass component and a distinctly winter-rainfall regime. Unlike the adjacent Namib Desert and Gariep Desert, where the perennial vegetation is mainly confined to moisture-concentrating areas, such as rocky crags or depressions or water-storing sands, in the CB Nama-Karoo, much of the vegetation of the open rain-fed plains is perennial. Contrary to the neighbouring Arid Savanna regional biomes, where the perennial woody vegetation occurs mainly in the form of small trees or tall shrubs, the woody perennial vegetation in the CB Nama-Karoo contains few such lofty elements, and smaller shrubs together with perennial grasses predominate (Fig. 7.4g and h). Another consequence of the lower rainfall in Nama-Karoo is the generally inadequate fuel load to sustain fires that are more commonly found in the Arid Savanna biome. It has, nevertheless, been shown that in some higher rainfall parts of Nama-Karoo in South Africa, sporadic fires can occur and can change the cover and composition of the dwarf shrub component, at least in the short term (du Toit et al. 2015a, b).

We distinguish five regional biomes within the CB Nama-Karoo: Kaokoland Pediment Karoo, Namaland-Bushmanland Karoo, Upper Karoo, Lower Karoo, and Escarpment Karroid

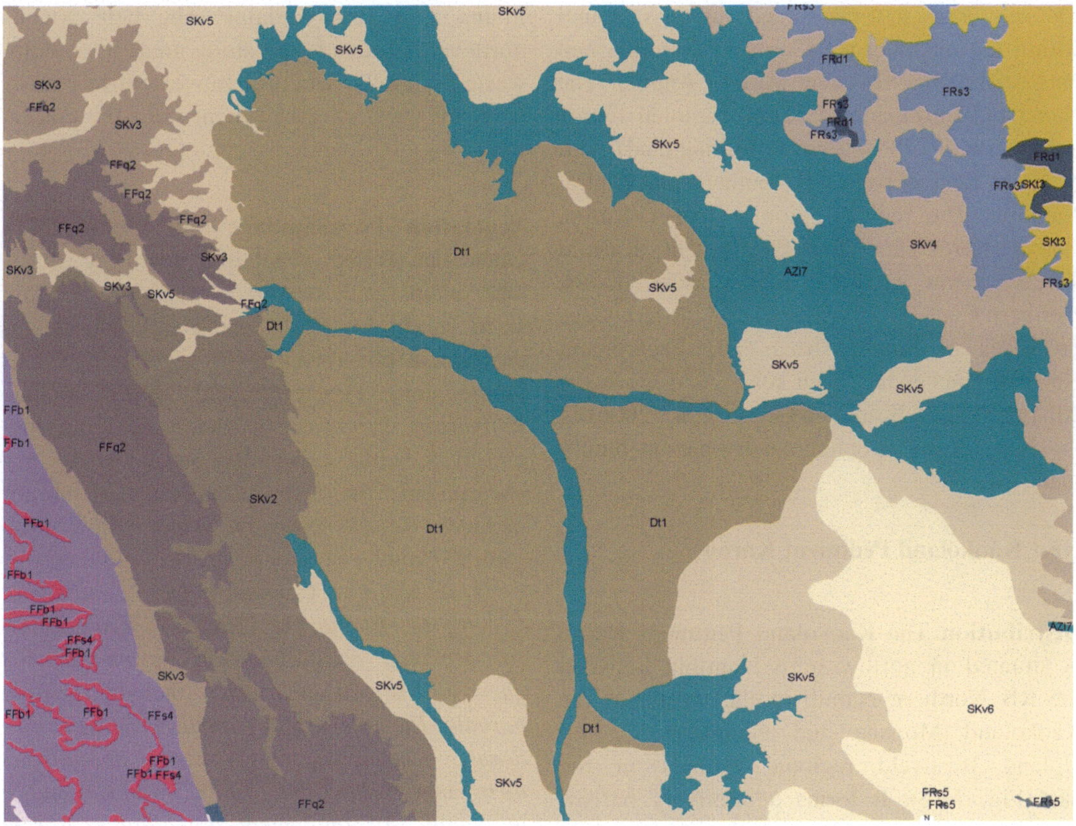

Fig. 7.10 The map of the Tankwa Desert and the surrounding vegetation types (the codes of those match the mapping concepts of the VegMap2006 South Africa; Mucina and Rutherford 2006)

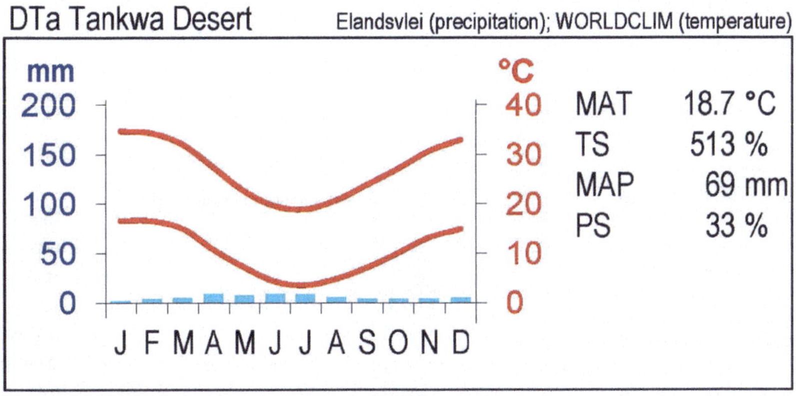

Fig. 7.11 The climate diagram of the Elandsvlei Farm climatic station located in the Tankwa/Tanqua Basin representing the regional biome Dt1 Tankwa Desert. *Source* Original climatic data of the Elandsvlei climatic station no. 86079 (32.318195° S, 19.556837° E), measured over more than 40 years. Temperature data come from WorldClim (http://worldclim.org)

Thicket. The first-mentioned unit is found within the mapping area, only in Namibia and outside the area—in southwestern Angola. The Namaland-Bushmanland Karoo regional biome occurs in South Africa (Bushmanland) and occupies large expanses in southern and central Namibia. The regional biomes Upper Karoo, Lower Karoo and the Escarpment Karroid Thicket cover extensive inland areas of the Northern Cape, Western Cape, and Eastern Cape provinces of South Africa. The Escarpment Karroid Thicket has been reassigned from the Albany Thicket Biome of Hoare et al. (2006) to the Nama-Karoo Continental Biome on bioclimatic grounds.

NK1 Kaokoland Pediment Karoo

Distribution The Kaokoland Pediment Karoo is situated in northwestern Namibia between the RB Northern Namib in the west and the Kaokoland Mopane and Namibian Central Upland Bushveld regional biomes in the east (Fig. 7.12). It forms a relatively narrow belt from the Kunene River valley in the north, mainly along the pediment and lower slopes of the western Escarpment, but also extends to the upper slopes in the south. It includes the sparsely distributed settlements of Orupembe, Sesfontein, Palmwag, Uis and Usakos to the vicinity of Solitaire in the south. The unit also occurs on the marginal grassy plains of the Marienfluss Valley, Giribes Plain, Achab Plains, and Tinkas Plains. The belt extends northwards (outside the study area) into southwestern Angola, where it moves entirely away from the western escarpment onto the coastal plain.

Vegetation Physiognomy and Composition Open but varied vegetation featuring grassland plains and scattered small to medium-sized shrubs on plains and rocky hillsides is the typical physiognomy of vegetation of this biome unit. The prominent shrubs include *Euphorbia damarana* on the lower-lying flatter areas, while especially on the rocky dry Escarpment, the rich assemblage of shrubs includes, *Colophospermum mopane* (low-shrub form), *Commiphora giessii*, *C. glaucescens*, *C. kraeuseliana*, *C. multijuga*, *C. saxicola*, *C. virgata*, *Euphorbia virosa*, *Maerua schinzii*, *Moringa ovalifolia*, and *Sesamothamnus guerichii.* Important grasses include *Aristida parvula*, *Enneapogon desvauxii*, *Eragrostis nindensis*, *Kaokochloa nigrirostris*, *Stipagrostis hirtigluma*, *S. hochstetteriana*, *S. obtusa*, and *S. uniplumis.* *Aizoanthemum galenioides*, *Boscia foetida*, *Euphorbia glanduligera*, *Justicia cleomoides*, *Kissenia capensis*, *Petalidium variabile*, and *Tetraena cylindrifolia* are also prominent. Fairy circles of *Stipagrostis giessii* or *S. hochstetteriana* occur in places on the plains, often close to the border with the Northern Namib Desert regional biome. The narrow belt contains common intrusions of savanna elements

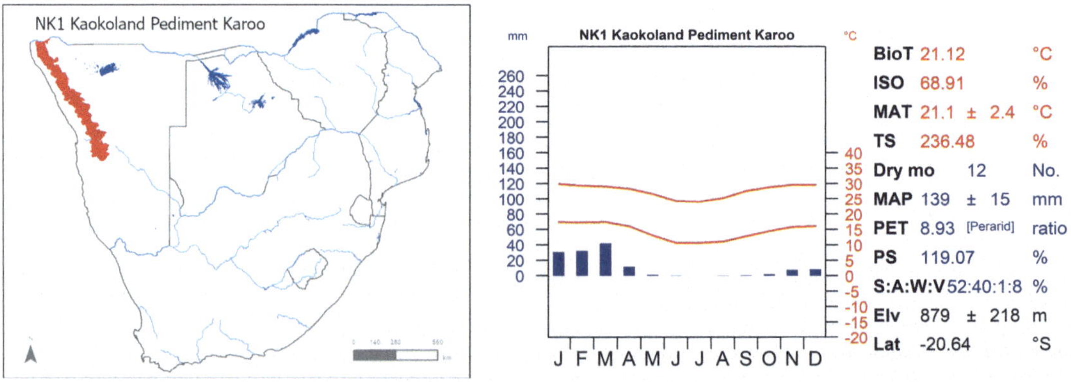

Fig. 7.12 The distribution area and the climatic diagram of the regional biome NK1 Kaokoland Pediment Karoo

from the east and desert elements from the west. The intermediate status between savanna and desert has been well-noted. This ecotonal character of the belt has also been reflected by fossil pollen spectra from the Late Holocene. Major inselbergs, such as the Brandberg, can also contain numerous floristic elements of the savanna vegetation in the interior (e.g. *Dombeya rotundifolia* and *Senegalia hereroensis*). This regional biome includes most of Jürgens's (1991) 'Damaraland-Kaokoland Domain', one of the phytogeographic subdivisions of the 'Karoo-Namib Region'.

Climate The MAP of 139 mm ranges from around 80 mm in the west to between 180 and 200 mm in the east and is concentrated in summer with very dry winters. This regional biome has a higher MAT and considerably warmer winters than the other Nama-Karoo regional biomes. The probability of frost is generally very low. Fog can occasionally extend from the coast into the lower altitude western parts of the zone. Westerly sea breezes can occur throughout the year, and occasional hot easterly berg winds occur in winter (Fig. 7.12).

Topography, Geology and Soils Topography is characterised by gently sloping plains with many intervening rock outcrops, including inselbergs, and the rocky slopes of substantial parts of the west-facing escarpment. The belt is traversed by several ephemeral east–west running river courses, sometimes with associated canyon formations. Very variable substrates range from extensive sand sheets, calcrete, saline soils, gravel, and other plains of superficial deposits to rocky ridges and dykes, spurs, and slopes of the western Escarpment and a particularly wide size range of inselbergs. Broadly, geology includes Huab and Epupa Metamorphic Complexes, schists of the Swakop Group, basalts of the Etendeka Group, sandstone of the Hakos Group, and numerous outcrops of Damara granites. Inselbergs, such as the Brandberg and the Spitzkoppe, consist of granites, whereas the Rostock inselberg comprises granitic gneisses and other rocks. Apart from the many rock outcrops, soils are mainly lithic and eutric leptosols, eutric regosols, and petric calcisols.

Sources Jürgens (1991, 2013), Irish (1994), Cowlishaw and Davies (1997), Becker (2000), Becker and Getzin (2000), Burke (2001b, 2002a, b, c, 2003, 2005b), Mendelsohn et al. (2002), van Rooyen et al. (2004), Gil-Romera et al. (2007), Jürgens et al. (2013), Rohde et al. (2019), Becker et al. (2021), Strohbach (2021), Getzin et al. (2022), De Cauwer et al. (2023).

NK2 Namaland-Bushmanland Karoo

NKb 1 Lower Gariep Broken Veld (Dayaram et al. 2019)
NKb 2 Blouputs Karroid Thornveld (Dayaram et al. 2019)
NKb 3 Bushmanland Arid Grassland (Dayaram et al. 2019)
NKb 4 Bushmanland Sandy Grassland (Dayaram et al. 2019)
NKb 5 Kalahari Karroid Shrubland (Dayaram et al. 2019)
NKb 6 Bushmanland Basin Shrubland (Dayaram et al. 2019)

Distribution About two-thirds of this large regional biome is found north of the Gariep River, comprising mainly the interior, Namaland, plain of southern Namibia from around Rehoboth to the vicinity of Warmbad in the south. It extends into South Africa from the Upington area, encompassing the entire, extensive Bushmanland Plain, southwards to the vicinities of Gamoep, Sakrivier and Vanwyksvlei in the Northern Cape Province. In Namibia and the Northern Cape north of the Gariep River, it remains west of the extensive Kalahari Dunefield regional biome. The western perimeter in Namibia is formed by the Pronamib (Vornamib) plains and associated escarpment mountains, including the Naukluft, Zaris, and Tiras Mountains, Huib-Hoch Plateau, and numerous inselbergs such as Loskop and Chowagasberg. Other notable mountains further inland include the Schwarzrand, Hunsberge, Brukkaros, Klein Karasberge, and Groot Karasberge. This regional biome is partially bisected by the CB Gariep Desert along the lower Gariep River valley (Fig. 7.13).

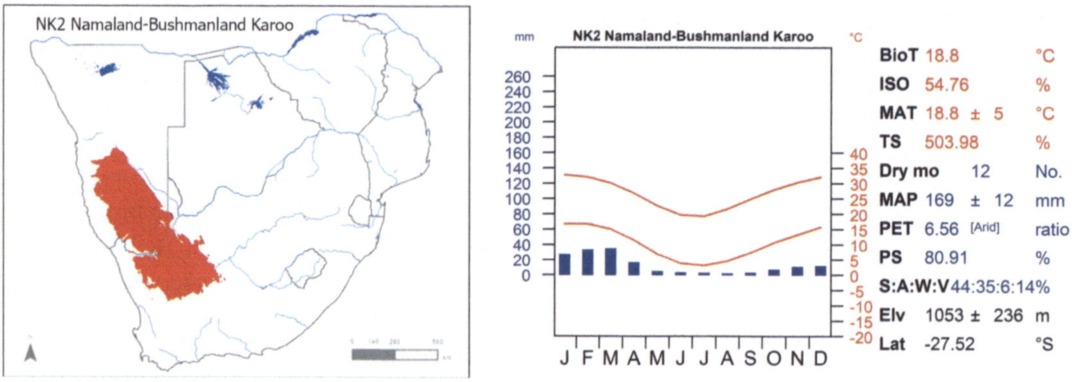

Fig. 7.13 The distribution area and the climatic diagram of the regional biome NK2 Namaland-Bushmanland Karoo

Vegetation Physiognomy and Composition The dominant vegetation physiognomy is low to medium height, sparse to open grassy shrubland. On the plains, the proportion of sand, gravel and calcrete governs the ratio of grass to shrub, with sandy plains, sometimes extensive, tending to support almost pure stands of perennial grass. In contrast, the shrub component increases on gravel and calcrete. Numerous common species, in both Namaland and Bushmanland, include the dominant shrub *Rhigozum trichotomum* as well as *Aptosimum spinescens*, *Barleria rigida*, *Boscia foetida*, *Caroxylon* spp., *Dicoma capensis*, *Eriocephalus ambiguus*, *Hermannia spinosa*, *Lycium bosciifolium*, *Parkinsonia africana*, *Phaeoptilum spinosum*, *Tetragonia arbuscula*, *Xerocladia viridiramis*, and *Zygophyllum simplex*. *Aloidendron dichotomum* is a conspicuous feature on rocky outcrops. Important, widespread grasses include *Aristida adscensionis*, *Enneapogon cenchroides*, *E. desvauxii*, *Eragrostis nindensis*, *Schmidtia kalahariensis*, *Stipagrostis brevifolia*, *S. ciliata*, *S. hochstetteriana*, *S. obtusa*, and *S. uniplumis*. Some important shrub species predominantly found on the Namaland plains are *Catophractes alexandri* on calcretes, *Vachellia nebrownii* in lower-lying areas (e.g. shallow washes), *Leucosphaera bainesii* in stony habitats, and *Senegalia mellifera* on deeper and often sandy soils. Conversely, some shrubs, such as *Berkheya annectens*, *Osteospermum spinescens*, and *Pentzia spinescens*, occur predominantly in the Bushmanland region. In the western escarpment zone of southern Namibia, other species, in addition to many of the above-listed widespread species, include *Amphiasma divaricatum*, *Commiphora glaucescens*, *C. tenuipetiolata*, *Cyphostemma bainesii*, *Euphorbia virosa*, *Justicia cleomoides*, and *Myrothamnus flabellifolius*. Also, on deeper western sandy flats of this Pronamib zone *Stipagrostis obtusa*, *S. uniplumis* and other grasses dominate. *S. ciliata* forms the edge of the circular bare patches (fairy circle) features. In areas of southwestern Namaland, *Euphorbia gregaria* is dominant over extensive areas (e.g., north of the Gariep Desert continental biome), whereas, in southeastern Namaland, *E. avasmontana* dominates several areas, including in the Gariep River valley and Karas Mountain complex.

Climate MAP is 169 mm, peaking in the summer and early autumn. Winters are dry, although light winter rains can occur closer to the Southern Namib Desert regional biome. MAP falls close to a minimum of 70–80 mm over a wide front where this regional biome adjoins desert climates associated with its lengthy border with the regional biomes of Namib Sand Sea Desert, Southern Namib Desert, and Gariep Desert. Summers are hot, with maximum summer temperatures higher than any other Nama-Karoo regional biome. Winter days are warm but cold at night, with frosts occurring, except for the low-lying plains of the Pronamib (Fig. 7.13).

Topography, Geology and Soils Extensive plains dominate the topography, often gently sloping and sometimes dissected by a dense network of shallow washes in the more arid parts. Some moderately steep hills and steep inselbergs, rugged escarpment elements and mountains are also included, a few exceeding an elevation of 2000 m above sea level. Rugged, broken rocky terrain surrounds the canyon-like feature on the Gariep River. Many large pans are embedded within the area, especially the Bushmanland Basin. The geology of this regional biome is highly varied. In southern Namibia, the Nama Group is widespread and includes the sandstones and shales of the Fishriver Subgroup with the addition of limestones in the Schwarzrand and Kubis Subgroups. Much of the remaining area in southern Namibia and in the Bushmanland Basin comprises the tillites and other sedimentary rocks of the Dwyka and Ecca Formations of the Karoo Sequence. Basalt outcrops of the Kalkrand Formation of the same sequence are important in the north of the regional biome. In contrast, outcrops of granite and gneiss are common in the Namaqualand Metamorphic Complex occurring further south. The Sinclair Sequence in the west contains a wide range of rock types, including granites, quartzites, gabbro, norite, rhyolite, and basalt. The geology in the vicinity of the Gariep River, in addition to the metamorphosed sediments and outcrops of the ultra-metamorphic rocks of the Namaqualand Metamorphic Complex, includes carbonates and cherts of the Campbell Group, banded iron, and amphibolite of the Asbestos Hills Subgroup, quartzites and gneisses of the Korannaland Supergroup as well as the Riemvasmaak Gneiss. In many areas of the regional biome, localised dolerite outcrops, in the form of sills and dykes, occur as well. Soils are poorly developed, skeletal, usually coarse-grained, well-drained, sandy, and gravelly; they are generally shallow, often due to widely distributed shallow-lying hard calcrete beds, with surface calcrete outcrops also a common feature. However, the depth of sand can vary significantly on the occasional, extensive sandy plains.

Sources Giess (1971), Werger (1978), Kent (1980), Acocks (1988), Irish (1994), Palmer and Hoffman (1997), Burke (1997, 2001a, b, 2002a, b, 2013, 2020), Strohbach (2001), Rutherford et al. (2003), van Rooyen et al. (2004), Mucina et al. (2006b), Jürgens et al. (2010), Dreber et al. (2011), Strohbach and Jankowitz (2012), Hanke et al. (2014), Cramer et al. (2017), van der Merwe (2020), Getzin et al. (2022).

NK3 Upper Karoo

NKu 1 Western Upper Karoo (Dayaram et al. 2019)
NKu 2 Upper Karoo Hardeveld (Dayaram et al. 2019)
NKu 3 Northern Upper Karoo (Dayaram et al. 2019)
NKu 4 Eastern Upper Karoo (Dayaram et al. 2019)

Distribution This regional biome encompasses a large, relatively high-elevation continuous plain lying between the Great Escarpment and the Bushmanland Basin in the Williston, Fraserburg, Carnarvon, and Victoria West areas in the west, extending eastwards to the Orange River on the southern boundary of the Free State Province and including the areas of Colesburg, Middelburg and Cradock, all north of the Great Escarpment. It extends further north through the Britstown, De Aar, and Philipstown regions to the vicinity of Prieska and Hopetown. It continues north of the Orange River, mainly in the Koffiefontein area of the southwestern Free State (Fig. 7.14). Excluded are rocky hills and ridges of the Dry Highveld Koppie Grassland regional biome scattered throughout the belt of Upper Karoo, lying immediately south and southwest of the Orange River and extending as far as the Richmond area in the southwest.

Vegetation Physiognomy and Composition The dominant vegetation is open dwarf shrubland with a greater or lesser grass component, often depending on the local grazing regime. Some taller shrubs often occur on the koppies. Important shrubs and forbs include *Aizoon plinthoides*, *Aptosimum spinescens*, *Caroxylon* spp., *Chrysocoma ciliata*, *Eriocephalus ericoides*, *E. spinescens*, *Euryops lateriflorus*, *Felicia muricata*, *Gnidia polycephala*, *Helichrysum lucilioides*, *Indigofera alternans*,

Limeum aethiopicum, *Lycium cinereum*, *Oedera humilis*, *Osteospermum spinescens*, *Pentzia calcarea*, *P. globosa*, *P. incana*, *P. spinescens*, *Phymaspermum parvifolium*, *Polygala leptophylla*, *Pteronia glauca*, *P. sordida*, *Rhigozum obovatum*, *Ruschia intricata*, *Searsia burchellii*, *Tetragonia arbuscula*, and *Thesium hystrix*. Important grasses include *Aristida adscensionis*, *A. congesta*, *A. diffusa*, *Cenchrus ciliata*, *Enneapogon desvauxii*, *E. scoparius*, *Eragrostis bergiana*, *E. bicolor*, *E. lehmanniana*, *E. obtusa*, *E. truncata*, *Sporobolus fimbriatus*, *Stipagrostis ciliata*, *S. obtusa*, *Tragus berteronianus*, and *T. koelerioides*. Large shrubs or small trees of *Senegalia mellifera* occur in parts of the far north of this regional biome. The northeastern boundary between the Upper Karoo and the adjacent dry grassland in the Free State has been much debated. Despite earlier claims that the original state of the north-eastern-most parts of Upper Karoo was dry grassland prior to the introduction of domestic grazing stock by European settlers and that parts of dry grassland in the Free State were being invaded by shrubby elements of the Upper Karoo, it is now accepted that the border between Upper Karoo and dry grassland is dynamic. This boundary can shift, often in a fragmented fashion, with a light grazing history, good rainfall seasons, and occasional shrub-inhibiting fire promoting a transformation from Upper Karoo to dry grassland. Conversely, heavy sustained grazing and prolonged drought periods favour transforming parts of the dry grasslands to Upper Karoo vegetation.

Climate The MAP of 310 mm varies from under 200 mm in the west to over 400 mm in the east. Rain falls mainly in summer and autumn, with only 10% falling in winter. Summer temperatures are hot, whereas winter nights are cold with frequent, often heavy, frost. MAT is the lowest of all the Nama-Karoo regional biomes (Fig. 7.14).

Topography, Geology and Soils The landscape comprises open plains dotted with occasional small rocky koppies and more prominent flat-topped mesas. More extensive, steep, and rugged topography is associated with the rocky Great Escarpment along the southern perimeter of this regional biome. The area is covered by shales, mudstones, and sandstones of the Beaufort Group (Adelaide and Tarkastad Subgroups) and of the Ecca Group (Waterford, Volksrust, and Prince Albert Formations), also including Dwyka Group diamictites in parts of the north, all of which fall within the Karoo Supergroup. Intrusive sills and dykes of dolerite (Jurassic Karoo Dolerite Suite) are common in many areas. The lithic and cumulic soils are usually weakly developed, freely drained, and shallow to medium depth. Shallow lithosols tend to be patchily distributed, whereas deeper soils are commonly found in the north, often together

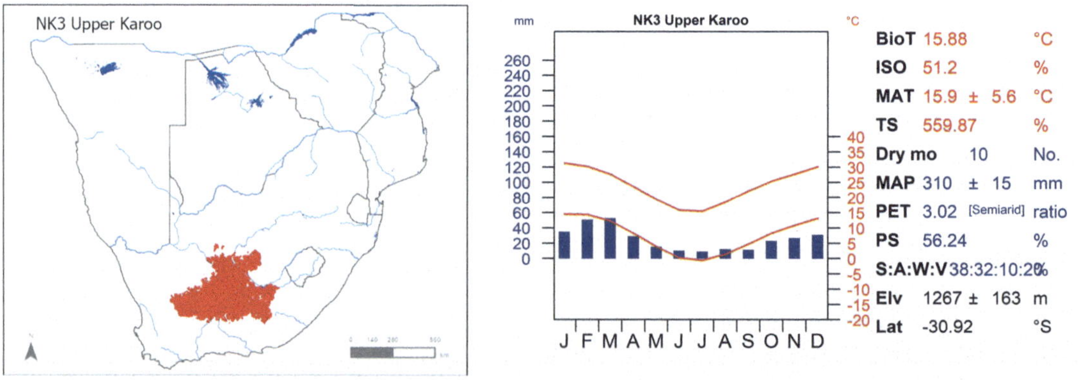

Fig. 7.14 The distribution area and the climatic diagram of the regional biome NK3 Upper Karoo

with calcrete formation. Higher nutrient-status soils can develop on the dolerite.

Sources Acocks (1953, 1988), Werger (1973a, b), van der Walt (1980), Werger (1980), Palmer (1988, 1990, 1991a, b, c), Jooste (1989), Bousman and Scott (1994), Lloyd and Badenhorst (1995, 1996), O'Connor and Roux (1995), du Toit (1996, 2010, 2019), Low and Rebelo (1996), Rubin and Palmer (1996), Hoffman et al. (1999), Brown and Bezuidenhout (2000, 2005), Pienaar (2002), Rutherford et al. (2003, 2012), Pienaar et al. (2004), Mucina et al. (2006b), Clark et al. (2009, 2011), Masubelele et al. (2013, 2014), du Toit et al. (2015a, b, 2018), Kraaij et al. (2017), Martin (2017), van der Merwe et al. (2018, 2021), du Toit and O'Connor (2020), Petersen et al. (2020), van der Merwe (2020), Arena (2021), Blanckenberg (2021), Milton et al. (2021), Harmse et al. (2022), Arena et al. (2023), Rybchak et al. (2024).

NK4 Lower Karoo

AT 3 Groot Thicket (Mucina and Rutherford 2006)
NK 11 Gamka Karoo (Dayaram et al. 2019)
NK 12 Eastern Lower Karoo (Dayaram et al. 2019)
NK 13 Lower Karoo Gwarrieveld (Mucina and Rutherford 2006)
SKv 12 Willowmore Gwarrieveld (Mucina and Rutherford 2006)
SKv 14 Steytlerville Karoo (Dayaram et al. 2019)

Distribution This regional biome encompasses vast inland stretches of karoo scrub south of the Great Escarpment, spanning Laingsburg in the west and Pearston in the east (Fig. 7.15). It occurs mainly in the Western Cape Province and, to a lesser extent, in the Eastern Cape Province of South Africa. This Karoo area has been classified in both VegMap2006 and VegMap2018 as NKl 1 Gamka Karoo, NKl 2 Eastern Lower Karoo and NKl 3 Lower Karoo Gwarrieveld. In the current work, it also encompasses the arid to semiarid karoo shrublands and low thickets (formerly classified as an Albany Thicket unit—AT 3 Groot Thicket) on the steep, mainly south-facing flanks of mountains north of the Steytlerville area. Two formerly SK units (SKv 12 and SKv 14) found in piedmonts in broad intra-mountain valleys were reclassified as part of the CB Nama-Karoo and incorporated into the Lower Karoo regional biome. The RB Lower Karoo lies, on average, at a lower elevation than that of the other regional biomes of the Nama-Karoo.

Vegetation Physiognomy and Composition The dominant vegetation physiognomy is defined by low scrub, both non-succulent (*Chrysocoma*, *Eriocephalus*, *Pentzia*, *Rhigozum*) and succulent shrubs of the families Aizoaceae, Crassulaceae, and Zygophyllaceae (*Augea*, *Crassula*, *Drosanthemum*, *Malephora*, *Ruschia*, *Zygophyllum*), alternating with patches of drought-resistant 'white' grasses belonging to the genera *Aristida* and *Stipagrostis*. Groups of tall shrubs and solitary scattered trees (*Boscia oleoides*, *Euclea undulata*, *Pappea capensis*, and *Schotia afra*) are also common. *Portulacaria afra* occurs on steep slopes in some vegetation units (especially in Groot Thicket) of this regional biome.

Climate Bioclimatically, the Lower Karoo regional biome straddles a cool-temperate and warm-temperate climate. It is pronouncedly arid at the drier end of the typical semidesert climate. The prominent rainfall peak is in the late warm season (late summer to early autumn), with a subtending maximum in the early warm season (late spring or early summer). When the climate of several component vegetation types is examined, a bimodal pattern becomes discernible (see Fig. 7.2 in Mucina et al. 2006b). This is one of the reasons that led to the re-classification described above into this regional biome of the two easternmost former Succulent Karoo units, which show a similar bimodal rainfall regime (see Fig. 5.65 in Mucina et al. 2006a; see also Fig. 7.15).

Topography, Geology and Soils The landscapes supporting this regional biome are irregular to slightly undulating plains, often interrupted by some dolerite dykes, butts and mesas in the north and east. The soils of very shallow and stony soils of the Glenrosa and/or Mispah forms

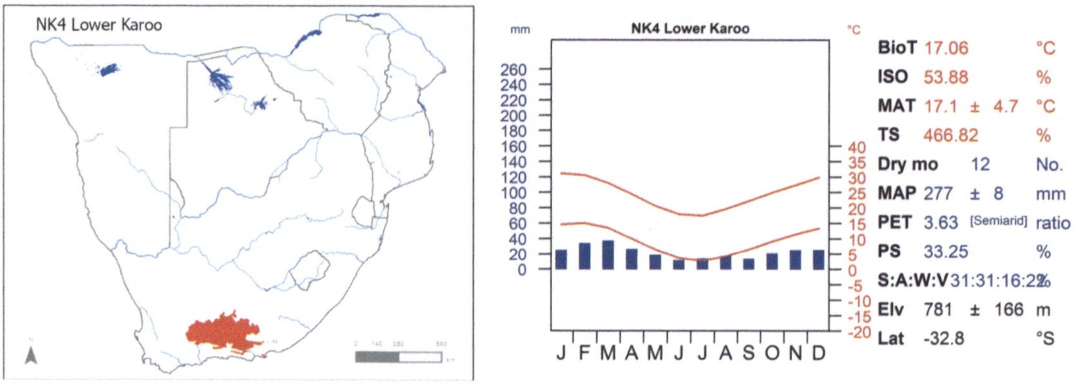

Fig. 7.15 The distribution area and the climatic diagram of the regional biome NK4 Lower Karoo

develop over a plethora of sedimentary mudstones and sandstones of the Beaufort Group (Adelaide Subgroup) with some Ecca sediments and Dwyka tillites in the south and with some Jurassic dolerite intrusions in the north. Some areas are built from shales, sandstone and siltstone of the Bokkeveld Group (mostly the Traka Subgroup as well as the Ceres Subgroup, both Devonian) as well as quartzites of the slightly younger Witteberg Group (all of the Cape Supergroup). These support clayey soils, as are the soils derived from the dolerite dykes. For more details, see the descriptions of the vegetation units classified within this regional biome (Hoare et al. 2006; Mucina et al. 2006a, b).

Sources Acocks (1953, 1979, 1988), Palmer (1988, 1989, 1991a, c), Hoffman (1989a, b), Esler (1993), Esler and Cowling (1995), Milton et al. (1995), du Toit (1996), Low and Rebelo (1996), Rubin and Palmer (1996), Wiegand and Milton (1996), Settele et al. (1999), Cowling and Heijnis (2001), Lloyd et al. (2002), Vlok and Euston-Brown (2002), Vlok et al. (2003), Hoare et al. (2006), Mucina et al. (2006a; b), Masubelele et al. (2013), Martin (2017), Schmidt et al. (2019), Saayman et al. (2021).

NK5 Escarpment Karroid Thicket

AT 13 Eastern Cape Escarpment Thicket (Mucina and Rutherford 2006)
AT 14 Camdebo Escarpment Thicket (Mucina and Rutherford 2006)

Distribution This regional biome is located on mainly south-facing slopes of the Great Escarpment of the Eastern Cape Province, South Africa. In the west, it forms an arc from Bruintjieshoogte in the east via the Coetzeeberg Mountains and Graaff-Reinet as far as Kamdebooberg and Aberdeen. In the east, Amathole occurs on the Winterberg and Swaershoek Mountains piedmonts and on the foothills south and west of Cradock (Fig. 7.16). For further details on the distribution of this biome, see the descriptions of the AT 13 and AT 14 vegetation units in Hoare et al. (2006).

Vegetation Physiognomy and Composition The vegetation of this regional biome has a tall-scrub, thicket-like physiognomy with dominating *Olea europaea* subsp. *africana*, *Scutia myrtina*, and *Vachellia karroo* in the AT 13, or *Euclea undulata*, *Pappea capensis*, *Portulacaria afra*, *Gymnosporia polyacantha*, and *V. karroo* in the AT 14, both with species-rich C4-grass understorey, including *Aristida adscensionis*, *A. congesta*, *Cenchrus ciliaris*, *Digitaria eriantha*, *Enneapogon desvauxii*, *Eragrostis lehmanniana*, *E. obtusa*, and *Heteropogon contortus*.

Climate Most precipitation occurs in the warm season, with considerably less in the winter months (Fig. 7.16). There is a tendency towards a bimodal pattern, with a central maximum in March and a somewhat lesser maximum in

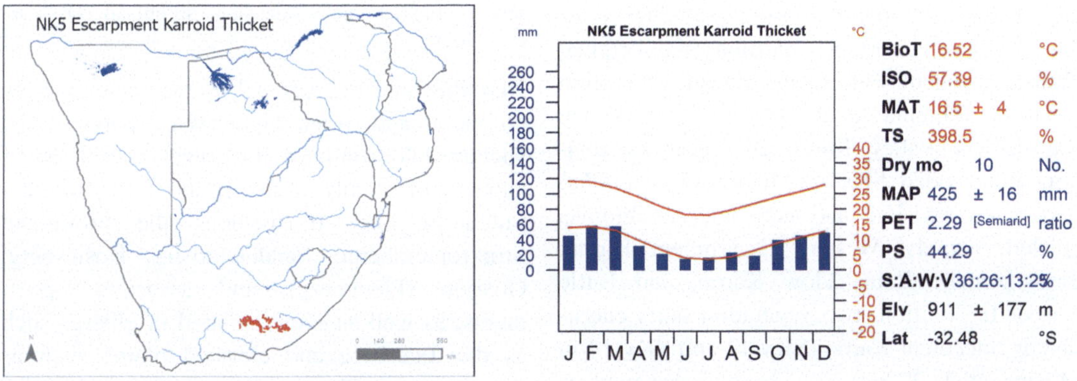

Fig. 7.16 The distribution area and the climatic diagram of the regional biome NK5 Escarpment Karroid Thicket

November. MAP is higher than that of the other regional biomes of the Nama-Karoo and ranges from about 400 to 700 mm per annum on the southern side of the escarpment and from 310 to 400 mm per annum on the northern side around Cradock, increasing with elevation. The bioclimate is warm-temperate, semiarid, with a high incidence of frost. Bioclimatic analyses of the Albany Thicket units (as defined by Hoare et al. 2006) as well as application of further bioclimatic criteria, including precipitation seasonality as reflected by the climate diagrams in both Mucina et al. (2006b) and Hoare et al. (2006) as well as the more significant occurrence of frost, have resulted in the extraction of the former Albany Thicket units AT 13 and AT 14 (as coded in Hoare et al. 2006) to create the new regional biome called Escarpment Karroid Thicket.

Topography, Geology and Soils The topography of this regional biome is formed by rugged, steeply sloping escarpment and mountain slopes. The shallow, skeletal soils are predominantly Mispah with fine-grained, nutrient-poor silts or more nutrient-rich red clays. They are derived from mudstones and arenite of the Permian Adelaide Subgroup of the Karoo Supergroup, interrupted in places by dykes of Jurassic Karoo dolerites.

Sources Acocks (1953, 1988), Palmer (1988, 1989, 1991a, b), Lloyd et al. (2002), Robertson and Palmer (2002), Vlok and Euston-Brown (2002), Vlok et al. (2003), Brown and Bezuidenhout (2005), Hoare et al. (2006), Masubelele et al. (2013).

7.6 Continental Biome Succulent Karoo (SK)

The Succulent Karoo biome is a broadly coastal semidesert famous for its extraordinary succulent flora. Shrubs and dwarf shrubs, mainly leaf succulents, are dominant. In contrast to Nama-Karoo, grasses are uncommon in Succulent Karoo and usually contribute little to cover (Fig. 7.4f). The South African section (hence most of the territory covered by this biome; see Fig. 7.1c and 7.3d) has been described and delimited in detail by Mucina et al. (2006b). Here, apart from updated descriptions, we present two changes. One is spatial, which involves recognising this biome in southern Namibia. The other is conceptual, as it involves splitting the VegMap concept of SKv 5 Tanqua Karoo into two—the Tanqua Karoo (in new delimitation) and a new—CB Tankwa Desert (described above).

The CB Succulent Karoo consists of eight regional biomes (Table 7.1). Most of these regional biomes are based on the bioregions carrying the same names as described in Mucina et al. (2006a), except for the Richtersveld Bioregion (SKr in Mucina et al. 2006a), which has been extended into southern Namibia and named the regional biome Richtersveld-Aus Upland Karoo. In addition, the original Transescarpment Succulent Karoo Bioregion (SKt in Mucina et al. 2006a) and Valley Rainshadow Valley Karoo Bioregion (vegetation units coded SKv in Mucina et al. 2006a) have been revisited and new regional biomes created, namely, Western Escarpment Karoo, Transmontane Rainshadow Karoo, and Little Karoo. In addition, two vegetation units classified as Succulent Karoo (SKv 12 and SKv 14 in Mucina et al. 2006a) have been reclassified as Nama-Karoo units.

Succulent Karoo is a member of the GB Subtropical Coastal Hot Semidesert global biome (see also Mucina 2023). The West Coast Arid Strandveld continental ecotone biome (see Chap. 9) is a member of the global zonoecotone ZE Semidesert–Mediterranean Coastal Ecotone, which straddles the CB Succulent Karoo and CB Fynbos.

SK1 Richtersveld-Aus Upland Karoo

SKr 1 Central Richtersveld Mountain Shrubland (Dayaram et al. 2019)
SKr 2 Upper Annisvlakte Succulent Shrubland (Dayaram et al. 2019)
SKr 3 Goariep Mountain Succulent Shrubland (Dayaram et al. 2019)
SKr 4 Lekkersing Succulent Shrubland (Dayaram et al. 2019)
SKr 5 Vyftienmyl se Berge Succulent Shrubland (Dayaram et al. 2019)
SKr 6 Stinkfonteinberge Eastern Apron Shrubland (Dayaram et al. 2019)
SKr 7 Northern Richtersveld Scorpionstailveld (Dayaram et al. 2019)
SKr 8 Rosyntjieberg Succulent Shrubland (Dayaram et al. 2019)
SKr 9 Tatasberg Mountain Succulent Shrubland (Dayaram et al. 2019)
SKr 10 Die Plate Succulent Shrubland (Dayaram et al. 2019)
SKr 11 Rooiberg Quartz Vygieveld (Dayaram et al. 2019)
SKr 12 Kosiesberg Succulent Shrubland (Dayaram et al. 2019)
SKr 13 Southern Richtersveld Scorpionstailveld (Dayaram et al. 2019)
SKr 14 Southern Richtersveld Inselberg Shrubland (Dayaram et al. 2019)
SKr 15 Anenous Plateau Shrubland (Dayaram et al. 2019)
SKr 16 Umdaus Mountains Succulent Shrubland (Dayaram et al. 2019)
SKr 17 Eenriet Plains Succulent Shrubland (Dayaram et al. 2019)

Distribution This regional biome occurs mainly on the inland, often higher-lying parts of the Richtersveld south of the lower Gariep River valley and mainly north of the Port Nolloth–Steinkopf line. It includes the prominent Stinkfonteinberge, Vandersterberg, Kosiesberg, Goarieb (Ploegberg), and associated pediments, as well as some isolated inselbergs such as the Tatasberg and elevated plains such as the Anenous Plateau. The main settlements within the regional biome are Eksteenfontein and Lekkersing. Its distribution is more limited in southwestern Namibia north of the Gariep River, occurring mainly in the Namuskluftberge and Witputz area, north of Rosh Pinah, but also occurring on some isolated, elevated areas farther north, including the highlands in the vicinity of Aus at the northeastern extremity of this regional biome, as well as on some major inselbergs embedded within the RB Southern Namib Desert, such as the Aurus Mountains, Heioab Mountain and several scattered outcrops of the Klinghardt Mts (not all patches were mappable). In addition, it occurs on some lower altitude prominences, with southwest-facing, fog-catching slopes nearer the coast, such as the Boegoeberg and Sargdeckel (Fig. 7.17).

Vegetation Physiognomy and Composition The vegetation of this regional biome is dominated by low to medium-tall shrublands of moderate to dense cover, often with a preponderance of leaf succulents. However, stem succulents can also be important. Important species, generally limited to the extensive Richtersveld area south of the Gariep River in South Africa, include *Cephalophyllum regale*, *C. rigidum*, *Mesembryanthemum amplectens*, *Ruschia leucosperma*, and *Tylecodon buchholzianus*. Those important species shared with both the Richtersveld and southwestern Namibia, include *Aizoon fruticosum*, *Aloe pearsonii*, *A. microstigma*, *Aloidendron ramosissimum*, *Asparagus capensis*, *Augea capensis*, *Cheiridopsis robusta*, *Cotyledon orbiculata*,

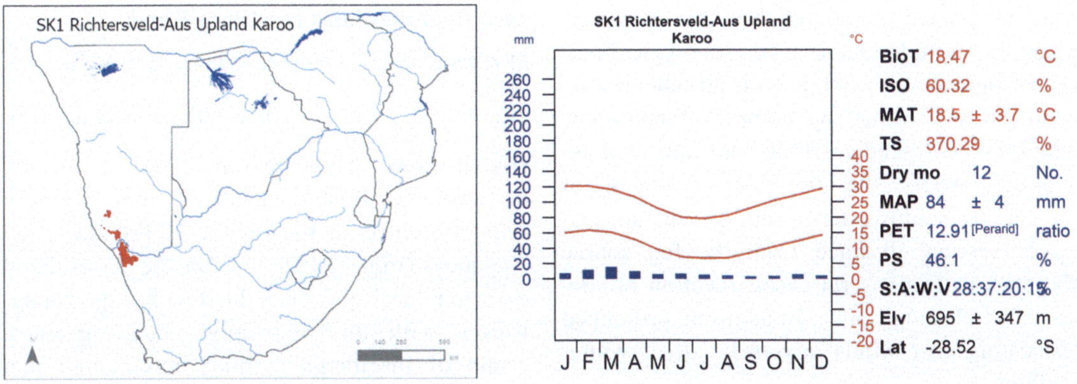

Fig. 7.17 The distribution area and the climatic diagram of the regional biome SK1 Richtersveld-Aus Upland Karoo

Crassula muscosa, *Didelta carnosa*, *D. spinosa*, *Euphorbia ephedroides*, *E. gummifera*, *E. mauritanica*, *E. rhombifolia*, *Mesembryanthemum guerichianum*, *M. hypertrophicum*, *M. pseudoschlichtianum*, *Montinia caryophyllacea*, *Othonna furcata*, *Osteospermum microcarpum*, *Pachypodium namaquanum*, *Portulacaria namaquensis*, *Roepera foetida*, *Searsia populifolia*, *Stoeberia frutescens*, *Tetraena prismatocarpa*, *Trachyandra muricata*, *Tylecodon paniculatus*, and *T. wallichii*. In Namibia, apart from the core area of the Namuskluftberg and its surrounds, some isolated inselbergs, such as the Heioab and Aurus Mountains, can also support relatively dense species-rich vegetation cover, including additional localised species, for example, *Conophytum klinghardtense*, *Crassula aurusbergensis*, *Tylecodon schaeferianus*, and *T. aurusbergensis*. The relatively large, isolated Aus patch is more marginal and diffuse, probably owing to its climatically peripheral position. It lacks many species typical of this regional biome to the south, and it often constitutes the approximate northeastern limit of some typical taxa such as *Aloe microstigma Crassula ausensis* subsp. *ausensis*, *Hereroa hesperantha*, *Mesembryanthemum hypertrophicum*, *Monsonia drudeana*, and *Oxalis luederitzii*. Despite this thinning out of species from the southwest, the Aus area is one of the most species-rich in the whole of Namibia, helped in part by the presence of species endemic to the Aus area, such as *Ruschia deminuta* and *Oxalis ausensis*, but a larger extent by partial merging with many species from the closely neighbouring eastern interior, e.g., *Aloe hereroensis* and *Wahlenbergia denticulata*.

Climate Rain falls predominantly in winter in the southern parts, although the proportion of winter rainfall decreases steeply towards the northeast in Namibia. Fog can be an important water source for plants in areas of the regional biome near the sea and at lower altitudes on the west-facing side of the Richtersveld. Summers are hot but cooler at higher altitudes in the mountains. Winters are warm to mild. Strong winds prevail nearer the coast. The MAP of 84 mm given by the global model in this study for this regional biome (Fig. 7.17) is an underestimate, probably owing to the lack of formal rain stations. It especially underestimates rainfall in the higher-lying parts of the regional biome where MAP up to 200 mm, or greater, has been previously credibly estimated for several VegMap2006 vegetation types of the Richtersveld, e.g. Central Richtersveld Mountain Shrubland, Goariep Mountain Succulent Shrubland and Kosiesberg Succulent Shrubland. MAP between about 80 and 120 mm is commonly expected on the lower-lying plains, with the modelled MAP of 84 mm limited to those areas and patches of the regional biome nearer the neighbouring desert margins. The climate that supports this regional biome on the number of inselbergs, ridges and some associated plains in Namibia is determined by multiple factors where the typical pattern of increase in rainfall with altitude can be significantly modified by complex topography, including local wind corridors and rain shadow-creating barriers, the direction of slopes for intercepting southwestern wind-driven rain and fog harvesting, distance from the fog source (the sea), and in several cases, location of specific sand blasting routes. In addition, prolonged desiccating east winds adversely affect winter seedlings.

Topography, Geology and Soils Much of the area is characterised by exceptionally rugged terrain, including steep mountain complexes, rocky ridges, deep ravines, diverse outcrops, inselbergs, as well as some enclosed, sloping plains, plateaus, and pediment slopes. The maximum elevation of over 1700 m is attained on the Aus plateau and associated ridges, compared to mostly under 1200 m in the Richtersveld. Geology varies greatly, including a spectrum of metavolcanic and metasediments. Quartzites, granites, schists, and gneisses are most common, with dolomite, limestone, sandstone, shale, and tillite present in some places. Alluvial and aeolian overburdens and superficial calcrete deposits are common in terrain with gentle gradients. Main stratigraphic units include the Vioolsdrif Suite, Stinkfontein Formation, and Namaqualand Metamorphic Complex. Soils are generally weakly developed, shallow, and skeletal but strongly variable over short distances in hilly terrain. Deeper soils occur on the plains but often with underlying calcareous duripans and quartz-rich top layers in places. Soil textures are commonly loamy sands, gravel sands, silty loams, and silty sands, particularly on heuweltjies.

Sources Williamson (1997, 2000), Craven (2002, 2009), Burke (2004, 2020), Burke et al. (2004), Hendricks et al. (2005), Craven and Vorster (2006), Mucina et al. (2006a), Jürgens et al. (2010), Snijman (2013), Burke and Loots (2020), Konje et al. (2021).

SK2 Bushmanland Inselberg Karoo

SKr18 Bushmnaland Inselberg Shrubland (Dayaram et al. 2019)

SKr19 Aggenys Gravel Vygieveld (Dayaram et al. 2019)

Distribution This regional biome occurs in northern Bushmanland in the Northern Cape Province in the region of Pofadder and Aggeneys (Fig. 7.18). It can be considered extrazonal as it lies embedded within the Nama-Karoo, with this vegetation occurring on a group of inselbergs (solitary mountains) that are separated from the core of the Succulent Karoo—Namaqualand) by expansive flat sandy plains. These inselbergs mark the easternmost extent of the Succulent Karoo in Bushmanland and lie scattered in an east–west orientation towards the edges of the Namaqualand hill ridges in the west. Although this regional biome occurs in a transitional winter/summer-rainfall region, it is distinct from the surrounding Bushmanland Arid Grassland (Nama-Karoo) vegetation due to specific substrate conditions promoting the dominance of dwarf leaf-succulent species.

Vegetation Physiognomy and Composition The vegetation is generally characterised by a variety of succulents and sparse grassy undergrowth (*Aristida*, *Eragrostis*, *Stipagrostis*), although the vegetation of the inselbergs and surrounding plains are markedly different. On the flat plains, the vegetation is sparse and low-growing, dominated by small to dwarf leaf-succulents with a perennial graminoid element. These plains contain many local endemics, especially various dwarf succulents, incl. iconic *Conophytum* and *Lithops*. The inselbergs host shrublands with a succulent (including *Adromischus*, *Euphorbia*, *Tylecodon*, and *Ruschia*) and a non-succulent herbs (mainly Asteraceae) component on the southern slopes. Many of the succulents on the inselbergs are local endemics, particularly members of the families of the Aizoaceae, Asphodelaceae, and Crassulaceae families, as well as some intriguing species of Didiereaceae (*Ceraria*) and

Anacampserotaceae (*Anacampseros*). The northern slopes of the inselbergs present much hotter, harsher conditions and are inhabited by sparse vegetation belonging to the Dg 9 East Gariep Rocky Desert vegetation unit.

Climate Rainfall is low and erratic, typically occurring with thunderstorms concentrated in late summer and early autumn between February and April. This regional biome has the highest proportion of summer rainfall compared to the other regional biomes of Succulent Karoo. The MAP of 150 mm is greater than previous estimates by Mucina et al. (2006a). Temperatures range widely, with a mean maximum of 38 °C in February and a mean minimum of −3 °C in July. Frost is relatively frequent, occurring for about 20 (range 10–30) days a year (Fig. 7.18).

Topography, Geology and Soils The topography is flat to slightly sloping plains surrounding distinct inselbergs and smaller koppies. The plains comprise gneiss from the Little Namaqualand and Hoogor Suites and quartzite from the Bushmanland and Geelvloer Groups. Soils here range from loamy to sandy and are shallow and skeletal, with various gravel patches being a common feature across the landscape. Quartz gravel patches are widespread but can mostly be found at inselbergs' foothills (and some summits). Calcrete gravel patches can be found on open plains where erosion has uncovered a calcrete layer. Patches of feldspar gravel occur on open plains overlying pink gneiss, although rare. The inselbergs in this region are made up of high-grade metamorphic rocks formed during the Namaqualand Metamorphic Event. They comprise clastic sediments and volcanic and intrusive rocks that date back to the Mokolian Age.

Sources Anderson and van Heerden (2000), Desmet (2000a, b), Fick (2000), Hammer (2002), Schmiedel (2002a, b), and Mucina et al. (2006a).

SK3 Namaqualand Hardeveld Karoo

SKn 1 Namaqualand Klipkoppe Shrubland (Dayaram et al. 2019)
SKn 2 Namaqualand Shale Shrubland (Dayaram et al. 2019)
SKn 3 Namaqualand Blomveld (Dayaram et al. 2019)
SKn 4 Namaqualand Heuweltjieveld (Dayaram et al. 2019)
SKn 6 Kamiesberg Mountains Shrubland (Dayaram et al. 2019)

Distribution This regional biome stretches from the northwestern corner of the Western Cape Province (around Nuwerus) in the south to around Steinkopf in the Richtersveld region in the Northern Cape Province in the north (Fig. 7.19). The western foothills of the Namaqualand section of the Escarpment mark the seaward boundary, while the inland boundary is shared mainly with the Bushmanland and Knersvlakte regions. This region is topographically

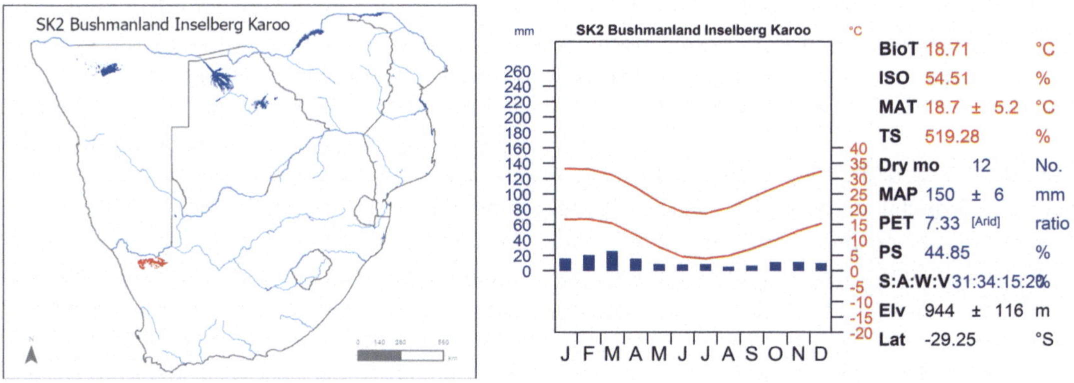

Fig. 7.18 The distribution area and the climatic diagram of the regional biome SK2 Bushmanland Inselberg Karoo

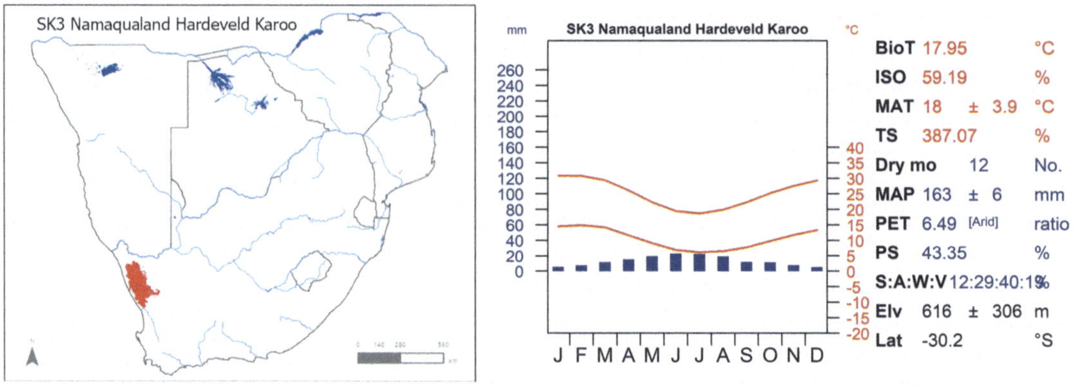

Fig. 7.19 The distribution area and the climatic diagram of the regional biome SK3 Namaqualand Hardeveld Karoo

heterogeneous, covering valleys, flats, undulating hills, and steep slopes. Elevation ranges between 100 and 1260 m.

Vegetation Physiognomy and Composition Physiognomy is open to dense shrublands dominated by a mixture of succulent and non-succulent dwarf to medium shrubs. Dominant succulent shrubs include the genera *Crassula*, *Didelta*, *Drosanthemum*, *Euphorbia*, *Galenia*, *Lampranthus*, *Leipoldtia*, *Ruschia*, *Zygophyllum*, and common non-succulent shrubs include *Eriocephalus*, *Dodonaea*, *Hermannia*, *Lebeckia*, and *Pteronia*. Dwarf succulents are often abundant in rocky areas. The charismatic kokerboom *Aloidendron dichotomum* is generally seen on some north-facing granodiorite slopes. Annual herbs and geophytes provide spectacular mass flower displays in years with sufficient rainfall. On the undulating plains of the Namaqualand escarpment's western foothills, mosaics of low succulent shrublands can be found on and among rounded ancient termitaria known as 'heuweltjies'.

Climate Winter rainfall is concentrated between May and September, with MAP ranging from around 120 mm at lower elevations to 230 mm at higher elevations. The low rainfall along the west-facing escarpment is frequently supplemented by coastal fog, and dew can generally occur throughout the winter. Frost is common throughout most of this regional biome, occurring approximately 7–13 days a year but with significant interannual variability. Drought periods (< 100 mm of yearly rainfall) can occur sporadically for one or two successive years. There are occasional thunderstorms in late summer (February to April). Mean daily temperatures range around 18 °C with hot summers (mean maximum: 30 °C) and cold winters (mean minimum 5 °C) (Fig. 7.19).

Topography, Geology and Soils The landscape is characterised by flat, smooth, undulating plains between hills and mountains of Mokolian granite and gneiss. Imposing scenes of sloping rock sheets, enormous domes (bornhardts), and disintegrating castle koppies (tors) can be seen throughout parts of this region. Some of the smooth hills in the north of this regional biome bear shallow soils over the quartzite of the Kuibis Subgroup and deep soils over the shales of the Schwarzrand Subgroup. Most of the soil is yellow brown to brown, fine to coarse loamy sands. Deep red loamy soils are found over the granites and gneisses on the western foothills of the Namaqualand escarpment. Heuweltjies (ancient, abandoned termitaria) are particularly common in this regional biome, especially on the foothills of rocky outcrops. Heuweltjies are found on base-rich substrates, and their soils are generally far more fertile than the surrounding soils. The differences in physical and chemical soil properties between heuweltjies and their surroundings create a characteristic patchy pattern of environmental heterogeneity over the landscape, reflecting the sharp distinctions between plant communities over small spatial scales.

Sources van Rooyen et al. (1979a, 1996, 2010, 2018b), van Rooyen and Grobbelaar (1982), le Roux (1984), Theron et al. (1993), Steyn et al. (1996), Rösch et al. (1997a, b, c), Rösch (2001), van Rooyen (2002), Helme and Desmet (2006), Mucina et al. (2006a), Todd and Hoffman (2009), Broodryk (2010), Luther-Mosebach et al. (2012), Hanke et al. (2014), Schmiedel et al. (2016, 2023), Nenzhelele (2017), Nenzhelele et al. (2018), Schmiedel and Oldeland (2018a, b), Bell et al. (2021), Samuels et al. (2022), Geldenhuys et al. (2023)

SK4 Namaqualand Sandveld

SKs 1 Richtersveld Coastal Duneveld (Dayaram et al. 2019)
SKs 3 Southern Richtersveld Yellow Duneveld (Dayaram et al. 2019)
SKs 4 Richtersveld Sandy Coastal Scorpionstailveld (Dayaram et al. 2019)
SKs 5 Richtersveld Red Duneveld (Dayaram et al. 2019)
SKs 6 Oograbies Plains Sandy Grassland (Dayaram et al. 2019)
SKs 7 Namaqualand Strandveld (Dayaram et al. 2019)
SKs 8 Namaqualand Coastal Duneveld (Dayaram et al. 2019)
SKs 9 Namaqualand Inland Duneveld (Dayaram et al. 2019)
SKs 10 Riethuis-Wallekraal Quartz Vygieveld (Dayaram et al. 2019)
SKs 11 Namaqualand Arid Grassland (Dayaram et al. 2019)
SKs 12 Namaqualand Spinescent Grassland (Dayaram et al. 2019)
SKs 13 Klawer Sandy Shrubland (Dayaram et al. 2019)
SKs 14 Namaqualand Heuweltjie Strandveld (Dayaram et al. 2019)

Distribution This regional biome occurs along the Atlantic Ocean coast of Namaqualand (Northern and Western Cape Provinces), forming a broad coastal belt on the western side of the Namaqualand Escarpment (Fig. 7.20). It runs from around the Holgat River near Alexander Bay in the north to around Donkins Bay (near Doringbaai) in the south. It penetrates up to approximately 40 km inland in some areas. Most of this region occurs below elevations of 300 m, although some patches of SKs 11 Namaqualand Arid Grassland can occur up to 840 m.

Vegetation Physiognomy and Composition This regional biome supports a variety of habitats, with substrate conditions (including soil depth and type), topography and vegetation age being some of the major drivers of habitat variability. The deep white sands in the coastal areas north of Kleinsee host a variety of communities dominated by succulent subshrubs with some graminoids. While the stabilised sand sheets support relatively homogenous communities dominated by *Stoeberia beetzii*, habitats on active dunes are heterogeneous (i.e., high beta diversity) due to variability in slope aspect and the stage of sedimentation or deflation. S. utilis often dominate *dune crests*, while interdune valleys commonly support *Cephalophyllum ebracteatum*, *Cheiridopsis robusta*, *Mesembryanthemum arenosum*, and *M. pseudoschlichtianum.* Further inland, the red dune fields of the Richtersveld support tall grasses and shrubs, with a prevalence of *Lebeckia cinerea* thickets. The coastal peneplains further south (Kleinsee to Donkins Bay) are generally comprised of species-rich shrublands with many low succulents (*Cephalophyllum*, *Didelta*, *Othonna*, *Ruschia*, *Tetragonia*, *Tripteris*, *Zygophyllum*) and non-succulent shrubs (*Eriocephalus*, *Lebeckia*, *Pteronia*, *Salvia*). Slightly inland, there are two patches of active dune fields that support tall, non-succulent shrublands with some grasses (*Ehrharta*) and restioids (*Willdenowia*). Grassy shrublands also occur on flat sandy plains about 25 km east of Port Nolloth, dominated by the spinifex-like *Cladoraphis spinosa* and *Stipagrostis ciliata*. Scattered patches of arid grassland dominated by *S. ciliata* and *S. obtusa* occur near Soebatsfontein and south of Kliprand. The southeastern corner of the regional biome is also characterised by spinescent grasslands dominated by the *C. spinosa* with some sparse shrubs and dense, tall *Montinia caryophyllacea* shrublands with a grassy understorey. Local endemic Aizoaceae species like *Jacobsenia*, *Monilaria*, and *Dicrocaulon* dominate quartz patches on the somewhat saline skeletal soils in the Wallekraal area.

Climate The region is arid with substantial seasonal winter rainfall (the highest winter rainfall proportion in the Succulent Karoo). MAP is generally lowest at the coast, ranging from 50 to 100 mm and increasing to roughly 80–150 mm inland, although a small area in the southeastern extreme may receive almost 200 mm. Sea fog is frequent throughout most of this region, and dew occurs in some inland areas throughout the winter, both acting as important supplementary water sources in the arid landscape. Temperatures generally range from mean minima of 5–10 °C in the winter to mean maxima of 25–30 °C in the summer. Frost occurs rarely. There are occasional to frequent strong southerly winds and sand blasting in the northern coastal areas (Fig. 7.20).

Topography, Geology and Soils The landscape varies from generally flat coastal plains to the gently undulating foothills of the escarpment. The northernmost stretches are characterised by flat, stable sand sheets and some gentle hills, with soils comprised of wind-blown marine sands, white to yellow. From around Kleinsee southwards, the coastal peneplains comprise semi-mobile sand plains and dune fields (some are mobile, while others are stabilised), with some active angular dune plumes originating from river mouths. The soils here are made up of Quaternary aeolian fine sands, white to grey or yellowish to red, lying over Tertiary marine sediments and granite gneisses. The habitats that lie further inland are fragmented and with variable topography, including areas of flat sand sheets and fossil dune aggregates in the northern parts of this region, dwarf succulent shrublands on plains characterised by white quartz pavement, and higher-lying grasslands and shrublands overlying deep coarse red sands. Biological soil crusts cover large areas of the Richtersveld coastal plain.

Sources Milton (1978), van Jaarsveld (1981), Boucher and le Roux (1993), Jürgens (1991, 2004), de Villiers et al. (1994, 1999, 2001, 2003), Jürgens et al. (1997), Desmet and Cowling (1999), Eccles et al. (1999, 2001), de Villiers (2000), Eccles (2000), Oguz (2000), Stöcker (2000), van Wyk and Smith (2001), Schmiedel (2002a, b), Boucher (2003), Nußbaum (2003), Oguz et al. (2004), Mucina et at. (2006a), Ndemoina (2006), van Rooyen et al. (2010), Pauw (2011), Schmiedel and Oldeland (2018b).

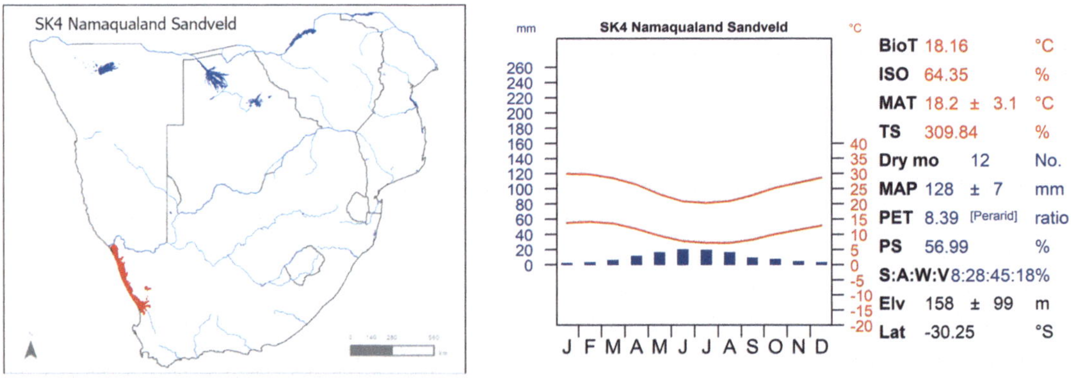

Fig. 7.20 The distribution area and the climatic diagram of the regional biome SK4 Namaqualand Sandveld

SK5 Knersvlakte Karoo

SKk 1 Northern Knersvlakte Vygieveld (Dayaram et al. 2019)
SKk 2 Central Knersvlakte Vygieveld (Dayaram et al. 2019)
SKk 3 Knersvlakte Quartz Vygieveld (Dayaram et al. 2019)
SKk 4 Knersvlakte Shale Vygieveld (Dayaram et al. 2019)
SKk 5 Vanrhynsdorp Gannabosveld (Dayaram et al. 2019)
SKk 6 Knersvlakte Dolomite Vygieveld (Dayaram et al. 2019)
SKk 7 Citrusdal Vygieveld (Dayaram et al. 2019)
SKk 8 Piketberg Quartz Succulent Shrubland (Dayaram et al. 2019)
SKk 9 Kobee Succulent Shrubland (Dayaram et al. 2019)
SKn 7 Southern Namaqualand Quartzite Klipkoppe Shrubland (Dayaram et al. 2019)

Distribution The Knersvlakte occurs in the northwestern corner of the Western Cape (and slightly into the Northern Cape), covering the low-lying southern region of Namaqualand (Fig. 7.21). The main body of this regional biome is situated between the Hardeveld Karoo and the Bokkeveld Escarpment to the east, extending from the Vanrhynsdorp region in the south to the upper reaches of the Sout River to the north. Several patches of succulent shrubland belonging to this regional biome occur further south in the Olifants River valley and on quartz patches near the Piketberg Mts. The elevation is mostly around 300 m with little elevational range.

Vegetation Physiognomy and Composition Knersvlakte supports over 50 species specially adapted to living on quartz fields, with the majority being endemic, making this region an important centre of endemism in the Succulent Karoo. The region is characterised by high edaphic heterogeneity that results in rapid species turnover across small spatial scales (within a few meters in some cases). The vegetation consists of open shrublands dominated by low succulents, with *Ruschia*, *Drosanthemum*, *Leipoldtia* and *Salsola* being the most common genera. Marvellous flower displays by various annuals and geophytes are typical following winters with sufficient rainfall. Heuweltjies in some areas are dominated by *Stoeberia frutescens* or *Lampranthus uniflorus*. The patches in the Olifants River valley comprise medium to tall shrublands dominated by *Euphorbia mauritanica* and various Aizoaceae. Low Salicornia and Aizoaceae shrublands characterise the patches in the vicinity of Piketberg Mt.

Climate Rainfall is strongly seasonal, occurring over mild winters, while summers are hot and dry. Mean minimum winter and mean maximum summer temperatures are generally around 5–10 °C and > 30 °C, respectively. The MAP is approximately 194 mm but much lower (110–130 mm) throughout most of this regional biome. It can be substantially higher (over 160 mm) in some parts due to localised orographic effects around the Escarpment and mountains in the region. Some areas that include patches of this regional biome may experience even higher rainfall, such as the Olifants

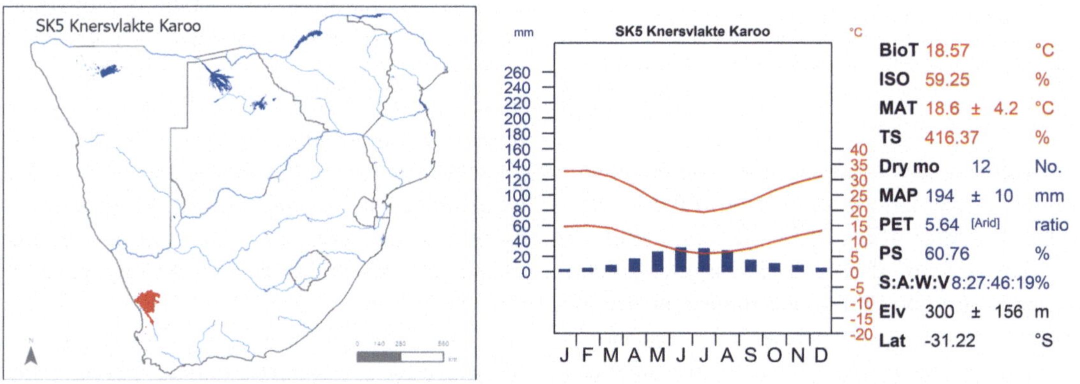

Fig. 7.21 The distribution area and the climatic diagram of the regional biome SK5 Knersvlakte Karoo

River valley (MAP: 316 mm) and the Piketberg Mountains area (MAP: 264 mm). Occasional fog and dew occur in some parts. Frost can occur in winter throughout most of the region but is rare (ca. 3–5 days a year) (Fig. 7.21).

Topography, Geology and Soils The landscape undulates smoothly and is characterised by a layer of quartz gravel on the soil surface. Most of the region overlies alluvium, clastic sediments and calcrete of the Knersvlakte Subgroup (Vanrhynsdorp Group) and schists of the Gariep Supergroup. The patches across the Olifants River valley overlie the shale and quartzite of the Cape Supergroup, and the patches around the Piketberg Mountains comprise soft shales that have weathered to reveal a layer of quartzitic rubble. Soils vary from shallow to moderately deep, sandy to loamy and acidic to alkaline soils.

Sources Milton (1978), Greeff (1987), Knight et al. (1989), Boucher (1990, 1991, 2003), Steinschen et al. (1996), van Jaarsveld (1998a, b), Schmiedel and Jürgens (1999), Schmiedel (2002a, b), Klak (2003), Manning and Goldblatt (2003), Ellis et al. (2006), Mucina et al. (2006a), Zedda and Rambold (2009), Schmiedel et al. (2012, 2015), Eibes et al. (2022), Zhigila et al. (2024).

SK6 Western Escarpment Karoo

SKn 5 Platbakkies Succulent Shrubland (Dayaram et al. 2019)
SKt 1 Western Bushmanland Klipveld (Dayaram et al. 2019)
SKt 2 Hantam Karoo (Dayaram et al. 2019)
SKt 3 Roggeveld Karoo (Dayaram et al. 2019)
SKv 6 Koedoesberge-Moordenaars Karoo (Dayaram et al. 2019)

Distribution This regional biome occurs mainly in the Northern Cape and partly in the Western Cape Provinces (Fig. 7.22). It lies along the western edge of the Great Escarpment, delineating a large portion of the Succulent Karoo's inland boundary. The northernmost extreme is marked by piedmonts between Bushmanland and the Hardeveld near Gamoep. It extends in a southeasterly direction across the northwestern plains of Bushmanland, the Onder-Bokkeveld and Hantam region (surrounding Loeriesfontein and Calvinia) and over the Roggeveld. The southern extreme of this regional biome is separated from the main area and covers the Koedoesberge and Moordenaars Karoo areas in the Laingsburg region. This area includes the Doekberg area and piedmonts of the Elandsberg near the Gamkapoort Dam. The elevation of the regional biome ranges from less than 500 m in the western parts of the Hantam Karoo to 1680 m in the Roggeveld Karoo.

Vegetation Physiognomy and Composition Shrublands (often sparse) are dominated by dwarf or low succulents, particularly Aizoaceae (e.g., *Antimima*, *Caroxylon*, *Drosanthemum*, *Eberlanzia*, *Galenia*, *Mesembryanthemum*, *Ruschia*), and low microphyllous shrubs, especially Asteraceae (e.g., *Eriocephalus*, *Pentzia*, *Pteronia*). There is also a grassy component (*Aristida*, *Stipagrostis*, *Ehrharta*) throughout most of the region, which is most striking in the transitional area between the Succulent Karoo and Nama-Karoo (especially the Roggeveld). While the shrubs are phenologically adapted to the predominant winter rainfall, the C4 grasses fare well in the late summer rainfall parts of the region. Spectacular displays of flowering annuals and geophytes can be found in the spring.

Climate Most of this region experiences strongly seasonal winter rainfall. MAP is 208 mm, but some areas, like the arid Western Bushmanland Klipveld, are far drier (MAP 90 mm) with erratic rainfall. Mean annual potential evaporation is extreme throughout most regions (> 2550 mm). MAT is 16.1 °C, the lowest of Succulent Karoo's regional biomes. Mean minimum temperatures in winter are below 5 °C, whereas mean maximum temperatures in summer exceed 30 °C. Frost is a common occurrence (13–40 days a year) due to the region's high elevation and inland position. Within this regional biome, the Roggeveld area falls between the summer and winter rainfall regimes of Southern Africa, which, when averaged over the whole area, results in rainfall peaking bimodally in March and June. This

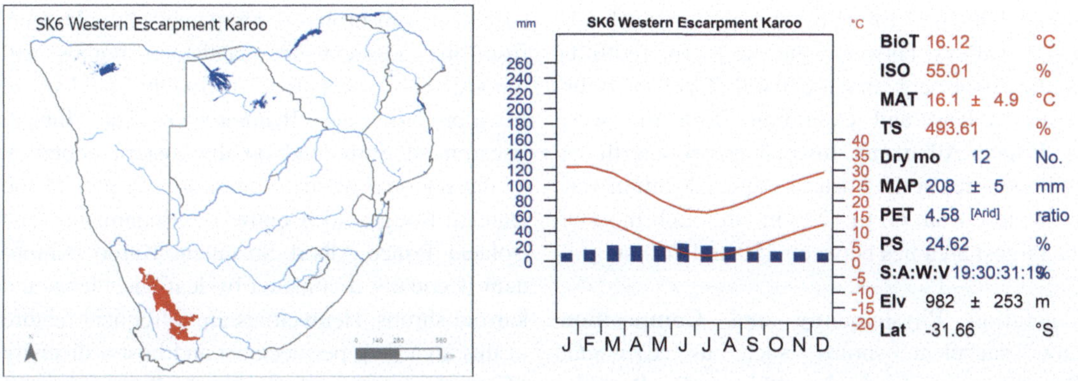

Fig. 7.22 The distribution area and the climatic diagram of the regional biome SK6 Western Escarpment Karoo

Roggeveld area has a relatively high MAP of 234 mm despite being somewhat influenced by a rain shadow. The Roggeveld is known as one of the coldest areas in South Africa, with frequent cold days in autumn and winter and regular frost (56 days a year on average) (Fig. 7.22).

Topography, Geology and Soils This regional biome is characterised by plains, low koppies and mountain slopes lying predominantly over Karoo Supergroup deposits, mainly sandstone, shale, and mudstone of the Ecca and Beaufort Groups, and diamictites and tillite of the Dwyka Group. The northernmost area overlies gneiss from the Little Namaqualand Suite and metasediments and metavolcanics from the Bushmanland Group. In this area, unique silcrete caps from the Bitterfontein Formation lie above weathered gneiss, which has resulted in quartz gravel plains. Different gravel plains to the north and the east (around the boundary with Bushmanland) comprise rounded rocks, boulders and calcrete nodules associated with palaeo-river terraces.

Sources Acocks (1953, 1988), Lloyd (1985, 1989a, b), Snijman and Perry (1987), Manning and Goldblatt (1997), van Wyk and Smith (2001), Petersen (2003), Mucina et al. (2006a), van der Merwe and Hoffman (2019)

SK7 Transmontane Rainshadow Karoo

SKv 1 Doringrivier Quartzite Karoo (Dayaram et al. 2019)
SKv 2 Swartruggens Quartzite Karoo (Dayaram et al. 2019)
SKv 3 Agter-Sederberg Shrubland (Dayaram et al. 2019)
SKv 4 Tanqua Escarpment Shrubland (Dayaram et al. 2019)
SKv 5 Tanqua Karoo p.p. (Dayaram et al. 2019)
SKv 7 Robertson Karoo (Dayaram et al. 2019)
SKv 13 Prince Albert Succulent Karoo (Dayaram et al. 2019)

Distribution The main area of this regional biome lies to the south of the Hantam Karoo in the Northern Cape Province, stretching down into the Western Cape Province (Fig. 7.23). This area extends from the Cederberg in the west across the Tanqua Basin (where it surrounds the Tankwa Desert continental biome) to the westerly slopes of the Roggeveld Escarpment and Klein Roggeveldberge in the east. In the northwest, there is also a stretch from around Doringbos across the Knolvlei area (north of the Pakhuis Pass) to the Doring River valley where the Doring and Olifant Rivers meet. A small portion of this regional biome, the Robertson Karoo, lies separately to the south of the main area in the broad valley around the Breede River, roughly between Worcester, Villiersdorp, and Robertson. To the east, another portion (Prince Albert Succulent Karoo) lies far removed from the main area of this regional biome. This part forms a band along broad colluvial valleys between the northern foothills of the Swartberg and the parallel ridges lying further inland and extending from the west of Prince Albert to Bosmanspoort (north of Willowmore) in the Eastern Cape. Elevation varies widely from 60 to 1220 m, although most of the largest area lies between 450 and 700 m.

Vegetation Physiognomy and Composition Low succulent shrubs such as *Crassula*, *Drosanthemum*, *Euphorbia*, *Phiambolia*, *Ruschia*, *Tylecodon*, and *Zygophyllum* are the most common life forms throughout this regional biome. Non-succulent shrubs like *Asparagus*, *Berkheya*, *Felicia*, and *Pteronia* are relatively common in some areas. The regional biome's largest area, the plains of the Tanqua Basin, supports sparse, low, succulent shrubland and may resemble a barren desert near the boundary with the Tankwa Desert continental biome in extremely dry years. In wetter parts, especially to the east and south, vegetation is denser and may be locally dominated by *Stipagrostis ciliata* or *S. obtusa* grasses on sandy soils. Annual herbs (especially Asteraceae) become apparent with adequate rainfall, as do grasses and geophytes to a lesser degree. On the quartzites around the Doring River valley in the northwest, low succulent shrubs co-occur with tall non-succulent shrubs such as *Berkheya fruticosa*, *Diospyros ramulosa*, and *Montinia caryophyllacea*). The quarzitic sandstone bordering the western edge of the Tanqua Basin hosts succulent subshrubs accompanied by elements from the adjacent dry sandstone fynbos and renosterveld vegetation, including *Dodonaea*, *Elytropappus*, and *Willdenowia*. The Tanqua Escarpment Shrubland on the eastern boundary of the regional biome's main area is part of the Hantam-Roggeveld Centre of Endemism. The isolated Prince Albert Succulent Karoo component is equally dominated by leaf succulents and karroid shrubs. Heuweltjies are a frequent feature in this area (~ 2 per hectare) and host a diversity of salt-tolerant succulents, especially members of the Aizoaceae family.

Climate The main area of this regional biome (the Tanqua Basin and surrounds) experiences strong seasonal winter rainfall, while the Prince Albert Succulent Karoo portion to the east experiences equinoctial rainfall. Rainfall is highly spatially variable due to the localised influences of different rain shadows, with MAP ranging from about 90 mm (bordering the Tankwa Desert) up to 350 mm in other parts. The mean annual evaporation potential can also be very high (over 2500 mm in some areas). Temperatures can occasionally drop below 0 °C in winter and reach up to 40 °C in summer. The incidence of frost is relatively low in the main area (up to roughly 15 days a year) but notably higher in the Tanqua Escarpment Shrubland and Prince Albert Succulent Karoo (26–30 days a year). Intermittent north-westerly warm bergwinds can occur in winter (Fig. 7.23).

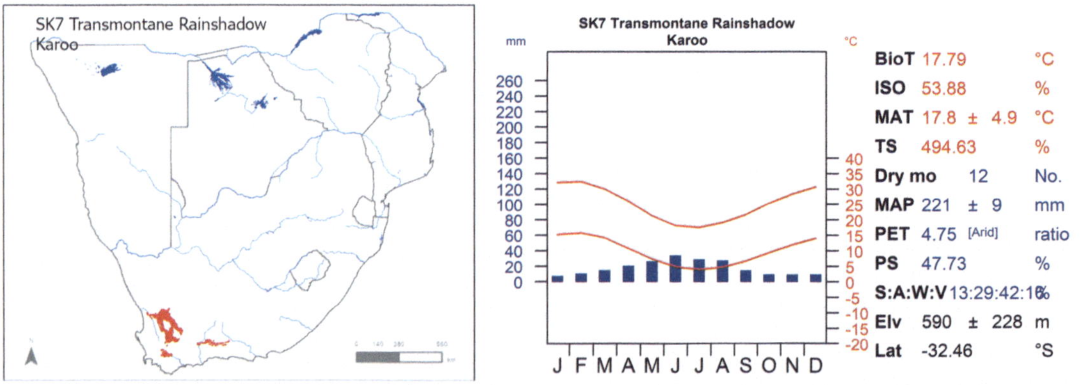

Fig. 7.23 The distribution area and the climatic diagram of the regional biome SK7 Transmontane Rainshadow Karoo

Topography, Geology and Soils The landscape's topography varies from flat plains to slightly undulating slopes and hills to deep incised valleys and river canyons. The largest area, the Tanqua Karoo, is a gently undulating plain interspersed by isolated dolerite buttes (in the north), level sheet-washes and incised channels of ephemeral rivers. Most of the area lies over sedimentary rocks from groups belonging to the Karoo Supergroup, including the Ecca Group (mudrock, shales), Dwyka Group (sandstone, diamictite), and the Beaufort Group (mudrock). Siltstone, shale, mudstone, and sandstone from the Bokkeveld Group (Cape Supergroup) are also common. The northwestern corner around the Doring River valley lies mainly over the Table Mountain Group quartzites. The western edge of the regional biome overlies the quartzitic sandstone of the Witteberg Group. Soils are loamy to sandy-loamy and vary in depth. Small quartz patches can be found in the southern parts of the Tanqua Basin. It has been shown, experimentally, that at least a few typical dominants of this regional biome can potentially extend well into surrounding sandstone fynbos at considerably higher and wetter elevations but are probably excluded from doing so by competitive effects and/or the periodic fire regime associated with fynbos. Importantly, they do not appear to be edaphically limited here.

Sources Olivier (1966, 1979), Joubert (1968), Lane (1977), Acocks (1979, 1988), Jürgens (1986), Boshoff (1989), Midgley and Musil (1990), Milton et al. (1990a, b, 1992, 1994, 1995), Smitheman and Perry (1990), Milton (1992), Esler (1993), Mackay (1994), Esler and Cowling (1995), Mackay and Zietsman (1996), Wiegand and Milton (1996), Milton et al. (1997), Lechmere-Oertel (1998), Lechmere-Oertel and Cowling (1999, 2001), Rubin (1998), Settele et al. (1999), Bossdorf et al. (2000), Oliver (2000), Cowling and Heijnis (2001), van Wyk and Smith (2001), Beukes and Ellis (2003), Boucher (2003), Chesselet et al. (2003), Klak (2003), Mucina et al. (2006a), Schmiedel and Mucina (2006), Rahlao et al. (2008), Rutherford and Powrie (2010), Seymour et al. (2010), Esler et al. (2015), Cramer et al. (2016), van der Merwe et al. (2016), Milton et al (2023), Samuels et al. (2023).

SK8 Little Karoo

AT 2 Gamka Thicket (Mucina and Rutherford 2006)
SKv 8 Western Little Karoo (Dayaram et al. 2019)
SKv 9 Western Gwarrieveld (Dayaram et al. 2019)
SKv 10 Little Karoo Vygieveld (Dayaram et al. 2019)
SKv 11 Eastern Little Karoo (Dayaram et al. 2019)

Distribution This regional biome marks the southern and eastern limits of the Succulent Karoo. It occurs in the Little Karoo basin in the eastern part of the Western Cape Province, with a marginal area in the Eastern Cape Province (Fig. 7.24). It lies between the Witteberg and Swartberg Mts in the north and the Outeniqua Mts in the south. The largest cohesive parts of this vegetation are concentrated in the area between Anysberg and Van Wyksdorp and the area surrounding the stretch between Calitzdorp and Oudtshoorn. The eastern and northeastern parts of this vegetation occur as isolated narrow bands and patches on lower mountain slopes and river valleys. Elevation ranges from 200 to 1000 m.

Vegetation Physiognomy and Composition The region hosts mosaics of dense shrublands ranging from low to medium height. Shrublands are dominated by low succulents, especially Aizoaceae (*Drosanthemum*, *Gibbaeum*, and *Ruschia*) and Crassulaceae (*Cotyledon*, *Crassula*, *Tylecodon*). However, a variety of non-succulent shrubs like *Chrysocoma*, *Euclea*, *Lycium*, *Nymania*, *Pappea*, *Pteronia*, *Rhigozum*, and *Searsia* are also common. In some areas, magnificent spring flower displays of annuals and geophytes occur in years with adequate rainfall. The vegetation on the mountain slopes and valleys to the east and northeast is peculiar as it has an intricate spatial and floristic connection with the Albany Thicket, Nama-Karoo, and Fynbos biomes. These areas are characterised by low succulent thicket, often dominated by dense stands of *Portulacaria afra*, with various non-succulent shrubs also being common. Grasses

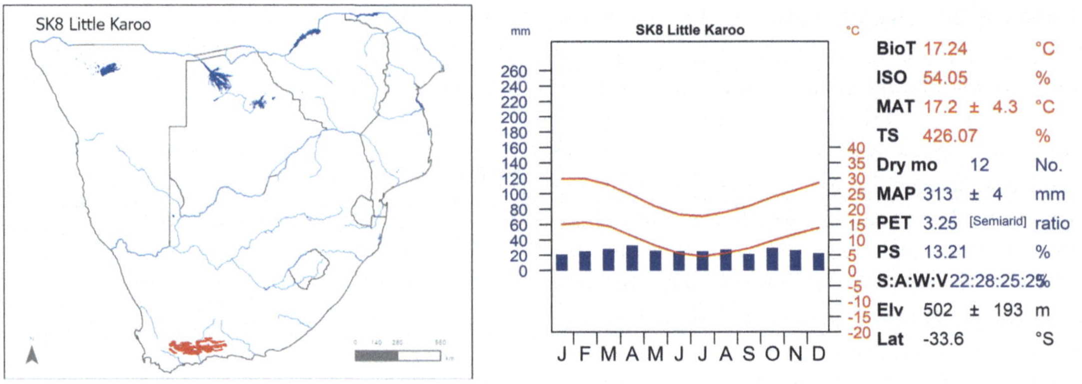

Fig. 7.24 The distribution area and the climatic diagram of the regional biome SK8 Little Karoo

are generally sparse but may become abundant following sufficient rains.

Climate Rainfall is relatively non-seasonal (especially compared to the rest of the Succulent Karoo), with several minor peaks between March and November. However, summer months are notably drier, suggesting a tendency towards a winter rainfall regime (especially in the more western parts). The MAP of 313 mm is the highest of the regional biomes of the Succulent Karoo. However, its MAP can range widely from less than 200 mm to over 500 mm on the south-facing foothills to the east. Temperatures can regularly reach >30 °C in summer, while winters are generally mild. Frost occurs between 7 to 14 days a year on average (Fig. 7.24).

Topography, Geology and Soils The landscape mainly comprises flat to slightly undulating plains and includes piedmonts, hills, steep foothills, and valleys. This regional biome lies mainly over shales, siltstones and mudstones of the Bokkeveld Group and some sandstones of the Table Mountain, Wittenberg, and Uitenhage Groups. Soils vary in depth and texture but are mainly loamy-sandy, loamy-clayey, and moderately deep. Quartz gravel patches occur over shales associated with weathered quartz veins, especially at the northern foothills of the Langeberg Mts. The patchy vegetation to the east and northeast are confined to areas with deep, gravelly soils over the conglomerate of the Enon Formation (Uitenhage Group).

Sources Muir (1929), Acocks (1953, 1988), Schmiedel and Jürgens (1999, 2002, 2004), van Wyk and Smith (2001), Lloyd et al. (2002), Schmiedel (2002a, b, 2004), Vlok and Euston-Brown (2002), van Jaarsveld and van Wyk (2003), Vlok et al. (2003), Cleaver et al. (2005), Mucina et al. (2006a), Schmiedel and Mucina (2006), Nongwe (2008), Vorster (2017), Hoffman et al. (2009), Thompson et al. (2009), Saayman et al. (2016), Kirsten et al. (2023).

References

Acocks JPH (1953) Veld types of South Africa. Mem Bot Surv S Afr 28:1–192

Acocks JPH (1979) The flora that matched the fauna. Bothalia 12:673–709

Acocks JPH (1988) Veld types of South Africa, 3rd edn. Mem Bot Surv S Afr 57:1–146

Anderson T, van Heerden A (2000) Preliminary findings of a botanical reconnaissance of inselbergs similar to Gamsberg for the Gamsberg Zinc Project. McGregor Museum, Kimberley

Arena G (2021) Expanding grasslands? Structural biome shifts in the dryland rangelands of the eastern Karoo revealed through long-term observation of climate, vegetation and land use change. Thesis, Department of Biological Sciences, University of Cape Town, Cape Town

Arena G, Hoffman MT, van der Merwe H et al (2023) Expansion of the Grassland Biome in the eastern Karoo corresponds with changes in rainfall and livestock numbers. Afr J Range Forage Sci 40:1–19

Aushiku NNL, Buschbeck L, Christian A et al (2015) Changes in rainfall and biomass in the Namib Sand Sea Dune Ecosystem over 52 years. J Namib Sci Soc 63:219–237

Balinsky BI (1962) Patterns of animal distribution on the African continent. Ann Cape Prov Mus 11:299–310

Becker T (2000) Muster der Vegetation und ihre Determinanten in einem desertifikationsgefährdeten Raum im Nordwesten Namibias (Kaokoland). Thesis, Universität zu Köln, Köln

Becker T, Getzin S (2000) The fairy circles of Kaokoland (North-West Namibia) origin, distribution, and characteristics. Basic Appl Ecol 1:149–159

Becker RW, Swanepoel W, van Jaarsveld E et al (2021) Euphorbias of three high mountaintops in the Kaokoveld Centre of Endemism in northern Namibia and southern Angola. Euphorbia World 17(3):15–27

Bell WD, Hoffman MT, Vernon Visser V (2021) Regional land degradation assessment for dryland environments: the Namaqualand Hardeveld bioregion of the Succulent Karoo biome as a case-study. Land Degrad Develop 32:2287–2302

Bester SP, Klopper RR, Steyn HM et al (2012) New plant records for Tankwa Karoo National Park. Koedoe 54(1):Art#1066. https://doi.org/10.4102/koedoe.v54i1.1066

Beukes PC, Ellis F (2003) Soil and vegetation changes across a Succulent Karoo grazing gradient. Afr J Range Forage Sci 20:11–19

Blanckenberg C (2021) The effects of land use change, from small livestock farming to protected area, on vegetation and mammal communities in the SKA region of the Karoo, South Africa. Thesis, Department of Biological Sciences, University of Cape Town, Cape Town

Boshoff CR (1989) Structure, distribution and phenology of perennial plant species in the Worcester Veld Reserve, in the arid winter rainfall region of the southwestern Cape. Thesis, Department of Botany, University of Cape Town, Cape Town

Bossdorf O, Schurr O, Schumacher J (2000) Spatial patterns of plant association in grazed and ungrazed shrublands in the semi-arid Karoo, South Africa. J Veg Sci 11:253–258

Boucher C (1990) Report on inspection and study of some proposed Olifants River Dam sites. Ecological Research Report 45. Department of Botany, University of Stellenbosch, Stellenbosch

Boucher C (1991) Vegetation survey of Rondegat Farm, Clanwilliam. Ecological Research Report 8. Department of Botany, University of Stellenbosch, Stellenbosch

Boucher C (2003) Western Cape Olifants/Doring River Irrigation Study. Botanical Report. PGWC Report 259/2004/2. Department of Economic Affairs, Agriculture and Tourism, Provincial Government of the Western Cape Province, Cape Town

Boucher C, le Roux A (1993) Dry coastal ecosystems of the South African West Coast. In: van der Maarel E (ed) Ecosystems of the world 2B. Dry coastal ecosystems: Africa, America, Asia and Oceania. Elsevier, Amsterdam, pp 75–88

Bousman B, Scott L (1994) Climate or overgrazing?: the palynological evidence for vegetation change in the eastern Karoo. S Afr J Sci 90:575–578

Boyer DC (1989) Some characteristics of the plant communities of three dunes situated across a climatic gradient in the Namib Desert. Madoqua 16:141–148

Broodryk NL (2010) Long-term monitoring of vegetation dynamics in the Goegap Nature Reserve, Namaqualand, South Africa. Thesis, Department of Plant Science, University of Pretoria, Pretoria

Brown LR, Bezuidenhout H (2000) The phytosociology of the De Rust section of the Mountain Zebra National Park, Eastern Cape. Koedoe 43(1):1–18

Brown LR, Bezuidenhout H (2005) The vegetation of the farm Ingleside and Welgedacht of the Mountain Zebra National Park, Eastern Cape. Koedoe 48(2):23–42

Burke A (1997) The impact of large herbivores on floral composition and vegetation structure in the Naukluft Mountains, Namibia. Biodivers Conserv 6:1203–1217

Burke A (2001a) Classification and ordination of plant communities of the Naukluft Mountains, Namibia. J Veg Sci 12:53–60

Burke A (2001b) Determinants of inselberg floras in arid Nama Karoo landscapes. J Biogeogr 28:1211–1220

Burke A (2002a) Properties of soil pockets on arid Nama Karoo inselbergs—the effect of geology and derived landforms. J Arid Environ 50:219–234

Burke A (2002b) Are Namibian inselbergs conservation islands? A floral perspective. S Afr J Sci 98:560–562

Burke A (2002c) Plant communities of a central Namib inselberg landscape. J Veg Sci 13:483–492

Burke A (2003) How special are Etendeka mesas? Flora and elevation gradients in an arid landscape in northwest Namibia. J Arid Environ 55:747–764

Burke A (2004) A preliminary account of patterns of endemism in Namibia's Sperrgebiet—the succulent karoo. J Biogeogr 31:1613–1622

Burke A (2005a) Endemic plants of the arid succulent karoo in Namibia: towards hypotheses for their evolution. Ecography 28:171–180

Burke A (2005b) Vegetation types of mountain tops in Damaraland, Namibia. Biodivers Conserv 14:1487–1506

Burke A (2007) Plant endemism in the central Namib desert. Evol Ecol Res 9:283–297

Burke A (2011) Managing biodiversity in the /Ai-/Ais Hotsprings Game Park. EnviroScience and Strengthening the Protected Area Network Project (SPAN), Windhoek

Burke A (2013) Plant species of the Naute Game Park—an annotated inventory. Dinteria 33:3–11

Burke A (2015) Rare plants in the Succulent-Nama Karoo ecotone in southern Africa. Afr J Ecol 53:93–102

Burke A (2020) Criteria for biodiversity special value zones in the Sperrgebiet—plant endemism and species richness measures in practice. Namib J Environ Sect A 4:28–40

Burke A, Loots S (2020) Plant endemics of the Tsau ǁKhaeb (Sperrgebiet) National Park. Namib J Environ Sect A 4:62–70

Burke A, Mannheimer C (2004) Plant species of the Sperrgebiet (Diamond Area 1). Dinteria 29:79–109

Burke A, Juergens N, Seely MK (1998) Floristic affinities of an inselberg archipelago in the southern Namib desert—relict of the past, centre of endemism or nothing special? J Biogeogr 25:311–317

Burke A, Cooper T, Noli D (2004) Initial results determining spatial and temporal patterns of rainfall in the Sperrgebiet, south-west Namibia. Namib Sci Soc J 52:117–122

Chase BM, Boom A, Carr AS (2023) A ca. 39,000-year record of vegetation and climate change from the margin of the Namib Sand Sea. Quater Res 116:1–11. https://doi.org/10.1017/qua.2023.29

Chesselet P, Hammer S, Oliver I (2003) *Brianhuntleya*, a new genus endemic to the Worcester-Robertson Karoo, South Africa. Bothalia 33:160–164

Clark VR, Barker NP, Mucina L (2009) The Sneeuberg: a new centre of floristic endemism on the Great Escarpment, South Africa. S Afr J Bot 75:196–238

Clark VR, Barker NP, Mucina L (2011) A phytogeographic assessment of the Nuweveldberge, South Africa. S Afr J Bot 77:147–159

Cleaver G, Brown LR, Bredenkamp GJ (2005) The phytosociology of the Vermaaks, Marnewicks and Buffelsklip valleys of the Kammanassie Nature Reserve, Western Cape. Koedoe 48:1–16

Cowling RM, Heijnis CE (2001) The identification of Broad Habitat Units as biodiversity entities for systematic conservation planning in the Cape Floristic Region. S Afr J Bot 67:15–38

Cowlishaw G, Davies JG (1997) Flora of the Pro-Namib desert Swakop River catchment, Namibia: community classification and implications for desert vegetation sampling. J Arid Environ 36:271–290

Cramer MD, Barger NN, Tschinkel WR (2017) Edaphic properties enable facilitative and competitive interactions resulting in fairy circle formation. Ecography 40:1139–1250

Cramer MD, von Holt J, Khomo L et al (2016) Evidence for aeolian origins of *heuweltjies* from buried gravel layers. S Afr J Sci 112(1/2):Art. #2015-0025. https://doi.org/10.17159/sajs.2016/20150025

Craven P (2002) Phytogeography of Namibia: A taxon approach to the spermatophyte flora. Thesis, Department of Botany, Stellenbosch University, Stellenbosch

Craven P (2009) Phytogeographic study of the Kaokoveld Centre of Endemism. Thesis, Department of Botany and Zoology, Stellenbosch University, Stellenbosch

Craven P, Vorster P (2006) Patterns of plant diversity and endemism in Namibia. Bothalia 36:175–189

Dayaram A, Harris LR, Grobler BA et al (2019) Vegetation Map of South Africa, Lesotho and Swaziland 2018: a description of changes since 2006. Bothalia 49(1):a2452. https://doi.org/10.4102/abc.v49i1.2452

De Cauwer V, Becker RW, Gomes A et al (2023) The mountain top flora and vegetation of the remote Ovahimba Highlands in the Kaokoveld Centre of Endemism: a reconnaissance. Trans Roy Soc S Afr 78:109–122. https://doi.org/10.1080/0035919X.2023.2211040

de Villiers AJ (2000) Seed bank dynamics of the Strandveld Succulent Karoo. Thesis, Department of Botany, University of Pretoria, Pretoria

de Villiers AJ, van Rooyen MW, Theron GK (1994) Comparison of two methods for estimating size of the viable seed bank of two plant communities in the Strandveld of the west coast, South Africa. S Afr J Bot 60:81–84

de Villiers AJ, van Rooyen MW, Theron GK (2001) Seed bank phytosociology of the Strandveld Succulent Karoo, South Africa: a pre-mining benchmark survey for rehabilitation. Land Degrad Develop 12:119–130

de Villiers AJ, van Rooyen MW, Theron GK (2003) Similarity between the soil, seed bank and the standing vegetation in the Strandveld Succulent Karoo, South Africa. Land Degrad Develop 14:527–540

de Villiers AJ, van Rooyen MW, Theron GK et al (1999) Vegetation diversity of the Brand-se-Baai coastal dune area, West Coast, South Africa: a pre-mining benchmark survey for rehabilitation. Land Degrad Develop 10:207–224

Desmet P (2000a) Laying out options for a national park in the Inselberg region of northern Bushmanland. Institute for Plant Conservation, University of Cape Town, Cape Town, South Africa

Desmet P (2000b) The succulents of Northern Bushmanland: their distribution and implications for conservation. Aloe 37:32–35

Desmet PG, Cowling RM (1999) Biodiversity, habitat and range-size aspects of a flora from a winter-rainfall desert in north-western Namaqualand, South Africa. Plant Ecol 142:23–33

Dreber N, Oldeland J, van Rooyen GMW (2011) Species, functional groups and community structure in seed banks of the arid Nama Karoo: grazing impacts and implications for rangeland restoration. Agric Ecosyst Environ 141:399–409

du Toit JCO (2010) Description of very wet to very dry periods at Grootfontein, eastern Karoo, 1889–2012. Groot Agric 14(1):42

du Toit JCO (2019) Drivers of vegetation change in the eastern Karoo. Thesis, School of Life Sciences, University of KwaZulu-Natal, Pietermaritzburg

du Toit JCO, O'Connor TG (2020) Long-term influence of season of grazing and rainfall on vegetation in the eastern Karoo, South Africa. Afr J Range Forage Sci 37:159–171

du Toit JCO, van den Berg L, O'Connor TG (2015a) Fire effects on vegetation in a grassy dwarf shrubland at a site in the eastern Karoo, South Africa. Afr J Range Forage Sci 32:13–20

du Toit JCO, O'Connor TG, van den Berg L (2015b) Photographic evidence of fire-induced shifts from dwarf-shrub- to grass-dominated vegetation in Nama-Karoo. S Afr J Bot 101:148–152

du Toit JCO, Ramaswiela T, Pauw MJ (2018) Interactions of grazing and rainfall on vegetation at Grootfontein in the eastern Karoo. Afr J Range Forage Sci 35:267–276

du Toit PCV (1996) Development of a model to estimate grazing index values for Karoo plant species. Thesis, Department of Botany, University of Pretoria, Pretoria

Eccles NS (2000) Plant ecology of Namaqualand Strandveld: community structure and dynamics of a unique winter-rainfall desert ecosystem. Thesis, Department of Botany, University of Cape Town, Cape Town

Eccles NS, Esler KJ, Cowling RM (1999) Spatial pattern analysis in Namaqualand desert plant communities: evidence for general positive interactions. Plant Ecol 142:71–85

Eccles NS, Lamont B, Esler KJ et al (2001) Relative performance of clumped vs. experimentally isolated plants in a South African winter-rainfall desert community. Plant Ecol 155:219–227

Eckardt FD, Soderberg K, Coop LJ et al (2013) The nature of moisture at Gobabeb, in the central Namib Desert. J Arid Environ 93:7–19

Eibes PM, Schaffrath F, Oldeland J et al (2022) Testing the concept of edaphism for the quartz island flora of the Knersvlakte, South Africa. S Afr J Bot 151:555–564

Ellis AG, Weis AE, Gaut BS (2006) Evolutionary radiation of "stone plants" in the genus *Argyroderma* (Aizoaceae): unravelling the effects of landscape, habitat, and flowering time. Evolution 60:39–55

Esler KJ (1993) Vegetation patterns and plant reproductive processes in the succulent Karoo. Thesis, Department of Botany, University of Cape Town, Cape Town

Esler KJ, Cowling RM (1995) The comparison of selected life-history characteristics of Mesembryanthema a species occurring on and off Mima-like mounds (heuweltjies) in semi arid southern Africa. Vegetatio 116:41–50

Esler KJ, von Staden L, Midgley GF (2015) Determinants of the Fynbos/Succulent Karoo biome boundary: insights from a reciprocal transplant experiment. S Afr J Bot 101:120–128

Fick A (2000) The Gamsberg zinc mine project. Aloe 37:14–17

Geldenhuys C, van der Merwe H, van Rooyen MW (2023) Vegetation response to grazing and drought (13 yr) in a conservation area in the Succulent Karoo, South Africa. J Arid Environ 219:105093. https://doi.org/10.1016/j.jaridenv.2023.105093

Getzin S, Holch S, Yizhaq H et al (2022) Plant water stress, not termite herbivory, causes Namibia's fairy circles. Persp Plant Ecol Evol Syst 57:125698. https://doi.org/10.1016/j.ppees.2022.125698

Giess W (1971) A preliminary vegetation map of South West Africa. Dinteria 4:1–114

Giess W (1981) Die in der Zentralen Namib von Südwestafrika/Namibia festgestellten Pflanzenarten und ihre Biotope. Dinteria 15:13–73

Gil-Romera G, Scott L, Marais E et al (2007) Late Holocene environmental change in the northwestern Namib Desert margin: new fossil pollen evidence from hyrax middens. Palaeogeogr Palaeoclimatol Palaeoecol 249:1–17

Gottlieb TR, Eckardt FD, Venter ZS et al (2019) The contribution of fog to water and nutrient supply to *Arthraerua leubnitziae* in the central Namib Desert, Namibia. J Arid Environ 161:35–46

Greeff LS (1987) 'n Voorlopige ondersoek na die plantegroei van die plaas Kalkgat op die Knersvlakte. BSc(Hons) Project, Department of Botany, University of Stellenbosch, Stellenbosch

Günster A (1993) Microhabitat differentiation among serotinous plants in the Namib Desert. J Veg Sci 4:585–590

Hammer SA (2002) Dumpling and his wife. New views of *Conophytum* N.E.Br. East Anglia Engraving, London

Hanke W, Böhner J, Dreber N et al (2014) The impact of livestock grazing on plant diversity: an analysis across dryland ecosystems and scales in southern Africa. Ecol Appl 24:1188–1203

Harmse CJ, Gerber H, van Niekerk A (2022) Evaluating several vegetation indices derived from Sentinel-2 imagery for quantifying localized overgrazing in a semi-arid region of South Africa. Rem Sens 14:1720. https://doi.org/10.3390/rs14071720

Helme N, Desmet P (2006) A description of the endemic flora and vegetation of the Kamiesberg uplands, Namaqualand, South Africa. CEPF/SKEP, Cape Town

Hendricks HH, Bond WJ, Midgley JJ et al (2005) Plant species richness and composition along livestock grazing intensity gradients in a Namaqualand (South Africa) protected area. Plant Ecol 176:19–33. https://doi.org/10.1007/s11258-003-0009-6

Henschel JR, Lancaster N (2013) Gobabeb—50 years of Namib Desert research. J Arid Environ 93:1–6

Henschel JR, Seely MK (2000) Long-term growth patterns of *Welwitschia mirabilis*, a long-lived plant of the Namib Desert (including a bibliography). Plant Ecol 150:7–26

Hoare DB, Mucina L, Rutherford MC et al (2006) Albany Thicket Biome. In: Mucina L, Rutherford MC (eds) The vegetation of South Africa, Lesotho and Swaziland. SANBI, Pretoria, pp 540–567

Hoffman MT (1989a) A preliminary investigation of the phenology of subtropical thicket and karroid shrubland in the lower Sundays River Valley, SE Cape. S Afr J Bot 55:586–597

Hoffman MT (1989b) Vegetation studies and the impact of grazing in the semi-arid Eastern Cape. Thesis, Department of Botany, University of Cape Town. Cape Town

Hoffman MT, Cousins B, Meyer T et al (1999) Historical and contemporary land use and the desertification of the Karoo. In: Dean WRJ, Milton SJ (eds) The karoo: ecological patterns and processes. Cambridge University Press, Cambridge, pp 257–273

Hoffman MT, Madden CF, Erasmus K et al (2009) The impact of indigenous ungulate herbivory over five years (2004–2008) on the vegetation of the Little Karoo, South Africa. Afr J Range Forage Sci 26:169–179

Irish J (1994) The biomes of Namibia, as determined by objective categorization. Navors Nasion Mus Bloemfontein 10:549–592

Jacobsen NHG, Moss H (1987) A contribution to the flora of the northern Namib. Dinteria 19:27–68

Jacobson PJ, Jacobson KM (2013) Hydrologic controls of physical and ecological processes in Namib Desert ephemeral rivers: implications for conservation and management. J Arid Environ 93:80–93

Jooste JF (1989) A study of phytosociology and small mammals of the Rolfontein Nature Reserve, Cape Province. Thesis, Department of Botany, University of Stellenbosch, Stellenbosch

Joubert JGV (1968) Die ekologie van die weiveld van die Robertson-Karoo. Thesis, Department of Botany, University of Stellenbosch, Stellenbosch

Jürgens N (1986) Untersuchungen zur Ökologie sukkulenter Pflanzen des südlichen Afrika. Mitt Inst Allgem Bot Hamburg 21:139–365

Jürgens N (1991) A new approach to the Namib region, 1: Phytogeographic subdivision. Vegetatio 97:21–38

Jürgens N (1997) Floristic biodiversity and history of African arid regions. Biodivers Conserv 6:495–514

Jürgens N (2004) A first classification of the vegetation of the Richtersveld (RSA) and directly adjacent regions in Namibia and South Africa. Schumannia 4:149–180

Jürgens N (2013) The biological underpinnings of Namib Desert fairy circles. Science 339:1618–1621

Jürgens N, Burke A, Seely MK et al (1997) Desert. In: Cowling RM, Richardson DM, Pierce SM (eds) Vegetation of southern Africa. Cambridge University Press, Cambridge, pp 189–214

Jürgens N, Desmet PG, Rutherford MC et al (2006) Desert Biome. In: Mucina L, Rutherford MC (eds) The vegetation of South Africa, Lesotho and Swaziland. SANBI, Pretoria, pp 300–323

Jürgens N, Haarmeyer DH, Luther-Mosebach J et al (2010) Biodiversity in southern Africa. Volume 1: Patterns at local scale—the BIOTA Observatories. Klaus Hess Publishers, Göttingen

Jürgens N, Oldeland J, Hachfeld B et al (2013) Ecology and spatial patterns of large-scale vegetation units within the central Namib Desert. J Arid Environ 93:59–79

Jürgens N, Oncken I, Oldeland J et al (2021) *Welwitschia*: Phylogeography of a living fossil, diversified within a desert refuge. Sci Rep 11:1–14

Kaseke KF, Wang L, Wanke H et al (2016) An analysis of precipitation isotope distributions across Namibia using historical data. PLoS ONE 11(5):e0154598. https://doi.org/10.1371/journal.pone.0154598

Kent LE (comp) (1980) Stratigraphy of South Africa. Part 1. Lithostratigraphy of the Republic of South Africa, South West Africa/Namibia and the Republics of Bophuthatswana, Transkei and Venda. Geological Survey Handbook 8. Department of Mineral and Energy Affairs, Pretoria

Kers LE (1967) The distribution of *Welwitschia mirabilis* Hook.F Svensk. Bot Tidskr 61:97–125

Kirsten T, Hoffman MT, Bell WD et al (2023) A regional, remote sensing-based approach to mapping land degradation in the Little Karoo, South Africa. J Arid Environ 219:105066. https://doi.org/10.1016/j.jaridenv.2023.105066

Klak C (2003) New combinations, a new genus and five new species in the Aizoaceae. Bradleya 21:107–120

Knight RS, Rebelo AG, Siegfried WR (1989) Plant assemblages on Mima-like earth mounds in the Clanwilliam district, South Africa. S Afr J Bot 55:465–472

Konje MM, Muoria PK, Wabuyele E et al (2021) Rainfall and soils, not grazing intensity, determine the composition and productivity of annual plants in a biodiverse arid winter rainfall region. J Nat Sci Res 12(10):35–48. https://doi.org/10.7176/JNSR/12-10-04

Kraaij T, Young C, Bezuidenhout H (2017) Growth-form responses to fire in Nama-Karoo Escarpment Grassland, South Africa. Fire Ecol 13(3):85–94. https://doi.org/10.4996/fireecology.130308594

Lalley JS, Viles HA (2005) Terricolous lichens in the northern Namib Desert of Namibia: distribution and community composition. Lichenologist 37:77–91

Lane SB (1977) An ecological study of the Tanqua-Doring Karoo based on a structural-physiognomic vegetation survey. Botany 111 Project Report, Department of Botany, University of Cape Town, Cape Town

le Roux A (1984) 'n Fitososiologiese studie van die Hester Malan-natuurreservaat. Thesis, Department of Botany, University of Pretoria, Pretoria

Lechmere-Oertel RG (1998) The environmental determinants of the fynbos/succulent karoo boundary in the Matjiesrivier Nature Reserve, Western Cape. Thesis, Department of Botany, University of Cape Town, Cape Town

Lechmere-Oertel RG, Cowling RM (1999) Predicting the distribution of fynbos and succulent karoo biome boundaries and plant communities using general linear models and geographic information systems. S Afr J Bot 65:89–96

and water vapour and elemental accumulation by lichens explain their relative abundance in a coastal desert. J Arid Environ 82:71–80
Lechmere-Oertel RG, Cowling RM (2001) Abiotic determinants of the fynbos/succulent karoo boundary, South Africa. J Veg Sci 12:75–80
Livingstone I (2013) Aeolian geomorphology of the Namib sand sea. J Arid Environ 93:30–39
Lloyd JW (1985) A plant ecological study of the Farm "Vaalputs", Bushmanland, with special reference to edaphic factors. Thesis, Department of Botany, University of Cape Town, Cape Town
Lloyd JW (1989a) Discriminant analysis and ordination of vegetation and soils on the Vaalputs radioactive waste disposal site, Bushmanland, South Africa. S Afr J Bot 55:12–136
Lloyd JW (1989b) Phytosociology of the Vaalputs radioactive waste disposal site, Bushmanland, South Africa. S Afr J Bot 55:372–382
Lloyd JW, Badenhorst M (1995) Vegetation and soils: synthesis of baseline datasets, inventories and maps for Rolfontein Nature Reserve. Scientific Services Section, Northern Cape Nature Conservation Service, Kimberley
Lloyd JW, Badenhorst M (1996) Vegetation and soils: synthesis of baseline datasets, inventories and maps for Dornkloof Nature Reserve. Scientific Services Section, Northern Cape Nature Conservation Service, Kimberley
Lloyd JW, van den Berg E, van Wyk E et al (2002) Patterns of degradation and degradation in the Thicket Biome. Terrestrial Ecology Research Unit Report 39. University of Port Elizabeth, Port Elizabeth
Loris K, Schieferstein B (1992) Ecological investigations on lichen fields of the Central Namib. Vegetatio 98:113–128
Low AB, Rebelo A(T)G (eds) (1996) Vegetation of South Africa, Lesotho and Swaziland. A companion to the vegetation map of South Africa, Lesotho and Swaziland. Department of Environmental Affairs and Tourism, Pretoria
Luther-Mosebach J, Dengler J, Schmiedel U et al (2012) A first formal classification of the Hardeveld vegetation in Namaqualand, South Africa. Appl Veg Sci 15:401–431. https://doi.org/10.1111/j.1654-109X.2011.01173.x
Mackay C (1994) The application of integrated remotely sensed and geographic data to facilitate rangeland mapping and condition assessment in the Ceres Karoo region of southern Africa. Thesis, Department of Geography, University of Stellenbosch, Stellenbosch
Mackay CH, Zietsman HL (1996) Assessing and monitoring rangeland condition in extensive pastoral regions using satellite remote sensing and GIS techniques: an application to the Ceres Karoo region of South Africa. Afr J Range Forage Sci 13:100–112
Manning J, Goldblatt P (1997) South African Wild Flower Guide 9. Nieuwoudtville, Bokkeveld Plateau & Hantam. Botanical Society of South Africa, Kirstenbosch
Manning J, Goldblatt P (2003) A new species and new combinations in *Drimia* (Urgineoideae). Bothalia 33:109–111
Maphangwa KW, Musil CF, Raitt L et al (2012) Differential interception and evaporation of fog, dew
Martin K (2017) Potential eco-physiological and phytosociological impacts of fracking on the vegetation of the Karoo, Eastern Cape, South Africa. Thesis, Faculty of Science, Nelson Mandela University, Port Elizabeth
Masubelele ML, Hoffman MT, Bond W et al (2013) Vegetation change (1988–2010) in Camdeboo National Park (South Africa), using fixed-point photo monitoring: the role of herbivory and climate. Koedoe 55(1):Art.#1127. https://doi.org/10.4102/koedoe.v55i1.1127
Masubelele ML, Hoffman MT, Bond WJ et al (2014) A 50 year study shows grass cover has increased in shrublands of semi-arid South Africa. J Arid Environ 104:43–51
McAuliffe JR, McFadden LD, Hoffman MT (2018) Role of aeolian dust in shaping landscapes and soils of arid and semi-arid South Africa. Geosciences 8:171. https://doi.org/10.3390/geosciences8050171
Mendelsohn J, Jarvis A, Roberts C et al (2002) Atlas of Namibia: a portrait of the land and its people. David Philip, Cape Town
Middleton N, Thomas D (1997) World atlas of desertification. Arnold, London
Midgley GF, Musil CF (1990) Substrate effects of zoogenic soil mounds on vegetation composition in the Worcester-Robertson valley, Cape Province. S Afr J Bot 56:158–166
Milton SJ (1978) Plant communities of Andriesgrond, Clanwilliam District. Bolus Herbarium, Department of Botany, University of Cape Town, Cape Town
Milton SJ (1990a) Above-ground biomass and plant cover in a succulent shrubland in the southern Karoo, South Africa. S Afr J Bot 56:587–589
Milton SJ (1990b) Life styles of plants in four habitats in an arid Karoo shrubland. S Afr J Ecol 1:63–70
Milton SJ (1992) Studies of herbivory and vegetation change in Karoo shrublands. Thesis, Faculty of Science, University Cape Town, Cape Town
Milton SJ (1994) Growth, flowering and recruitment of shrubs in grazed and in protected rangelands in the arid Karoo, South Africa. Vegetatio 111:17–27
Milton SJ (1995) Spatial and temporal patterns in the emergence and survival of seedlings in arid Karoo shrubland. J Appl Ecol 32:145–156
Milton SJ, Dean WRJ, Kerley GIH (1992) Tierberg karoo research centre: history, physical environment, flora and fauna. Trans Roy Soc S Afr 48:15–46
Milton SJ, Dean WRJ, Marincowitz CP et al (1995) Effects of the 1990/91 drought on rangeland in the Steytlerville Karoo. S Afr J Sci 91:78–84
Milton SJ, Yeaton RI, Dean WRJ et al (1997) Succulent karoo. In: Cowling RM, Richardson DM, Pierce SM (eds) Vegetation of southern Africa. Cambridge University Press, Cambridge, pp 131–166
Milton SJ, Henschel JR, van der Merwe H et al (2021) Environmental baseline review of the core area and

surrounds of the Square Kilometre Array (SKA). SAEON Report and Appendices. SAEON, Pretoria. https://doi.org/10.15493/saeon.arid.10000001

Milton SJ, Petersen H, Namba G et al (2023) Drought as a driver of vegetation change in Succulent Karoo rangelands, South Africa. Afr J Range Forage Sci 40:181–195. https://doi.org/10.2989/10220119.2021.1992501

Moisel A, Moll EJ (1981) A Braun-Blanquet survey of the vegetation of the Welwitschia Plain. Dinteria 15:3–11

Mucina L (2017) *Caroxylon* (Chenopodiaceae s.str.) in continental southern Africa and Madagascar: a preliminary nomenclatural synopsis and biogeographic considerations. Phytotaxa 312:151–178

Mucina L (2023) Biomes of the Southern Hemisphere. Springer Nature, Cham

Mucina L, Rutherford MC (eds) (2006) The vegetation of South Africa, Lesotho and Swaziland. SANBI, Pretoria

Mucina L, Jürgens N, le Roux A et al (2006a) Succulent Karoo Biome. In: Mucina L, Rutherford MC (eds) The vegetation of South Africa, Lesotho and Swaziland. SANBI, Pretoria, pp 220–299

Mucina L, Rutherford MC, Palmer AR et al (2006b) Nama-Karoo Biome. In: Mucina L, Rutherford MC (eds) The vegetation of South Africa, Lesotho and Swaziland. SANBI, Pretoria, pp 324–347

Mucina L, Rutherford MC, Powrie LW et al (2006c) Inland azonal vegetation. In: Mucina L, Rutherford MC (eds) The vegetation of South Africa, Lesotho and Swaziland. SANBI, Pretoria, pp 616–657

Muir J (1929) The vegetation of the Riversdale Area, Cape Province. Mem Bot Surv S Afr 13:1–82

Mvondo F, Dauteuil O, Guillocheau F (2011) The Fish River canyon (Southern Namibia): a record of Cenozoic mantle dynamics? Compt Rend Geosci 343:478–485

Ndeimona A (2006) Mycorrhiza re-establishment on post-mined rehabilitated areas of the Brand se Baai Succulent Karoo vegetation. Thesis, Department of Botany & Zoology, University of Stellenbosch, Stellenbosch

Nenzhelele E (2017) Long-term impacts of livestock grazing in the Succulent Karoo: a 20-year study of vegetation change under different grazing regimes in Namaqualand. Thesis, Department of Biological Sciences, University of Cape Town, Cape Town

Nenzhelele E, Todd SW, Hoffman MT (2018) Long-term impacts of livestock grazing and browsing in the Succulent Karoo: a 20-year study of vegetation change under different grazing regimes in Namaqualand. Afr J Range Forage Sci 35:277–287

Nongwe N (2008) A systematic conservation assessment of habitat transformation and degradation in the Little Karoo, South Africa. Thesis, Department of Botany, University of Cape Town, Cape Town

Nußbaum S (2003) Ecological studies on the vegetation of a semi-arid desert following a climatic gradient (Richtersveld, South Africa). Thesis, Botanisches Institut und Botanischer Garten, Universität zu Köln, Köln

O'Connor TG, Roux PW (1995) Vegetation changes (1949–71) in a semi-arid, grassy dwarf shrubland in the Karoo, South Africa: influence of rainfall variability and grazing by sheep. J Appl Ecol 32:612–626

Oguz I (2000) Vegetationsökologische Transektanalyse der Sand-Ökosysteme im westlichen Richtersveld (Republik Südafrika). Thesis, Botanisches Institut und Botanischer Garten, Universität zu Köln, Köln

Oguz I, Stöcker B, Jürgens N (2004) Succulent plant communities and edaphical factors along a coast-inland-transect in the Sandveld of the Northern Richtersveld. Schumannia 4:133–149

Oliver I (2000) A unique desert garden in the winter rainfall area of South Africa. Aloe 37:28–30

Olivier MC (1966) Die plantegroei en flora van die Worcesterse Veldreservaat. Thesis, Department of Botany, University of Stellenbosch, Stellenbosch

Olivier MC (1979) An annotated systematic check list of the Angiospermae of the Worcester Veld Reserve. J S Afr Bot 45:49–62

Osman R, Masubelele M, Khomo L (2019) Soil and vegetation differences across ecological boundaries in an arid South African ecosystem. Afr J Ecol 57:344–352

Palmer AR (1988) Vegetation ecology of the Camdebo and Sneeuberg regions of the Karoo Biome, South Africa. Thesis, Department of Botany, Rhodes University, Grahamstown

Palmer AR (1989) The vegetation of the Karoo Nature Reserve, Cape Province. I. A phytosociological reconnaissance. S Afr J Bot 55:215–230

Palmer AR (1990) Vegetation survey of the Carnarvon Experimental Station, Cape Province. Agricultural Research Council, Grassland Institute, Pretoria

Palmer AR (1991a) A syntaxonomic and synecological account of the vegetation of the eastern Cape midlands. S Afr J Bot 57:76–94

Palmer AR (1991b) Vegetation/environment relationships in the central area of the Cape Midlands, South Africa. Coenoses 6:29–38

Palmer AR (1991c) The potential vegetation of the upper Orange River, South Africa: concentration analysis and its application to rangeland assessment. Coenoses 6:131–138

Palmer AR, Hoffman MT (1997) Nama-karoo. In: Cowling RM, Richardson DM, Pierce SM (eds) Vegetation of southern Africa. Cambridge University Press, Cambridge, pp 167–188

Pauw MJ (2011) Monitoring ecological rehabilitation on coastal mineral sands in Namaqualand, South Africa. Thesis, Department of Botany & Zoology, Stellenbosch University, Stellenbosch

Petersen A (2003) A vegetation study of Paulshoek: considering the relevant biophysical parameters and the impact of different management strategies. Thesis, Department of Botany, University of Cape Town, Cape Town

Petersen H, Jack SL, Hoffman MT et al (2020) Patterns of plant species richness and growth form diversity in critical habitats of the Nama-Karoo Biome, South Africa. S Afr J Bot 135:201–211

Pienaar E (2002) Vegetation on and adjacent to mesas in the Nama-karoo, South Africa—characteristics and comparisons. Thesis, Department of Botany, University of Stellenbosch, Stellenbosch

Pienaar E, Esler KJ, Mucina L (2004) Vegetation of mesas in the southeastern Nama-Karoo, South Africa. S Afr J Bot 70:540–558

Pokorny L, Riina R, Mairal M et al (2015) Living on the edge: timing of Rand Flora disjunctions congruent with ongoing aridification in Africa. Front Genet 6:154. https://doi.org/10.3389/fgene.2015.00154

Qiao N, Wang L (2022) Satellite observed vegetation dynamics and drivers in the Namib sand sea over the recent 20 years. Ecohydrology 15(3):e2420. https://doi.org/10.1002/eco.2420

Rahlao SJ, Hoffman MT, Todd SW et al (2008) Long-term vegetation change in the Succulent Karoo, South Africa following 67 years of rest from grazing. J Arid Environ 72:808–819

Robertson MP, Palmer AR (2002) Predicting the extent of succulent thicket under current and future climate scenarios. Afr J Range Forage Sci 19:21–28

Robinson ER (1976) Phytosociology of the Namib Desert Park. Thesis, University of Natal, Pietermaritzburg

Rohde RF, Hoffman MT, Durbach I et al (2019) Vegetation and climate change in the Pro-Namib and Namib Desert based on repeat photography: insights into climate trends. J Arid Environ 165:119–131

Rösch H (2001) The identification and description of the management units of the Goegap Nature Reserve. Koedoe 44:17–30

Rösch H, van Rooyen MW, Theron GK (1997a) Community-level competition between five Namaqualand pioneer plant species. S Afr J Bot 63:1–3

Rösch H, van Rooyen MW, Theron GK (1997b) Competitive effect and response of ten Namaqualand ephemeral plant species at two nutrient levels. S Afr J Bot 63:210–215

Rösch H, van Rooyen MW, Theron GK (1997c) Predicting competitive interactions between pioneer plant species by using plant traits. J Veg Sci 8:489–494

Rubin F (1998) The physical environment and major plant communities of the Tankwa-Karoo National Park. Koedoe 41(2):61–94

Rubin F, Palmer AR (1996) The physical environment and major plant communities of the Karoo National Park, South Africa. Koedoe 39(2):25–52

Rutherford MC, Powrie LW (2010) Severely degraded rangeland—implications for plant diversity from a case study in Succulent Karoo, South Africa. J Arid Environ 74:692–701

Rutherford MC, Westfall RH (1986) Biomes of southern Africa: an objective categorization. Mem Bot Surv S Afr 54:1–98

Rutherford MC, Mucina L, Powrie LW (2003) Nama-karoo veld types revisited: a numerical analysis of original Acocks' field data. S Afr J Bot 69:52–61

Rutherford MC, Powrie LW, Husted LB (2012) Plant diversity consequences of a herbivore-driven biome switch from grassland to Nama-Karoo shrub steppe in South Africa. Appl Veg Sci 15:14–25

Rybchak O, du Toit J, Delorme J-P (2024) Livestock grazing and biodiversity: effects on CO_2 exchange in semi-arid Karoo ecosystems, South Africa. Sci Total Environ 910:168517. https://doi.org/10.1016/j.scitotenv.2023.168517

Saaed MWB, Jacobs SM, Masubelele ML et al (2018) The composition of the seedbank and its role in ecosystem dynamics and rehabilitation potential in the arid Tankwa Karoo region, South Africa. Afr J Range Forage Sci 35:351–361

Saaed MW, Jacobs SM, Masubelele ML et al (2020) Fifteen-year resilience against further degradation of Succulent Karoo vegetation in South Africa. J Arid Environ 178:1–8

Saaed MWB, Jacobs SJ, Masubelele ML et al (2022) Does the landscape functionality approach provide insight into rangeland conditions? A case study in the Tanqua Karoo region, South Africa. PeerJ 10:e13305. https://doi.org/10.7717/peerj.13305

Saayman N, Morris C, Botha H (2016) Livestock grazing has minimal effect on the species richness and diversity of semi-arid shrubland in the Western Little Karoo, South Africa. Afr J Range Forage Sci 33:225–233

Saayman N, Morris CD, Swart R et al (2021) Drivers of change in vegetation and soil in the Gamka Karoo: grazing versus rainfall. Afr J Range Forage Sci 39:310–319. https://doi.org/10.2989/10220119.2021.1947373

Samuels MI, Allsopp N, Hoffman MT (2022) Stocking density and distribution in relation to vegetation in a biodiverse semi-arid pastoral system in South Africa. Rangel J 44:1–12. https://doi.org/10.1071/RJ21046

Samuels MI, Saaed M, Jacobs S et al (2023) Vegetation structure and composition at different elevational intervals in the arid Tankwa Karoo National Park, South Africa. Veg Classif Surv 4:115–126. https://doi.org/10.3897/VCS.86310

Schmidt A, Kerley G, Watson L (2019) Threshold changes in the structure and composition of bush clumps in piospheres in an arid thicket mosaic, South Africa. J Arid Environ 169:19–28

Schmiedel U (2002a) *Conophytum* and the southern African quartz fields. In: Hammer SA (ed) Dumpling and his wife. New views of *Conophytum* N.E.Br. East Anglia Engraving, London, pp 348–362

Schmiedel U (2002b) The quartz fields of southern Africa flora, phytogeography, vegetation, and habitat ecology. Thesis, Botanisches Institut, Universität zu Köln, Köln

Schmiedel U (2004) The phytogeography of the obligate quartz field flora of southern Africa. Schumannia 4:181–205

Schmiedel U, Jürgens N (1999) Community structure on unusual habitat island: quartz-fields in the Succulent Karoo, South Africa. Plant Ecol 142:57–69

Schmiedel U, Jürgens N (2002) Untersuchungen zur Steuerung der Lebensformzusammensetzung der Quarzflächen-Vegetation im ariden südlichen Afrika. Ber Reinhold-Tüxen-Ges 14:45–58

Schmiedel U, Jürgens N (2004) Habitat ecology of southern African quartz fields: studies on the thermal properties near the ground. Plant Ecol 170:153–166

Schmiedel U, Mucina L (2006) Vegetation of quartz fields in the Little Karoo, Tanqua Karoo and eastern Overberg (Western Cape Province, South Africa). Phytocoenologia 36:1–44

Schmiedel U, Oldeland J (2018a) Vegetation dynamics in the Namaqualand Hardeveld—observations from 17 years of annual monitoring. In: Revermann R, Krewenka KM, Schmiedel U et al (eds) Climate change and adaptive land management in southern Africa—assessments, changes, challenges, and solutions. Klaus Hess Publishers, Göttingen, pp 450–457

Schmiedel U, Oldeland J (2018b) Vegetation responses to seasonal weather conditions and decreasing grazing pressure in the arid Succulent Karoo of South Africa. Afr J Range Forage Sci 35:303–310

Schmiedel U, Dengler J, Etzold S (2012) Vegetation dynamics of endemic-rich quartz fields in the Succulent Karoo, South Africa, in response to recent climatic trends. J Veg Sci 23:292–303

Schmiedel U, Kühne N, Twerski A et al (2015) Small-scale soil patterns drive sharp boundaries between succulent "dwarf" biomes (or habitats) in the arid Succulent Karoo, South Africa. S Afr J Bot 101:129–138

Schmiedel U, Röwer IU, Luther-Mosebach J et al (2016) Effect of grazing on vegetation and soil of the heuweltjieveld in the Succulent Karoo, South Africa. Acta Oecol 77:27–36

Schmiedel U, Hanke W, Mayer C et al (2023) Vegetation dynamics over 18 years in response to seasonal climatic conditions and land use in the Kamiesberg mountains of the succulent Karoo, South Africa. J Arid Environ 219:105085. https://doi.org/10.1016/j.jaridenv.2023.105085

Schultz C (2006) Remote sensing the distribution and spatiotemporal changes of major lichen communities in the Central Namib Desert. Thesis, Universität Kaiserlautern, Kaiserlautern

Settele J, Hoffmann I, Jahn R et al (eds) (1999) Rangeland management in the southern Karoo (South Africa): conflicts of landuse and environmental conservation (report of a scientific students excursion). Umweltforschungszentrum, Leipzig

Seymour CL, Milton SJ, Joseph GS et al (2010) Twenty years of rest returns grazing potential, but not palatable plant diversity, to Karoo rangeland, South Africa. J Appl Ecol 47:859–867. https://doi.org/10.1111/j.1365-2664.2010.01833.x

Smitheman J, Perry P (1990) A vegetation survey of the Karoo National Botanic Garden Reserve, Worcester. S Afr J Bot 56:525–541

Snijman DA (ed) (2013) Plants of the Greater Cape Floristic Region. Volume 2: The Extra Cape flora. SANBI, Pretoria

Snijman D, Perry P (1987) A floristic analysis of the Nieuwoudtville Wild Flower Reserve, northwestern Cape. S Afr J Bot 53:445–454

Steinschen AK, Görne A, Milton SJ (1996) Threats to the Namaqualand flowers: outcompeted by grass or exterminated by grazing. S Afr J Sci 92:237–242

Steyn HM, van Rooyen N, van Rooyen MW et al (1996) The prediction of phenological stages in four Namaqualand ephemeral species using thermal unit indices. Isr J Plant Sci 44:147–160

Steyn HM, Bester SP, Bezuidenhout H (2013) An updated plant checklist for Tankwa Karoo National Park, South Africa. S Afr J Bot 88:247–251

Stöcker B (2000) Vegetationsökologische Untersuchungen zweier Landschaftsausschnitte in den Dünenfeldern des nördlichen Sandveldes (Republik Südafrika). Thesis, Botanisches Institut, Universität zu Köln, Köln

Strohbach B (2021) Vegetation survey of the Khomas Hochland in central-western Namibia: syntaxonomical descriptions. Bothalia 51(2):a4. https://doi.org/10.38201/btha.abc.v51.i2.4

Strohbach BJ (2001) Vegetation survey of Namibia. J Namib Sci Soc 49:93–123

Strohbach BJ, Jankowitz WJ (2012) Phytosociology of the farm Haribes in the Nama-Karoo biome of southern Namibia. Koedoe 54(1):Art.#1038. https://doi.org/10.4102/koedoe.v54i1.1038-2

Theron GK, van Rooyen MW, van Rooyen N (1993) Skilpad Wildflower Reserve: botanical survey. Report to WWF-SA. Department of Botany, University of Pretoria, Pretoria

Thompson M, Vlok J, Rouget M et al (2009) Mapping grazing-induced degradation in a semi-arid environment: a rapid and cost effective approach for assessment and monitoring. Environ Manage 43:585–596

Todd SW, Hoffman MT (2009) A fence line in time demonstrates grazing-induced vegetation shifts and dynamics in the semiarid Succulent Karoo. Ecol Appl 19:1897–1908

Van Damme P (1991) Plant ecology of the Namib Desert. Afrika Focus 7:355–400

van der Merwe H (2009) Patterns of plant diversity in the Hantam–Tanqua–Roggeveld Subregion of the Succulent Karoo, South Africa. Thesis, Department of Plant Science, University of Pretoria, Pretoria

van der Merwe H (2010) Wild flowers of the Roggeveld and Tanqua. H van der Merwe, Calvinia

van der Merwe H (2020) Vegetation of the Square Kilometre Array (SKA), South Africa: a baseline to measure local and global change. S Afr J Bot 132:22–29 van der Merwe H, Hoffman MT (2019) Vegetation of Akkerendam Nature Reserve, Northern Cape: Delineation and dynamics over 100 years. Bothalia 49(1):a2401. https://doi.org/10.4102/abc.v49i1.2401

van der Merwe H, van Rooyen MW (2011a) Guiding conservation efforts in the Hantam–Tanqua–Roggeveld, Succulent Karoo (South Africa) using diversity parameters. Koedoe 53(1):Art#1018. https://doi.org/10.4102/Koedoe.v52i.1018

van der Merwe H, van Rooyen MW (2011b) Species-area relationships in the Hantam-Tanqua-Roggeveld, Succulent Karoo, South Africa. Biodivers Conserv 20:1183–1201

van der Merwe H, van Rooyen MW, van Rooyen N (2008) Vegetation of the Hantam-Tanqua-Roggeveld subregion, South Africa, Part 2: Succulent Karoo Biome related vegetation. Koedoe 50:160–183. https://doi.org/10.4102/koedoe

van der Merwe H, Bezuidenhout H, Bradshaw PL (2015) Landscape unit concept enabling management of a large conservation area: a case study of Tankwa Karoo National Park, South Africa. S Afr J Bot 99:44–53

van der Merwe H, Rahlao S, Hein L et al (2016) Vegetation dynamics after experimental fire disturbance in the arid Succulent Karoo, South Africa. Afr J Range Forage Sci 33:131–135

van der Merwe H, du Toit JCO, van den Berg L (2018) Impact of sheep grazing intensity on vegetation at the Arid Karoo Stocking Rate Trial after 27 years, Carnarvon, South Africa. J Arid Environ 155:36–45. https://doi.org/10.1016/j.jaridenv.2018.02.005

van der Merwe H, Milton SJ, Dean WRJ (2021) Developing an environmental research platform in the Karoo at the Square Kilometre Array. S Afr J Sci 117(11/12):Art. #10511. https://doi.org/10.17159/sajs.2021/10511

van der Walt PT (1980) A phytosociological reconnaissance of the Mountain Zebra National Park. Koedoe 23:1–32

van Jaarsveld E (1981) A preliminary report on the vegetation of the Richtersveld with specific reference to the trees and shrubs of the area. Trees S Afr 33:58–84

van Jaarsveld E (1998a) *Diplosoma retroversum* (Kensit) Schwantes. Aloe 35:23

van Jaarsveld E (1998b) *Tylecodon peculiaris* (Crassulaceae), a new species from the Northern Knersvlakte region (Western Cape). Aloe 35:10–11

van Jaarsveld EJ, van Wyk AE (2003) Two new cremnophilous taxa from semi-arid regions in South Africa. Bothalia 33:115–118

van Rooyen MW (2002) Management of the old field vegetation in the Namaqua National Park, South Africa: conflicting demands of conservation and tourism. Geogr J 168:211–223

van Rooyen MW, Grobbelaar N (1982) Saadbevolkings in die grond van die Hester Malannatuurreservaat in die Namakwalandse Gebroke Veld. S Afr J Bot 1:41–50

van Rooyen MW, Theron GK, van Rooyen N (1996) Skilpad Wildflower Reserve. Veld Flora 82:40–42

van Rooyen MW, Grobbelaar N, Theron GK (1979a) Phenology of the vegetation in the Hester Malan Nature Reserve in the Namaqualand broken veld: 2. The therophyte population. J S Afr Bot 45:433–452

van Rooyen MW, Theron GK, Grobbelaar N (1979b) Phenology of the vegetation in the Hester Malan Nature Reserve in Namaqualand broken veld: 1. General observation. J S Afr Bot 45:279–293

van Rooyen MW, Theron GK, van Rooyen N et al (2004) Mysterious circles in the Namib Desert: review of hypotheses on their origin. J Arid Environ 57:467–485

van Rooyen MW, Henstock R, van Rooyen et al (2010) Plant diversity and flowering displays on old fields in the arid Namaqua National Park, South Africa. Koedoe 52(1):Art.#1004. https://doi.org/10.4102/koedoe.v52i1.1004

van Rooyen MW, le Roux A, van der Merwe H et al (2018) Long-term vegetation change (>20 years) in the plains habitat on the Goegap Nature Reserve, Succulent Karoo, South Africa. Afr J Range Forage Sci 35:289–302. https://doi.org/10.2989/10220119.2018.1498802

van Wyk AE, Smith GF (2001) Regions of floristic endemism in southern Africa. A review with emphasis on succulents. Umdaus Press, Hatfield

Verdcourt B (1969) The arid corridor between the north-east and southwest areas of Africa. Palaeoecol Afr 4:140–144

Vlok JHJ, Euston-Brown, DIW (2002) Subtropical Thicket Ecosystem Planning Project (STEP). Biological Survey Report (plants and birds). Terrestrial Ecology Research Unit, University of Port Elizabeth, Port Elizabeth

Vlok JHJ, Euston-Brown DIW, Cowling RM (2003) Acocks's Valley Bushveld 50 years on: new perspectives on the delimitation, characterization and origin of subtropical thicket vegetation. S Afr J Bot 69:27–51

Vorster L (2017) Current status and impact (2004–2015) of indigenous ungulate herbivory on the vegetation of Sanbona Wildlife Reserve in the Little Karoo. Thesis, Department of Biological Sciences University of Cape Town, Cape Town

Werger MJA (1973a) An account of the plant communities of Tussen die Riviere Game Farm, Orange Free State. Bothalia 11:165–176

Werger MJA (1973b) Phytosociology of the Upper Orange River Valley, South Africa. Thesis, Catholic University of Nijmegen, Nijmegen

Werger MJA (1978) The Karoo-Namib region. In: Werger MJA, van Bruggen AC (eds) Biogeography and ecology of southern Africa. Dr W. Junk Publishers, The Hague, pp 231–299

Werger MJA (1980) A phytosociological study of the Upper Orange River Valley. Mem Bot Surv S Afr 46:1–98

Wiegand T, Milton SJ (1996) Vegetation change in semi-arid communities: simulating probabilities and time scales. Vegetatio 125:169–183

Williamson G (1997) Preliminary account of the floristic zones of the Sperrgebiet (Protected Diamond Area) in southwest Namibia. Dinteria 25:1–68

Williamson G (2000) Notes on the vegetation and climate of the Richtersveld National Park. Aloe 37(4):83–87

Yeaton RI (1988) Structure and function of the Namib dune grasslands. Characteristics of the environmental gradients and species distributions. J Ecol 76:744–758

Zedda L, Rambold G (2009) Diversity and ecology of soil lichens in the Knersvlakte (South Africa). Bryologist 112:19–29

Zhigila DA, Elliott TL, Schmiedel U et al (2024) Do phylogenetic community metrics reveal the South African quartz fields as terrestrial-habitat islands? Ann Bot 133:833–885 https://doi.org/10.1093/aob/mcae027

8 Drakensberg: Southern African Subtropical Alpine Zone

Ladislav Mucina, Michael C. Rutherford, and James L. Tsakalos

Abstract

The Subtropical Alpine Zone is formed by two widely scattered archipelagos of small patches of ecosystems confined to habitats above the timberline, both in the Southern and Northern Hemisphere. In Southern Africa, it is limited to the highest elevations (above the climatic timberline) encountered in the Drakensberg/uKhahlamba massif, particularly in Lesotho and much less in South Africa. This biome is a well-studied high-elevation grassland containing a high number (and cover) of low shrubs, especially where disturbed by long-term grazing. Only one regional biome (DAG1 Drakensberg Alpine Grassland) falls within the continental biome Drakensberg Alpine Grassland.

Keywords

Drakensberg · Lesotho · South Africa · Timberline

L. Mucina (✉) · J. L. Tsakalos
Harry Butler Institute, Murdoch University, 90 South Street, Murdoch, 6150 Perth, Australia
e-mail: Ladislav.Mucina@murdoch.edu.au

L. Mucina
Department of Geography and Environmental Studies, Stellenbosch University, Private Bag X1, Matieland, Stellenbosch 7602, South Africa

M. C. Rutherford
High Peak, Derbyshire, United Kingdom

Department of Botany and Zoology, Stellenbosch University, Private Bag X1, Matieland, Stellenbosch 7602, South Africa

J. L. Tsakalos
Plant Diversity and Ecosystems Management Unit, School of Biosciences and Veterinary Medicine, University of Camerino, Via Pontoni 5, Camerino 62032, Italy

The 'Afroalpine' Drakensberg Vegetation in Global Context of Alpine Ecosystems

The ecosystems above the timberline are often called 'alpine' since they were reminiscent, especially to the European colonial botanists and ecologists, of the high elevations of the Alps—the towering mountain range of Central Europe. This term underwent a few modifications, leading to coining terms such as 'afro-alpine' (Hauman 1933; Hedberg 1955) in Africa, for instance. It was initially intended to apply to the 'alpine' region of the African tropics but was later copied (possibly not in a correct sense; Carbutt and Edwards 2015) by South African vegetation ecologists (e.g., Killick 1963; Mucina et al. 2006 for mountain ranges far from the Tropics per se). This section follows the concept of an 'alpine' ecosystem found at high elevations, separated from other shrub- or tree-dominated vegetation by the climatic timberline at its low-elevation limit (Fig. 8.1).

L. Mucina and M. C. Rutherford (eds.), *Zonal Biomes of Southern Africa*, Biome Ecology 2,
https://doi.org/10.1007/978-3-031-63731-5_8

The subdivision of the alpine para-biome follows Mucina (2023). In Southern Africa, this para-biome is represented by the Subtropical Alpine Zone, which contains only one global biome called Southern Subtropical Alpine Biome, with one continental biome in Africa—the Drakensberg Alpine Grassland. This biome has been described under the name Drakensberg Grassland by Mucina et al. (2006). Its global position has been elucidated by Mucina (2023). Here, we shall limit ourselves only to a few biogeographic and palaeobotanical notes.

The close biogeographical connections between the Cape flora and that of the Drakensberg have been the subject of several studies (Hilliard and Burtt 1987; Carbutt and Edwards 2001, 2006, 2012, 2015). This connection, maintained by the Great Escarpment (Clark et al. 2011a, b), has been shown to reach as far north as the uplands of south-central Africa and to the mountains of

Fig. 8.1 High-elevation Drakensberg alpine grassland—a member of the Subtropical Alpine Zone (zonobiome). **a** Wet grasslands dominated by *Merxmuellera drakensbergensis* (Poaceae) on shallow soils on basalt at an elevation of about 2880 m, top of Sani Pass, Lesotho. **b** African Montane Grassland—dense grasslands at an elevation of around 2000 m on shallow soil and grass-sod mats over Clarens Sandstone in the Golden Gate Highlands National Park near Phuthaditjhaba, eastern Free State, South Africa, with scattered *Protea roupelliae* (Proteaceae) low trees. Photo credits: Both images by L. Mucina

Ethiopia and Cameroon. Phylogenetic analyses of the directionality of the plant migrations along the Great Escarpment suggest that the unidirectional Cape to Cairo route is the most probable scenario for the migrations (Carbutt and Edwards 2001; Gehrke and Linder 2009; Clark et al. 2011a, b). These patterns were found, for instance, in *Disa* (Bytebier et al. 2007), *Pentaschistis* (Galley et al. 2007), *Euryops* (Devos et al. 2010), *Cotula* (A. Jakoet et al., unpublished data), and finally also in *Erica* (Pirie et al. 2011; but see McGuire and Kron 2005). Dating estimates show that the migration into tropical East Africa has occurred in the last 17 Myr, consistent with the Mio-Pliocene formation of the mountains in this area.

The Drakensberg grasslands might have served as a Pleistocene refugium (Carbutt and Edwards 2001), rendering the Drakensberg Range a source of austral afrotemperate taxa also represented in the Cape, such as *Satyrium, Kniphofia,* and *Aloe* (Galley et al. 2007; Ramdhani et al. 2008). Of the recent palaeoecological sources featuring this biome, the papers by Norström et al. (2018) and Kylander et al. (2021) deserve attention.

This continental biome consists of only one regional biome—the DAG1 Drakensberg Alpine Grassland (see Table 8.1).

DAG1 Drakensberg Alpine Grassland

Gd10 Drakensberg Afroalpine Heathland (Mucina and Rutherford 2006)
Gd6 Drakensberg-Amathole Afromontane Fynbos (Mucina and Rutherford 2006)
Gd7 uKhahlamba Basalt Grassland (Mucina and Rutherford 2006)

Distribution This alpine regional biome occurs in the study region, mainly in Lesotho and very marginally in KwaZulu-Natal and Free State Provinces of South Africa. Here, it is found at the highest plateaus, embedded mountain ridges and adjacent upper steep slopes of the Great Escarpment (in this region, also known as uKhahlamba), mainly above an altitude of about 2,900 m. It includes the highest mountain in Southern Africa (Thabana Ntlenyana: 3482 m) (Fig. 8.2).

Vegetation Physiognomy and Composition Vegetation physiognomy is dominated by grasslands and low scrub-dominated areas. The main components of the vegetation cover are grasses *Merxmuellera drakensbergensis* and *Tenaxia disticha,* and low shrubs and herbs such as *Helichrysum praecurrens, H. sessilioides, H. trilineatum,* and many more. *Chrysocoma ciliata* is common in disturbed areas. Patches of montane fynbos-like

Table 8.1 A system of zonobiomes ZB, global biomes GB, continental biomes CB, and regional biomes RB for the Zonobiome A2 in Southern Africa

Rank	Code	Biome	Zonality	Climatic diagnosis
Zonobiome A2 Subtropical Alpine Zone				
ZB	A2	Subtropical Alpine Zone	Zonal	Alpine subtropical; arid to subhumid summer-rainfall
GB	SAZs	Southern Subtropical Alpine Zone	Zonal	Alpine subtropical; arid to subhumid summer-rainfall
CB	DAG	Drakensberg Alpine Grassland	Zonal	
RB	DAG1	Drakensberg Alpine Grassland	Zonal	

The codes of the zonobiomes follow Mucina (2023)

1–3 m tall scrub containing genera such as *Arrowsmithia, Cliffortia, Erica, Euryops, Helichrysum, Passerina, Protea, Restio*, and *Widdringtonia* also occur. On basalt slopes, terraced tussocks are dominated by *Bromus speciosus, Cymbopogon nardus, Festuca caprina, Pentameris tysonii, Rendlia altera*, and *Themeda triandra*. These grasslands are very rich in herbs and low shrubs. The vegetation of this regional biome has been the subject of attention of vegetation ecologists and biogeographers since the fifties of the last century.

Climate The entire region is dominated by summer rainfall but subject to some precipitation from cold fronts in winter. MAP ranges between 800 mm to nearly 3000 mm, with an average of around 1000 mm. There is a rapid decline in MAP from near (2–3 km) the edge of the Escarpment to the interior at high elevations in Lesotho. The mean annual temperature of about 7.4 °C is lower than that of other regional biomes. Hot, dry winds are common from July to October. Morning summer mists are also frequent (Fig. 8.2).

Topography, Geology and Soils The alpine zone of the Lesotho Drakensberg is a rolling plateau with steep slopes on the edges. Geologically, the area is formed by basalts of the Drakensberg Group (Karoo Supergroup). Soils are mollisols, indicating an udic moisture regime and frigid temperature regime. Deep nutrient-rich soils are formed on less steep slopes, while basalt outcrops usually do not carry any fine soil, and instead, shallow pockets of basalt rubble occur. Frost action is important in alpine soil formation (Mokuku 1991). The freezing and thawing of the soil heave the soil material, gradually removing finer soil particles downslope, mainly from the existing micro-erosion terraces (Mucina et al. 2006).

Sources Story (1952), Acocks (1953, 1988), Roberts (1961, 1966), Killick (1963, 1978a, b, 1994), Herbst (1971), Jacot Guillarmod (1971), Herbst and Roberts (1974a, b), White (1978, 1983), Hilliard and Burtt (1987), du Preez and Bredenkamp (1991), Mokuku (1991), Morris et al. (1993), Hill (1996), Carbutt and Edwards (2001, 2006, 2012, 2015), van Wyk and Smith (2001), Kobisi (2005), Mucina et al. (2006), Brand et al. (2015, 2019), Morris (2017), Seleteng-Kose et al. (2021), Harrison et al. (2022), Mucina (2023).

DAG1 Drakensberg Alpine Grassland

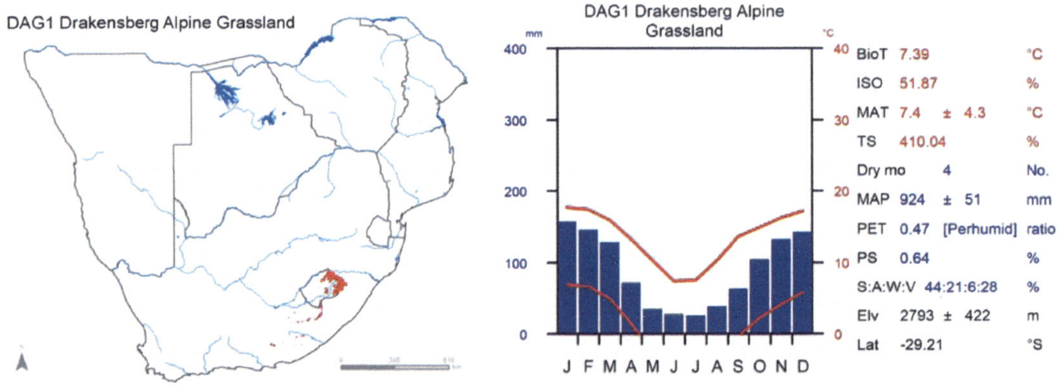

Fig. 8.2 The distribution area and the climatic diagram of the regional biome DAG1 Drakensberg Alpine Grassland

References

Acocks JPH (1953) Veld types of South Africa. Mem Bot Surv S Afr 28:1–192

Acocks JPH (1988) Veld types of South Africa, 3rd edn. Mem Bot Surv S Afr 57:1–146

Brand RF, Collins N, du Preez PJ (2015) A phytosociology survey and vegetation description of inselbergs in the uKhahlamba-Drakensberg Park world heritage site, South Africa. Koedoe 57(1):Art.#1233. https://doi.org/10.4102/koedoe.v57i1.1233

Brand RF, Scott-Shaw CR, O'Connor TG (2019) The alpine flora on inselberg summits in the Maloti-Drakensberg Park, KwaZulu-Natal. South Africa. Bothalia 49(1):a2386. https://doi.org/10.4102/abc.v49i1.2386

Bytebier B, Bellstedt DU, Linder HP (2007) A molecular phylogeny for the large African orchid genus *Disa*. Mol Phylogen Evol 43:75–90

Carbutt C, Edwards T (2001) Cape elements on high-altitude and edaphic islands: historical aspects and preliminary phytogeography. Syst Geogr Plants 71:1033–1061

Carbutt C, Edwards T (2006) The endemic and near-endemic angiosperms of the Drakensberg Alpine Centre. S Afr J Bot 72:105–132

Carbutt C, Edwards TJ (2012) An Eden in exile: outliers of the Cape Floristic Region. An Eden in exile: outliers of the Cape Floristic Region. PlantLife 41&42:4–11

Carbutt C, Edwards TJ (2015) Reconciling ecological and phytogeographical spatial boundaries to clarify the limits of the montane and alpine regions of sub Sahelian Africa. S Afr J Bot 98:64–75

Clark VR, Barker NP, Mucina L (2011a) Taking the scenic route—the southern Great Escarpment (South Africa) as part of the Cape to Cairo floristic highway. Plant Ecol Divers 4:313–328

Clark VR, Barker NP, Mucina L (2011b) The Great Escarpment of southern Africa: a new frontier for biodiversity exploration. Biodivers Conserv 20:2543–2561

Devos N, Barker NP, Nordenstam B (2010) A multi-locus phylogeny of *Euryops* (Asteraceae, Senecioneae) augments support for the "Cape to Cairo" hypothesis of floral migrations in Africa. Taxon 59:57–67

du Preez PJ, Bredenkamp GJ (1991) Vegetation classes of the southern and eastern Orange Free State (Republic of South Africa) and the highlands of Lesotho. Navors Bloemfontein Mus 7:477–526

Galley C, Bytebier B, Bellstedt DU (2007) The Cape element in the Afrotemperate flora: from Cape to Cairo? Proc Roy Soc Lond B Biol Sci 274:535–543

Gehrke B, Linder HP (2009) The scramble for Africa: pan-temperate elements on the African high mountains. Proc Roy Soc Lond B Biol Sci 276:2657–2665

Harrison R, van Tol J, Amiotte Suchet P (2022) Hydropedological characteristics of the Cathedral Peak research catchments. Hydrology 9:189. https://doi.org/10.3390/hydrology9110189

Hauman L (1933) Esquisse de la végétation des hautes altitudes sur le Ruwenzori. Bull Acad Roy Sci Lettr Beaux-Arts Belg 5:602–616, 702–717, 900–917

Hedberg O (1955) Altitudinal zonation of the vegetation on the East African mountains. Proc Linn Soc Lond 165:134–136

Herbst SN (1971) 'n Ekologiese plantopname van 'n gedeelte van die Oxbow-opvanggebied, Lesotho. Thesis, University of the Orange Free State, Bloemfontein

Herbst SN, Roberts BR (1974a) Quantitative ecological relationships in the Alpine grassland of Lesotho. Proc Grassl Soc S Afr 9:61–66

Herbst SN, Roberts BR (1974b) The alpine vegetation of the Lesotho Drakensberg: a study in quantitative floristics at Oxbow. J S Afr Bot 40:257–267

Hill TR (1996) Description, classification and ordination of the dominant vegetation communities, Cathedral Peak, KwaZulu-Natal Drakensberg. S Afr J Bot 62:263–269

Hilliard OM, Burtt BL (1987) The botany of the southern Natal Drakensberg. National Botanic Gardens, Kirstenbosch

Jacot Guillarmod A (1971) Flora of Lesotho (Basutoland). J. Cramer, Lehre

Killick DJB (1963) An account of the plant ecology of the Cathedral Peak area of the Natal Drakensberg. Mem Bot Surv S Afr 34:1–178

Killick DJB (1978a) Notes on the vegetation of the Sani Pass area of the southern Drakensberg. Bothalia 12:537–542

Killick DJB (1978b) The Afro-alpine region. In: Werger MJA, Van Bruggen AC (eds) Biogeography and ecology of southern Africa. Dr. W. Junk, The Hague, pp 515–560

Killick DJB (1994) Drakensberg Alpine region. In: Davis SD, Heywood VH, Hamilton AC (eds) Centres of plant diversity. A guide and strategy for their conservation. Volume 1. IUCN Publications, Cambridge, pp 257–259

Kobisi K (2005) Preliminary checklist of the plants of Lesotho. SABONET Rep 34:1–84

Kylander ME, Holm M, Fitchett J et al (2021) Late glacial (17,060–13,400 cal yr BP) sedimentary and paleoenvironmental evolution of the Sekhokong Range (Drakensberg), southern Africa. PLoS ONE 16(3):e0246821. https://doi.org/10.1371/journal.pone.0246821

McGuire AF, Kron KA (2005) Phylogenetic relationships of European and African Ericas. Int J Plant Sci 166:311–318

Mokuku C (1991) Classification of the alpine plant communities of Mafika-Lisiu. Thesis, Department of Biological Sciences, University of Zimbabwe, Harare

Morris C (2017) Historical vegetation—environment patterns for assessing the impact of climatic change in the mountains of Lesotho. Afr J Range Forage Sci 34:45–51

Morris CD, Tainton NM, Boleme S (1993) Classification of the eastern alpine vegetation of Lesotho. Afr J Range Forage Sci 10:47–58

Mucina L (2023) Biomes of the Southern Hemisphere. Springer Nature, Cham

Mucina L, Rutherford MC (eds) (2006) The vegetation of South Africa, Lesotho and Swaziland. SANBI, Pretoria

Mucina L, Hoare DB, Lötter MC et al (2006) Grassland Biome. In: Mucina L, Rutherford MC (eds) The vegetation of South Africa, Lesotho and Swaziland. SANBI, Pretoria, pp 348–437

Norström E, Bringensparr C, Fitchett JM et al (2018) Late-Holocene climate and vegetation dynamics in eastern Lesotho highlands. Holocene 28:1483–1494

Pirie MD, Oliver EGH, Bellstedt DU (2011) A densely sampled ITS phylogeny of the Cape flagship genus *Erica* L. suggests numerous shifts in floral macromorphology. Mol Phylogen Evol 61:593–601

Ramdhani S, Barker NP, Baijnath H (2008) Exploring the Afromontane centre of endemism: *Kniphofia* Moench (Asphodelaceae) as a floristic indicator. J Biogeogr 35:2258–2273

Roberts BR (1961) Preliminary notes on the vegetation of Thaba ‘Nchu. J S Afr Bot 27:241–251

Roberts BR (1966) The ecology of Thaba ’Nchu—a statistical study of vegetation/habitat relationships. Thesis, Department of Pasture Sciences, University of Natal, Pietermaritzburg

Seleteng-Kose L, Kobisi K, Pool-Stanvliet R et al (2021) A rapid biodiversity assessment of Lesotho’s first proposed biosphere reserve: a case study of Bokong Nature Reserve and Tšehlanyane National Park. Bothalia 51(2):a6. https://doi.org/10.38201/btha.abc.v51.i2.6

Story RA (1952) Botanical survey of the Keiskammahoek district. Mem Bot Surv S Afr 27:1–228

van Wyk AE, Smith GF (2001) Regions of floristic endemism in southern Africa. A review with emphasis on succulents. Umdaus Press, Hatfield

White F (1978) The Afromontane Region. In: Werger MJA, van Bruggen AC (eds) Biogeography and ecology of southern Africa. Dr. W. Junk Publishers, The Hague, pp 463–513

White F (1983) Vegetation of Africa: a descriptive memoir to accompany the UNESCO/AETFAT/UNSO vegetation map of Africa. UNESCO, Paris

Ecotonal Biomes of Southern Africa

9

Ladislav Mucina, Mervyn C. Lötter, Michael C. Rutherford and James L. Tsakalos

Abstract

This chapter presents continental ecotonal biomes (CEBs) that straddle zonobiomes E1, E2, S1, S2, T3, and A4. Zonobiome E2 (tropical seasonal zone) creates the largest (and ecologically most complex) zonoecotonal units—either when straddling the zonobiome T3 Warm-Temperate Zone (AMG African Montane Grassland units, including those representing the iconic Highveld) or S2 Hot Arid Zone (DHG Dry Highveld Grassland, incl. the arid grasslands marginal to Kalahari savanna). Endemic to South Africa is an intricate Albany Thicket, here interpreted as an ecotonal biome straddling zonobiomes S2 and E1 (subtropical section of the Equatorial Zone). Cape Valley Thicket (CVT) is found at the interface between S1 (Ethesial Zone), E1, and T3 zonobiomes. We hypothesise that this globally unique combination of bioclimates carries a relict signature. Several vegetation types found along the West Coast represent a transition between the zonobiomes S1 and S2 (WCS West Coast Arid Strandveld)—a bioclimatic situation repeated in all five global regions supporting the S1 Ethesial Zonobiome supporting the mediterranean-type ecosystems. The high elevations of the Cape Fold Mountains are home to a transitional CEB Cape Altimontane Fynbos, involving the S1 zonobiome and the austroalpine section of the A4 Antarctic Zone.

Keywords

Albany Thicket biome · Cape arid strandveld · Cape Valley thicket · Grassland · Highveld · Manica Highlands · Mozambique · South Africa · West Coast · Zimbabwe · Zonoecotones

L. Mucina (✉) · J. L. Tsakalos
Harry Butler Institute, Murdoch University, 90 South Street, Murdoch, Perth 6150, Australia
e-mail: Ladislav.Mucina@murdoch.edu.au

L. Mucina
Department of Geography and Environmental Studies, Stellenbosch University, Private Bag X1, Matieland 7602, Stellenbosch, South Africa

M. C. Lötter
Mpumalanga Tourism and Park Agency, P/Bag X1088, Lydenburg 1120, South Africa

School of Animal, Plants and Environmental Sciences, University of the Witwatersrand, Private Bag 3, Wits, Johannesburg 2050, South Africa

M. C. Rutherford
High Peak, Derbyshire, United Kingdom

Department of Botany and Zoology, Stellenbosch University, Private Bag X1, Matieland 7602, Stellenbosch, South Africa

J. L. Tsakalos
Plant Diversity and Ecosystems Management Unit, School of Biosciences and Veterinary Medicine, University of Camerino, Via Pontoni 5, Camerino 62032, Italy

L. Mucina and M. C. Rutherford (eds.), *Zonal Biomes of Southern Africa*, Biome Ecology 2,
https://doi.org/10.1007/978-3-031-63731-5_9

This chapter focuses on the continental ecotonal biomes (CEBs) resulting from spatial interactions (transitions) among most of the subcontinent's zonobiomes (E1, E2, S1, S2, T3), and A4 that is represented only in the Southern Hemisphere (see Table 2.1 in this book; also Mucina 2023). With so many zonobiomes converging in Southern Africa, a complex mosaic of ecotonal biomes appears to have emerged in the distant past and survives today.

The zonobiome E2, represented by the subtropical savanna woodlands and grasslands, is undisputedly the largest zonobiome in Southern Africa where it intergrades with the Warm-Temperate Zone (T3) and it forms two large-scale ecotonal biomes that dominate south-central Southern Africa. These ecotones are classified as CEB African Montane Grassland and CEB Dry Highveld Grassland. Formerly, they were viewed as forming a heterogenous rankless biome known to South African ecologists as the Grassland Biome (Mucina et al. 2006a). Southern Africa is also home to a complex of arid ecosystems, involving Arid Savanna and several semideserts (Succulent Karoo, Nama-Karoo). The semideserts belong to the zonobiome S2, while the Arid Savanna (in the region represented by the iconic Kalahari 'desert') is a member of the zonobiome E2. Interestingly, zonobiome S2 is involved in even more types of ecotonal biomes than is the case with zonobiome E2. The largest of these ecotonal biomes is the CEB Albany Thicket (where the zonobiomes S2 and E1 come into contact), followed by CEB Southern Cape Valley Thicket and the CEB West Coast Arid Strandveld (both involving the zonobiome S2 and the zonobiome S1) and the CAP Cape Altimontane Fynbos involving S1 zonobiome and the austroalpine section of the Antarctic zonobiome.

In this chapter, we shall delimit the ecotonal biomes of Southern Africa and present their regional-biome structure (Table 9.1).

9.1 Continental Ecotonal Biome African Montane Grassland (AMG)

ZE Subtropical–Warm-Temperate Wet Ecotone E2(savm)–T3

Several types of grasslands have a temperate (warm-temperate) climate in the Southern Hemisphere (see Mucina 2023). These mainly, and in southern Africa exclusively, occur at higher elevations and hence can be called 'upland grasslands' or 'montane grasslands'. From a point of view, they are pure (tree-less) grasslands despite having a climate that can support tree growth (Fig. 9.1; see also Bond 2019; Higgins et al. 2023, 2024). Woody shrubland vegetation found within the confines of these vast grassland-dominated landscapes is embedded either in sheltered (extrazonal) habitats, such as depressions and gullies or in soil-specific (azonal) and less frost-affected habitats, such as rocky ridges and localised, usually rocky hillocks also known as koppies (see Mucina et al. 2006a: 351). These grasslands experience cold winters but warm and high-precipitation summers. They are dominated by C4 grasses, although some C3 grasses and numerous other herbaceous species also occur (Fig. 9.1).

If we accept a recent functional definition of savanna (biome) as a 'grassland dominated by C4 plants' (Lehmann et al. 2011) the C4-dominated Highveld grasslands suggest a savanna affiliation. Indeed, Backéus (1992) did not hesitate to use the term 'humid savanna' for some African high-elevation grasslands. However, these 'savannas' are clearly different from the woody savannas that are dealt with in Chap. 3 of this book. Primarily, the bioclimate of the treeless Highveld mesa is different from the typical savanna biome units such as Miombo, Mopane, and Lowveld. The CART modelling analyses (Rutherford et al. 2006)

Table 9.1 A system of zonoecotones (ZE), continental ecotonal biomes (CEB), and regional ecotonal biomes (REB) for the zonal non-forest vegetation of Southern Africa. The codes of the ecotonal biomes follow Mucina (2023)

Rank	Code	Biome	Zonality	Azonality type	Climatic Diagnosis
Zonoecotone E2(savm)–T3 Subtropical Warm-Temperate Wet Ecotone					
ZE		**Subtropical Warm-Temperate Wet Ecotone**	Zonoecotonal		**Warm-temperate; predictable high summer rainfall**
CEB	**AMG**	**African Montane Grassland**	Ecotonal		**Warm-temperate; predictable high summer rainfall**
REB	AMG1	Mesic Southern Highveld Sandy Grassland	Ecotonal		
REB	AMG2	Mesic Northern Highveld Sandy Grassland	Ecotonal		
REB	AMG3	Mesic Highveld Clayey Grassland	Ecotonal		
REB	AMG4	Northern Drakensberg Montane Grassland	Ecotonal		
REB	AMG5	Southern Drakensberg Montane Grassland	Ecotonal		
REB	AMG6	Dry Sub-Escarpment Grassland	Ecotonal		
REB	AMG7	Wet Sub-Escarpment Grassland	Ecotonal		
REB	AMG8	Eastern Highlands Grassland	Ecotonal		
REB	AMG9	Eastern Dolomite Grassland	Ecotonal/intrazonal	Peinobiome	
REB	AMG10	Western Dolomite Grassland	Ecotonal/intrazonal	Peinobiome	
REB	AMG11	Northern Quartzite Sourveld	Ecotonal/intrazonal	Peinobiome	
REB	AMG12	Mesic Montane Shrubland	Ecotonal/intrazonal	Lithobiome	
Zonoecotone [E2(sava)–[E2(savm)–T3]–S2 Subtropical Semiarid Seasonal–Semidesert Ecotone					
ZE		**Subtropical Semiarid Seasonal–Semidesert Ecotone**	Zonoecotonal		**Warm-temperate; predictable low summer rainfall**
CEB	**DHG**	**Dry Highveld Grassland**	Ecotonal		**Warm-temperate; predictable low summer rainfall**
REB	DHG1	Dry Highveld Sandy Grassland	Ecotonal		
REB	DHG2	Dry Highveld Clayey Grassland	Ecotonal		
REB	DHG3	Dry Highveld Koppie Shrubland	Ecotonal/intrazonal	Lithobiome	
Zonoecotone S2–E1(sub) Subtropical Semiarid–Semidesert Ecotone					
ZE		**Subtropical Semiarid–Semidesert Ecotone**	Zonoecotonal		**Subtropical; subarid bi-seasonal**
CEB	**AT**	**Subtropical Albany Thicket**	Ecotonal		**Subtropical; subarid bi-seasonal**
REB	AT1	Albany Thicket	Ecotonal		
REB	AT2	Albany Calcrete Bontveld	Ecotonal/intrazonal	Peinobiome	
REB	AT3	Albany Strandveld	Ecotonal/intrazonal	Psammobiome	
REB	AT4	Sundays Dry Thicket	Ecotonal		

(continued)

Table 9.1 (continued)

Rank	Code	Biome	Zonality	Azonality type	Climatic Diagnosis
Zonoecotone S2(sdesc)–S1 Semidesert–Ethesial Coastal Ecotone					
ZE		**Semidesert–Ethesial Coastal Ecotone**	Zonoecotonal		**Warm-temperate-subtropical transitional; low winter rainfall**
CEB	**WCS**	**West Coast Arid Strandveld**	Ecotonal		**Warm-temperate-subtropical transitional; low winter rainfall**
REB	WCS1	West Coast Arid Sandy Strandveld	Ecotonal/intrazonal	Psammobiome	
REB	WCS2	West Coast Arid Granite Strandveld	Ecotonal/intrazonal	Lithobiome	
REB	WCS3	West Coast Arid Limestone Strandveld	Ecotonal/intrazonal	Peinobiome	
Zonoecotone S1–A4 Ethesial–Austro-Alpine Ecotone					
ZE		**Ethesial–Austro-Alpine Ecotone**	Zonoecotonal		**Alpine warm-temperate; ethesial seasonal winter-rainfall**
CEB	**CAP**	**Cape Altimontane Fynbos**	Ecotonal		**Alpine warm-temperate; ethesial seasonal winter-rainfall**
REB	CAP1	Western Altimontane Fynbos	Ecotonal/intrazonal	Peinobiome	
REB	CAP2	Eastern Altimontane Fynbos	Ecotonal/intrazonal	Peinobiome	
Zonoecotone T3–[S2–E1(sub)] Semiarid Warm-Temperate Ecotone					
ZE		**Semiarid Warm-Temperate Ecotone**	Zonoecotonal		**Warm-temperate; aseasonal low rainfall**
CEB	**CVT**	**Southern Cape Valley Thicket**	Ecotonal		**Warm-temperate; aseasonal low rainfall**
REB	CVT1	Southern Cape Valley Thicket	Ecotonal/relict		

using five macro-climatic variables (assumed to be informative in terms of eco-physiological response) suggest that the bioclimatic signature of the Highveld grasslands is of zonoecotonal nature—transitional between the climate of the zonobiome E2 (Tropical Seasonal Zone) and zonobiome T3 (Warm-Temperate Zone). In South Africa, the upland grasslands are adjacent to lowland savannas in many areas, with an abrupt boundary between them that could be termed a savanna–grassland 'treeline'. Both systems are dominated by C4 grasses and burn regularly, yet fire-tolerant savanna trees do not survive in the grasslands.

Mucina et al. (2006a) suggested that the origins of the South African 'Grassland Biome', should be sought in the coincidence of Late Miocene global cooling that might have initiated the development of modern, grass-rich biomes. Continental uplift, which started in the Early Miocene and culminated in significant vertical movement of up to 600–900 m in the southeastern regions of the Southern African subcontinent during the Pliocene, moved a large area into colder, higher altitudes more suitable for grasslands than for (sub)tropical savanna woodlands (Partridge 1997; Mucina et al. 2006a).

Fig. 9.1 Physiognomy of high-elevation (warm-temperate) grasslands dominated by C4-grasses of southern Africa. **a** Mesic Highveld Grassland represented by vegetation type Paulpietersburg Moist Grassland along the old Piet Retief (eMkhondo)–Wakkerstroom road (northern KwaZulu-Natal, South Africa) interrupted by drainage lines supporting shrublands with *Leucosidea sericea* (Rosaceae). **b** The rolling landscape of wet sub-escarpment grasslands (CEB African Montane Grassland) near Underberg, KwaZulu-Natal, South Africa, rich in *Helichrysum* spp., Asteraceae. **c** High-elevation (ca. 2200 m), species-rich grasslands of the Tsetserra region (Zimbabwe–Mozambique border), supporting many endemic taxa to this region. **d** Western part of the Nyanga Massif on the Zimbabwe–Mozambique border, supporting grassland rich in perennial herbs. **e** Dry Highveld Grassland typical of the koppie-dominated landscapes near Springfontein, southern Free State, South Africa, with prominent grasses *Hyparrhenia hirta* and several *Eragrostis* species. **f** Dry Highveld grassland in the Mountain Zebra National Park near Craddock, Eastern Cape, South Africa. Photo credits: **a** & **c**: M. C. Lötter; **b**, **e** & **f**: L. Mucina; **d**: V. R. Clark

Among the hypotheses attempting to explain the origin and persistence of these grasslands, Bredenkamp et al.'s hypothesis (2002) deserves particular attention. These authors argued that the South African grassland is a climatically controlled ecosystem in which cooler conditions at high altitudes are one of the major driving forces that prevent the colonisation of trees of a generally tropical origin. Wakeling et al. (2012) submitted this hypothesis to a test. They tested experimentally whether frost in the dormant season or slow growth in the growing season in conjunction with frequent fires may explain the tree-less state of grasslands. Wakeling et al. (2012) found that savanna trees grew more slowly in cooler upland grassland sites compared with lower-elevation warm savanna sites and, under frequent fire regimes, would be prevented from reaching maturity. This may be true globally for similar grasslands where tree growth can occur and partly explain the lack of trees in grasslands. Another partial explanation has also been offered by Wakeling et al. (2010). They suggested that tree seedling growth on the leached soils of the Highveld grasslands was reduced due to the restricted nutrient availability of the leached soils. They observed that the grasslands of South Africa (both subtropical and? temperate) occur on a range of surfaces, from the ancient peneplains of the African surface to recently dissected areas (Partridge 1997). The African surface is stable and has undergone gentle lowering rather than large-scale dissection—remnants of this surface support highly leached soils.

On the other hand, South African savannas occur largely on younger land surfaces (King 1978; Partridge 1997). In the upland region, higher rainfall and an older land surface may have produced more leached soils with lower nutrient availability (King 1978), possibly resulting in slower plant growth. The cooler temperatures of high elevation grassland climates may also influence the nitrogen and phosphorus mineralisation rates (Lewis 1986; Miller and Cramer 2004), which may further constrain the availability of these elements in grassland relative to savanna sites.

In high-elevation grasslands, slower growth might result in juvenile trees taking too long to escape the frequent fires to become adults. Frost sensitivity of juvenile trees undoubtedly plays a crucial role as well. Thus, the interaction between frequent fires, low-nutrient soils, and frequent frost might explain the tree-less nature of these grasslands.

The above general explanations on the origin and maintenance of the 'Grassland Biome' will need revisiting three biome concepts, namely the zonal DAG Drakensberg Alpine Grassland (zonobiome A2) and both continental ecotonal biomes AMG African Montane Grassland and DHG Dry Highveld Grassland.

The mesic Highveld grasslands (the flagship of the AMG African Montane Grassland) are a global 'regular anomaly'—these are C4-dominated grasslands in a warm-temperate thermo-climatic envelope. On the global scale, the C4-dominated warm-temperate grasslands include (a) pampa and 'campo limpo' ('campo do sul') of northern Argentina, Uruguay, and southern Brazil (see also Mucina 2023), (b) West Indian Ghat shola grasslands, and (c) Southern (Short Grass) Prairie of the western Mississippi basin of the United States; the latter two are located in the Northern Hemisphere and therefore not handled here in detail.

Besides the South African Highveld, Manica Highlands, and Gorongosa (all within the MBSA), the AMG African Montane Grassland occurs further north in Malawi and southern Tanzania. In the MBSA, the African Montane Grassland comprises 12 regional ecotonal biomes, of which eight are zonal/ecotonal and four are intrazonal/ecotonal. The regional ecotonal biomes AMG1 Mesic Southern Highveld Sandy Grassland and AMG2 Mesic Northern Highveld Sandy Grassland AMG1 on the one hand and AMG3 Mesic Highveld Clayey Grassland on the other, correspond roughly to the Mesic Highveld Grassland Bioregion (Mucina and Rutherford 2006). These regional biomes are contrasted by prevalent sandy vs. clayey soils. The Highveld grasslands found on and just below the Great Escarpment experience relatively high precipitation due to orographic rainfall originating from moisture propelled towards the Great Escarpment by the easterly flows that leave the tablelands above the Great Escarpment in a partial rain shadow (see Cadman et al. 2013). These areas constitute two regional ecotonal biomes—AMG4 Northern Drakensberg Montane Grassland and AMG5 Southern Drakensberg Montane Grassland. The regional biomes AMG6 Dry Sub-Escarpment Grassland and AMG7 Wet Sub-Escarpment Grassland correspond roughly to the Gs Sub-Escarpment Grassland Bioregion (Mucina and Rutherford 2006). The occurrence of grassland physiognomy at lower (sub-Escarpment) elevations might be attributed to cold-air flows from the Escarpment. The last zonal/ecotonal unit within AMG is AMG8 Eastern Highlands Grassland, geographically distinct from the bulk of the AMG and located in the Manica Highlands, straddling the borders of Zimbabwe and Mozambique. The remaining four REBs are either intrazonal peinobiomes (AMG9 Eastern Dolomite Grassland, AMG10 Western Dolomite Grassland, and AMG11 Northern Quartzite Sourveld) or lithobiomes (AMG12 Mesic Montane Shrubland).

The correspondence between these regional ecotonal biomes and the formal vegetation units of the VegMap2006, VegMap2018, and VegMapMoz is given in the respective descriptions of the regional ecotonal biomes (below).

The CEB African Montane Grassland encompasses 12 regional ecotonal biomes (Table 9.1):

AMG01 Mesic Southern Highveld Sandy Grassland

Gd 3 Stormberg Plateau Grassland (Dayaram et al. 2019)
Gm 1 Zastron Moist Grassland (Dayaram et al. 2019)
Gm 4 Eastern Free State Sandy Grassland (Dayaram et al. 2019)

Distribution This regional ecotonal biome ranges from the Stormberg Plateau in the Eastern Cape Province, northwards into the Free State in the surrounds of Zastron, the western and northwestern foot slopes of the Maloti and the Low Drakensberg is known as East Free State sandy grasslands, as well as parts of the lowlands of eastern Lesotho (Fig. 9.2).

Vegetation Physiognomy and Composition The vegetation of this regional biome is a C4-dominated grassland, with frequent patchy occurrence of dwarf-shrub synusia. The common grass dominants include *Aristida congesta, A. junciformis* subsp. *galpinii, Cymbopogon pospischilii, Digitaria argyrograpta, D. monodactyla, Elionurus muticus,* numerous *Eragrostis* species (*E. capensis, E. chloromelas, E. curvula, E. plana, E. racemosa*), *Harpochloa falx, Heteropogon contortus, Hyparrhenia hirta, Tribolium purpureum, Microchloa caffra, Monocymbium ceresiiforme, Setaria sphacelata, Tenaxia disticha, Themeda triandra, Trisetopsis imberbis,* and *Tristachya leucothrix*. The dwarf shrubs are represented by *Felicia filifolia* and *Chrysocoma ciliata*. Herbs and subshrubs of the genera *Barleria, Berkheya, Dicoma, Helichrysum,* and *Vernonia* s.l. (incl. *Hillardiella*).

Climate Unlike its northern biome counterpart (AMG02, see below), this unit is cool-temperate with a mean annual temperature of 14.1 °C. Summers are usually hot, while winter can be bitterly cold, and frost is common. The seasonality of precipitation varies from spring–autumn bimodal (Gd 3 Stormberg Plateau Grassland) to strictly summer-rainfall pattern. The MAP of 692 mm varies from 450 mm in parts of the south to over 800 mm in the north (Fig. 9.2).

Topography, Geology and Soils The landscapes supporting this regional biome are flat to undulating, broken in places by sandstone outcrops forming extensive terraces. Generally, the soils are shallow, containing variable amounts of soil and overlying mudstones and sandstones of the Elliot and Molteno Formations (Karoo Supergroup). Relatively deep sandy layers occur on the sandstone of the Tarkastad Subgroup. Typical soil forms present on the sandstone terraces are Clovelly and Avalon. Mudstones, sandstones and shale of the Beaufort Group also occur.

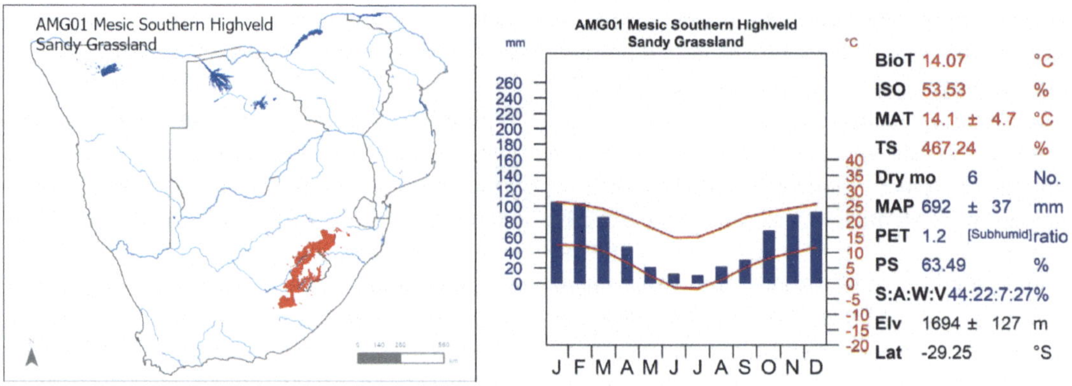

Fig. 9.2 The distribution area and the climatic diagram of the regional ecotonal biome AMG01 Mesic Southern Highveld Sandy Grassland. The construction of the climatic diagrams and the distribution maps of the biomes is described in Chap. 1

Sources Acocks (1953, 1988), Scheepers (1975), Werger (1980), Potgieter (1982), du Preez (1991), du Preez and Bredenkamp (1991), Eckhardt (1993), Fuls (1993), Kay et al. (1993), Eckhardt et al. (1995), Smit et al. (1995c), Malan (1998), Malan et al. (1999), Hoare and Bredenkamp (2001), Mucina et al. (2006a), Dingaan et al. (2013, 2016, 2017), Dingaan and Tsubo (2019), Daemane et al. (2022).

AMG02 Mesic Northern Highveld Sandy Grassland

Gm 10 Egoli Granite Grassland (Dayaram et al. 2019)
Gm 11 Rand Highveld Grassland (Dayaram et al. 2019)
Gm 12 Eastern Highveld Grassland (Dayaram et al. 2019)
Gm 15 Paulpietersburg Moist Grassland (Dayaram et al. 2019)

Distribution This regional biome includes the grasslands of the eastern North West and Gauteng Provinces (Rand and Egoli grasslands), extending further east to include the Mpumalanga Province and northern KZN grasslands as far as Paulpietersburg (Fig. 9.3).

Vegetation Physiognomy and Composition The vegetation is a typical, dense Highveld grassland dominated by C4 grasses of the genera *Aristida* (*A. aequiglumis, A. canescens, A. congesta, A. junciformis* subsp. *galpinii*), *Eragrostis* (*E. capensis, E. chloromelas, E. curvula, E. plana, E. racemosa, E. sclerantha*), *Sporobolus* (*S. africanus, S. pectinatus*), and *Tristachya* (*T. biseriata, T. leucothrix, T. rehmannii*). Other dominants include *Andropogon schirensis, Brachiaria serrata, Cynodon dactylon, Diheteropogon amplectens, Digitaria monodactyla, Elionurus muticus, Harpochloa falx, Heteropogon contortus, Hyparrhenia hirta, Loudetia simplex, Melinis repens* subsp. *repens, Microchloa caffra, Panicum natalense, Rendlia altera, Schizachyrium sanguineum, Setaria sphacelata*, and locally also *Ctenium concinnum* and *Monocymbium ceresiiforme*. *Themeda triandra* is a common and often dominating element, especially in well-managed grasslands. Locally, a high diversity of herbs (mainly of the family Asteraceae, mainly *Berkheya* and *Helichrysum*) is typical. Of species of other families, *Justicia anagalloides* and *Pollichia campestris* may dominate in places.

Climate In thermic terms, the climate of this regional biome is transitional between cool-temperate and warm-temperate (the strongest warm-temperate signal is observed with the Gm 15 Paulpietersburg Moist Grassland). The mean annual temperature is 16.1 °C. It has a strongly seasonal precipitation pattern, with most of the rainfall occurring in summer, with a MAP of 792 mm, from about 550 mm in the west to over

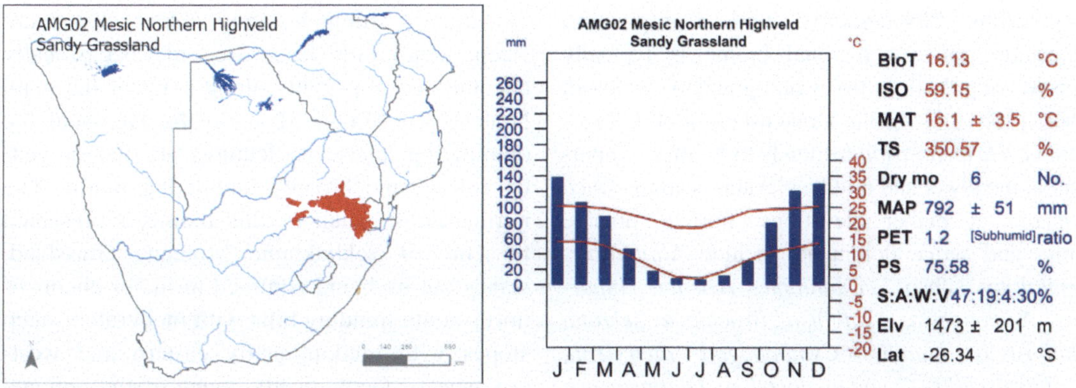

Fig. 9.3 The distribution area and the climatic diagram of the regional ecotonal biome AMG02 Mesic Northern Highveld Sandy Grassland

900 mm in the east. The incidence of frost is high, especially in the western, more continental, section of the biome (Fig. 9.3).

Topography, Geology and Soils The landscape of this regional biome is shaped by moderately undulating and extensive sloping plains and low hills interrupted by a series of ridges. The geologic substrate is variable and includes mafic and sedimentary rocks. Archaean granite (Halfway House Granite at the core of the Johannesburg Dome) supports leached, shallow, coarsely-grained, sandy soil of the Glenrosa form that is poor in nutrients. Here a small group of ultramafics is also found. Quartzite ridges of the Witwatersrand Supergroup and the Pretoria Group, as well as the Selons River Formation of the Rooiberg Group (the last two are of the Transvaal Supergroup), and support soils of various quality (shallow Glenrosa and Mispah forms, especially on rocky ridges). Red to yellow sandy soils are formed over the arenites of the Madzaringwe Formation of the Karoo Supergroup (see Mucina et al. 2006a).

Sources Acocks (1953, 1988), Bredenkamp (1975, 1976, 1977), Bezuidenhout (1988), Turner (1989), Bezuidenhout and Bredenkamp (1991b), Coetzee (1993), Eckhardt (1998), Coetzee et al. (1994, 1995), Eckhardt et al. (1996a, c), Smit et al. (1997), Camp (1999a), Burgoyne et al. (2000), Grobler (2000), Grobler et al. (2002), Ellery et al. (2001), Reddy et al. (2001), Grobler et al. (2006), Mucina et al. (2006a), Siebert (2011), Mochesane (2018), Pule et al. (2018), Mokgakane (2021), Mokgakane et al. (2021), Oosthuysen et al. (2023), van der Walt et al. (2024).

AMG03 Mesic Highveld Clayey Grassland

Gd 8 Lesotho Highland Basalt Grassland (Dayaram et al. 2019)
Gd 9 Western Lesotho Basalt Grassland (Dayaram et al. 2019)
Gm 3 Eastern Free State Clay Grassland (Dayaram et al. 2019)
Gm 6 Frankfort Highveld Grassland (Dayaram et al. 2019)
Gm 8 Soweto Highveld Grassland (Dayaram et al. 2019)
Gm 9 Tsakane Clay Grassland (Dayaram et al. 2019)
Gm 13 Amersfoort Highveld Clay Grassland (Dayaram et al. 2019)
Gm 16 KaNgwane Montane Grassland (Dayaram et al. 2019)
Gm 19 Sekhukhune Montane Grassland (Dayaram et al. 2019)

Distribution This is a very extensive regional biome covering most of Lesotho, parts of northeastern Eastern Cape, eastern Free State, southern Gauteng and Mpumalanga and marginal reaching into southern and northwestern KZN and western eSwatini. The average elevation is 1845 m, with lowest and highest elevation extremes occurring at about 900 m and 2900 m, respectively (Fig. 9.4).

Vegetation Physiognomy and Composition Vegetation of this regional biome is typically species-rich (both in herbs and grasses) in closed, short grasslands. In the Drakensberg and its piedmonts, *Themeda triandra* tends to be more important at the lower and middle elevations and *Festuca caprina* at higher elevations. Further notable grass and sedge dominants include *Alloteropsis semialata* subsp. *eckloniana, Aristida bipartita, A. congesta, A. diffusa, Brachiaria serrata, Bulbostylis humilis, Cymbopogon dieterlenii, C. pospischilii, Cynodon dactylon, Digitaria eriantha, Dracoscirpoides falsa, Elionurus muticus, Eragrostis chloromelas, E. curvula, E. plana, E. planiculmis, E. racemosa, Harpochloa falx, Heteropogon contortus, Hyparrhenia hirta, Koeleria capensis, Microchloa caffra, Panicum natalense, Pentameris oreodoxa, Poa binata, Setaria sphacelata, Tenaxia disticha, Trachypogon spicatus, Tristachya leucothrix*, and many more. At higher elevations (Lesotho Plateau), shrublands dominated by *Chrysocoma ciliata, Leucosidea sericea, Passerina montana,* and *Pentzia cooperi* are very common. Herbs are also very common and diverse; of the dominant genera *Anthospermum, Gnidia, Helichrysum*, and *Senecio* stand out.

Climate This regional biome experiences summer rainfall, averaging 717 mm, which varies from below 600 mm in places to 1250 mm in parts of the KaNgwane area. Much of the rainfall comes in the form of convectional summer storms. Thermically, this is a cool-temperate region with maximum summer temperatures reaching over 30 °C and the lowest winter temperatures as low as −10 °C. Gm 9 Tsakane Clay Grassland has an intermediate position between a cool-temperate and warm-temperate climate. Frost occurs regularly, and the Gm 3 Eastern Free State Clay Grassland is one of the coldest regions of the Highveld (Fig. 9.4).

Topography, Geology and Soils The landscapes supporting this regional biome show dramatic features especially in the southern areas of the biome in Lesotho where often deep valleys separate many plateaus and high ridges of mountains. Although many valley slopes are steep, major cliff faces are only occasionally encountered, especially along parts of the main Drakensberg/Maloti Mts. On the Highveld the dominating landscape features are flat or gently rolling and broadly undulating plains. The northernmost patch of this biome, represented by Gm 19 Sekhukhune Montane Grassland, stands out by being confined to major chains of north–south trending hills with moderately steep slopes with predominantly eastern and western aspects. Geologically, much of the regional biome is underlain by Jurassic basaltic lava flows of the Drakensberg Group (especially in the southern areas). Jurassic dolerite dykes add to the geological variability and also support clayey soils. The major geological and pedological feature—the domination of geology giving rise to heavy, clayey soils—is the primary motivation for the recognition of this regional biome. The Karoo Supergroup is represented by mudstones and sandstones of the Adelaide, Madzaringwe, Tarkastad and Volksrust Formations. The soils derived from the basalt have fairly even proportions of coarse sand, fine sand, silt, clay and organic matter. In Gm 9 Tsakane Clay Grassland, the most significant rock is the basaltic lava of the Klipriviersberg Group (Ventersdorp Supergroup). Gm 16 KaNgwane Montane Grassland is found mainly on the granite of the Mpuluzi Granite (Randian Erathem), which also gives rise to melanic soils. Gm 19 Sekhukhune Montane Grassland geologically stands out by the domination of mafic intrusive rocks of the Upper and Main Zones of the Rustenburg Layered Suite, which is economically the most important part of the Bushveld Igneous Complex (Vaalian Erathem). Gabbro and norite are the dominant rocks here, giving rise to vertic, melanic or red-structured Arcadia, Mayo, Milkwood, Mispah, Shortlands, and Steendal soils.

Sources Acocks (1953, 1988), Herbst and Roberts (1974a, b), Bredenkamp (1975, 1976, 1977), Scheepers (1975), Bredenkamp and Theron (1978, 1980), Killick (1978a, b), Potgieter (1982), Wieland (1982), Martin

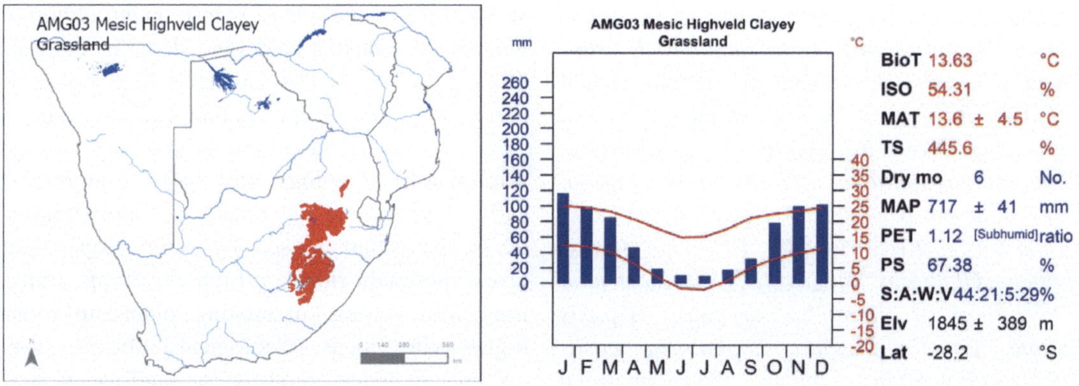

Fig. 9.4 The distribution area and the climatic diagram of the regional ecotonal biome AMG03 Mesic Highveld Clayey Grassland

(1986), Hilliard and Burtt (1987), Seaman (1987), Backéus (1988), Morris (1990, 1994), Bezuidenhout and Bredenkamp (1991a), Breytenbach (1991), du Preez (1991), Mokuku (1991), Breytenbach (1991), Breytenbach et al. (1992, 1993a, b, c), Myburgh et al. (1992), Eckhardt (1993), Eckhardt et al. (1993a, b), Fuls (1993), Fuls et al. (1993c), Kay et al. (1993), Morris et al. (1993), Coetzee et al. (1995), Smit et al. (1995a, c), Bester (1998), de Frey (1999), Smit (1992), Willis et al. (1999), Anonymous (2000), Siebert (2001), Siebert et al. (2001, 2002a, b, c, 2003), van Wyk and Smith (2001), Bredenkamp and Brown (2003), Kobisi (2005), Mucina et al. (2006a), Brand et al. (2011), Seleteng-Kose et al. (2021), Rahlao et al. (2023), Stephen et al. (2023).

AMG04 Northern Drakensberg Montane Grassland

Gm 14 Wakkerstroom Montane Grassland (Dayaram et al. 2019)
Gm 17 Barberton Montane Grassland (Dayaram et al. 2019)
Gm 25 Woodbush Granite Grassland (Dayaram et al. 2019)
Gm 30 Steenkampsberg Montane Grassland (Dayaram et al. 2019 & VegMap 2012)
Gm 31 Long Tom Pass Montane Grassland (Dayaram et al. 2019 & VegMap 2012)
Gs 1 Northern Zululand Mistbelt Grassland (Dayaram et al. 2019)

Distribution This regional biome occurs in the form of three isolated areas along the Great Escarpment. The first straddles Mpumalanga and KwaZulu-Natal Provinces, but with the Gs 1 Northern Zululand Mistbelt Grassland is confined to KZN. The second and the largest are found mainly on Mpumalanga, with outliers in eSwatini, while the third is confined to southern Limpopo (Gm 25 Woodbush Granite Grassland). The average elevation in this regional biome is about 1600 m (spanning 760 and 2200 m), indicating the high-elevation preference of this biome unit (Fig. 9.5).

Vegetation Physiognomy and Composition As a rule, the vegetation of this biome is dominated by short grasslands rich in herbs and low shrubs. The dominant grassy element includes *Alloteropsis semialata* subsp. *eckloniana, Andropogon schirensis, Ctenium concinnum, Cymbopogon caesius, Digitaria tricholaenoides, Diheteropogon amplectens, D. filifolius, Eragrostis chloromelas, E. plana, E. racemosa, Eulalia villosa, Festuca costata, Harpochloa falx, Heteropogon contortus, Hyparrhenia hirta, Koeleria capensis, Loudetia densispica, L. simplex, Microchloa caffra, Monocymbium ceresiiforme, Paspalum scrobiculatum, Rendlia altera, Stipa dregeana, Themeda triandra, Trachypogon spicatus,* and *Tristachya leucothrix*. Among the herbs (and low shrubs), many species of *Helichrysum* (*H. adenocarpum, H. melanacme, H. spiralepis, H. splendidum, H. subluteum, H. wilmsii*, etc.) are prominent, joined by *Acalypha wilmsii, Argyrolobium tuberosum, Asparagus devenishii, Asplenium sandersonii, Cephalanthus natalensis, Cliffortia linearifolia, C. nitidula* subsp. *pilosa, C. repens, Erica drakensbergensis, Erica woodii, Eriosema kraussianum, Gladiolus longicollis, Syncolostemon subvelutinus, Hilliardiella aristata, Hypoxis costata, Lobelia flaccida, Lopholaena coriifolia, Otholobium wilmsii, Phymaspermum acerosum, P. athanasioides, Protea parvula, Psammotropha myriantha, Psoralea latifolia, Rubus ludwigii, Searsia discolor, Selago atherstonei, Sopubia cana*, and *Xysmalobium acerateoides*. In places, pockets of tall scrub also occur, with *Buddleja salviifolia, Leucosidea sericea*, and *Protea roupelliae* subsp. *roupelliae* being prominent.

Climate The rainfall is typically concentrated in the summer months; in the southern regions of this biome, it falls in the mid-summer, while in the central and northern regions, it is concentrated in the early-summer months. The MAP span (per vegetation units involved) is broad—from 700 to 1500 mm (883 mm in terms of total average). Because of its position on the Great Escarpment, the climate of this regional biome experiences orographic rain as well as an increased frequency of mist. Apart from the Wakkerstroom area, frost is infrequent, although winters are usually cold (and dry). In thermo-bioclimatic terms, this unit straddles cool-temperate and warm-temperate realms (Fig. 9.5).

Topography, Geology and Soils Topographically, this biome is linked to mountainous landscapes characterised by steep slopes and broad mountain ridges at high elevations, transiting into rolling, undulating plains in other higher-lying areas. Geological substrate varies and includes: mudstones, sandstones, and shale of the Madzaringwe, Pietermaritzburg and Volksrust Formations (Karoo Supergroup) intruded by voluminous Jurassic dolerite dykes and sills (Gm 14 Wakkerstroom Montane Grassland and Gs 1 Northern Zululand Mistbelt Grassland), Barberton Supergroup schists, gneiss, felspathic quartzites and various lavas of the Figtree, Moodies, and Onverwacht Formations (Gm 17 Barberton Montane Grassland), Archaean granite, gneiss and greenstone basement: Turfloop granite (Randian) and relicts of Goudplaats gneiss (Swazian Erathem) and occasional dolerite dykes or sills, and quartz veins (Gm 25 Woodbush Granite Grassland), the Pretoria Group, with the Dullstroom, Steenkampsberg, Lakenvlei, Vermont, Magaliesberg, Silverton, Strubenkop, Daspoort, Hekpoort, and Timeball Hill Formations (Gm 30 Steenkampsberg Montane Grassland and Gm 31 Long Tom Pass Montane Grassland). Depending

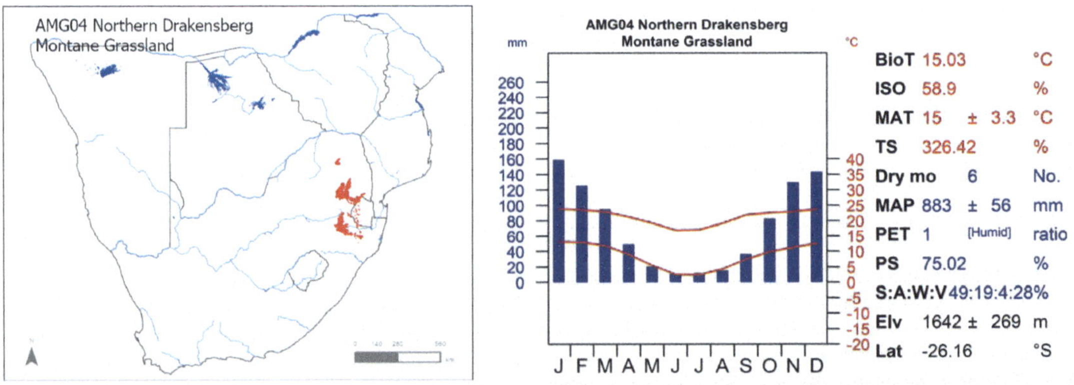

Fig. 9.5 The distribution area and the climatic diagram of the regional ecotonal biome AMG04 Northern Drakensberg Montane Grassland

on geology, the soil is mainly shallow to deep, well-drained, sandy to clay loams, either dystrophic and/or mesotrophic.

Sources Acocks (1953, 1988), Codd (1968), Deal et al. (1989a, b), Bloem et al. (1993), Scott-Shaw et al. (1996), Eckhardt et al. (1997), Camp (1999a), Scott-Shaw (1999), Stalmans et al. (1999), van Wyk and Smith (2001), Lötter et al. (2002), Mucina et al. (2006a), Lötter and Burrows (2012a, b), Kremer-Köhne et al. (2022).

AMG05 Southern Drakensberg Montane Grassland

Gd 4 Southern Drakensberg Highland Grassland (Dayaram et al. 2019)
Gd 5 Northern Drakensberg Highland Grassland (Dayaram et al. 2019)

Distribution The region covered by this regional biome forms a lengthy, relatively narrow belt of mountain slopes that follows the contour of the steeper slopes of the lower-elevation Drakensberg from the Clarens area of the eastern Free State in the north, extending into KZN along the Little Berg, turning sharply south at Giants Castle, through the lower reaches of the Drakensberg in the southern KZN, to the escarpment slopes near Maclear, Elliot, and Dordrecht in the Eastern Cape, where it widens into an intricate network of valley slopes, including those of the broad valley of the Kwaai River near Barkly East. The elevation ranges from about 1450 to 2050 m (Fig. 9.6).

Vegetation Physiognomy and Composition These are relatively dense tussock grasslands, rich in forbs and having a shrub component. Widely scattered small trees of *Protea caffra* and occasionally *P. roupelliae* occur on the slopes, whereas occasional trees of *Cussonia paniculata* and *Greyia sutherlandii* can be prominent near sandstone cliffs and other rocky areas. Important graminoids include *Alloteropsis semialata* subsp. *eckloniana, Andropogon appendiculatus, Aristida junciformis, Catalepis gracilis, Cenchrus sphacelatus, Diheteropogon filifolius, Elionurus muticus, Eragrostis caesia, E. chloromelas, E. planiculmis, E. racemosa, Festuca caprina, Harpochloa falx, Heteropogon contortus, Hyparrhenia hirta, Juncus exsertus, Koeleria capensis, Microchloa caffra, Monocymbium ceresiiforme, Poa binata, Rendlia altera, Setaria sphacelata, Themeda triandra, Trachypogon spicatus, Trisetopsis imberbis,* and *Tristachya leucothrix.* Important forbs include *Acalypha depressinervia, A. punctata, Disa versicolor, Eriosema kraussianum, Gladiolus crassifolius, Helichrysum adenocarpum, H. allioides, H. auronitens, H. glomeratum, H. oreophilum, H. rugulosum, Lobelia erinus, L. flaccida, L. vanreenensis, Orthochilus foliosus, Pteridium aquilinum, Salvia stenophylla, Satyrium longicauda, Selago galpinii, Senecio erubescens,* and *S. latifolius.* Important low shrubs and subshrubs include *Anthospermum rigidum* subsp. *pumilum, Chrysocoma ciliata, Erica caffrorum, E. drakensbergensis, Euryops candollei, Felicia filifolia, F. muricata, Helichrysum asperum, H. hypoleucum, H. splendidum, Lasiosiphon kraussianus, Passerina montana, Pentzia cooperi, Protea simplex, P. subvestita,* and *Senecio burchellii.* In the northern areas, there is generally no sharp border between this regional biome and the grasslands of the Drakensberg foothills at lower altitudes, nor at higher elevations between grasslands on the basalt-capped terrace of the Little Berg.

Climate This is a humid area with a MAP of 747 mm; the water supply is strongly seasonal, with rain concentrated in summer, mainly through intense convective thunderstorms. Winters are dry, although cyclonic rain can sometimes occur due to cold fronts from the southwest in winter. Whereas winter frost is frequent, snowfalls are occasional in the north but more frequent in the south at higher elevations. MAT is 13.5 °C. Around the highest elevations of this regional biome, frost action plays a significant part in forming terracettes (horizontally arranged crescentic scars in the soil surface). Temperatures are mild to warm in summer with consistently freezing winter nights, but winter day temperatures vary from mild to very cold depending on prevailing frontal conditions (Fig. 9.6).

Topography, Geology and Soils Steeply sloping mountainous areas with some prominent sandstone cliffs, spurs and buttresses dominate the landscapes of this regional biome. The geological substrate is predominantly built of sandstones of the Clarens Formation and sandstones, siltstones, and mudstones of the Elliot Formation (both formations of the Stormberg Group, Karoo Supergroup), with basaltic lava flows of the Drakensberg Group embedded. In the north, an extensive shelf forms the Little Berg below the Great Escarpment and is capped by basalt, which weathers away to reveal Clarens Sandstone in places, most notably at the edge of the shelf, where it forms some dramatic cliffs. Soils are leached, acidic, fine-grained, and well-drained and vary from deep to shallow to very shallow on sandstone pavements. Humic and oxidic (ferrallitic) soils are more common at higher elevations.

Sources Acocks (1953, 1988), Killick (1963), Edwards (1967), Granger (1976), Everson (1985), Hilliard and Burtt (1987), Acocks (1988), Adcock (1990), Du Preez and Bredenkamp (1991), Hill (1996), Bester (1998), Camp (1999b), Hoare and Bredenkamp (2001), Mucina et al. (2006a), Gordijn et al. (2018), Gordijn and O'Connor (2023), Nsibande and Nkosi (2023).

AMG06 Dry Sub-Escarpment Grassland

Gd 1 Amathole Montane Grassland (Dayaram et al. 2019)
Gs 14 Mthatha Moist Grassland (Dayaram et al. 2019)
Gs 15 Tsomo Grassland (Dayaram et al. 2019)
Gs 16 Queenstown Thornveld (Dayaram et al. 2019)
Gs 17 Tarkastad Montane Shrubland (Dayaram et al. 2019)
Gs 18 Bedford Dry Grassland (Dayaram et al. 2019)
Gs 19 Dry Coast Hinterland Grassland (Dayaram et al. 2019)

Distribution Most of the area occupied by this biome is found in the north-western and western-central regions of Eastern Cape Province, where it is the main grassland unit of the lower elevations of the Amathole, Winterberg, and Kologha mountain ranges as well as in the mountains just north of Somerset East (Bosberg). It is also found on broken veld between Stutterheim and Komga, plains between Mthatha and Butterworth, and in the vicinity of Queenstown, Tarkastad, and Sterkstroom. Marginally it reaches into the Northern Cape Province and KZN. The geographic core of the biome lies at mid-elevations between 750 and 1600 m, reaching extremes of 450 mm and 1780 m locally (Fig. 9.7).

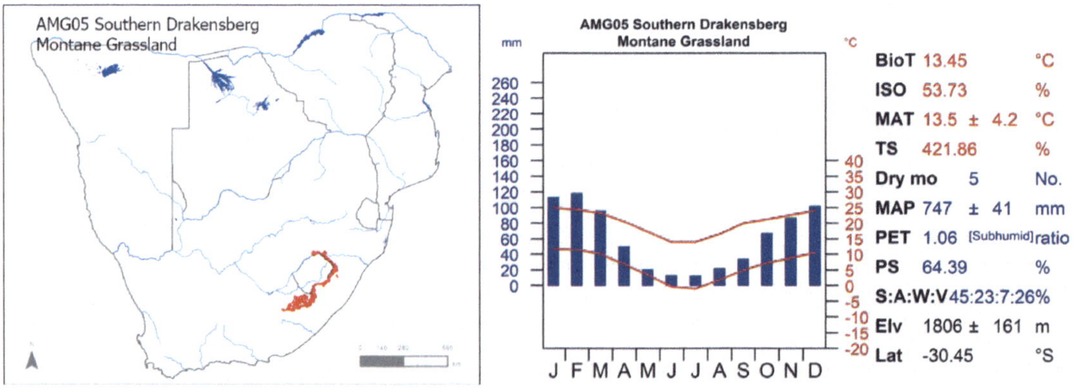

Fig. 9.6 The distribution area and the climatic diagram of the regional ecotonal biome AMG05 Southern Drakensberg Montane Grassland

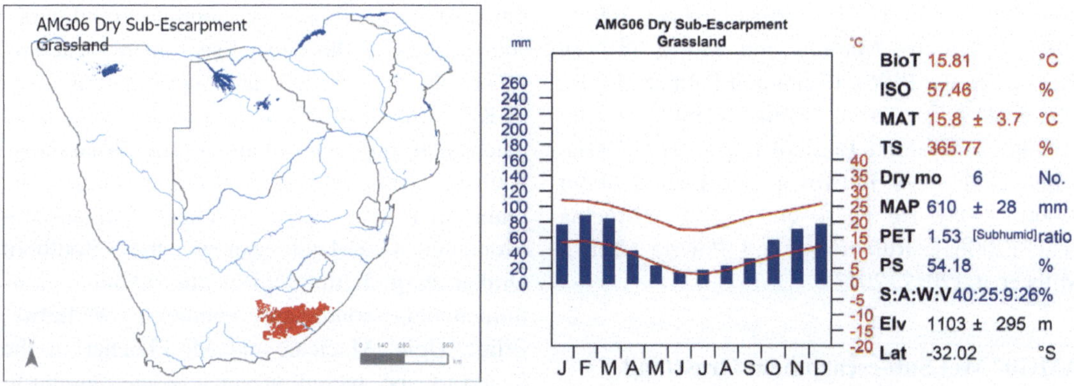

Fig. 9.7 The distribution area and the climatic diagram of the regional ecotonal biome AMG06 Dry Sub-Escarpment Grassland

Vegetation Physiognomy and Composition The grasslands are usually poor in species, giving this biome its physiognomic identity. Locally, they dominated by a variety of grasses, including *Alloteropsis semialata, Aristida congesta, Elionurus muticus, Eragrostis chloromelas, E. curvula, Eustachys paspaloides, Cymbopogon pospischilii, Cynodon dactylon, Digitaria argyrograpta, Heteropogon contortus, Microchloa caffra, Paspalum dilatatum, Sporobolus africanus, Sporobolus africanus, Themeda triandra, Tragus berteronianus,* and *Tristachya leucothrix*. Asteraceae and Fabaceae are prominent among many species of forbs. In thornvelds (such as found in the Gs 16 Queenstown Thornveld and Gs 18 Bedford Dry Grassland vegetation units), besides the prominent *Vachellia karroo*, low shrubs such as *Asparagus laricinus, Atriplex semibaccata* var. *appendiculata, Felicia filifolia* subsp. *filifolia, F. muricata, Helichrysum asperum* var. *albidulum,* and *H. dregeanum* are common. *Euryops floribundus* dominates, especially in overgrazed pastures. Patches of Gs 17 Tarkastad Montane Shrubland are also part of this biome. The vegetation physiognomy is a typical low, semi-open, mixed shrubland with 'white' grasses and dwarf shrubs. Here, *Aloe ferox, Diospyros austro-africana,* and *Euryops annae* are prominent, while the grassy understorey is composed of the same elements as in pure grassland vegetation types of this biome. The Gs 19 Dry Coast Hinterland Grassland (formerly part of the Ngongoni Veld) is perhaps the most nutrient-poor unit, dominated by unpalatable Ngongoni grass (*Aristida junciformis*), associated with low species diversity. In good condition dominated by *Themeda triandra* and *Tristachya leucothrix*. Being the lowest elevation unit of this biome, in valleys grades into SVs 3 KwaZulu-Natal Hinterland Thornveld and SVs 7 Bisho Thornveld.

Climate Only the westernmost Amathole and Bedford regions have a bimodal rainfall pattern with spring and late summer peaks. The rest of the unit experiences summer rainfall. Precipitation varies between 310 mm (Bedford grassland) and about 1000 mm (Amathole and Mthatha grasslands), with 610 mm being the average for the entire area occupied by the biome. Summers are warm to hot but winters are cold, with frequent frost in some western parts (Fig. 9.7).

Topography, Geology and Soils The topography of landscapes of this biome is characterised by low mountain ranges and moderately undulating plains and hills. Geology is dominated by sediments (mudstones and sandstones) of the Karoo Supergroup (Beaufort, Dwyka) and occasional intrusions of Jurassic dolerites forming numerous dykes and sills. The soils are moderately deep (loamy or clayey-loamy) but also shallow and skeletal in places.

Sources Acocks (1953, 1988), Palmer (1988, 1989, 1991a, b), Martens and Morris (1994), Martens et al. (1996), Rubin and Palmer (1996), Hoare (1997, 2002), Smits (1997), Camp (1999a), Hoare and Bredenkamp (1999, 2001), Smits et al. (1999), Brown and Bezuidenhout (2000, 2005), De Klerk et al. (2003), Mucina et al. (2006a), Rutherford and Powrie (2011), Mills et al. (2012, 2013).

AMG07 Wet Sub-escarpment Grassland

Gd 2 Amathole Mistbelt Grassland (Dayaram et al. 2019)
Gs 3 Low Escarpment Moist Grassland (Dayaram et al. 2019)
Gs 4 Northern KwaZulu-Natal Moist Grassland (Dayaram et al. 2019)
Gs 6 KwaZulu-Natal Highland Thornveld (Dayaram et al. 2019)
Gs 7 Income Sandy Grassland (Dayaram et al. 2019)
Gs 8 Mooi River Highland Grassland (Dayaram et al. 2019)
Gs 9 Midlands Mistbelt Grassland (Dayaram et al. 2019)
Gs 10 Drakensberg Foothill Moist Grassland (Dayaram et al. 2019)
Gs 11 Southern KwaZulu-Natal Moist Grassland (Dayaram et al. 2019)
Gs 12 East Griqualand Grassland (Dayaram et al. 2019)
Gs 13 Mabela Sandy Grassland (Dayaram et al. 2019)
Gs 20 Moist Coast Hinterland Grassland (Dayaram et al. 2019)

Distribution This is one of the largest regional biomes within the confined of South Africa, and this makes it a subject of future detailed studies, which may lead to its split into several units. As a rule, it occupies southeast (and partly south) areas of the Great Escarpment, reaching especially in Transkei and southwestern areas of KZN. South of this area, it occurs as isolated patches at high elevations of the Winterberg, Katberg, Amathole, and Kologha ranges. Its main area abuts REB Southern Drakensberg Montane Grassland below the Southern Drakensberg. It thus covers the inland regions immediately south and southeast of Indwe, Elliot, Ugie, Maclear, and Mt Fletcher in the Eastern Cape, including much of the Qumbu to Mt Ayliff areas. It extends into KZN, including the Matatiele, Kokstad, Underberg and Ixopo areas. It continues seamlessly into the north-central of KZN, where it is found in the large region from Ladysmith, Winterton, Estcourt, and Colenso, between Mooi River and Greytown, between Pomeroy and Babanago, and Vryheid, Paulpietersburg, Louwsburg, Newcastle, and Dundee—the elevations within the area of span 760 and 2080 m (Fig. 9.8).

Vegetation Physiognomy and Composition The vegetation of this regional biome is dominated by short or dense-tussock grasslands, sometimes with a dwarf-shrubby component and dwarf shrubland on exposed rocky areas. Savannoid woodlands with scattered *Vachellia sieberiana* var. *woodii* and small pockets, also with *Vachellia karroo* complex, are also found (especially in the northwestern KZN). The dominating graminoids include *Alloteropsis*

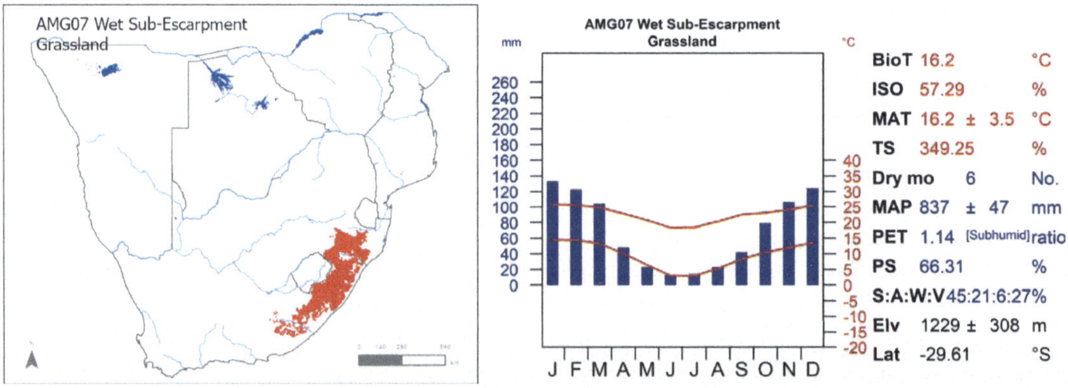

Fig. 9.8 The distribution area and the climatic diagram of the regional ecotonal biome AMG07 Wet Sub-Escarpment Grassland

semialata subsp. *ecklonianа, Andropogon appendiculatus, A. schirensis, Aristida junciformis* subsp. *galpinii* and subsp. *junciformis, Brachiaria serrata, Catalepis gracilis, Cynodon dactylon, C. incompletus, Cyperus obtusiflorus, Digitaria monodactyla, D. tricholaenoides, Diheteropogon filifolius, Elionurus muticus,* many *Eragrostis* species (*E. caesia, E. capensis, E. chloromelas, E. gummiflua, E. planiculmis, E. racemosa), Festuca caprina, F. costata, Helictotrichon turgidulum, Heteropogon contortus, Karroochloa purpurea, Loudetia simplex, Microchloa caffra, Monocymbium ceresiiforme, Panicum ecklonii, P. gilvum, Paspalum scrobiculatum, Pennisetum sphacelatum, Rendlia altera, Sporobolus africanus, Tenaxia disticha, Themeda triandra, Trachypogon spicatus,* and *Tristachya leucothrix.* These grasslands have a high diversity of asteraceous herbs of genera such as *Helichrysum (H. appendiculatum, H. aureum, H. asperum, H. simillimum, H. odoratissimum, H. nudifolium, H. anomalum*) and *Senecio* (e.g., *Senecio coronatus*).

Climate The term 'wet' in the name of this regional biome relates to the relatively high mean annual rainfall that characterises these grasslands, but also by the influence of orographic precipitation contributed by ascending, water-laden air masses towards higher elevations of the sub-escarpment mountain ranges, creating a mistbelt in some (e.g., KwaZulu-Natal Midlands or on Amathole Mts). The westernmost reaches of the distribution area of this regional biome (occupied by the Gd 2 Amathole Mistbelt Grassland) have a Bimodal spring–autumn rainfall pattern, although rain can fall at any time of the year, and it increases from west to east as well as with increasing elevation. The rest of the biome experiences summer rainfall because of the influence of the cyclonic Easterly flows. The (average) MAP of 837 mm spans 600 mm and 950 mm but can reach as much as 1300 mm (Midlands). Thermically, this regional biome straddles the cool-temperate and warm-temperate bioclimatic realms. Frost is frequent but of a moderate nature (Fig. 9.8).

Topography, Geology and Soils The topography of this extensive regional biome is characterised by mid-elevation mountain ranges with moderately undulating slopes below peaks, often strewn by ridges of rock or boulder outcrops. Part of the area (especially in the northwestern KZN) occurs on gently sloping valley bottoms. Geology is dominated by sedimentary rocks of the Karoo Sequence, in particular by sandstones, siltstones, and mudstones of the Dwyka, Ecca, Beaufort, and Stormberg Groups. In places, basaltic lava flows of the Drakensberg Group interrupt the sedimentary geological landscapes. Where sandstones dominate, the soils are deep, sandy and fine-grained. The soil typology varies considerably in this area. Red and black heavy soils are derived from dolerites and basalts.

Sources Bews (1917), Story (1952), Acocks (1953, 1988), Killick (1963, 1978b), Edwards (1967), Colvin (1984), Smit (1992), Smit et al. (1992, 1993a, 1993b, 1995a, 1995b, 1995c), Hurt et al. (1993), Fynn and O'Connor (2005), Fynn et al. (2005), Adams (1996), Eckhardt et al. (1996a, b), Lütge et al. (1996), Morris and Tainton (1996), Scott-Shaw et al. (1996), Hoare (1997), Roos and Allsopp (1997), Eckhardt (1998), Robesson (1998), Camp (1999a, b, c), Hoare and Bredenkamp (1999), Scott-Shaw (1999), KZN Daea (2003), Mucina et al. (2006a), O'Connor et al. (2011), Mills et al. (2013), Joubert et al. (2017), Morris and Scott-Shaw (2019), Dabengwa et al. (2022), Findlay et al. (2022), Chonco et al. (2023), Gokul et al. (2023), Jewitt et al. (2023), Ndamane et al. (2023).

AMG08 Eastern Highlands Grassland

Ge2 Gorongosa Montane Grassland (Lötter et al. 2023)
Ge3 Manica Montane Grassland (Lötter et al. 2023)

Distribution The Eastern Highlands Grasslands are found at high elevations (summit plateaus and sourvelds), usually above 1300 m with a mean of 1635 m, located along the border area between Zimbabwe and Mozambique, from the just north of the Haroni River in

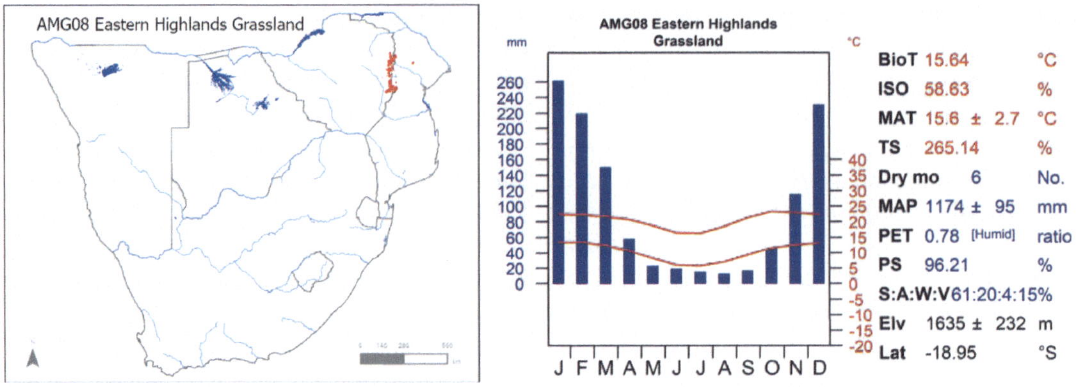

Fig. 9.9 The distribution area and the climatic diagram of the regional ecotonal biome AMG08 Eastern Highlands Grassland

the south, northwards through Rotanda and Manica to the Choa and Nyanga Mountains (Inyanga Highlands) in the north, with an eastern outlier on the summit of Mt Gorongosa in Mozambique (Fig. 9.9).

Vegetation Physiognomy and Composition The regional biome is dominated by grasslands on mountain summits, gentle to steep slopes, either as large expanses or as glades in a matrix of montane forests. Rocky outcrops often show an increase in shrubs and bush clumps due to the absence of fire. In the grasslands with nutrient-deficient quartzite soils, *Loudetia simplex* is dominant, together with *Elionurus muticus, Melinis nerviglumis, Monocymbium ceresiiforme, Panicum ecklonii, Rhytachne rottboellioides, Sporobolus festivus,* and *Trachypogon spicatus*. On gneiss- and schist-derived soils *Themeda triandra, Tristachya leucothrix,* and *Bulbostylis contexta* prevail. Other important grasses include *Andropogon schirensis, Bewsia biflora, Digitaria maitlandii, Eragrostis racemosa, Eulalia villosa, Ischaemum polystachyum, Panicum ecklonii, Setaria sphacelata,* and *Trichopteryx dregeana*. The diverse herbaceous layer contains many forb species, such as *Aeschynomene rhodesiaca, Aloe inyangensis, Anthospermum vallicola, Asparagus chimanimaniensis, Crotalaria collina, Eriosema psoraleoides, Exacum zombense, Gladiolus crassifolius, Haumaniastrum venosum, Helichrysum buchananii, Hilliardiella aristata, Indigofera hedyantha, Jamesbrittenia carvalhoi, Kniphofia splendida, Lotus wildii, Pachycarpus chirindensis, Pearsonia sessilifolia, Rhynchosia monophylla, Sopubia mannii, Sphenostylis erecta* subsp. *obtusifolia,* and *Tephrosia dasyphylla*. The grasslands support a rich non-graminoid flora of pyrophytic shrublets, forbs and geophytes. The fire-tolerant tree *Protea caffra* subsp. *gazensis* occurs in patches within the grasslands, together with other woody species such as *Erica benguelensis, Hypericum revolutum, Kotschya thymodora,* and tree fern *Alsophila dregei*. The common bracken fern *Pteridium aquilinum* forms extensive colonies.

Climate The climate of this regional biome is typical of a summer rainfall area with very limited rain in winter. MAP averages 1174 mm, the highest of the African Montane Grassland regional biomes within the MBSA. Mist is common and it constitutes an extensive unmeasured addition to precipitation. Frost is very infrequent (Fig. 9.9).

Topography, Geology and Soils The mountainous landscape of this unit is dominated by summits, ridges and adjacent, often steep, slopes. The dominant geology is mica schists and aluminous saccharoidal quartzites; some granites and granitoids are also found. The mountains of Inyanga comprise an upper sill of dolerite and sandstone, with the harder dolerite forming cliffs and ridges. These support the formation of shallow, nutrient-poor, sandy soils on the nutrient-poor lithology and more nutrient-rich and clayey soil on the dolerite.

Sources Pedro and Barbosa (1955), Wild and Barbosa (1967), Stalmans and Beilfuss (2008), Timberlake et al. (2016), Lötter et al. (2023).

AMG09 Eastern Dolomite Grassland

Gm 22 Northern Escarpment Dolomite Grassland (Dayaram et al. 2019)

Gm 26 Wolkberg Dolomite Grassland (Dayaram et al. 2019)

Distribution This pedobiome occurs in two discrete patches recognised by Mucina et al. (2006a) as separate Vegetation units (see the synonymy). In Limpopo Province, it occurs along the summit of the Strydpoort and Wolkberg Mountains, from the Lekgalameetse Nature Reserve in the east to Mogodi in the west. In Mpumalanga Province, it spans a narrow high-elevation belt between Abel Erasmus Pass and Blyde River (Vaalhoek) in the north. It reaches as far south as the vicinity of Kaapsehoop. The elevation span is 1000–1840 m (mean of 1349 m) (Fig. 9.10).

Vegetation Physiognomy and Composition The dominant vegetation is species-rich grasslands (rich in forbs) with occasional scrub pockets also occurring. Of the dominating grasses these deserve mentioning: *Andropogon schirensis, Brachiaria serrata, Cymbopogon caesius, C. nardus, Diheteropogon amplectens, Elionurus muticus, Eragrostis capensis, Heteropogon contortus, Hyparrhenia filipendula, Loudetia simplex, Monocymbium ceresiiforme, Rendlia altera, Setaria sphacelata, Schizachyrium sanguineum, Themeda triandra, Trichopteryx dregeana,* and *Tristachya leucothrix.*

Climate The climate of this regional biome is characterised by summer rainfall, with the annual mean of 937 mm varying from about 550–1000 mm, increasing southwards to around 1400 mm. Winters are dry. Most of this unit occurs in a mistbelt, with a consequential increase in precipitation. In thermo-bioclimatic terms, this is a warm-temperate climate with a low frequency of frost. The mean annual temperature of 16.8 °C is among the highest for all regional biomes of the African Montane Grassland (Fig. 9.10).

Topography, Geology and Soils These high-elevation landscapes, formed by broad mountain ridges, are built by Malmani dolomites of the Chuniespoort Group (Transvaal Supergroup), which overlies the Black Reef Quartzite Formation. Soils usually have a high pH, are rich in calcium and magnesium, and have low phosphorus status. Deep Hutton and Griffin soil forms are common (Mucina et al. 2006a).

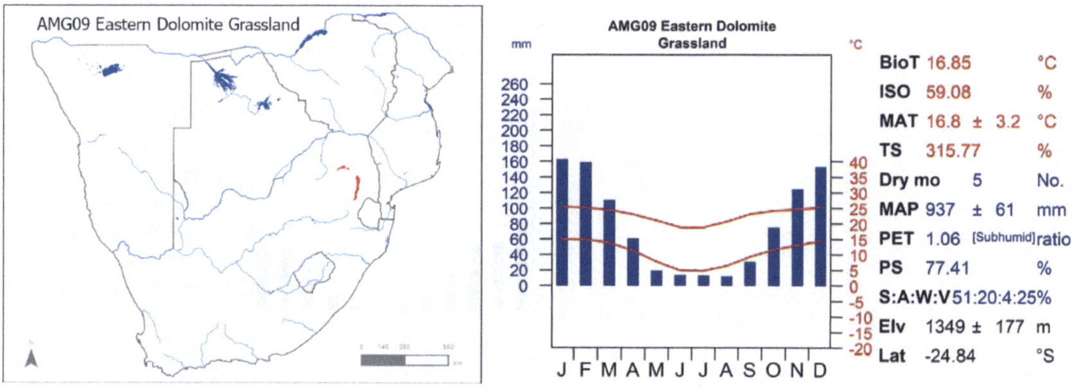

Fig. 9.10 The distribution area and the climatic diagram of the regional ecotonal biome AMG09 Eastern Dolomite Grassland

Sources Acocks (1953, 1988), Stalmans (1990), Matthews (1991), Matthews et al. (1991, 1992a, b, 1993, 1994), Stalmans et al. (1997), Siebert (2001), van Wyk and Smith (2001), Mucina et al. (2006a).

AMG10 Western Dolomite Grassland

Gh 12 Vaal Reefs Dolomite Sinkhole Woodland (Dayaram et al. 2019)
Gh 15 Carletonville Dolomite Grassland (Dayaram et al. 2019)

Distribution This regional pedobiome is limited to northeastern North West Province (especially around Potchefstroom, Ventersdorp, and Carletonville) and western Gauteng. Apert from a marginal occurrence in the Free State, the Vaal River forms the southern distribution limit of this biome. The mean elevation is 1504 m, spanning 1280–1620 m (Fig. 9.11).

Vegetation Physiognomy and Composition The vegetation is a grassland–woodland vegetation complex, with woodland clumps found typically around sinkholes, especially in places of dolomite outcrops. Here, shrubs and small trees of *Senegalia caffra, Vachellia erioloba, V. karroo, V. robusta* subsp. *clavigera, Asparagus suaveolens, Celtis africana, Diospyros lycioides* subsp. *lycioides, Ehretia rigida, Grewia flava, Gymnosporia heterophylla, Pavonia burchellii, Searsia lancea,* and *Sida dregei* are prominent. In the grassy undergrowth, *Aristida congesta, Brachiaria serrata, Cynodon dactylon, Digitaria eriantha, D. tricholaenoides, Diheteropogon amplectens, Eragrostis biflora, E. chloromelas, E. curvula, E. racemosa, Heteropogon contortus, Loudetia simplex, Schizachyrium sanguineum, Setaria sphacelata,* and *Themeda triandra* dominate.

Climate This is a warm-temperate, summer-rainfall region with MAP of 637 mm varying from less than 550 mm in the extreme west to around 700 mm in parts of the east. Its large temperature amplitude confirms its continental nature. Summer is hot; severe and frequent frost occurs in the dry winter (Fig. 9.11).

Topography, Geology and Soils The landscapes supporting this small-sized pedobiome are typically slightly undulating, dissected by prominent rocky chert ridges and incorporating sinkhole depressions. The pedobiome status of the biome is characterised by the almost exclusive domination of the dolomites of the Malmani Subgroup (Chuniespoort Group, Transvaal Supergroup). The soils are shallow (50–150 mm) and rocky, of Mispah, Glenrosa, and Hutton forms.

Sources Louw (1951), Acocks (1953, 1988), Morris (1973, 1976), Coetzee (1974), Coetzee and Werger (1975), van Wyk (1983), van Wyk and Bredenkamp (1986), Bezuidenhout and Bredenkamp (1990), Scoggins and Theron

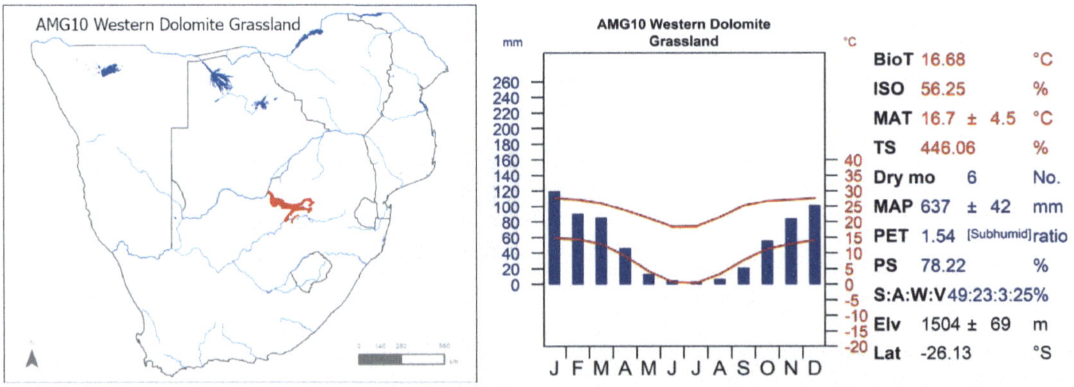

Fig. 9.11 The distribution area and the climatic diagram of the regional ecotonal biome AMG10 Western Dolomite Grassland

(1990), Bezuidenhout (1993), Bezuidenhout et al. (1994a, b, c, d, e, f), Bredenkamp et al. (1994), Grobler (2000), Hartmann (2001), Siebert and Siebert (2005), Grobler et al. (2006), Mucina et al. (2006a), Furniss (2008), Mochesane (2018), Pule et al. (2018, 2023a, b).

AMG11 Northern Quartzite Sourveld

Gm 23 Northern Escarpment Quartzite Sourveld (Dayaram et al. 2019)
Gm 24 Northern Escarpment Afromontane Fynbos (Dayaram et al. 2019)
Gm 27 Strydpoort Summit Sourveld (Dayaram et al. 2019)
Gm 28 Soutpansberg Summit Sourveld (Dayaram et al. 2019)
Gm 29 Waterberg-Magaliesberg Summit Sourveld (Dayaram et al. 2019)

Distribution The main distribution area forms an arch-like archipelago occurring along the high-altitude crests of the Northern Escarpment, from Haenertsburg in the north south-eastwards, then bending southwards past Blyde River Canyon, Graskop and as far south as the vicinity of Kaapsehoop (in Mpumalanga and Limpopo Provinces). The biome occurs in a series of isolated patches in high-lying areas of the Strydpoortberge, from Bewaarkloof in the east as far as the mountains just above Mokopane (Potgietersrus) in the west. To the north of this area, this regional biome is found in Soutpansberg and Blouberg, while to the south of this area, the high-elevated ridges and plateaus of the Magaliesberg and Waterberg (Gauteng and North West Provinces, respectively) form another geographic outlier. The mean elevation of 1443 m ranges between 1000 m (Mpumalanga Escarpment) and nearly 2100 m (Blouberg) (Fig. 9.12).

Vegetation Physiognomy and Composition The overall vegetation physiognomy of this regional biome is characterised by short, wiry, and often patchy (due to surface rocks) grasslands, dominated by C4 grasses such as *Aristida junciformis* subsp. *galpinii, Aristida transvaalensis, Diheteropogon amplectens, Eragrostis nindensis, Loudetia simplex, Melinis nervliglumis, Monocymbium ceresiiforme, Schizachyrium sanguineum, Panicum ecklonii, Trachypogon spicatus*, and *Tristachya leucothrix*. In places, the grasses are replaced by restios (*Restio schoenoides*) and sedges (*Bulbostylis burchellii, Scleria transvaalensis*). Sclerophyllous and succulent small trees and shrubs are a common sight (*Aloe arborescens, Combretum molle, Englerophytum magalismontanum, Lopholaena coriifolia, Protea caffra* subsp. *caffra, P. roupelliae subsp. roupelliae,* and *Senegalia caffra*). The extent of these bush clumps within the grasslands depends mainly on the extent of rock outcrops, rock boulders and the local soil depth. Due to frequent mist, the surface of rocks and woody vegetation support rich epilithic, epiphytic lichen, and bryophyte flora. In places, montane microphyllous fynbos scrub (*Anthospermum hispidulum, Cliffortia nitidula, C. serpyllifolia, Erica natalitia* var. *natalitia, Hypericum revolutum,* and *Passerina montana*) is formed.

Climate The climate of this archipelago of high-elevation summit ridge islands is characterised by summer rainfall with a MAP of 897 mm ranging from 530 mm in Gm 27 Strydpoort Summit Sourveld to more than 1400 m (Gm 24 Northern Escarpment Afromontane Fynbos). Some localities, such as in Gm 28 Soutpansberg Summit Sourveld, are suspected to have precipitation as high as 2000 mm, enhanced by regional orographic rainfall and frequent occurrence of mist. The temperature regime is indicative of a transition between the cool temperate and warm temperate bioclimates. Frost is infrequent (Fig. 9.12).

Topography, Geology and Soils Northern Quartzite Sourveld occurs in isolated patches of mainly quartzite ridges at high elevations. The terrain is very rocky and the landscape is characterised by rugged features such as steep east-facing cliffs. Fires are infrequent because of low-standing biomass (and poor accumulation of necromass). The predominant rock types are quartzite of the Black Reef Formation and the Wolkberg Group,

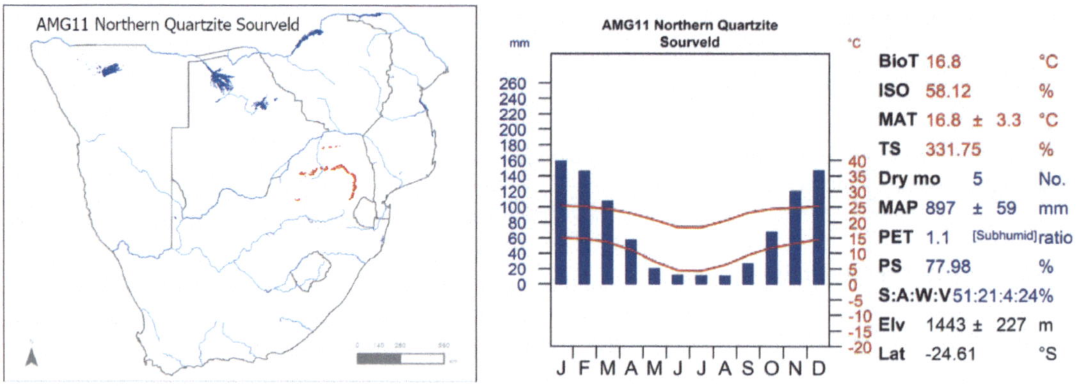

Fig. 9.12 The distribution area and the climatic diagram of the regional ecotonal biome AMG11 Northern Quartzite Sourveld

with occurrences on the westerly Timeball Hill Formation and other quarzitic formations of the Pretoria Group (Transvaal Supergroup). The acidic sandy soil in the region is derived from sandstone, quartzite, and shale, which are shallow and nutrient-poor due to eluviation. Therefore, this regional biome should be considered a peinobiome (see also Mucina 2018). In the Gm 27 Strydpoort Summit Sourveld, the rugged summit crests and adjacent steep rocky slopes consist of sandstone, quartzite, and shale of Mokolian Erathem (Soutpansberg Group, Wyllie's Poort Formation), giving rise to highly shallow, coarse-sand Glenrosa and Mispah soils. In Gm 29 Waterberg-Magaliesberg Summit Sourveld, these nutrient-poor soils are formed over sedimentary sandstone, quartzite, conglomerate, or shale of the Kransberg Subgroup (Waterberg Group, Mokolian Erathem).

Sources Acocks (1953, 1988), van der Schijff (1963), Smith (1972), Coetzee (1974, 1975), Coetzee and Werger (1975), Scholes (1979), van der Meulen (1979a, b), Westfall (1981), Deall (1985), Deal et al. (1989a, 1989b), Stalmans (1990), Matthews (1991), Matthews et al. (1991, 1992a, 1993, 1994), Stalmans and De Klerk (1991, 1992), Stalmans and Mentis (1993), Burgoyne et al. (2000), van Wyk and Smith (2001), Emery et al. (2002), Hahn (2002),van Staden (2002), Brown et al. (2005), van Staden and Bredenkamp (2005), Mucina et al. (2006a), Mills et al. (2012, 2013), Mucina (2018).

AMG12 Mesic Montane Shrubland

Gm 2 Senqu Montane Shrubland (Dayaram et al. 2019)
Gm 5 Basotho Montane Shrubland (Dayaram et al. 2019)
Gm 7 Northern Free State Shrubland (Dayaram et al. 2019)
Gs 5 Northern KwaZulu-Natal Shrubland (Mucina and Rutherford 2006)

Distribution The main distributional feature of this regional biome is its archipelago-like occurrence of solitary koppies, mainly concentrated along the foothills of Drakensberg and Maloti, mainly in the Free State and eastern Lesotho, and to a lesser extent also in the Eastern Cape, KZN, and Mpumalanga. More extensive patches are found on steep slopes of the Senqu River and its tributaries in the south and southwest of Lesotho. The mean elevation of 1736 m ranges from 1100 m (in northern KZN) to 1940 m (in Lesotho) (Fig. 9.13).

Vegetation Physiognomy and Composition The vegetation is characterised by a two-phase appearance—sparse to dense grassland overtopped by tall shrubland (locally forming thickets). The woody phase is dominated by evergreen shrubs such as *Buddleja salviifolia, Diospyros austro-africana, Euclea crispa* subsp. *crispa, Gymnopentzia buxifolia, Kiggelaria africana, Leucosidea sericea, Olea europaea* subsp. *africana, Rhamnus prinoides, Scutia myrtina,*

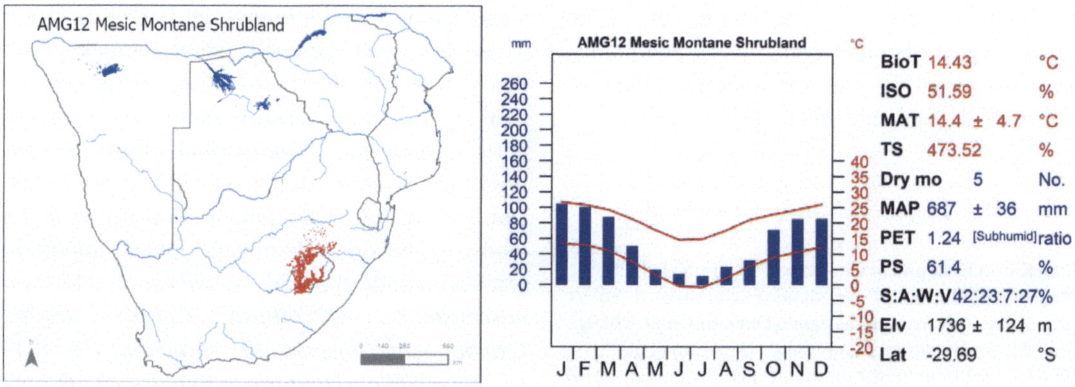

Fig. 9.13 The distribution area and the climatic diagram of the regional ecotonal biome AMG12 Mesic Montane Shrubland

Searsia burchellii, and *S. erosa*. In the grassy undergrowth C4 grasses dominate: *Andropogon appendiculatus, A. schirensis, Aristida congesta, Cymbopogon pospischilii, Cynodon dactylon, Elionurus muticus, Eragrostis chloromelas, E. plana, E. racemosa, Heteropogon contortus, Hyparrhenia hirta, Microchloa caffra, Setaria sphacelata, Themeda triandra,* and *Tristachya leucothrix.*

Climate The biome experiences summer rainfall, with a MAP of 687 mm, spanning approx. 600 mm to 800 mm. Due to high elevation, the thermic climate tends toward cool-temperate. Frost is frequent (Fig. 9.13).

Topography, Geology and Soils The dominant topographical features in this regional biome are steep, often boulder-strewn talus slopes and deep gullies. These occur often on prominent landscape inselberg—tafelbergs, koppies, and butts. Karoo Supergroup and sedimentary rocks of the Clarens, Elliot, and Molteno Formations are prominent geological substrates. In places, the dolerite dykes cut through the thick sandstones, resulting in the formation of sheltered ravines. In northern KZN, Madzaringwe Formation shale and sandstones are found in the east, and Adelaide Subgroup and Volksrust Formation mud rocks occur in the west. Notably, the intrusive dolerite dykes and sills give rise to heavy clayey soils.

Sources Müller DB (1986), Du Preez (1991), du Preez and Bredenkamp (1991), Fuls (1993), Fuls et al. (1993a, b), Smit et al. (1995b), Eckhardt et al. (1997), Eckhardt (1998), Malan (1998), Malan et al. (1998, 1999), Robesson (1998), Boucher and Tlale (1999a, b), Müller (2002), Mucina et al. (2006a), Brand et al. (2011), Mills et al. (2012, 2013), Daemane et al. (2022).

9.2 Continental Ecotonal Biome Dry Highveld Grassland (DHG)

ZE Subtropical–Warm-Temperate Dry Ecotone E2(sava)–[E2(savm)–T3]–S2

This ecotonal biome is a reincarnation of the Dry Highveld Grassland Bioregion (see Mucina and Rutherford 2006; Mucina et al. 2006a). Its complicated zonoecotonal formula indicates that it is a transitional entity involving Arid Savanna (SAVa) with which it shares a border in the west, mesic African Montane Grassland (E2(savm)/T3) bordering in the east, and finally, inland semidesert (Nama-Karoo, Zonobiome S2) with which shares the border in the south. It is a warm-temperate/subtropical grassland dominated by C4 grasses, among which the dominant subfamily is Chloridoideae (*Eragrostis*). The more mesic AMG African Montane Grassland is dominated by the subfamily Panicoideae.

Three regional biomes within the DHG biome were distinguished on a pedological basis—considering the dominating sandy vs clayey vs skeletal substrates (Table 9.1):

DHG1 Dry Highveld Sandy Grassland

Gh 1 Karoo Escarpment Grassland (Dayaram et al. 2019)
Gh 2 Aliwal North Dry Grassland (Dayaram et al. 2019)
Gh 3 Xhariep Karroid Grassland (Dayaram et al. 2019)
Gh 5 Bloemfontein Dry Grassland (Dayaram et al. 2019)
Gh 6 Central Free State Grassland (Dayaram et al. 2019)
Gh 10 Vaal-Vet Sandy Grassland (Dayaram et al. 2019)
Gh 11 Vredefort Dome Granite Grassland (Dayaram et al. 2019)
Gh 14 Western Highveld Sandy Grassland (Dayaram et al. 2019)

Distribution It comprises a broad belt stretching from the north near Lichtenburg and Ventersdorp in the North West Province, extending southwards through the central Free State Province, including the Kroonstad, Welkom and Bloemfontein areas, continuing as far as the vicinity of the Orange River in the southern Free State, as well as south of the river in the Aliwal North region in the Eastern Cape. Further south, it is limited to isolated higher-lying parts and disjunct mountains, notably on the Great Escarpment, including the summit area of the Nuweveld Mountains near Beaufort West, high mountains around Graaff Reinet and the north-facing slopes of the Winterberge near Tarkastad (Fig. 9.14).

Vegetation Physiognomy and Composition These are open, generally short tussock grasslands with an increased low shrub component in the more arid western and southern parts. *Themeda triandra* and *Aristida congesta* are widely dominant, although on the southern Great Escarpment *Tenaxia disticha* tends to become dominant. Other important grasses include *Anthephora pubescens, Aristida adscensionis, A. diffusa, Chloris virgata, Cymbopogon pospischilii, Cynodon dactylon, C. incompletus, Digitaria argyrograpta, D. eriantha, Elionurus muticus, Eragrostis curvula, E. cylindriflora, E, lehmanniana, E. obtusa, E. plana, E. superba, Fingerhuthia africana, Heteropogon contortus, Panicum coloratum, P. stapfianum, Setaria sphacelata, Tragus berteronianus, T. koelerioides,* and *Triraphis andropogonoides.* Important herbs include *Berkheya onopordifolia, Commelina africana, Conyza podocephala, Gazania krebsiana, Geigeria ornativa, Helichrysum lineare, Hermannia coccocarpa, H. depressa, Indigofera alternans, Jamesbrittenia aurantiaca,* and *Stachys spathulata.* Common geophytes include *Bulbine narcissifolia, Ledebouria marginata, Moraea pallida, Oxalis depressa,* and *Trachyandra saltii.* In the southwest drier grasslands, low shrubs common in the adjacent RB Upper Karoo include *Chrysocoma ciliata, Eriocephalus ericoides, E. spinescens, Felicia filifolia* subsp. *filifolia, F. muricata, Helichrysum lucilioides, Lycium cinereum, Nenax microphylla, Pentzia*

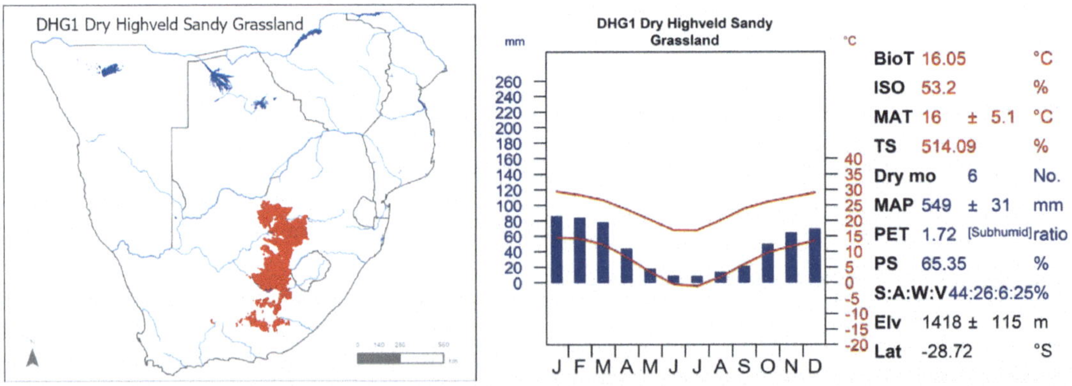

Fig. 9.14 The distribution area and the climatic diagram of the regional ecotonal biome DHG1 Dry Highveld Sandy Grassland

globosa, and *P. incana.* The instability of the boundary between the dry grassland and Upper Karoo has been examined in several studies, which have indicated that heavy sustained grazing or prolonged drought periods can (or already have) transform(ed) parts of these marginal grassland areas to karoo vegetation.

Climate The MAP is 549 mm but with a substantial rainfall gradient from a MAP of around 400 mm in the southwest to over 600 mm in the northeast and east of the regional biome. Rain falls mainly in summer and its shoulder months, with a little winter rain mainly in the south of the regional biome. Summer temperatures are warm to hot whereas winter nights are cold with frequent, and often heavy, frost (Fig. 9.14).

Topography, Geology and Soils Typical topographical features are the extensive, undulating open plains, with prominent outcrops of REB Dry Highveld Koppie Shrubland often embedded in the southern and central parts. In the most southerly parts, the landscape is dramatically different with the regional biome distributed along the upper reaches and summits of escarpment mountains, some higher than 2000 m. The extensive southern and central parts of the regional biome are dominated by mudstones and sandstones of the Adelaide and Tarkastad Subgroups of the Beaufort Group and the Volksrust Formation of the Ecca Group (all Karoo Supergroup). Jurassic dolerite intrusive sills and dykes, in the form of ridges and other outcrops (with the REB Dry Highveld Koppie Shrubland), also occur throughout these parts. Much of the northern part of this regional biome is covered by basaltic lavas of the Klipriviер Group and andesitic lavas of the Allanridge Formation (both Ventersdorp Supergroup) and to a lesser extent by granite and gneiss in the central area of the Vredefort Dome. Soils vary, ranging from aeolian sands to clayey soils, the latter mainly found in the bottomlands. However, widespread, shallow to deep aeolian and colluvial sand often overlie deeper clayey horizons. Particularly towards the more arid margins of the regional biome, calcrete can occur, whereas, on the southern mountains, soils are generally shallow lithosols.

Sources Louw (1951), Acocks (1953, 1988), Morris (1973, 1976), Werger (1973a, b, 1980), van der Walt (1980), du Preez (1986), Müller (1986, 2002), Bezuidenhout (1988, 1993), Bezuidenhout et al. (1994a, 1994b), Bredenkamp and Bezuidenhout (1990), Kooij et al. (1990b, 1992), Palmer (1991a, 1991b), Fuls et al. (1992), Malan (1993, 1998), Malan et al. (1994, 1995), Rubin and Palmer (1996), Hoare (1997), Dingaan (1996, 1999), Brown and Bezuidenhout (2000), Hoare and Bredenkamp (2001), Vukeya et al. (2003, 2021, 2022, 2023), Janecke et al. (2005), Mucina et al. (2006a), Daemane et al. (2010, 2012), Rutherford et al. (2012), Dingaan and du Preez (2013, 2017), Dingaan et al. (2013, 2016, 2017), Masubelele et al. (2013), Munyai et al. (2023), Novellie and Gaylard (2013), Dingaan and Tsubo (2019).

DHG2 Dry Highveld Clayey Grassland

Gh 9 Western Free State Clay Grassland (Dayaram et al. 2019)
Gh 13 Klerksdorp Thornveld (Dayaram et al. 2019)

Distribution This regional ecotonal biome is found on lower-lying or bottomland plains occurring in blocks embedded mostly within the REB Dry Highveld Sandy Grassland, from around Mafikeng and Ottosdal in the north to the pan veld areas of the western Free State in the vicinities of Wesselsbron, Bultfontein, and Dealesville in the south (Fig. 9.15).

Vegetation Physiognomy and Composition The physiognomy of vegetation is characterised by short open grassland with important widespread grasses, including *Aristida adscensionis, A. bipartita, A. congesta, Cynodon dactylon, Eragrostis curvula, E. lehmanniana, E. plana, Heteropogon contortus, Microchloa caffra, Panicum coloratum, Setaria sphacelata, Sporobolus discosporus, S. fimbriatus,* and *Themeda triandra.* The southern half of the regional biome, in which many azonal salt

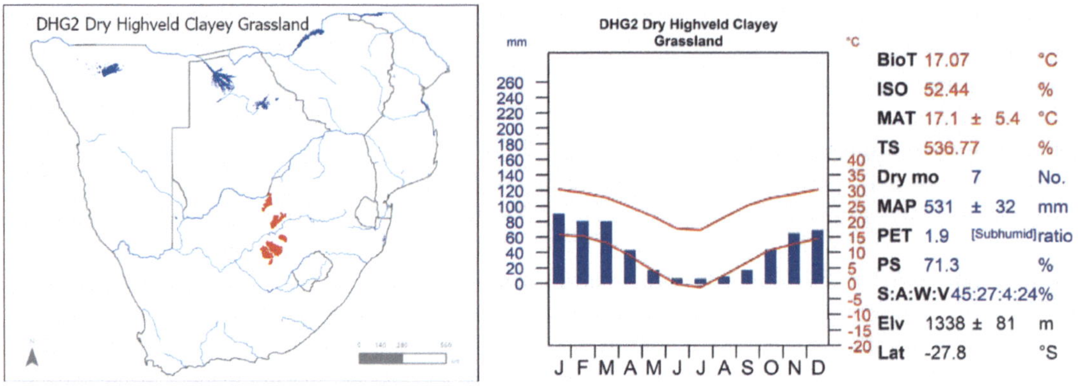

Fig. 9.15 The distribution area and the climatic diagram of the regional ecotonal biome DHG2 Dry Highveld Clayey Grassland

pans are embedded, comprises a species-poor grassland with a low shrub component, including *Amphiglossa triflora, Felicia muricata, Hermannia depressa, Lycium cinereum, Pentzia globosa,* and *P. incana.* The grassland in the northern section includes a variable cover of large thornveld patches dominated by mid-tall *Vachellia karroo*. However, other woody species, such as *Asparagus suaveolens, Diospyros lycioides, Grewia flava, Searsia lancea, S. pyroides,* and *Ziziphus mucronata,* can also occur.

Climate MAP is 531 mm, varying from drier in the south to wetter in the north of the regional biome. Rain falls predominantly in summer; winters are relatively dry. Summer temperatures are hot whereas winter nights are cold with frequent frost (Fig. 9.15).

Topography, Geology and Soils The typical topography is flats in the endorheic drainage panveld area, overlying sandstone, mudstone, and shale of the Volksrust Formation (Ecca Group) in the south. In the north, it occurs mainly on foot slopes of ridges and bottomlands with underlying shale, slate and quartzite of the Pretoria Group, as well as volcanics and sediments of the Ventersdorp Supergroup. Soils are moderately deep, often clayey, alluvium or colluvium, but range from shallow rocky soils to deep, clayey, and of duplex character.

Sources Louw (1951), Morris (1973, 1976), Bredenkamp and Bezuidenhout (1990), Kooij (1990), Kooij et al. (1990a, 1992), Bezuidenhout (1993), Bezuidenhout et al. (1994c, d), Mucina et al. (2006a).

DHG3 Dry Highveld Koppie Shrubland

Gh 4 Besemkaree Koppies Shrubland (Dayaram et al. 2019)
Gh 7 Winburg Grassy Shrubland (Dayaram et al. 2019)
Gh 8 Bloemfontein Karroid Shrubland (Dayaram et al. 2019)

Distribution This regional biome is formed by an archipelago of islands embedded in the eastern RB Upper Karoo from near Richmond and Middelburg in the southwest, through the southern and central Free State around Bloemfontein and Winburg, extending to more sparsely scattered occurrences embedded within the RB Mesic Highveld Clayey Grassland in the northeastern Free State (and including a few outliers between Volksrust and Standerton in Mpumalanga). It is particularly well-represented in the rugged terrain of the Gariep/Orange River valley between Aliwal North and Orania (Fig. 9.16).

Vegetation Physiognomy and Composition Medium-height, multi-stratal grassy shrublands are the typical vegetation of this regional biome. Often, there is a distinct dwarf-shrub component. Dwarf small-leaved karroid and succulent shrubs dominate localised rocky sheets with pockets of shallow soil even in some parts of the mesic northeast. Important shrubs include *Asparagus suaveolens, Buddleja saligna, Chasmatophyllum musculinum, Chrysocoma ciliata, Diospyros lycioides* subsp. *lycioides, D. pubescens, Eriocephalus ericoides, Euclea crispa, Euphorbia mauritanica, Felicia filifolia, Grewia occidentalis, Gymnosporia buxifolia, Helichrysum dregeanum, Olea europaea* subsp. *africana, Pentzia globosa, Searsia burchellii, S. ciliata,* and *S. erosa.* Important grasses include *Aristida adscensionis, A. congesta, A. diffusa, Cenchrus ciliaris, Cymbopogon caesius, C. pospischilii, Cynodon dactylon, C. incompletus, Digitaria eriantha, E. curvula, E. lehmanniana, E. micrantha, E. nindensis, E. obtusa, Eustachys paspaloides, Heteropogon contortus, Oropetium capense, Panicum stapfianum, Setaria lindenbergiana, S. sphacelata, Sporobolus fimbriatus, Themeda triandra,* and *Tragus koelerioides.*

Climate MAP is 454 mm, varying from less than 400 mm in the southwest to over 600 mm in the northeast, with rain falling predominantly in summer and early autumn. The northwestern slopes of the koppies are generally drier than the southeastern ones. Large dolerite boulders can result in microhabitats with a moister environment than surrounding areas. Summers are hot, whereas winters are cold but probably with less severe frost on the elevated koppies compared to frost on the surrounding plains (Fig. 9.16).

Topography, Geology and Soils The typical topography are hills, ridges and other rocky outcrops widely distributed on the plains. This regional biome is closely coupled with the geology. Jurassic dolerite intrusive sills and dykes form rocky koppies, ridges, and small escarpments, all embedded within Karoo Supergroup sediments. In places on the slopes of mesas and butts, the dolerite can be mixed with sandstone and mudstone of the Ecca and Beaufort Groups. Soils are lithosols, young soils on weathered rock, varying from thin layers of aeolian sand overlying dolerite sheets to better-developed soil in deeper pockets and larger crevices and other spaces between dolerite boulders.

Sources Potts and Tidmarsh (1937), Acocks (1953, 1988), Mostert (1958), Werger (1973a, b), Roussouw (1983), Werger (1983), Müller (1986, 2002), Jooste (1989), du Preez (1979, 1991), du Preez and Bredenkamp (1991), Malan (1993, 1998), Malan et al. (1994, 1995, 1998, 1999), Lloyd and Badenhorst (1995), Dingaan (1996, 1999), Dingaan and du Preez (2002, 2017), Botha (2003), Haasbroek (2003), Vukeya et al. (2003, 2021, 2022, 2023), Mucina et al. (2006a), Mills et al. (2012, 2013), Brown and du Preez (2014).

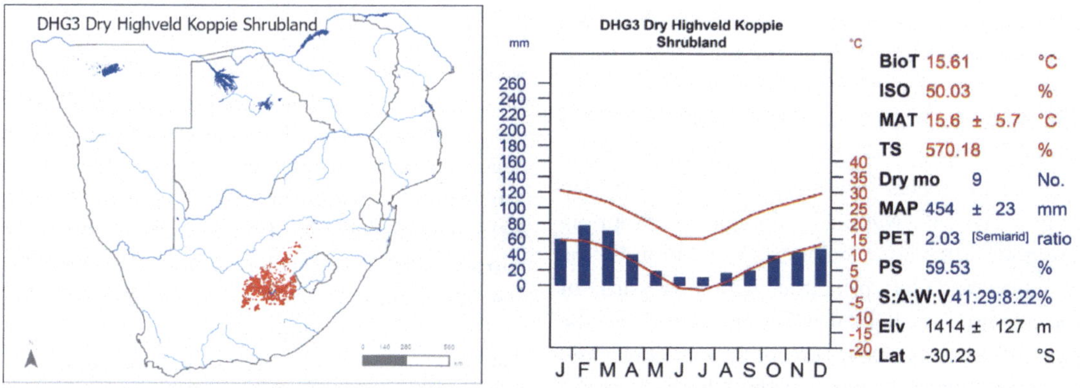

Fig. 9.16 The distribution area and the climatic diagram of the regional ecotonal biome DHG3 Dry Highveld Koppie Shrubland

9.3 Continental Ecotonal Biome Albany Thicket (AT)

ZE Subtropical Semiarid Ecotone S2–E1(sub)

9.3.1 Position of the Albany Thickets in the Southern African Context

The CEB Albany Thicket is physiognomically complex (see Fig. 9.17 a–e) and constitutes ecologically and biogeographically challenging vegetation and biome concepts. Hoare et al. (2006) summarised the history of the evolution of the (mis-)understanding of the structure and position of the Albany Thicket at both subcontinental (Southern Africa) and global scales. The milestones of this genesis of opinions include Rutherford and Westfall (1986), Scholes (1997), Vlok et al. (2003), and Cowling et al. (2005). In their unpublished yet well-accessible report, Vlok and Euston-Brown (2002 and further discussed by Vlok et al. 2003) presented a detailed vegetation system composed of many small-scale and poorly defined (lacking detail on the floristic and ecological distinction) vegetation units. This system has been revisited (simplified) by Hoare et al. (2006: Table 10.1). The detailed classification has served the regional research and nature-management programme, the Subtropical Thicket Ecosystem Planning (STEP) Project. Despite its problems, which we detail below, this classification made a VegMap comeback in 2018 when implemented in the updated (VegMap2018) version of the Albany Thicket coverage. The unfortunate part of this comeback rests in a few, yet profound, flaws which emanate from disregarding several basic theoretical tenets underpinning the VegMap South Africa (National Vegetation Mapping Programme). These include (1) violation of the zonality/azonality principle, (2) disregard of the essential ecological differences between the extant forests and their surrounding thickets, and (3) the failure to recognise the 'thicket-centric' bias in annexing numerous bioclimatically non-AT units.

To document the nature of Problems (1) and (2), we have prepared a spatial crosswalk (Table 9.2) between the original VegMap2006 vegetation mapping units covering the AT and the VegMap 2018 spatial coverage (see Dayaram et al. 2019; www.sanbi.org for the spatial coverage embedded within the VegMap 2018, and the associated descriptions of the 'newly' defined units).

Problem 1 *Disregard for the zonality/azonality principle*

This problem is demonstrated by the fact that two VegMap2006 azonal vegetation units (AZs 1 Algoa Dune Strandveld and AZs 2 Albany Dune Strandveld; see Mucina et al. 2006b) have been dissolved into a series of new zonal AT vegetation. These steps have led to the definition of several mixed (zonal/azonal) units in VegMap 2018, namely AT 41, AT 48, and 56.

Problem 2 *Mixing-up of concepts—forests vs thickets*

Numerous patches of subtropical forests are found within the AT realm (see Mucina et al. 2022). These (especially FOz 5 and FOz 6) were, in VM2018, virtually 'dissolved' and re-branded as AT units, carrying the misleading label 'Forest Thicket'.

Problem 3 *Inconsistent application of bioclimatic biome-delimitation criteria*

Research of several people involved in SKEP (e.g., Becker et al. 2015; Duker et al. 2015a, b) revealed the importance of the frost occurrence criterion driving the difference between the Albany Thickets and Nama-Karoo (NK). This critical criterium finds our support, too. However, it had been neglected in the re-definition of the extent of the 'Subtropical Albany Thicket' featured in VegMap 2018. The re-drawing of the borders of the 'new' Albany Thicket biome has failed to exclude the vegetation units AT 3, AT 4, AT 13, and AT 14—all showing high counts of frost days (13, 19, 16 and 19 days, resp.), but has also included other NK and Succulent Karoo (SK) units (see below) not fitting the subtropical bioclimate.

Fig. 9.17 Vegetation types of disparate physiognomy characterising the CEB Albany Thicket. **a** A Valley thicket dominated by tall succulent *Euphorbia triangularis*; Bushmans River near Kasouga, Eastern Cape Province of South Africa. **b** Low succulent scrub with *Aloe africana* (Asphodelaceae; in the background) and iconic ornamental *Strelitzia reginae* (Strelitziaceae; in the foreground); Ecca Pass, northeast of Grahamstown, Eastern Cape. **c** *Portulacaria afra* (Didiereaceae) is a locally dominating succulent shrub typical of the Albany Thicket, Greater Addo Elephant National Park near Paterson, Eastern Cape. **d** Succulent thicket dominated by *Portulacaria afra* on a steep slope of the Huisrivier Pass W of Calitzdorp, Western Cape Province of South Africa. **e:** Noorsveld is a vegetation type dominated by noors *Euphorbia caerulescens* near Jansenville, Eastern Cape. **f** View of the steep slopes of the Gouritz River, supporting Southern Cape Valley Thicket, from the bridge on N2 road over the Gouritz River, Western Cape. Photo credits: All images by L. Mucina

Table 9.2 Quantitative crosswalk between the vegetation units of VM2006 (Mucina and Rutherford 2006) and VM2018 (Dayaram et al. 2019) mapping coverages

	AT vegetation units (VegMap2006)													
	AT1	AT10	AT11	AT12	AT13	AT14	AT2	AT3	AT4	AT5	AT6	AT7	AT8	AT9
AT15			0										**100**	
AT16			0										**43**	
AT17												0	**95**	1
AT18		0	0								0	0	**93**	0
AT19								**78**	0	0	0		0	
AT20								**100**			0			
AT21				**98**										1
AT22				**99**	0									0
AT23			**86**	0										
AT24			**100**	0	0						0			0
AT25								**18**			0			
AT26									**47**		**53**			
AT27			0		**25**	**65**		9			0			
AT28			0	0	**91**	7								
AT29						**100**								
AT30		**100**	0										0	
AT31			**100**										0	0
AT32		0	**91**		0	8					0		0	0
AT33							**91**							
AT34							**99**							
AT35													0	**100**
AT36														
AT37	**95**													
AT38			0										0	**35**
AT39											0	**100**	0	0
AT40	1													
AT41											0		0	**61**
AT42											**100**			
AT43							**100**							
AT44								0			**100**	0		
AT45			0								0	0	**52**	**48**
AT46							**100**							
AT47		0	0								0		0	
AT48														
AT49			0			0		**19**	5	**22**	**51**			
AT50								0	**17**		**80**	0	0	0
AT51								0	**17**	0	**83**	0		0
AT52													**100**	0

(continued)

Table 9.2 (continued)

	AT vegetation units (VegMap2006)													
	AT1	AT10	AT11	AT12	AT13	AT14	AT2	AT3	AT4	AT5	AT6	AT7	AT8	AT9
AT53				**86**										0
AT54									**33**		9			0
AT55	0						0							
AT56			0	0										**87**
AT57								0	0		0	0		0
AT58							0	0	0		0			

	Non-AT vegetation units (VegMap2006)												
	AZs1	AZs2	FFd11	FFd9	FFh10	FFl3	FFs28	FOz5	NKl3	NKl4	SKv12	SKv9	SVs7
AT15													
AT16										0			***57***
AT17		0			2								0
AT18					3					0			0
AT19					6		3			0	0		
AT20	0												
AT21		0											0
AT22													0
AT23													***13***
AT24										0			0
AT25					0				***74***		2		
AT26	0						0						
AT27													
AT28										0			0
AT29													
AT30										0			
AT31		0			0					0			0
AT32										0			0
AT33											0		
AT34													
AT35													
AT36			***86***										
AT37				0		4						0	
AT38					3					1			***53***
AT39	0												
AT40				***29***		***51***							
AT41		***35***											
AT42													
AT43											0		
AT44	0												

(continued)

Table 9.2 (continued)

	Non-AT vegetation units (VegMap2006)												
	AZs1	AZs2	FFd11	FFd9	FFh10	FFl3	FFs28	FOz5	NKl3	NKl4	SKv12	SKv9	SVs7
AT45		0											0
AT46												0	
AT47					2					***97***			0
AT48	***87***												
AT49					0		0		0	0	0		
AT50	0				1		0						
AT51	0						0				0		
AT52													
AT53		0						***13***					
AT54					***17***		***30***						
AT55												***100***	
AT56		5						0					2
AT57	***88***		7										
AT58							0				***100***		

The VU codes AT15 through AT58 (in the first column) are the AT units of the VM2018. The values in the body of the matrix are % values of the overlap between the VM2018 and VM2006. The bold-highlighted cells mark the high (more than 10%) overlaps among the AT units; the bold and italics-highlighted cells mark the high (more than 5%) overlaps among the AT units and non-AT units. The values of 0 indicate values below 1%

9.3.2 Bioclimatic Analysis of the Albany Thicket Biome and Related Units

A bioclimatic analysis of the climatic space encompassing the Albany Thicket (AT) biome as defined by the VM06 mapping (Mucina and Rutherford 2006), hence including all AT vegetation units and the NKl 4 Albany Broken Veld—a Nama-Karoo (NK) unit, embedded geographically within the AT Albany Thicket, was undertaken to establish the bioclimatic identity of the Albany Thicket biome.

We employed a wide range of climatic variables (Bio1 through Bio19) derived from the CHELSA database (Karger et al. 2017a, b, 2020). The number of frost days, as defined and used in constructing the climatic diagrams in Mucina and Rutherford (2006), was also included as an additional variable. The AT vegetation unit (and the vegetation unit NKl 4) stratified the sampling points. One thousand points were distributed according to the size of the vegetation unit (i.e., if 50% of the total area, 500 points would fall in that type). The points were forced to be more than 1 km apart. Some units (AT 1 Southern Cape Valley Thicket, AT 10 Great Fish Noorsveld, AT 7 Coega Bontveld) were small in extent and hence scored fewer points. The number of points here was boosted by adding additional ones so that they achieved a minimum of 30.

Fifteen VUs were tentatively classified into five (A through E) informal groups (see Figs. 9.18, 9.19 and 9.20), defined as follows:

A: NKl 4 (the only original NK unit);
B: AT 2 and 3; two AT units occurring outside the core AT in the form of prolonged patches straddling NK, Succulent Karoo (SK) and Fynbos (FYN) biomes;
C: AT 13 and 14: original AT units closely following the Great Escarpment and occurring at the northern and northwestern edge of the AT proper;

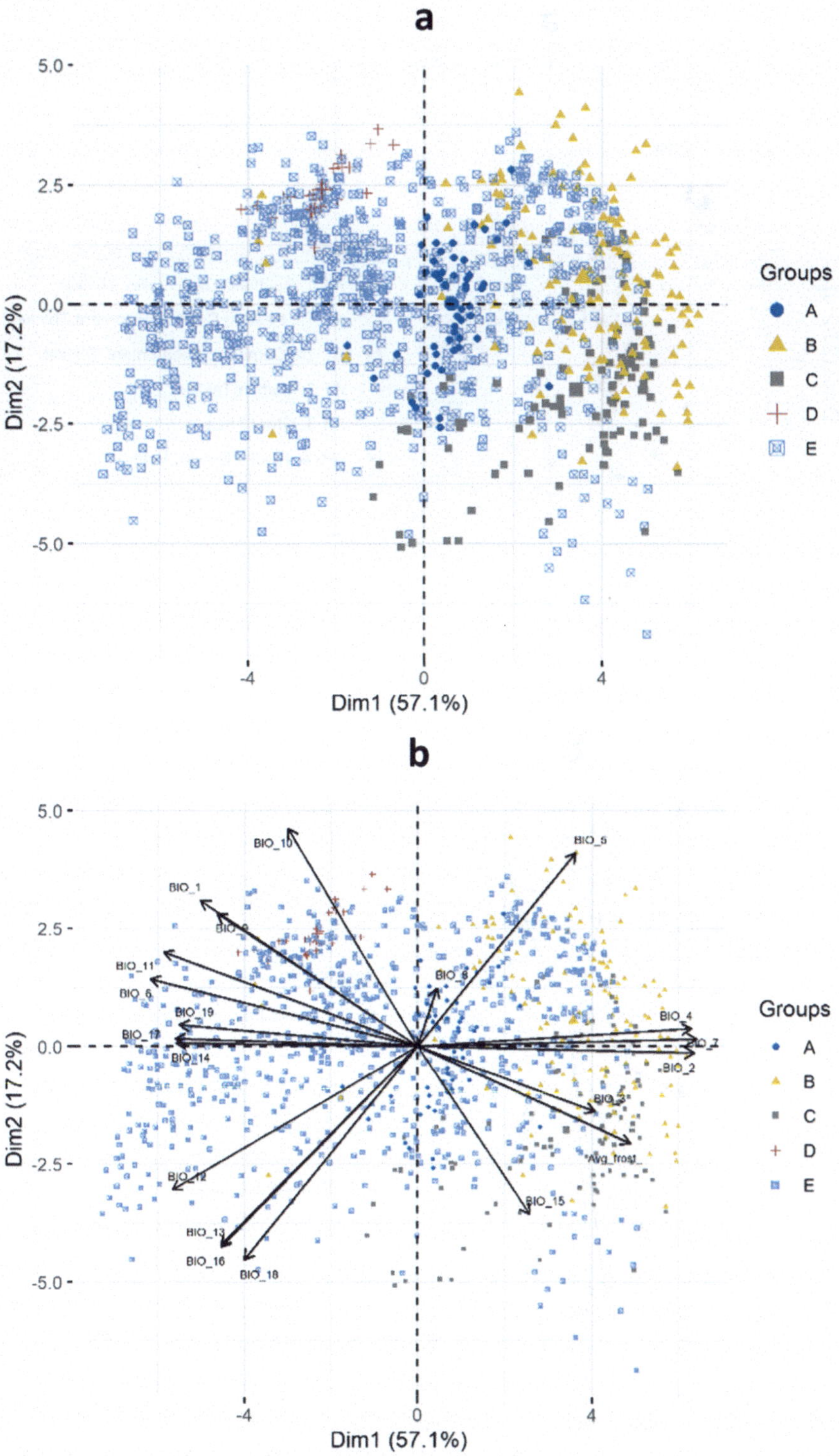

Fig. 9.18 Ordination diagrams featuring the classification of the sampled points as stratified into five informal groups A through E (A: NKl4; B: AT2, 3; C: AT13, 14; D: AT1; E: AT4, 5,6 7, 8, 9, 10, 11—all vegetation units of Mucina and Rutherford 2006). a: the classified points; b: biplot

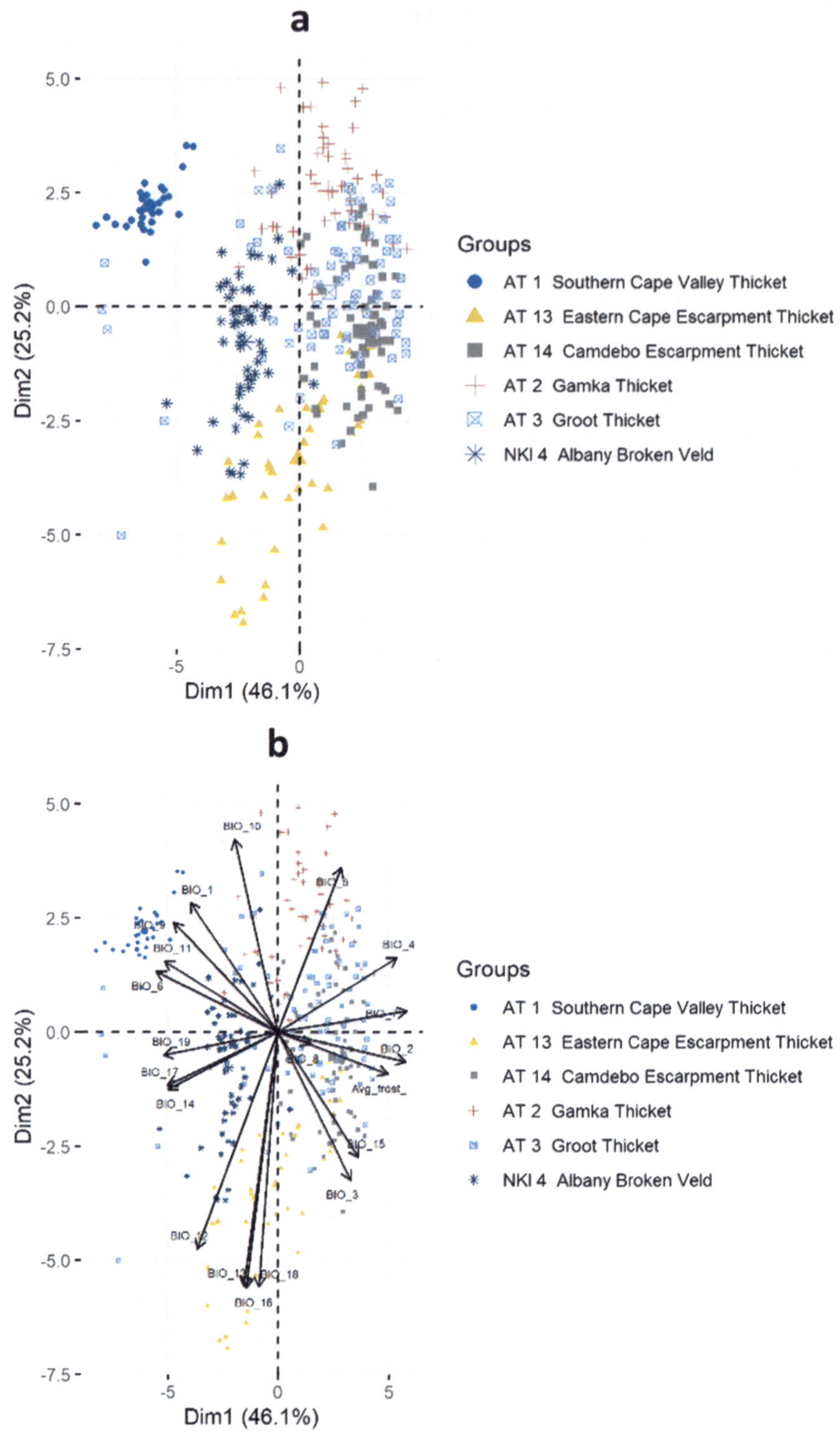

Fig. 9.19 Ordination diagrams featuring the classification of the sampled points of the informal groups A through D. The group E has been masked out. For the definition of the informal groups, see the key in the caption of Fig. 9.19. a: the classified points; b: biplot

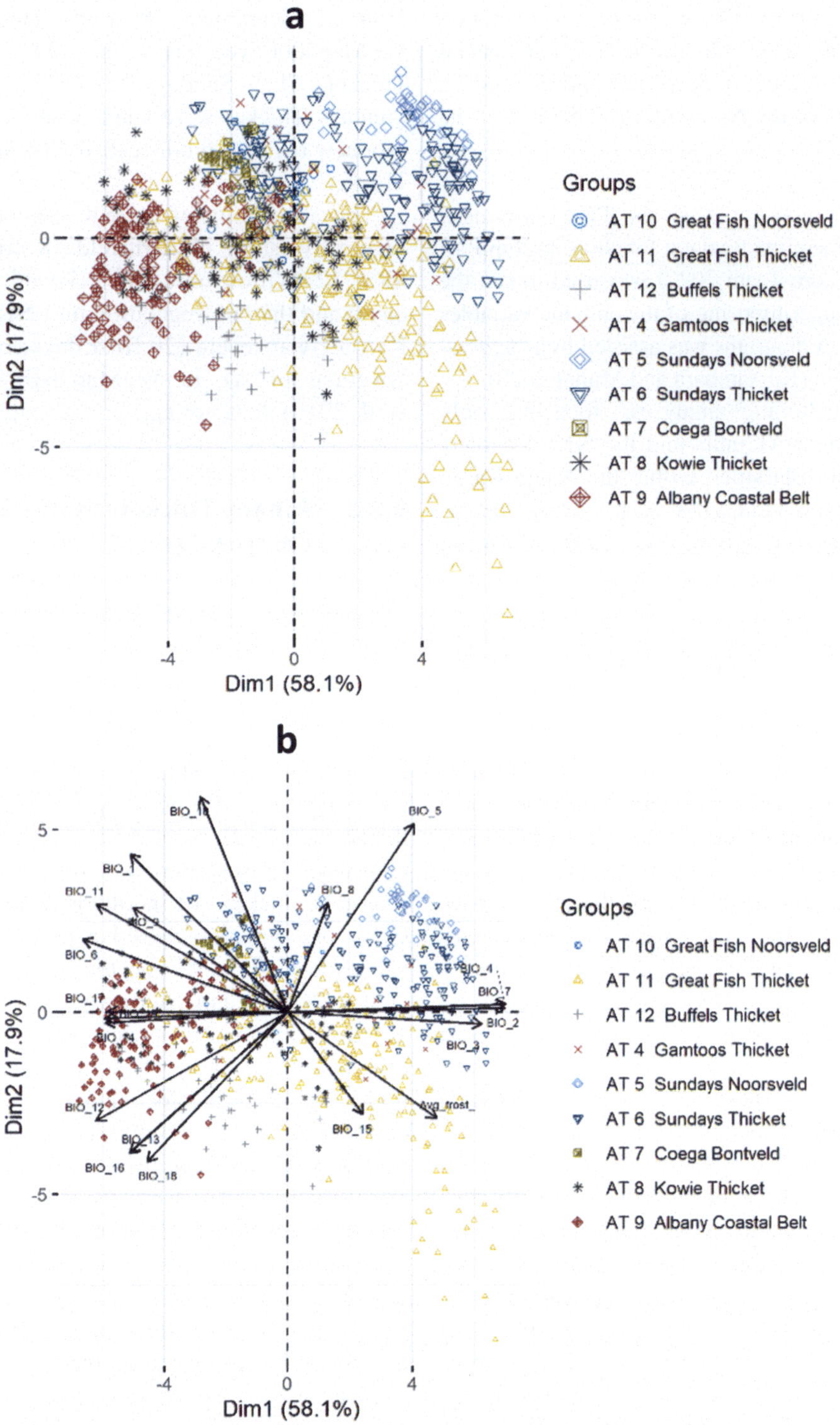

Fig. 9.20 Ordination diagrams featuring the sampled points of the vegetation units classified in group E. For the definition of this informal group, see the key in the caption to Fig. 9.19. a: the classified points; b: biplot

D: Southern Cape Valley Thicket—a vegetation unit embedded fully within FYN in special habitats (slopes of deep river valleys) and

E: the core of the AT, comprising the rest of the units.

Principal Component Analysis (PCA) was used to analyse the climatic data; function 'prcomp' of base R (R Core Team 2022) was used to run the PCA. The superimposing of the climatic variables onto the PCA diagrams was assisted by R's *factoextra* package (Kassambara and Mundt 2020).

Our bioclimatic analyses, focusing only on the original AT units and the NKl 4 Albany Broken Veld (classified within the Nama-Karoo biome; Mucina et al. 2006c), and using successive analytical steps, have revealed the following important issues:

(1) The NKl 4 Albany Broken Veld is bioclimatically embedded within the Albany Thicket biome, supported by its low counts of frost days (5).
(2) Four of the vegetation units (Gamka Thicket, Groot Thicket, Eastern Cape Escarpment Thicket, Camdebo Escarpment Thicket) showed marginal position in the ordination-modelled bioclimatic space driven by the frost frequency and the bioclimatic variables Bio2 (mean monthly diurnal range), Bio3 (isothermality), Bio4 (temperature seasonality), and Bio7 (annual temperature range). The latter four variables are indicative of increased thermic continentality.

Applying the results of this analysis, we reclassified the NKl 4 Albany Broken Veld as a member of the CEB Albany Thicket. Hence, here we agree with the VegMap2018 (Dayaram et al. 2019) and reclassify the NKl 4 into the CEB Albany Thicket. Likewise, the AT 2 Gamka Thicket was reclassified as a member of an SK biome. The remaining 'cold-climate' AT units (Groot Thicket, Eastern Cape Escarpment Thicket and Camdebo Escarpment Thicket) were reclassified as belonging to the CB Nama-Karoo. The last two units form a new regional biome, Escarpment Karroid Thicket. The VegMap2006 vegetation units AT 4 Gamtoos Thicket, AT 5 Sundays Noorsveld, and AT 6 Sundays Thicket were singled out as elements of a new regional biome called AT4 Sunday Dry Thicket.

The absorption of the SK vegetation units SKv 9 Western Gwarrieveld (8 frost days), SKv 12 Willowmore Gwarrieveld (21 frost days) and the NK vegetation unit NKl 3 Lower Karoo Gwarrieveld (20 frost days) by the new coverage of AT in VegMap2018 (Dayaram et al. 2019) does not appear warranted.

9.3.3 Albany Thickets in the Global Perspective

The comments on the global position of the Albany Thicket in Hoare et al. (2006) were provided by L. Mucina. Here, the Albany Thicket was analogised with the Chaco of Paraguay and Argentina and the succulent coastal thickets of southern Madagascar. These authors also suggested that possibly the Somali-Masai thickets of the Horn of Africa, vine thickets of Queensland (Australia) and the succulent-rich (coastal) thickets of northern Venezuela and Colombia 'could be viewed as part of the global subtropical succulent-rich thickets'. The Hoare et al.'s (2006) comparisons were based on physiognomy (domination of succulents) and, in part, also on climate since only similar thickets found under (sub)tropical bioclimatic conditions were considered. However, after the biomes of the Southern Hemisphere were revisited using a hemispheric (and indeed also global) magnifying glass (Mucina 2023), it became clear that the comparisons with those three selected thicket units were inaccurate.

Albany Thickets show a relationship with the true Subtropical Forests. All patches of subtropical scarp type (Albany Scarp Forest; see Mucina et al. 2022) are embedded within the Albany Thickets. However, there is a significant structural difference between AT and subtropical forests per se: AT has an important succulent element and a bimodal precipitation regime.

The CEB Albany Thicket was found (Mucina 2023) to be a unique ecotonal phenomenon [Subtropical Semiarid Ecotone S2–E1(sub)]. The thickets of the Horn of Africa and the vine thickets of Queensland, found across the entire Top End of Australia, belong to the family of Tropical Dry Forests (member of the zonobiome E2). Chaco is also a unique CEB, belonging to the Subtropical Seasonal Arid Ecotone [(E2(tdf)–S2)]. Finally, the Madagascan Thorny Thicket is a 'Passatwüste' (Germ. for 'trade-wind desert')—hence belonging to the ZE Tropical Coastal Semideserts Ecotone [(S2–E2(tdf)].

9.3.4 Subdivision of the Albany Thickets in Regional Ecotonal Biomes

After excluding vegetation units that do not meet our bioclimatic criteria, we are left with the REB Albany Thicket and REB Sundays Dry Thicket as the only two core units of the CEB Subtropical Albany Thicket. We also distinguish intrazonal REB Albany Calcrete Bontveld (peinobiome) and REB Albany Strandveld (psammobiome).

The former subdivisions of the Albany Thickets into subunits may have been intended to satisfy objectives set at the time but have nevertheless relied on poorly quantified accounts (Vlok et al. 2003), driven more by intuition (or 'expert opinion') than by rigorous data analyses (AT in VegMap2018; Dayaram et al. 2019). Our revisit of the AT is consistently the only one using bioclimatic (and pedological—in the case of an azonal regional biome) criteria. Although this approach is relatively straightforward, when regional biomes (or regional ecotonal biomes) are defined, they lose power at the level of the 'vegetation unit'. Here, more data-intensive analyses are needed to test former classifications.

In our regional (ecotonal) biome scheme, the CEB Albany Thicket contains four units, as described below.

AT1 Albany Thicket

AT 8 Kowie Thicket (Mucina and Rutherford 2006)
AT 9 Albany Coastal Belt (Mucina and Rutherford 2006)
AT 10 Great Fish Noorsveld (Mucina and Rutherford 2006)
AT 11 Great Fish Thicket (Mucina and Rutherford 2006)
AT 12 Buffels Thicket (Mucina and Rutherford 2006)
NKl 4 Albany Broken Veld (Dayaram et al. 2019)

Distribution This regional biome forms the core of the CEB Albany Thicket, which is limited to the confines of the Eastern Cape Province of South Africa. It is concentrated in the river basins of the Kowie, Great Fish, and Buffels Rivers (from the west to the east). It also comprises the Albany Broken Veld, formerly assigned to the CB Nama-Karoo (Mucina et al. 2006c), and the vegetation unit AT 9 Albany Coastal Belt (Hoare et al. 2006). The latter is a relatively recent (the early 1800s, marking the colonisation of the region by white farmers) vegetation created by the clearing and deforestation of the coastal succulent scrub and subtropical forests embedded within (Fig. 9.21).

Vegetation Physiognomy and Composition Due to the large extent and landscape variability, the original vegetation structure shows many faces, including tall and dense, mainly evergreen shrublands, spekboom (*Portulacaria afra*) thickets, and tall succulent thickets, found mainly on steep slopes of the incised rivers. The dominant tall shrubs and small trees include representatives of the genera *Apodytes, Canthium, Cussonia, Maytenus, Pappea, Ptaeroxylon, Schotia, Sideroxylon, Vepris*, and many more. Low-grown shrubs are also prominent and diverse including, for instance, *Allophylus, Azima, Carissa, Ehretia, Dicerothamnus, Euclea, Grewia, Gymnosporia, Hippobromus, Maerua, Muraltia, Mystroxylon, Olea, Pterocelastrus, Putterlickia, Searsia, Scolopia*, and *Scutia*. The most prominent tall euphorbias are *Euphorbia grandidens, E. tetragona, E. triangularis*, and tall *Aloe* component (*A. africana, A. arborescens, A. ferox, A. speciosa*). Lianas are very common in these thickets, rendering them poorly penetrable. Heavily disturbed areas of this regional biome include parts

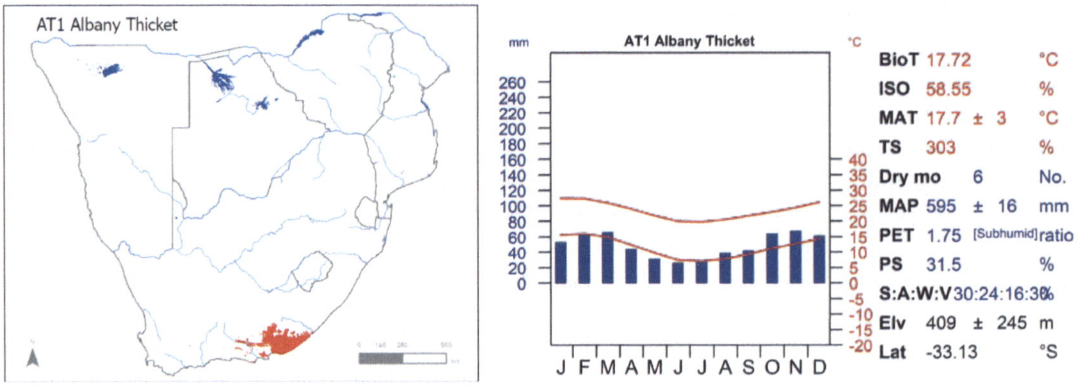

Fig. 9.21 The distribution area and the climatic diagram of the regional ecotonal biome AT1 Albany Thicket

dominated by low karoo shrublands as found in Albany Broken Veld or heavily utilised (dairy and meat production) coastal grasslands.

Climate Annual rainfall is bimodal, peaking in November and April, but remains relatively well-distributed through the year. The MAP of 595 mm varies from less than 400 mm in parts of Albany Broken Veld and Great Fish Noorsveld in the interior to between 800 and 900 mm in areas of Buffels Thicket and north-eastern parts of Albany Coastal Belt. Summers are warm to hot, and winters are cool. Frost occurrence is low, which is one of the critical defining characteristics of this regional biome. The mean annual number of frost days typically lies between 3 and 5 days but declines to zero days near the coast. In parts of the Great Fish Thicket and Kowie Thicket that lie farthest inland, the incidence of frost can increase markedly (Fig. 9.21).

Topography, Geology and Soils The landscape is remarkably varied, including steep slopes of highly dissected hills and river valleys, ridges and rills of low mountains and plateaus, and gentle to moderate undulating plains. The geological substrate is made up mainly of shale, sandstone, mudstone, arenite and tillite of the Beaufort, Ecca and Dwyka groups of the Karoo Super Group. Soil forms are mainly Glenrosa and Mispah, thus including many skeletal shallow soils, sometimes with a high base status. In the higher rainfall Buffels Thicket, fine-grained nutrient-poor silts and humus-rich, deep soils associated with forests are found.

Sources Dyer (1937), Acocks (1953, 1988), Palmer (1981), Pierce and Cowling (1984), Palmer et al. (1988), Everard (1987), La Cock (1992), Stuart-Hill (1992), Midgley and Cowling (1993), Stuart-Hill and Aucamp (1993), Low and Rebelo (1996), Evans et al. (1997), Judd (2001), Lloyd et al. (2002), Vlok and Euston-Brown (2002), Brink and Dold (2003), Vlok et al. (2003), Kamineth (2004), Hoare et al. (2006), Martin (2017).

AT2 Albany Calcrete Bontveld

AT17 Coega Bontveld (Mucina and Rutherford 2006)

Distribution This regional biome occupies a small area at altitudes below 400 m, north-east of Port Elizabeth just inland of Algoa Bay, where it occurs mainly around Coega but also in small patches in the Addo Elephant National Park (Zuurkop; Pentzhorn and Olivier 1974). It is limited to the confines of the Eastern Cape Province of South Africa (Fig. 9.22).

Vegetation Physiognomy and Composition The vegetation is a mosaic of dominating open grassland, with embedded pockets of low bush clumps. The species composition presents a

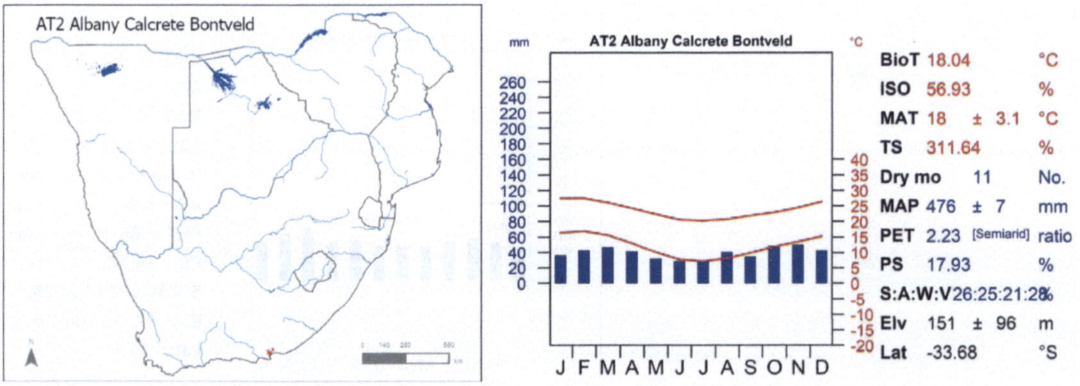

Fig. 9.22 The distribution area and the climatic diagram of the regional ecotonal biome AT2 Albany Calcrete Bontveld

mixture of fynbos, grassland and succulent karoo elements. The dominating grasses include *Aristida diffusa, Cynodon dactylon, C. incompletus, Eustachys paspaloides, Heteropogon contortus, Panicum maximum, Setaria sphacelata, Stipa dregeana, Tenaxia disticha,* and *Themeda triandra*. Low shrubs and herbs are numerous and diverse. The main component of the bush clumps is the tall shrub *Euclea undulata*,

Climate The region of this regional biome has a bimodal (tending to non-seasonal) rainfall regime, with maxima in March and October. MAP of 476 mm ranges from about 400 mm inland in the west to 550 mm in the east and closer to the coast. The incidence of frost is very low. Mean monthly maximum and minimum temperatures for the nearby Port Elizabeth are 32.1 °C and 3.4 °C for February and July, respectively (see Hoare et al. 2006) (Fig. 9.22).

Topography, Geology and Soils Undulating plains characterise the landscape, with some outcrops of Nanaga Formation limestone and calcareous paleo-dune fields of the Cenozoic Algoa Group. The soils are shallow, with increased clay content and naturally lime-rich. The limestone substrate (with its property to lower the levels of accessible phosphorus content in soil) is the reason to consider this regional biome azonal (peinobiome—a subtype of pedobiome).

Sources Acocks (1953, 1988), Pentzhorn and Olivier (1974), Vlok and Euston-Brown (2002), Vlok et al. (2003), Hoare et al. (2006), De Kock (2011), Carvalho et al. (2021).

AT3 Albany Strandveld

AZs 1 Algoa Dune Strandveld (Mucina and Rutherford 2006)
AZs 2 Albany Dune Strandveld (Mucina and Rutherford 2006)
FFd 11 Southern Cape Dune Fynbos (Mucina and Rutherford 2006)

Distribution This azonal coastal pedobiome (psammobiome) occurs in the form of a narrow strip along the Indian Ocean seaboard from the mouth of the Tsitsikamma River to the south of Kei Mouth, Eastern Cape Province of South Africa (Fig. 9.23).

Vegetation Physiognomy and Composition The original remnant vegetation is typically formed by tall (up to 5 m) dense thickets dominated by stunted trees, spiny and thorny shrubs of diverse height, accompanied by abundant lianas and sparse herbaceous and grassy undergrowth. The dominant woody elements comprise *Azima tetracantha, Brachylaena discolor, Chrysanthemoides monilifera, Cussonia thyrsiflora, Euclea racemosa* subsp. *racemosa, Maytenus procumbens, Metalasia muricata, Mystroxylon aethiopicum,*

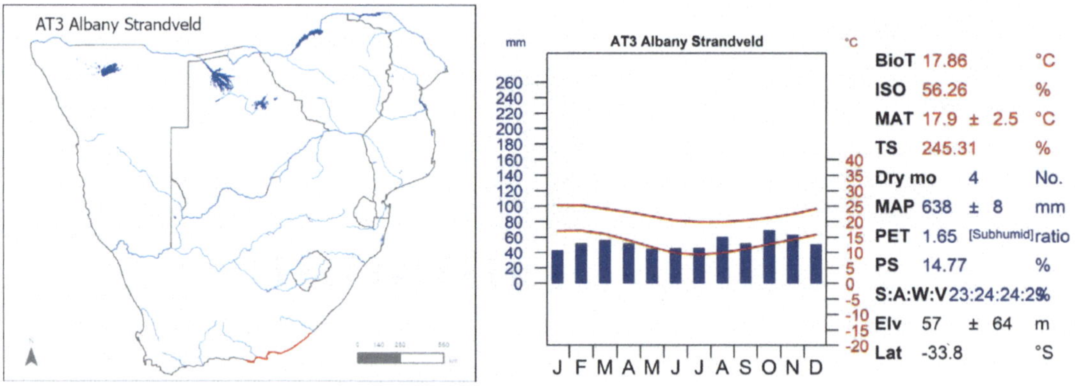

Fig. 9.23 The distribution area and the climatic diagram of the regional ecotonal biome AT3 Albany Strandveld

Pterocelastrus tricuspidatus, Schotia afra, Scutia myrtina, Searsia crenata, Sideroxylon inerme, and *Tarchonanthus littoralis*. Fynbos elements may also occur on crests of dunes with leached sands. The occurrence of bulbous geophytes and succulent herbs is an important feature of this coastal vegetation.

Climate The annual precipitation is bimodal (tending to aseasonality), with MAP around 600–700 mm. Temperatures are relatively equitable, with warm summers and mild winters. The mean daily maximum and minimum temperatures (for AZs 1 and AZs 2) are 25.1–26.4 °C and 8.3–8.8 °C for February and July, respectively, indicating a subtropical rather warm-temperate regime (Fig. 9.23).

Topography, Geology and Soils The main topographic featured here are slightly undulating aeolian sand dunes most of the time outside the influence of salt spray. The dune sands are of the Schelm Hoek Formation of the Algoa Group.

Sources Dyer (1937), Comins (1962), Taylor and Morris (1981), Lubke (1983), Olivier (1983), Cowling (1984), Cowling and Pierce (1985), Burns (1986), La Cock (1986), Young (1987), Lubke and De Villiers (1991), Talbot and Bate (1991), Taylor and Boucher (1993), Weisser and Cooper (1993), Avis and Lubke (1996), Vlok and Euston-Brown (2002), Vlok et al. (2003), Mucina et al. (2006b), Strydom et al. (2021).

AT4 Sundays Dry Thicket

AT 4 Gamtoos Thicket (Mucina and Rutherford 2006)
AT 5 Sundays Noorsveld (Mucina and Rutherford 2006)
AT 6 Sundays Thicket (Mucina and Rutherford 2006)

Distribution This regional biome is confined to the Eastern Cape Province of South Africa. It occupies the western, dry outreaches of the CEB Albany Thicket. It occurs mostly north of the Klein Winterhoek Mountains, including much of the Jansenville District and parts of the far-southern Pearston District and far-western Somerset East District, some patches south of this mountain range west of Kirkwood around Waterford and the Darlington Dam, surrounds of Uitenhage and the northern edge of Port Elizabeth into the lower Sundays River Valley to east of Colchester and northwards to the base of the Zuurberg Mountains and stretching west-wards north of the Groot Winterhoek Mountains to roughly the Kleinpoort longitude. It is also found, more marginally and to a limited fragmented extent, in the lower valley of the Gamtoos River to the south of the Winterhoek. The altitudinal span is 0–800 m (Fig. 9.24).

Vegetation Physiognomy and Composition The vegetation in this regional biome is a dense, usually up to 2 m tall succulent thicket consisting partly of a mosaic of noors (*Euphorbia caerulescens*) and low karoo shrub vegetation (dominated by *Pentzia incana* and *Rhigozum obovatum*).

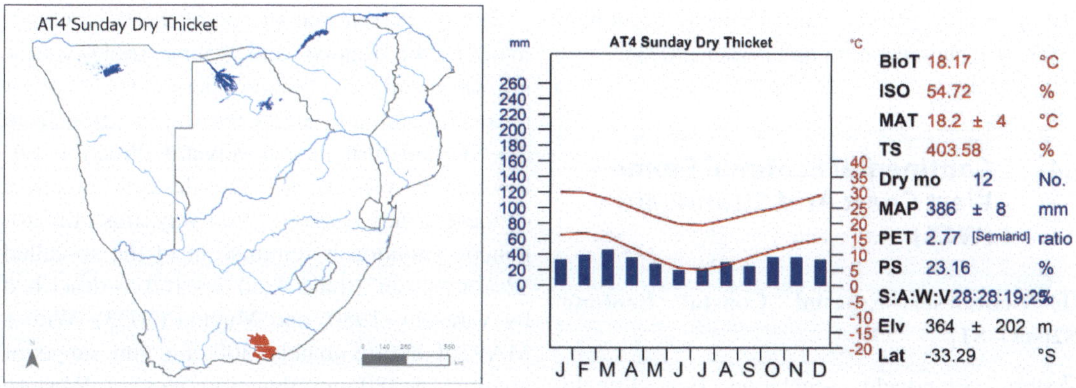

Fig. 9.24 The distribution area and the climatic diagram of the regional ecotonal biome AT4 Sundays Dry Thicket

Other important forms of vegetation physiognomy include tall, dense thickets, where trees, shrubs and succulents are common, with many spinescent species co-occurring. The succulent shrub *Portulacaria afra* can dominate north-facing ridge slopes. Solitary trees and shrub groups with *Boscia albitrunca, B. oleoides, Pappea capensis, Schotia afra, Vachellia karroo*, and tall shrubs *Euclea undulata, Grewia robusta, Gymnosporia polyacantha*, and *Searsia longispina* punctuate the vegetation landscape. The relative abundance of woody species present decreases with increasing aridity.

Climate The climate of this unit is slightly seasonal, with two less pronounced precipitation maxima in March and November, typical for the entire CEB Albany Thicket. MAP of 386 mm ranges from about 190 mm in the west to 480 mm in the SE on the coast near Port Elizabeth. The incidence of frost is only 5–8 days, ranging from 3 to 13 days of frost per year. Both the main vegetation units making up this regional biome are relatively arid, with an average precipitation below 350 mm. This is the leading reason why the Sundays Dry Thicket has been separated from the rest of the Albany Thicket regional ecotonal biomes. The smaller Gamtoos Thicket component has a wide range in MAP (180 mm in the northwest inland areas to an extreme of 850 mm on southeastern coastal sites), which nevertheless overlaps substantially with the range in MAP of the main areas above (Fig. 9.24).

Topography, Geology and Soils Undulating plains and flat low-lying foothills are the dominant topographic features. The substrate consists of shales, mudstones and sandstones of the Beaufort and Ecca Groups of the Karoo Supergroup in the north and of the Mesozoic Uitenhage Group in the south. Often, heavy, clayey, deep loamy soils are dominant. Some soils are derived from Sundays River and Kirkwood Formations (Mesozoic Uitenhage Group). In the Zuurberg Mountains, soils are more sandy, nutrient-poor, and derived from the Bokkeveld and Witteberg Groups (Cape Supergroup).

Sources Acocks (1953, 1988), Archibald (1955), van der Walt (1965), Pentzhorn and Olivier (1974), Pentzhorn et al. (1974), Aucamp (1976, 1979), Cowling (1983, 1984), Aucamp and Tainton (1984), Everard (1987), Hoffman (1989a, 1989b), Hoffman and Cowling (1990, 1991), La Cock (1992), Stuart-Hill (1992), Midgley and Cowling (1993), Stuart-Hill and Aucamp (1993), Du Toit and Blom (1995), Johnson (1998), Johnson et al. (1999), Kerley et al. (1999a, b), Sigwela (1999), Lombard et al. (2001), Todkill (2001), Cowling and Kerley (2002), Lloyd et al. (2002), Vlok and Euston-Brown (2002), Lechmere-Oertel (2003), Vlok et al. (2003), Sigwela et al. (2004), Hoare et al. (2006), Lechmere-Oertel et al. (2008), Rutherford et al. (2014), Landman et al. (2018, 2019), Martin

(2017), Parker (2017), Carvalho and Campbell (2023a, b), Nuttall-Smith and Parker (2023).

9.4 Continental Ecotonal Biome West Coast Arid Strandveld (WCS)

ZE Semidesert–Ethesial Coastal Ecotone [S2(scs)–S1]

Western Strandveld vegetation (see Rebelo et al. 2006 for a definition of 'strandveld' and the description of this bioregion and component vegetation types) was classified within the Fynbos Biome. The previous classification history, especially the Low and Rebelo's (1996) classification within 'Thicket Biome' (Dune Thicket) as well as later opinions placing this strandveld close to Succulent Karoo (Bergh et al. 2014), are indications of the uncertainty surrounding the placement of the Western Strandveld. Admittedly, the association of the Western Strandveld with Fynbos is more a matter of convenience and tradition than a result of applying rigorous bioclimatic and other (geology, soil) criteria.

The biome position of the Western Strandveld (spatially similar to F11 West Coast Strandveld Bioregion; Mucina and Rutherford 2006: the map) requires revision with a focus on the bioclimate and the underpinning climatic drivers. Disregarding the higher rainfall FS 6 through FS 9 vegetation units (Rebelo et al. 2006), the strandveld units on the West Coast are characterised by low MAP (approx. 300 mm with range 175–380 mm) and a strong winter rainfall regime. This precipitation pattern results from the rain-bearing Westerlies, which become weaker (less frequent rainfall) as one moves northwards from the Cape of Good Hope. The southernmost limit of the Western Strandveld is about 50 km north of Cape Town. The Westerlies do not encounter any barrier when crossing the West Coast and precipitate deeper inland—on the slopes of the Cape Fold Mountains. Besides the gentle topography, another climatic modifier enhances the aridity of the relevant stretch of the West Coast, namely, the regular and strong trade winds, known locally as 'Southeasters'. These occur in austral summer, affect the Cape region from the SE and turn in a northward direction following the West Coast. The desiccation effect of these winds is severe. In a way, this climatogenetic situation is reminiscent of the so-called 'Passatwüsten' (trade-wind deserts) as described by Vareschi (1980) and Mucina (2023). With a MAP of approximately 300 mm and range of about 175–380 mm this arid part of Western Strandveld does potentially qualify as a semidesert per se but this range also fully overlaps with several clearcut members of the Fynbos Biome (e.g., Bokkeveld Sandstone Fynbos FFs 1 with MAP 287 mm ranging 160–430 mm; Matjiesfontein Quartzite Fynbos with MAP of 318 mm ranging 150–450 mm; Leipoldtville Sand Fynbos with MAP of 263 ranging 130–450 mm; and numerous Renosterveld Vegetation Types) which confirms that (arid) Western Strandveld can be regarded as climatically associated with both Fynbos and Succulent Karoo, or, equivalently, be seen to lie in an intermediate position between the two. The (arid) Western Strandveld also stands out as climatically unique as the only area in the MBSA that experiences strong seasonality of winter rain (see Fig. 4.9; Rebelo et al. 2006), which in turn translates into a regime of arid summers that are exacerbated by the above-mentioned desiccating summer winds. On balance, we consider (arid) Western Strandveld as an ecotonal ecosystem, straddling the zonobiomes S1 and S2.

Interestingly, this transitional bioclimate fosters the emergence of a fascinating phenomenon—rich mass-flowering in late winter and spring (Fig. 9.25). Mucina (2023) has noticed that this occurs in all regions where the global analogues of the mediterranean-type climate border on arid hot ecosystems. To avoid confusion between the original nine-unit broad bioregion called Western Strandveld, we introduce a new name for our ecotonal unit—West Coast Arid Strandveld. In the global scheme of biomes, this unit is a continental ecotonal biome, which, in our area, we have subdivided

Fig. 9.25 Vegetation physiognomy of selected ecotonal biome units. **a** Wind-swept strandveld scrub with *Robsonodendron maritimum*, *Pterocelastrus tricuspidatus* (shrubby form) and numerous rutaceous shrubs of the general *Acmadenia* and *Agathosma*. Coastal cliffs near Koppie Alleen, De Hoop Nature Reserve. **b** Overberg strandveld (REB Southwestern Cape Strandveld) with dominating *Agathosma collina* (Rutaceae) and various species of *Helichrysum* (Asteraceae); coastal cliffs near Koppie Alleen, De Hoop Nature Reserve. **c** Low strandveld scrub on deep calcareous sand (Langebaan Dune Strandveld), a mixture of *Asparagus capensis* (Asparagaceae), *Euphorbia burmannii* (Euphorbiaceae), *Tetragonia fruticosa* (Aizoaceae), and several more succulent shrubs and herbs of the family Aizoaceae in the Postberg section of the West Coast National Park, near Langebaan, Western Cape. **d** A spring flush of annual flora (dominating *Dimorphotheca pluvialis*, Asteraceae) after late winter rains—a typical phenomenon of an ecotone straddling the West Coast semideserts and the Cape scrub; Jacobsbaai on the coast of the Atlantic Ocean, Western Cape, South Africa. Photo credits: All images by L. Mucina

into several regional ecotonal biomes based on geology and soil characteristics, namely, deep calcareous sands, limestone, and granite. Accordingly, we may distinguish REB West Coast Arid Sandy Strandveld, REB West Coast Limestone Strandveld and REB West Coast Granite Strandveld (Table 9.1), which are also all pedobiomes (psammobiome, peinobiome, and lithobiome, respectively).

The CEB West Coast Arid Strandveld consists of the following three regional ecotonal biomes:

WCS1 West Coast Arid Sandy Strandveld

FS 1 Lamberts Bay Strandveld (Dayaram et al. 2019)
FS 2 Saldanha Flats Strandveld (Dayaram et al. 2019)
FS 3 Langebaan Dune Strandveld (Dayaram et al. 2019)

Distribution This regional ecotonal biome forms a broad North–South trending coastal strip between Donkin Bay (north of Lambert's Bay) and Elands Bay (Verlorenvlei), penetrating deeply inland north of and along the Jakkals and Langvlei Rivers. This belt widens further south, also extending inland in a fragmented manner in the area south of Aurora, continuing south of the Berg River across the plains around Langebaanweg. Patches cover parts from Britannia Bay past Paternoster to Danger Bay near Saldanha Bay. The central belt continues south and from the surrounds of the Langebaan Lagoon as a very narrow strip along the West Coast seaboard as far south as Silverstroomstrand at Bokbaai (west of Atlantis) (Fig. 9.26).

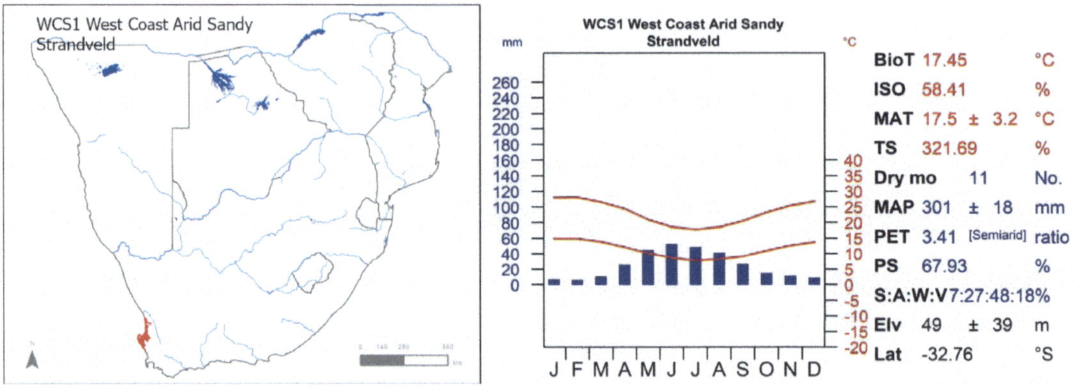

Fig. 9.26 The distribution area and the climatic diagram of the regional ecotonal biome WCS1 West Coast Arid Sandy Strandveld

Vegetation Physiognomy and Composition The vegetation cover is characterised by up to 2 m tall, dense shrubland composed of evergreen, sclerophyllous and fleshy, drought-deciduous shrubs, with a dense understory of low succulent shrubs; the vegetation physiognomy becomes more open on the sandy flats. Graminoids are important in places, and prominent annual herbaceous flora occurs in gaps and forms spectacular displays, especially after good rain in late winter. Important shrubs include *Asparagus capensis, Chrysanthemoides monilifera, Colpoon compressum, Eriocephalus africanus, E. racemosus, Euclea racemosa subsp. racemosa, Limonium capense, L. peregrinum, Metalasia muricata, Muraltia spinosa, Pterocelastrus tricuspidatus, Pteronia divaricata, Putterlickia pyracantha, Salvia africana-lutea, Searsia glauca,* and *S. laevigata*. Succulent shrubs include *Cotyledon orbiculata, Euphorbia burmannii, E. caput-medusae, E. mauritanica, Roepera morgsana, Tetragonia decumbens,* and *Tylecodon paniculatus*. Dominant herbs are *Carpobrotus acinaciformis, Cleretum bellidiforme, Dimorphotheca pluvialis, Oncosiphon suffruticosus, Senecio arenarius,* and *Zaluzianskya parviflora*. Graminoids include *Bromus pectinatus, Cladoraphis cyperoides, Ehrharta calycina, Thamnochortus spicigerus*, and *Willdenowia incurvata*.

Climate Rainfall is mainly cyclonic with the MAP of 301 mm, varying from approximately 125 mm in the north to 380 mm in the south, concentrated in winter, and summers are thus arid. The low monthly precipitation results in eleven months being characterised as dry. Fog and dew contribute to the moisture in summer and autumn. The temperature regime is typically warm-temperate. Frost and hail are rare phenomena. The winds tend to be strong southerly and southeasterly in summer and northerly in winter. Hot, dry, desiccating offshore winds occur in both winter and summer (Fig. 9.26).

Topography, Geology and Soils Series of old dunes, slightly undulating, consolidated sand-dune fields, and extensive sand flats are the typical topographic features of this regional biome. The substrate is mainly deep, Tertiary to Quaternary, white to pale red, calcareous, aeolian, sands and sandy loams. Local white sand of Pleistocene origin forms unstable blow-out dunes north of the mouths of the Verlorenvlei River at Elands Bay and the Jakkals River at Lambert's Bay. Recent calcareous sands of marine origin also line the coastal strip. On the flats, the calcareous substrate is covered by shallow lime-rich sandy soils. Here, the underlying geology is formed by a fossiliferous Pleistocene limestone hardpan of the Sandveld Group along an old marine terrace. It is often exposed in places where farmers rip the hardpan (Rebelo et al. 2006).

Sources Boucher and Jarman (1977), Boucher (1982, 1983, 1987), Liengme (1987), Bloemhoff and Craven (1990), Boucher and le Roux

(1993), Heydenrych (1995), van Rooyen (1981), Rebelo et al. (2006).

WCS2 West Coast Arid Granite Strandveld

FS 2 Saldanha Granite Strandveld (Dayaram et al. 2019)

Distribution This lithobiome is found mainly as granite domes (up to 180 m high) from Vredenburg to St Helena Bay (Western Cape Province of South Africa). In addition, it is limited to some small areas on the Langebaan Peninsula and on the opposite side of the lagoon towards the town of Langebaan and into the West Coast National Park. Also included is a series of coastal headlands between Paternoster and Saldanha (Fig. 9.27).

Vegetation Physiognomy and Composition Vegetation is dominated by low (in places medium-tall) scrub, containing some succulents (e.g., *Aloe glauca, Cleretum bellidiforme, Crassothonna floribunda, Drosanthemum floribundum, Euphorbia mauritanica, Lycium tetrandrum, Tetragonia fruticosa,* and *Tylecodon paniculatus*) and many herbaceous geophytes. The most prominent woody species include *Eriocephalus africanus, Limonium peregrinum, Pteronia divaricata, Searsia glauca* and many more (Rebelo et al. 2006: 199). In spring, under favourable circumstances (depending on timing and amount of rainfall), spectacular floral displays occur, mainly dominated by annual asteraceous herbs of the genera *Dimorphotheca, Oncosiphon, Senecio,* and *Ursinia.*

Climate This regional biome experiences a semi-arid climate (with 11 dry months), with an apparent rainfall concentration in autumn and winter (74%), provided by weakening (in the northward direction) rain-bearing Westerlies. These winds tend to be strong north-westerly in winter and southerly in summer (Rebelo et al. 2006). The MAP of 293 mm varies from about 250 mm in the north to 350 mm in the south. The thermic climate is warm-temperate, with mean daily maximum and minimum temperatures of 25.4 °C and 7.9 °C for February and July, respectively. Advective sea fog and dew contribute significantly to the moisture in summer and autumn. Frost is rare (Fig. 9.27).

Topography, Geology and Soils Rounded forms of granite sheets and high domes dominate the landscape of this regional biome. The undulating topography results from long-term erosion of the granite outcrops of the Vredenburg and Saldanha Batholiths (both Cape Granite Suite). The soils are shallow, coarse, sandy, and loamy.

Sources Boucher (1982), Liengme (1987), Rebelo et al. (2006).

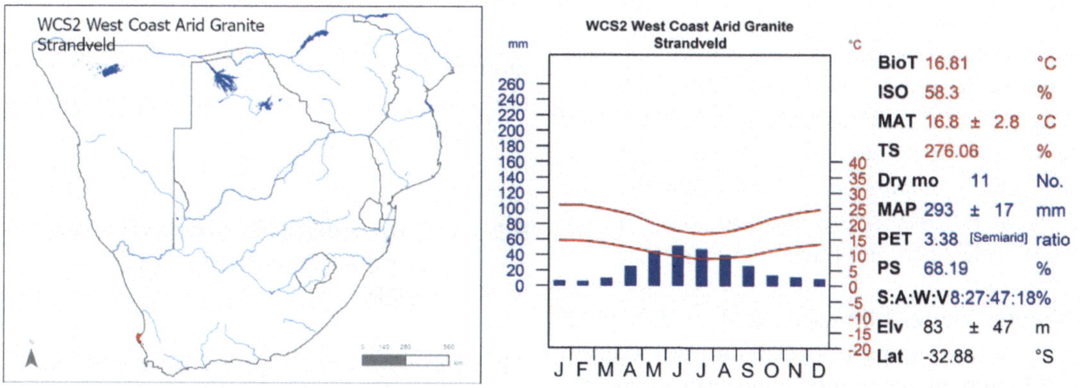

Fig. 9.27 The distribution area and the climatic diagram of the regional ecotonal biome WCS2 West Coast Arid Granite Strandveld

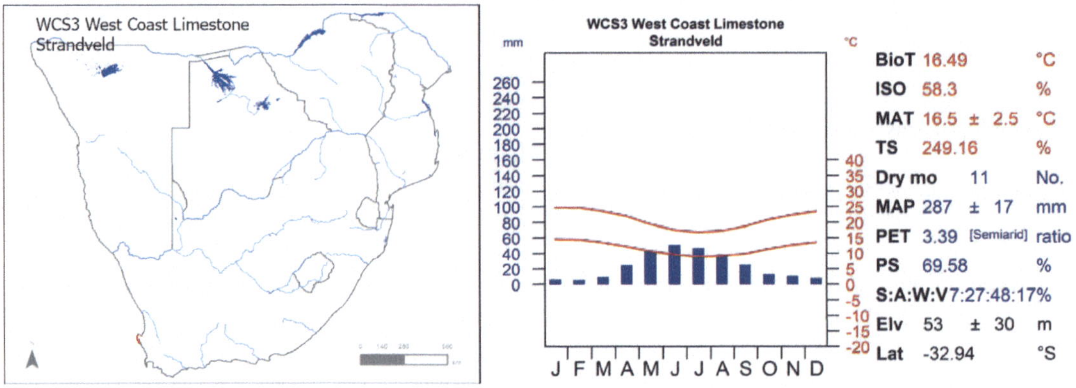

Fig. 9.28 The distribution area and the climatic diagram of the regional ecotonal biome WCS3 West Coast Arid Limestone Strandveld

WCS3 West Coast Arid Limestone Strandveld

FS 4 Saldanha Limestone Strandveld (Dayaram et al. 2019)

Distribution This regional ecotonal biome covers limited areas fringing the broad area from St Helena Bay to Saldanha Bay on the westernmost Atlantic Ocean seaboard of the Western Cape. It occurs as small patches south of Britannia Bay, a narrow belt and patches from Paternoster to Danger Bay, as well as several small patches in the vicinity of Saldanha Bay and Plankies Bay near the tip of the Langebaan peninsula (Fig. 9.28).

Vegetation Physiognomy and Composition The vegetation of this regional biome is characterised by low shrubland to mid-tall scrub, including *Euclea racemosa* subsp. *racemosa, Limonium capense, Muraltia spinosa, Searsia glauca* and other shrubs of the genera *Aspalathus, Clutia, Euryops, Hermannia, Lebeckia,* and many more. Succulent shrubs and herbs are very common (*Cleretum, Conicosia, Euphorbia, Mesembryanthemum, Roepera, Ruschia, Tetragonia*) and so are annual non-succulent herbs in spring.

Climate The region is semi-arid, with a MAP of 303 mm (ranging from 250 mm in the north to 350 mm in the south, concentrated mainly in the autumn and summer months (74%). Advective sea fog and dew contribute to the moisture balance in summer and autumn. Thermically, this is a warm-temperate region, and frost is very infrequent. Strong south-easterly winds are typical of the summer period, whereas northerly winds are more frequent in winter, especially between May and August (Fig. 9.28).

Topography, Geology and Soils The landscape comprises slightly undulating ridges and steeper coastal slopes with shallow sandy soil on hard-pan Tertiary limestone of the Sandveld Group. The limestone ridges are very stony, supporting soil only in pockets. The peinobiome nature of this regional biome is underpinned by the limestone substrate, which, due to the strong chemical bonding of phosphorus to calcium carbonate, causes low levels of available phosphorus for plants.

Sources Boucher and Jarman (1977), Boucher (1987), Bloemhoff and Craven (1990), Rebelo et al. (2006).

9.5 Continental Ecotonal Biome Cape Altimontane Fynbos (CAP)

ZE Ethesial–Austro-Alpine Ecotone S1–A4

The Cape Folded Mountains and the South Andes (roughly at the latitude of Santiago de

Chile) are the only regions in the Southern Hemisphere which support oromediterranean vegetation. Here, the mediterranean-type climate characterised by winter-rainfall maximum coincides with low (cool-temperate range) temperatures caused by high elevation. This ecosystem is a member of the global para-alpine biome, an informal biome unit comprising all ecosystems above the timberline.

Mucina (2023) has addressed the austro-oromediterranean biomes in the Southern Hemisphere in detail and provided arguments for identification of the Cape instalment of the southern oromediterranean to be recognised as an ecotone (at the level of a continental ecotonal biome) rather than a zonal biome per se. Here, we shall limit our discourse only to the subdivision of this continental ecotonal biome into regional ecotonal biomes. We recognise two regional ecotonal biomes (Table 9.1)—a western one based on the FFs 30 Western Altimontane Sandstone Fynbos and an eastern one based on the FFs 31 Swartberg Altimontane Sandstone Fynbos (Fig. 9.29). They are characterised by contrasting precipitation patterns. While the former regional ecotonal biome shows a typical ethesial (winter-rainfall maximum, summer drought) pattern, the latter reflects its geographic position in a region which experiences a large portion of the summer-rainfall, generating a quasi-year-round rainfall probability. Further information on these two regional ecotonal biomes can be found in Rebelo et al. (2006) and the literature cited therein.

The CEB Cape Altimontane Fynbos consists of the following two regional ecotonal biomes:

CAP1 Western Altimontane Fynbos

FFs 30 Western Altimontane Sandstone Fynbos (Mucina and Rutherford 2006)

Distribution This regional biome occurs in the form of an archipelago of small summit areas (peaks and adjacent ridges) at elevations above from about 1800 m in Cederberg, Groot Winterhoek and, most extensively, in Hex River Mts. patches are also found in the extreme western Langeberg and Du Toitsberge. Other localities are found in the vegetation unit FFs 30 description in Rebelo et al. (2006: 127–128) (Fig. 9.30).

Vegetation Physiognomy and Composition Vegetation in these high-elevation vegetation shrublands is dominated by low, open to medium dense restioid fynbos, with ericaceous and asteraceous fynbos also occurring locally. Interestingly, proteoid fynbos is generally absent. The dominating shrub genera are *Acmadenia, Aspalathus, Athanasia, Brunia, Cyclopia, Disparago, Dolichothrix, Erica, Euryops, Helichrysum, Lachnaea, Metalasia, Oedera, Passerina, Polyarrhena, Prismatocarpus, Protea, Selago, Spatalla, Syncarpha,* and *Thesium*. Among the graminoids (restios, sedges, and grasses), prominent genera are *Askidiosperma, Cannomois, Ehrharta, Elegia, Ficinia, Pentaschistis, Restio, Thamnochortus,* and *Willdenowia*.

Climate The MAP of 488 mm (following models as specified in Chap. 1) for this regional biome is much less than the previous estimate of 1384 mm (Rebelo et al. 2006). Resolving this difference requires establishing long-term rainfall stations at the high altitudes of this regional biome. Nevertheless, there is agreement that the MAP is concentrated in winter. The mean annual temperature is only 10 °C. Mean daily maximum and minimum temperatures are 22.9 °C and 0.1 °C for February and July, respectively. Microclimatic measurements on Waaihoek Peak at an elevation of about 1900 m (Boelhouwers 1998) revealed that diurnal frost cycles occur from May to September. Snow also occurs, and estimates suggest 31 days per year (Rebelo et al. 2006) (Fig. 9.30).

Topography, Geology and Soils Topography is typically dramatic for this high elevation. It is characterised by summit peaks, adjacent ridges and steep slopes, generally fragmented and localised but relatively extensive in the Hex River Mountains. Skeletal and rocky acidic lithosol soils derived from Ordovician

Fig. 9.29 **a** An early winter view of the Matroosberg in the Hexrivier Mts, Western Cape. The present snowline in the background indicates approximately the position of the lower boundary of the austro-oromediterranean scrub vegetation. **b** The summit area of the Rooiberg (2209 m) in the Hexrivier Mts, Western Cape. The relatively smooth boulder colluvium below the summit cliffs covers a shale band. **c** Summit and steep slopes of a peak in the Klein Swartberg with boulder fields containing pockets of nutrient-poor soils derived from sandstone. The oromediterranean vegetation is sparse and limited to the summit region of the mountain. **d** The high elevations in the Cedarberg characterised by a predominance of restios. In the rocky areas, sciophilous. Proteas are rare and are typically dwarf resprouting species exemplified by the endemic *Protea cryophila*. Cedarberg Tafelberg (1969 m), supporting the oromediterranean vegetation, visible in the background. Photo credits: **a** & **b**: L. Mucina; **c**: C. Boucher; **d**: D. Gwynne-Evans

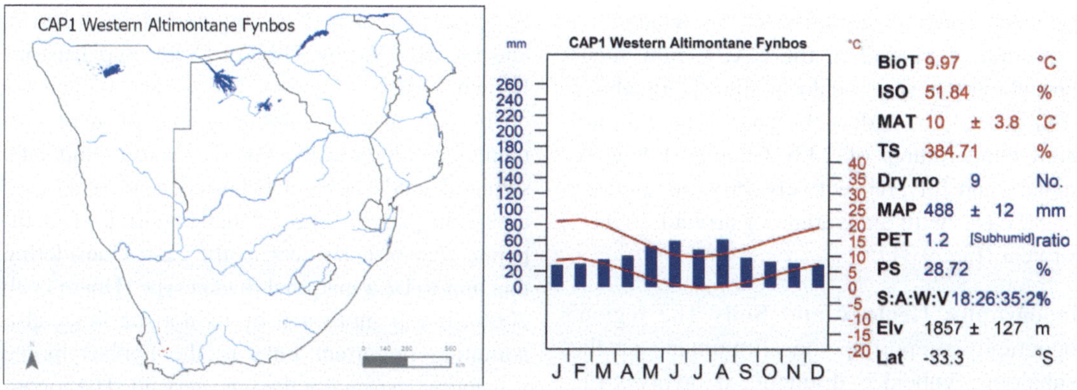

Fig. 9.30 The distribution area and the climatic diagram of the regional ecotonal biome CAP1 Western Altimontane Fynbos

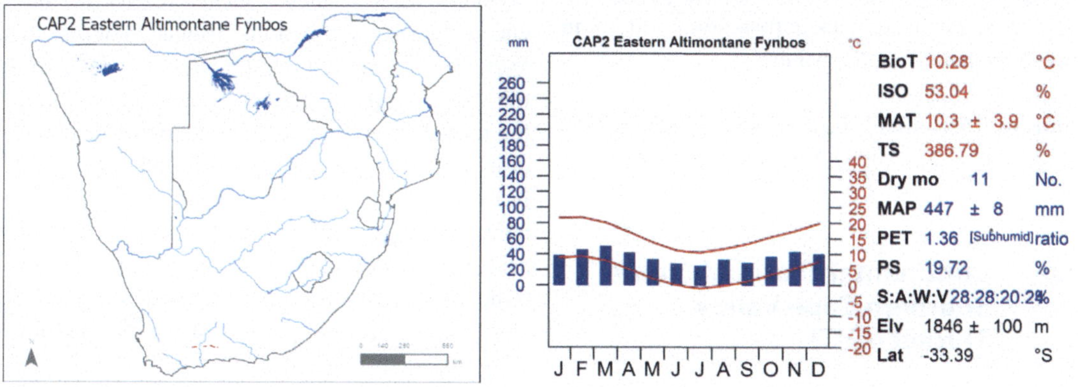

Fig. 9.31 The distribution area and the climatic diagram of the regional ecotonal biome CAP2 Eastern Altimontane Fynbos

sandstones of the Table Mountain Group (Cape Supergroup) dominate.

Sources Marloth (1902), Linder et al. (1993), McDonald et al. (1993), Taylor (1996), Boelhouwers (1998), Rebelo et al. (2006), CapeNature (2021).

CAP2 Eastern Altimontane Fynbos

FFs 31 Swartberg Altimontane Sandstone Fynbos (Mucina and Rutherford 2006)

Distribution This unit consists of a small number, yet sizable, patches of altimontane (subalpine) fynbos in the Western Cape Province at high elevations above 1800 m of the Groot Swartberg, Klein Swartberg, and Kammanassie mountain ranges. The full extent of the distribution is described in Rebelo et al. (2006: 128–129) (Fig. 9.31).

Vegetation Physiognomy and Composition The vegetation here is a typical low, open to medium dense restioid fynbos, with some more localised ericaceous and asteraceous fynbos. As in FFs 30 Western Altimontane Sandstone Fynbos, proteoid fynbos is generally absent.

Climate The climate of the regional biome CAP2 differs primarily from that of the CAP1 in terms of magnitude and seasonal distribution of MAP of 447 mm is lower than that of the CAP1 and, in contrast to the winter concentration of rainfall in the CAP1, rain occurs throughout the year, peaking slightly in March. MAT is 10.3 °C, with daily maximum and minimum temperatures of 23.6 °C and −1.1 °C for January and July, respectively. Frost incidence is 10–30 days (with an average of around 20 days) per year (Fig. 9.31).

Topography, Geology and Soils The region's topography supporting the altimontane patches (subalpine fynbos) is dramatic. It involves rugged high-elevation summit peaks and steep boulder slopes. Soils are shallow, skeletal, acidic lithosols derived from Ordovician sandstones of the Table Mountain Group (Cape Supergroup). Narrow shale bands run across some ridges in the Klein and Groot Swartberg.

Sources Marloth (1902), Linder et al. (1993), McDonald et al. (1993), Rebelo et al. (2006).

9.6 Continental Ecotonal Biome Southern Cape Valley Thicket (CVT)

ZE Semiarid Warm-Temperate Ecotone (S2–E1(sub))–T3

The position of the vegetation unit AT1 Southern Cape Valley Thicket (as defined by VegMap2006) has been a problem since the first days of coining this concept and classifying it within the Albany Thicket biome (see Hoare et al. 2006, incl. the relevant synonyms and the most detailed published description so far). Indeed, the vegetation typical of this unit has a strong floristic relationship with Albany Thickets. However, it is embedded within the Fynbos biome. It is, therefore, not surprising that in the past this ecosystem carried various, often conflicting, labels such as the Aloe Scrub (Muir 1929), Valley Bushveld and Coastal Renosterbosveld (Acocks 1953), South and South-west Coast Renosterveld (Low and Rebelo 1996), Gouritz Mesic Succulent Thicket (Cowling and Heijnis 2001), and Gouritz Valley Thicket (Vlok and Euston-Brown 2002).

To add another confusing classification element, bioclimatically, it is warm-temperate, semiarid, and shows a balanced year-round precipitation pattern, like the neighbouring afrotemperate forests it borders in the east. Considering this unit to be a marginal thicket type (Hoare et al. 2006), it was motivated by its floristic links, also admitting important links to the Fynbos biome vegetation, especially the renosterveld. The topography (mainly east-facing or west-facing steep slopes) and soil cover (shallow, well-drained and nutrient-poor soils) add to local aridity. The Southern Cape Valley Thicket region is influenced by two rain-bearing frontal systems—the Westerlies, bringing rainfall mainly in winter from the West, and the eastern anticyclonic front, bringing rainfall from the East. The topographic configuration (deeply incised valleys) might play a role in the relatively limited water availability in some areas, perhaps through the formation of small-scale rain shadows and by increasing microclimate temperatures on insolation-exposed steep valley slopes during certain periods of the day and possibly impeded air circulation.

Since the regional biomes are conceptually determined by the regional climate and soil/geological conditions, the CVT1 Southern Cape Valley Thicket cannot be classified unequivocally as Albany Thicket, Fynbos, or Afrotemperate Forest. However, the combination of its semiarid nature and equitable seasonal precipitation prompt us to consider the Southern Cape Valley Thicket as an ecotonal bioclimatic phenomenon, occupying the transition between CB Albany Thicket and CB Southern African Afrotemperate Forest climatic envelopes, as also expressed by its zonoecotonal code. This zonoecotone is globally unique. We also hypothesise that it may have relict origins, as it bears traces of previous dry periods.

CVT1 Southern Cape Valley Thicket

AT 1 Southern Cape Valley Thicket (Mucina and Rutherford 2006)

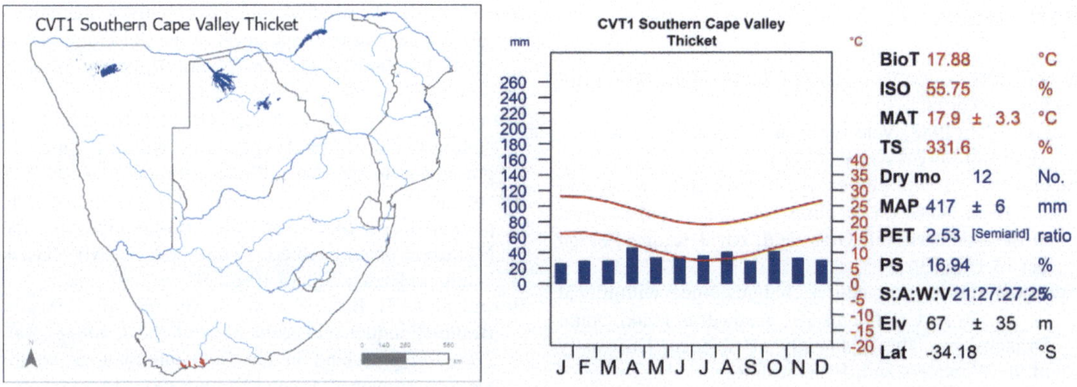

Fig. 9.32 The distribution area and the climatic diagram of the regional ecotonal biome CVT1 Southern Cape Valley Thicket

Distribution This regional ecotonal biome occurs in a series of isolated patches covering slopes, at elevations of 20–200 m, of incised rivers of the South Coast of the Western Cape Province, mainly along the Gouritz River Valley (between Herbertsdale and the Gouritzmond; Fig. 9.17f) as well as the Goekoe (between Riversdale and Still Bay), Groot Brak, Klein Brak and Kaaimans Rivers (here at its easternmost limit). Towards the west, this vegetation is found in the Duiwenhoek and Slang Rivers valleys and reaches its westernmost distribution limits in the Breede River Valley (Fig. 9.32).

Vegetation Physiognomy and Composition The vegetation is very heterogenous, characterised in places by dense thicket composed of sclerophyllous (often spinescent) evergreen shrubs (e.g., of the genera *Euclea, Grewia, Gymnosporia, Putterlickia, Searsia, Sideroxylon, Tarchonanthus*) with an important admixture of a microphyllous (partly ericoid) Cape shrub element (e.g., *Athanasia, Dicerothamnus, Oedera, Stoebe*) and a prominent (locally also dominant) succulent, rosulate tree *Aloe ferox*. The low shrub layer contains a high proportion of succulent shrubs (*Aloe, Crassula, Euphorbia, Ruschia*). Grasses are abundant in some favoured grazing areas.

Climate This unit has a non-seasonal rainfall with several minor relative maxima and the driest summer months, December to March. The MAP of 417 mm ranges from about 340 mm in the west (Swellendam) to about 940 mm in the east (Wilderness). Frost is infrequent. MAT falls within the warm-temperate range (17.9 °C). Mean monthly maximum and minimum temperatures for the nearby Riversdale weather station are 37.5 °C and 0.2 °C for February and July, respectively, and corresponding values on the coast for Mossel Bay (Cape St Blaize) are 29.6 °C and 7.4 °C for April and August, respectively (Hoare et al. 2006) (Fig. 9.32).

Topography, Geology and Soils The regional biome comprises steep slopes of deeply incised valleys of rivers flowing mainly in a north–south direction and dissecting the Southern Cape coastal peneplain. Shallow, loamy-clayey soils (mostly Glenrosa and Mispah) derived from siltstone and shales of the Bokkeveld Group are found in the western part of the area, and Jurassic Enon conglomerates and other clastic sediments of the Uitenhage Group in the eastern part of the area. On the hard Ordovician Table Mountain Sandstone, only on the Aasvogelberg range, there are rocky south-facing slopes with organic-rich soils (Rebelo et al. 1991).

Sources Muir (1929), Acocks (1953, 1988), Grobler and Marais (1967), Taylor (1970), Rebelo et al. (1991), Cowling and Heijnis (2001), Vlok and Euston-Brown (2002), Vlok et al. (2003), Hoare et al. (2006).

References

Acocks JPH (1953) Veld types of South Africa. Mem Bot Surv S Afr 28:1–192

Acocks JPH (1988) Veld types of South Africa, 3rd edn. Mem Bot Surv S Afr 57:1–146

Adams KM (1996) Influence of sward defoliation and soil disturbance on seedling emergence and survival in the Southern Tall Grassveld. Afr J Range Forage Sci 13:131–136

Adcock K (1990) Vegetation changes since withdrawal of fire from Catchment IX, Cathedral Peak, Natal Drakensberg. Thesis, Faculty of Science, University of the Witwatersrand, Johannesburg

Anonymous (2000) Biological diversity in Lesotho: a country study. National Environment Secretariat, Maseru

Archibald EEA (1955) An ecological survey of the Addo Elephant National Park. J S Afr Bot 20:137–154

Aucamp AJ (1976) The role of the browser in the bushveld of the eastern Cape. Proc Grassl Soc S Afr 11:135–138

Aucamp AJ (1979) Die produksiepotensiaal van die Valleibosveld as weiding vir Boer- en Angorabokke. Thesis, University of Pretoria, Pretoria

Aucamp AJ, Tainton NM (1984) Veld management in the Valley Bushveld of the Eastern Cape. Department of Agriculture Bulletin 41. Dohne Research Station, Stutterheim

Avis AM, Lubke RA (1996) Dynamics and succession of coastal dune vegetation in the Eastern Cape, South Africa. Landsc Urb Plan 34:237–254

Backéus I (1988) Mires in the Thaba-Putsoa Range of the Maloti, Lesotho. Stud Plant Ecol 17:1–89

Backéus I (1992) Distribution and vegetation dynamics of humid savannas in Africa and Asia. J Veg Sci 3:345–356

Becker CH, Coetsee C, Cowling RM et al (2015) The local landscape boundary between the Albany subtropical thicket and Nama-Karoo shrubland is not influenced by edaphic factors. S Afr J Bot 101:107–111

Bergh NG, Verboom GA, Rouget M et al (2014) Vegetation types of the Greater Cape Floristic Region. In: Allsopp N, Colville JF, Verboom GA et al (eds) Fynbos: ecology, evolution and conservation of a megadiverse region. Oxford University Press, Oxford, pp 1–25

Bester SP (1998) Vegetation and flora of the southern Drakensberg escarpment and adjacent areas. Thesis, Department of Botany, University of Pretoria, Pretoria

Bews JW (1917) The plant ecology of the Drakensberg Range. Ann Natal Mus 3:511–565

Bezuidenhout H (1988) 'n Platekologiese studie van die Mooirivieropvanggebied, Transvaal. Thesis, Department of Botany, Potchefstroom University for C.H.E., Potchefstroom

Bezuidenhout H (1993) Syntaxonomy and synecology of western Transvaal grasslands. Thesis, Department of Botany, University of Pretoria, Pretoria

Bezuidenhout H, Bredenkamp GJ (1990) A reconnaissance survey of the vegetation of the dolomitic region in the Potchefstroom–Ventersdorp–Randfontein area, South Africa. Phytocoenologia 18:387–403

Bezuidenhout H, Bredenkamp GJ (1991a) The vegetation of the Bc land type in the western Transvaal grassland, South Africa. Phytocoenologia 19:497–518

Bezuidenhout H, Bredenkamp GJ (1991b) Die plantgemeenskappe van die Ba-landtipe in die Mooirivieropvanggebied, Transvaal. S-Afr Tydsk Natuurwetensch Tegnol 10:85–92

Bezuidenhout H, Bredenkamp GJ, Theron GK (1994a) A Braun-Blanquet reclassification of the *Cymbopogon-Themeda* grassland in the Lichtenburg area, south-western Transvaal. S Afr J Bot 60:306–314

Bezuidenhout H, Bredenkamp GJ, Theron GK (1994b) A classification of the vegetation of the western Transvaal dolomite and chert grassland, South Africa. S Afr J Bot 60:152–161

Bezuidenhout H, Bredenkamp GJ, Theron GK (1994c) Phytosociological classes of the western Transvaal grassland, South Africa. Koedoe 37:1–18

Bezuidenhout H, Bredenkamp GJ, Theron GK (1994d) Syntaxonomy of the vegetation of the Fb land type in the western Transvaal grassland, South Africa. S Afr J Bot 60:72–80

Bezuidenhout H, Bredenkamp GJ, Theron GK (1994e) The vegetation syntaxa of the Ba land type in the western Transvaal grassland, South Africa. S Afr J Bot 60:214–224

Bezuidenhout H, Bredenkamp GJ, Theron GK et al (1994f) A Braun-Blanquet reclassification of the Bankenveld Grassland in the Lichtenburg area, south-western Transvaal. S Afr J Bot 60:297–305

Bloem KL, Theron GK, van Rooyen N (1993) Grassland communities of the Verlorenvlei Nature Reserve in the North-eastern Sandy Highveld, Transvaal. S Afr J Bot 59:273–280

Bloemhoff A, Craven A (1990) Gedetailleerde plantkundige opname van moontlike padbelyning deur 'n gedeelte van die Weskus Nasionale Park en die voorkoms van skaars en bedreigde plante daarin. BSc(Hons) Project, Department of Botany, University of Stellenbosch, Stellenbosch

Boelhouwers JC (1998) Environmental controls on soil frost activity in the Western Cape mountains, South Africa. Earth Surf Process Landsform 23:211–221

Bond WJ (2019) Open ecosystems. Ecology and evolution beyond the forest edge. Oxford University Press, Oxford

Botha AM (2003) A plant ecological study of the Kareefontein Private Nature Reserve, Free State Province. Thesis, Department of Plant Sciences, University of the Free State, Bloemfontein

Boucher C (1982) Some thoughts about the West Coast Strandveld vegetation. Veld Flora 68(4):108–109

Boucher C (1983) Floristic and structural features of the coastal foreland vegetation south of the Berg River, Western Cape Province, South Africa. Bothalia 14:669–674

Boucher C (1987) A phytosociological study of transects through the Western Cape coastal foreland, South Africa. Thesis, Department of Botany, University of Stellenbosch, Stellenbosch

Boucher C, Jarman ML (1977) The vegetation of the Langebaan area, South Africa. Trans Roy Soc S Afr 42:241–272

Boucher C, le Roux A (1993) Dry coastal ecosystems of the South African west coast. In: van der Maarel E (ed) Ecosystems of the world. 2B. Dry coastal ecosystems. Africa, America and Oceania. Elsevier, Amsterdam, pp 75–88

Boucher C, Tlale S (1999a) Lesotho Highlands Water Project. Consulting services for the establishment and monitoring of the instream requirements for river courses downstream of LHWP dams. Task 2 Report Series. Lesotho Highlands Development Authority, Maseru

Boucher C, Tlale S (1999b) Riparian and instream vegetation, Lesotho. In: Starter document for Instream Flow Requirement Workshop. Ecological Research Report 42. Department of Botany, University of Stellenbosch, Stellenbosch

Brand RF, Brown LR, du Preez PJ (2011) The Grassland vegetation of Platberg, eastern Free State, South Africa. Koedoe 53(1):Art. #1027. https://doi.org/10.4102/koedoe.v53i1.1027

Bredenkamp GJ (1975) Plant communities of the Suikerbosrand Nature Reserve, Transvaal. S Afr J Sci 71:30–31

Bredenkamp GJ (1976) Vegetation units for management of the grasslands of the Suikerbosrand Nature Reserve. S Afr J Wildl Res 6:113–122

Bredenkamp GJ (1977) The grasses of the Suikerbosrand Nature Reserve: their habitat preferences and synecological significance. Proc Grassl Soc S Afr 12:135–139

Bredenkamp GJ, Bezuidenhout H (1990) The phytosociology of the Faan Meintjes Nature Reserve in the western Transvaal grassland. S Afr J Bot 56:54–64

Bredenkamp GJ, Brown LR (2003) A reappraisal of Acocks' Bankenveld: origin and diversity of vegetation types. S Afr J Bot 69:7–26

Bredenkamp GJ, Theron GK (1978) A synecological account of the Suikerbosrand Nature Reserve, Part I: the phytosociology of the Witwatersrand geological system. Bothalia 12:513–529

Bredenkamp GJ, Theron GK (1980) A synecological account of the Suikerbosrand Nature Reserve, Part II: The phytosociology of the Ventersdorp geological system. Bothalia 13:199–216

Bredenkamp GJ, Bezuidenhout H, Joubert AF et al (1994) The vegetation of the Boskop Dam Nature Reserve, Potchefstroom. Koedoe 37:19–33

Bredenkamp GJ, Spada F, Kazmierczak E (2002) On the origin of northern and southern hemisphere grasslands. Plant Ecol 163:209–229

Breytenbach PJJ (1991) A phytosociological study of the south-eastern Transvaal Highveld. Thesis, Department of Botany, University of Pretoria, Pretoria

Breytenbach PJJ, Myburgh WJ, Theron GK et al (1992) The phytosociology of the Villers-Grootvlei area, South Africa. 1. Physical environment and the plant communities of the Bb Land Type. S Afr J Bot 58:239–249

Breytenbach PJJ, Myburgh WJ, Theron GK et al (1993a) The phytosociology of the Villers-Grootvlei area, South Africa. 2. The plant communicates of the Ba Land Type. S Afr J Bot 59:218–226

Breytenbach PJJ, Myburgh WJ, Theron GK et al (1993b) The phytosociology of the Villers-Grootvlei area, South Africa. 3. The plant communities of the Ib Land Type. S Afr J Bot 59:227–234

Breytenbach PJJ, Myburgh WJ, Theron GK et al (1993c) The phytosociology of the Villers-Grootvlei area, South Africa. 4. Plant communities of the Ea land type. S Afr J Bot 59:235–246

Brink E, Dold AP (2003) *Drimia acarophylla* (Hyacinthaceae), a new species from Eastern Cape, South Africa. S Afr J Bot 69:396–400

Brown LR, Bezuidenhout H (2000) The phytosociology of the De Rust section of the Mountain Zebra National Park, Eastern Cape. Koedoe 43:1–18

Brown LR, Bezuidenhout H (2005) The vegetation of the farm Ingleside and Welgedacht of the Mountain Zebra National Park, Eastern Cape. Koedoe 48:23–42

Brown LR, Du Preez PJ (2014) Threats of mining and urbanisation on a vulnerable ecosystem in the Free State, South Africa. In: Grillo O (ed) Biodiversity—The dynamic balance of the planet. IntechOpen, pp 229–348. https://doi.org/10.5772/57589

Brown LR, Marais H, Henzi SP et al (2005) Vegetation classification as the basis for baboon management in the Bourke's Luck Section of the Blyde Canyon Nature Reserve, Mpumalanga. Koedoe 48:71–92

Burgoyne PM, Krynauw S, Smith GF (2000) *Frithia*—up close and personal. Aloe 37:38–42

Burns MER (1986) A synecological study of the East London coast dune forests. Thesis, Department of Botany, Rhodes University, Grahamstown

Cadman M, de Villiers C, Lechmere-Oertel R et al (compilers) (2013) Grasslands Ecosystem Guidelines: landscape interpretation for planners and managers. SANBI, Pretoria

Camp K (1999a) The bioresource groups of KwaZulu-Natal. Mistbelt. Cedara Report N/A/99/13. KwaZulu-Natal Department of Agriculture, Cedara

Camp K (1999b) The bioresource groups of KwaZulu-Natal. Highland and montane. Cedara Report N/A/99/14. KwaZulu-Natal Department of Agriculture, Cedara

Camp K (1999c) The Bioresource Groups of KwaZulu-Natal. Tall Grassveld. Cedara Report No. N/A/99/15. KwaZulu-Natal Department of Agriculture, Cedara

CapeNature (2021) Hexriver Complex: Protected Area Management Plan 2021–2031. CapeNature, Cape Town

Carvalho SL, Campbell EE (2023a) Biodiversity-based evaluation of thicket thresholds at different levels of

elephant impact—towards assessment and monitoring of thicket vegetation. Ecol Inform 77:102197. https://doi.org/10.1016/j.ecoinf.2023.102197
Carvalho SL, Campbell EE (2023b) Using elephant impacts to gauge the biodiversity integrity of Valley Thicket in the Eastern Cape, South Africa. S Afr J Bot 153:53–61
Carvalho SL, Watson JJ, Campbell EE (2021) Calcrete Bontveld bushclumps are de novo thicket islands. S Afr J Bot 139:79–82
Chonco N, Slotow R, Tsvuura Z et al (2023) Ecosystem resilience of a South African mesic grassland with change from rotational to continuous grazing. Diversity 15:1187. https://doi.org/10.3390/d15121187
Codd LE (1968) The South African species of *Kniphofia* (Liliaceae). Bothalia 9:363–513
Coetzee BJ (1974) A phytosociological classification of the vegetation of the Jack Scott Nature Reserve. Bothalia 11:329–347
Coetzee BJ (1975) A phytosociological classification of the Rustenburg Nature Reserve. Bothalia 11:561–580
Coetzee BJ, Werger MJA (1975) A west-east vegetation transect through Africa south of the Tropic of Capricorn. Bothalia 11:539–560
Coetzee JP (1993) Phytosociology of the Ba and Ib land types in the Pretoria–Witbank–Heidelberg area. Thesis, Department of Botany, University of Pretoria, Pretoria
Coetzee JP, Bredenkamp GJ, van Rooyen N et al (1994) An overview of the physical environment and vegetation units of the Ba and Ib land types of the Pretoria–Witbank–Heidelberg area. S Afr J Bot 60:49–61
Coetzee JP, Bredenkamp GJ, van Rooyen N (1995) The phytosociology of the grasslands of the Ba and Ib land types in the Pretoria–Witbank–Heidelberg area. S Afr J Bot 61:123–133
Colvin IS (1984) Survey of the conservation worthy areas in Natal. The Natal Mistbelt of Ngongoni Veld. Internal Report No. 35. Natal Parks Board, Pietermaritzburg
Comins DM (1962) The vegetation of the districts of East London and King William's Town, Cape Province. Mem Bot Surv S Afr 33:1–32
Cowling RM (1983) Phytochorology and vegetation history in the southeastern Cape, South Africa. J Biogeogr 10:393–419
Cowling RM (1984) A syntaxonomic and synecological study in the Humansdorp region of the Fynbos Biome. Bothalia 15:175–227
Cowling RM, Heijnis CE (2001) The identification of Broad Habitat Units as biodiversity entities for systematic conservation planning in the Cape Floristic Region. S Afr J Bot 67:15–38
Cowling RM, Kerley GIH (2002) Impacts of elephants on the flora and vegetation of subtropical thicket in the Eastern Cape. In: Kerley G, Wilson S, Massey A (eds) Elephant conservation and management in the Eastern Cape. Terrestrial Ecological Research Unit Report 35. University of Port Elizabeth, Port Elizabeth, pp 55–72
Cowling RM, Pierce SM (1985) Southern Cape coastal dunes—an ecosystem lost? Veld Flora 71(4):99–103
Cowling RM, Procheş Ş, Vlok JHJ (2005) On the origin of southern African subtropical thicket vegetation. S Afr J Bot 71:1–23
Daemane ME, Cilliers SS, Bezuidenhout H (2010) An ecological study of the plant communities in the proposed Highveld National Park, in the peri-urban area of Potchefstroom, South Africa. Koedoe 52(1):Art. #708. https://doi.org/10.4102/koedoe.v52i1.708
Daemane ME, Cilliers SS, Bezuidenhout H (2012) Classification and description of the vegetation in the Spitskop area in the proposed Highveld National Park, North West Province, South Africa. Koedoe 54(1):Art. #1020. https://doi.org/10.4102/koedoe.v54i1.1020
Daemane ME, Ramoelo A, Adelabu S et al (2022) Plant community assemblages and environmental drivers in the mountainous grassland ecosystem of South Africa. Phytocoenologia 51:263–274
Dabengwa AN, Gillson L, Bond WJ (2022) Resilience modes of an ancient mountain valley grassland in South Africa indicated by palaeoenvironmental methods. Environ Res Lett 16:055002. https://doi.org/10.1088/1748-9326/abdf87
Dayaram A, Harris LR, Grobler BA et al (2019) Vegetation Map of South Africa, Lesotho and Swaziland 2018: A description of changes since 2006. Bothalia 49(1):a2452. https://doi.org/10.4102/abc.v49i1.2452
Deall GB (1985) A plant-ecological study of the Mpumalanga Escarpment in the Sabie area. Thesis, Department of Botany, University of Pretoria, Pretoria
Deal GB, Scheepers JC, Schutz CJ (1989a) The vegetation ecology of the eastern Transvaal escarpment in the Sabie area. 1 Physical Environment. Bothalia 19:53–67
Deal GB, Scheepers JC, Schutz CJ (1989b) The vegetation ecology of the eastern Transvaal escarpment in the Sabie area. 2. Floristic Classification. Bothalia 19:69–89
De Klerk J, Brown LR, Bezuidenhout H (2003) Plant communities of the Ebenaeser section of the Mountain Zebra National Park. Koedoe 46:1–13
de Frey WH (1999) Phytosociology of the Mpumalanga high altitude grasslands. Thesis, Department of Botany, University of Pretoria, Pretoria
De Kock R (2011) Bushclump rehabilitation of Grassridge Bontveld after strip mining. Thesis, Department of Botany, University of Port Elizabeth, Port Elizabeth
Dingaan MNV (1996) A phytosociological study of the 'Valley of Seven Dams', Bloemfontein. BSc(Hons) Project, Department of Botany and Genetics, University of the Orange Free State, Bloemfontein
Dingaan M (1999) The phytosociology of the natural open spaces in Bloemfontein, Free State. Thesis, Department of Botany and Genetics, University of the Orange Free State, Bloemfontein

Dingaan MNV, Tsubo M (2019) Improved assessment of pasture availability in semi-arid grassland of South Africa. Environ Monit Assess 191:733. https://doi.org/10.1007/s10661-019-7918-x

Dingaan MNV, du Preez PJ (2002) The phytosociology of the succulent dwarf shrub communities that occur in the 'Valley of Seven Dams' area, Bloemfontein, South Africa. Navors Nasion Mus Bloemfontein 18:33–48

Dingaan MNV, du Preez PJ (2013) Grassland communities of urban open spaces in Bloemfontein, Free State, South Africa. Koedoe 55(1):Art. #1075. https://doi.org/10.4102/koedoe.v55i1.1075

Dingaan MNV, du Preez PJ (2017) Floristic composition and species diversity of urban vegetation in Bloemfontein, Free State, South Africa. Bothalia 47(1):a2244. https://doi.org/10.4102/abc.v47i1.2244

Dingaan M, Walker S, Newby T et al (2013) Plant species richness, soil properties and grazing effects in semiarid grassland of South Africa. S Afr J Bot 86:174. https://doi.org/10.1016/j.sajb.2013.02.131

Dingaan MNV, Walker S, Tsubo M et al (2016) Influence of grazing on plant diversity-productivity relationship in semi-arid grassland of South Africa. Appl Ecol Environ Res 14:1–13. https://doi.org/10.15666/aeer/1404_001013

Dingaan MNV, Tsubo M, Walker S et al (2017) Soil chemical properties and plant species diversity along a rainfall gradient in semi-arid grassland of South Africa. Plant Ecol Evol 150:35–44

Duker R, Cowling RM, Du Preez DR et al (2015a) Frost, *Portulacaria afra* Jacq., and the boundary between the Albany Subtropical Thicket and Nama-Karoo biomes. S Afr J Bot 101:112–119

Duker R, Cowling RM, Du Preez DR et al (2015b) Community-level assessment of freezing tolerance: frost dictates the biome boundary between Albany subtropical thicket and Nama-Karoo in South Africa. J Biogeogr 42:167–178

du Preez PJ (1986) Ekologie van die boomgemeenskappe van die Vredefort Distrik, Oranje-Vrystaat. Thesis, University of the Orange Free State, Bloemfontein

du Preez PJ (1991) A syntaxonomical and synecological study of the vegetation of the south-eastern Orange Free State and related areas with special reference to Korannaberg. Thesis, Department of Botany and Genetics, University of the Orange Free State, Bloemfontein

du Preez PJ, Bredenkamp GJ (1991) Vegetation classes of the southern and eastern Orange Free State (Republic of South Africa) and the highlands of Lesotho. Navors Bloemfontein Mus 7:477–526

Du Toit PCV, Blom CD (1995) Diet selection by sheep and goats in the Noorsveld. Afr J Range Forage Sci 12:27–37

Dyer RA (1937) The vegetation of the divisions of Albany and Bathurst. Mem Bot Surv S Afr 17:1–138

Eckhardt HC (1993) A synecological study of the north-eastern Orange Free State. Thesis, Department of Botany, University of Pretoria, Pretoria

Eckhardt HC (1998) Vegetation ecology of the northern KwaZulu-Natal grasslands. Thesis, Centre for Wildlife Management, University of Pretoria, Pretoria

Eckhardt HC, van Rooyen N, Bredenkamp GJ (1993a) The vegetation of the north-eastern Orange Free State, South Africa: the physical environment and the plant communities of the Ea land type. Bothalia 23:117–127

Eckhardt HC, van Rooyen N, Bredenkamp GJ (1993b) An overview of the vegetation of the Vrede–Memel–Warden area, north-eastern Orange Free State. S Afr J Bot 59:391–400

Eckhardt HC, van Rooyen N, Bredenkamp GJ (1995) The grassland communities of the slopes and plains of the north-eastern Orange Free State. Phytocoenologia 25:1–21

Eckhardt HC, van Rooyen N, Bredenkamp GJ (1996a) Plant communities and species richness of the *Agrostis lachnantha–Eragrostis plana* Wetlands of northern KwaZulu-Natal. S Afr J Bot 62:306–315

Eckhardt HC, van Rooyen N, Bredenkamp GJ (1996b) Species richness and plant communities of the *Helichrysum rugulosum–Hyparrhenia hirta* Low-altitude grassland of northern KwaZulu-Natal. S Afr J Bot 62:296–305

Eckhardt HC, van Rooyen N, Bredenkamp GJ (1996c) The plant communities and species richness of the *Alepidea longifolia–Monocymbium ceresiiforme* high altitude grassland of northern KwaZulu Natal. Koedoe 39:53–68

Eckhardt HC, van Rooyen N, Bredenkamp GJ (1997) Plant communities of the forests, woodlands and thickets in northern KwaZulu-Natal. Koedoe 40:91–112

Edwards D (1967) A plant ecological survey of the Tugela River basin. Town and Regional Planning Commission, Pietermaritzburg

Ellery WN, Balkwill K, Ellery K et al (2001) Conservation of the vegetation on the Melville Ridge, Johannesburg. S Afr J Bot 67:261–273

Emery AJ, Lötter MC, Williamson SD (eds) (2002) Determining the conservation value of land in Mpumalanga. Mpumalanga Parks Board, Nelspruit

Evans N, Avis T, Palmer AR (1997) Changes to the vegetation of the mid-Fish River valley, Eastern Cape, South Africa, in response to land-use, as revealed by a direct gradient analysis. Afr J Range Forage Sci 14:68–74

Everard DA (1987) A classification of the subtropical transitional thicket in the eastern Cape, based on syntaxonomic and structural attributes. S Afr J Bot 51:329–340

Everson CS (1985) Ecological effects of fire in the Montane grasslands of Natal. Thesis, Department of Grassland Sciences, University of Natal, Pietermaritzburg

Findlay N, Manson A, Cromsigt JPGM et al (2022) Long-term frequent fires do not decrease topsoil carbon and nitrogen in an Afromontane grassland. Afr J Range Forage Sci 39:44–55

Fuls ER (1993) Vegetation ecology of the northern Orange Free State. Thesis, Department of Botany, University of Pretoria, Pretoria

Fuls ER, Bredenkamp GJ, van Rooyen N (1992) The plant communities of the undulating grasslands of the Vredefort–Kroonstad–Lindley–Heilbron area, northern Orange Free State. S Afr J Bot 58:224–230

Fuls ER, Bredenkamp GJ, van Rooyen N (1993a) Grassland communities of rocky outcrops in the northern Orange Free State. S Afr J Bot 59:370–376

Fuls ER, Bredenkamp GJ, van Rooyen N (1993b) Low thicket communities of rocky outcrops in the northern Orange Free State. S Afr J Bot 59:360–369

Fuls ER, Bredenkamp GJ, van Rooyen N et al (1993c) The physical environment and major plant communities of the Heilbron–Lindley–Warden–Villiers area, northern Orange Free State. S Afr J Bot 59:345–359

Furniss DG (2008) Can indices of Landscape Function Analysis (LFA) be derived from Ground-Based Spectroscopy? A case study from gold mines on the Highveld of South Africa, Thesis, School of Animal, Plant and Environmental Science, University of the Witwatersrand, Johannesburg

Fynn RWS, O'Connor TG (2005) Determinants of community organization of a South African mesic grassland. J Veg Sci 16:93–102

Fynn RWS, Morris CD, Edwards T (2005) Long-term compositional responses of a South African mesic grassland to burning and mowing. Appl Veg Sci 8:5–12

Gokul JK, Matcher G, Dames J et al (2023) Microbial community responses to alterations in historical fire regimes in montane grasslands. Diversity 15:818. https://doi.org/10.3390/d15070818

Gordijn PJ, Everson TM, O'Connor TG (2018) Resistance of Drakensberg grasslands to compositional change depends on the influence of fire-return interval and grassland structure on richness and spatial turnover. Persp Plant Ecol Evol Syst 34:26–36

Gordijn PJ, O'Connor TG (2023) The "eco-story" of a mountain range: the development of Socio-Ecological Disturbance Regimes in the northern Drakensberg and consequences for grassland-plant diversity. Trans Roy Soc S Afr 78:17–28. https://doi.org/10.1080/0035919X.2023.2175076

Granger JE (1976) The vegetation changes, some related factors and changes in the water balance following twenty years of fire exclusion in catchment IX, Cathedral Peak Forestry Research Station. Thesis, University of Natal, Pietermaritzburg

Grobler CH (2000) The vegetation ecology of urban open spaces in Gauteng. Thesis, Department of Botany, University of Pretoria, Pretoria

Grobler CH, Bredenkamp GJ, Brown LR (2002) Natural woodland vegetation and plant species richness of the urban open spaces in Gauteng. South Africa. Koedoe 45(1):19–34

Grobler CH, Bredenkamp GJ, Brown LR (2006) Primary grassland communities of urban open spaces in Gauteng, South Africa. S Afr J Bot 72:367–377

Grobler PJ, Marais J (1967) Die plantegroei van die Nasionale Bontebokpark, Swellendam. Koedoe 10:132–146

Haasbroek A (2003) Vegetation study of sensitive plant communities of dolerite outcrops, Bloemfontein. BSc(Hons) Project, Department of Plant Sciences, University of the Free State, Bloemfontein

Hahn N (2002) Endemic flora of the Soutpansberg. Thesis, University of Natal, Pietermaritzburg

Hartmann HEK (2001) The genus *Delosperma* in Gauteng. I. A new species in the white-flowered group: *Delosperma gautengense* H.E.K.Hartmann. Aloe 38:4–8

Herbst SN, Roberts BR (1974a) Quantitative ecological relationships in the Alpine grassland of Lesotho. Proc Grassl Soc S Afr 9:61–66

Herbst SN, Roberts BR (1974b) The alpine vegetation of the Lesotho Drakensberg: a study in quantitative floristics at Oxbow. J S Afr Bot 40:257–267

Heydenrych AJ (1995) 'n Evaluering van sommige plantkundige faktore wat kleinwild-digthede in die Weskus Nasionale Park beïnvloed. Thesis, Department of Botany, University of Stellenbosch, Stellenbosch

Higgins SI, Conradi T, Ongole S et al (2023) Changes in how climate forces the vegetation of southern Africa. Ecosystems 26:1716–1733

Higgins SI, Conradi T, Louw MA et al (2024) Biome change in southern Africa. In: von Maltitz GP, Midgley GF, Veitch J et al (eds) Sustainability of southern African ecosystems under global change. Springer, Cham, pp 369–405

Hill TR (1996) Description, classification and ordination of the dominant vegetation communities, Cathedral Peak, KwaZulu-Natal Drakensberg. S Afr J Bot 62:263–269

Hilliard OM, Burtt BL (1987) The botany of the southern Natal Drakensberg. National Botanic Gardens, Kirstenbosch

Hoare DB (1997) Syntaxonomy and synecology of the grasslands of the southern part of the Eastern Cape. Thesis, Department of Botany, University of Pretoria, Pretoria

Hoare DB (2002) Biodiversity and performance of grassland ecosystems in communal and commercial farming systems in South Africa. In: Proceedings of the FAO's Biodiversity and Ecosystem Approach in Agriculture, Forestry and Fisheries Event: 12–13 October, 2002. FAO, Rome, pp 10–27

Hoare DB, Bredenkamp GJ (1999) Grassland communities of the Amatola/Winterberg mountain region of the Eastern Cape, South Africa. S Afr J Bot 65:75–82

Hoare DB, Bredenkamp GJ (2001) Syntaxonomy and environmental gradients of the grasslands of the Stormberg/Drakensberg mountain region of the Eastern Cape, South Africa. S Afr J Bot 67:595–608

Hoare DB, Mucina L, Rutherford MC et al (2006) Albany Thicket Biome. In: Mucina L, Rutherford MC (eds) The vegetation of South Africa, Lesotho and Swaziland. SANBI, Pretoria, pp 540–567

Hoffman MT, Cowling RM (1990) Desertification in the lower Sundays River Valley, South Africa. J Arid Environ 9:105–117

Hoffman MT, Cowling RM (1991) Phytochorology and endemism along aridity and grazing gradients in the lower Sundays River Valley, South Africa: implications for vegetation history. J Biogeogr 18:189–201

Hoffman MT (1989a) A preliminary investigation of the phenology of subtropical thicket and karroid shrubland in the lower Sunday River Valley. S Afr J Bot 55:586–597

Hoffman MT (1989b) Vegetation studies and the impact of grazing in the semi-arid Eastern Cape. Thesis, Department of Botany, University of Cape Town, Cape Town

Hurt CR, Hardy MB, Tainton NM (1993) Identification of key grass species under grazing in the Highland Sourveld of Natal. Afr J Range Forage Sci 10:96–102

Janecke BB, du Preez PJ, Bredenkamp GJ (2005) A synoptic view on the grassland vegetation of Soetdoring Nature Reserve, Free State Province. S Afr J Bot 71:339–348

Jewitt D, Morris C, O'Connor T et al (2023) Dominant species of mid-elevation grasslands of the uKhahlamba-Drakensberg Park are predicted to be largely immune to climate change. J Mount Sci 20:2468–2486

Johnson CF (1998) Vulnerability, irreplaceability and reserve selection for the elephant-impacted flora of the Addo Elephant National Park, Eastern Cape, South Africa. Thesis, Department of Botany, Rhodes University, Grahamstown

Johnson CF, Cowling RM, Phillipson PB (1999) The flora of the Addo Elephant National Park, South Africa: are threatened species vulnerable to elephant damage? Biodivers Conserv 8:1447–1456

Joubert L, Pryke JS, Samways MJ (2017) Moderate grazing sustains plant diversity in Afromontane grassland. Appl Veg Sci 20:340–351

Judd RA (2001) The coastal grasslands of the Eastern Cape west of the Kei River. Thesis, Department of Botany, Rhodes University, Grahamstown

Kamineth AI (2004) The population dynamics and distribution of tree euphorbias in thicket. Thesis, Department of Botany, University of Port Elizabeth, Port Elizabeth

Karger DN, Conrad O, Böhner J et al (2017a) Climatologies at high resolution for the Earth land surface areas. Sci Data 4:170122. https://doi.org/10.1038/sdata.2017.122

Karger DN, Conrad O, Böhner J et al (2017b) Data from: climatologies at high resolution for the earth's land surface areas. Dryad Digit Reposit. https://doi.org/10.5061/dryad.kd1d4

Karger DN, Schmatz D, Dettling D et al (2020) High resolution monthly precipitation and temperature timeseries for the period 2006–2100. Sci Data 7:248. https://doi.org/10.1038/s41597-020-00587-y

Kassambara A, Mundt F (2020) *factoextra*: extract and visualize the results of multivariate data analyses. R Package Version 1.0.7. https://CRAN.R-project.org/package=factoextra

Kay C, Bredenkamp GJ, Theron GK (1993) The plant communities of the Golden Gate Highlands National Park in the north-eastern Orange Free State. S Afr J Bot 59:442–449

Kerley GIH, Boshoff AF, Knight MH (1999a) Ecosystem integrity and sustainable landuse in the Thicket Biome, South Africa. Ecosyst Health 5:104–109

Kerley GIH, Tongway D, Ludwig JA (1999b) Effects of goat and elephant browsing on soil resources in Succulent Thicket, Eastern Cape, South Africa. In: Proceedings of the VIth International Rangeland Congress 1, Townsville, pp 116–117

Killick DJB (1963) An account of the plant ecology of the Cathedral Peak area of the Natal Drakensberg. Mem Bot Surv S Afr 34:1–178

Killick DJB (1978a) Notes on the vegetation of the Sani Pass area of the southern Drakensberg. Bothalia 12:537–542

Killick DJB (1978b) The Afro-alpine region. In: Werger MJA, van Bruggen AC (eds) Biogeography and ecology of southern Africa. Dr W. Junk, The Hague, pp 515–560

King L (1978) The geomorphology of central and southern Africa. In: Werger MJA, van Bruggen AC (eds) Biogeography and ecology of Southern Africa. Dr W. Junk Publishers, The Hague, pp 3–18

Kobisi K (2005) Preliminary checklist of the plants of Lesotho. SABONET Rep 34:1–84

Kooij MS (1990) A phytosociological study of the north-western Orange Free State. Thesis, Department of Botany, University of Pretoria, Pretoria

Kooij MS, Bredenkamp GJ, Theron GK (1990a) The vegetation of the north-western Orange Free State, South Africa. 2. The D Land Type. Bothalia 20:241–248

Kooij MS, Bredenkamp GJ, Theron GK (1990b) The vegetation of the B land type in the north-western Orange Free State. S Afr J Bot 56:309–318

Kooij MS, Scheepers JC, Bredenkamp GJ et al (1992) The vegetation of the Kroonstad area: a description of the grassland communities. S Afr J Bot 58:155–164

Kremer-Köhne S, Thompson DI, Witkowski ETF (2022) Vegetation structure and composition associated with *Aloe lettyae* in dwindling grassland fragments: Fine-scale baseline data to underpin conservation efforts. S Afr J Bot 147:1156–1163

KZN DAEA (2003) Bioresource Unit Inventory Program. Version 5.007.D.3003.P.03. Report. Natural Resource Section, KwaZulu-Natal Department of Agriculture and Environment Affairs, Cedara

La Cock G (1986) The Boknes–Boesmansriviermond coast: an ecological study and management proposals. BSc(Hons) Project, Department of Botany, Rhodes University, Grahamstown

La Cock G (1992) The conservation status of Subtropical Transitional Thicket, and regeneration through seeding of shrubs in the Xeric Succulent Thicket of the

Eastern Cape. Eastern Cape Scientific Services, Cape Nature Conservation, Port Elizabeth

Landman M, Kloppers K, Kerley GIH (2018) Settling the browser–grazer debate for African buffalo in grass-limited Eastern Cape thicket, South Africa. Koedoe 60(1):a1465. https://doi.org/10.4102/koedoe.v60i1.1465

Landman M, Mgqatsa N, Cromsigt JPGM et al (2019) Elephant effects on treefall and logfall highlight the absence of megaherbivores in coarse woody debris conceptual frameworks. For Ecol Manage 438:57–62

Lechmere-Oertel RG (2003) The effect of goat browsing on ecosystem patterns and processes in succulent thicket, South Africa. Thesis, Department of Botany, University of Port Elizabeth, Port Elizabeth

Lechmere-Oertel RG, Kerley GIH, Mills AJ et al (2008) Litter dynamics across browsing-induced fenceline contrasts in succulent thicket, South Africa. S Afr J Bot 74:651–659

Lehmann CER, Archibald SA, Hoffmann WA et al (2011) Deciphering the distribution of the savanna biome. New Phytol 191:197–209

Lewis OAM (1986) Plants and nitrogen. Edward Arnold Publishers, London

Liengme CA (1987) West Coast Strandveld, a Mediterranean-type shrubland. Thesis, Department of Botany, University of Cape Town, Cape Town

Linder HP, Vlok JH, McDonald DJ et al (1993) The high altitude flora and vegetation of the Cape Floristic Region, South Africa. Opera Bot 121:247–261

Lloyd JW, Badenhorst M (1995) Vegetation and soils: synthesis of baseline datasets, inventories and maps for Rolfontein Nature Reserve. Scientific Services Section, Northern Cape Nature Conservation Service, Kimberley

Lloyd JW, van den Berg E, van Wyk E et al (2002) Patterns of degradation and degradation in the Thicket Biome. Terrestrial Ecology Research Unit Report 39. University of Port Elizabeth, Port Elizabeth

Lombard AT, Johnson CF, Cowling RM et al (2001) Protecting plants from elephants: botanical reserve scenarios within Addo Elephant National Park, South Africa. Biol Conserv 102:191–203

Lötter MC, Burrows JE (2012a) Gm 30 Steenkampsberg Montane Grassland. 2012 Vegetation Map—Biodiversity BGIS (sanbi.org). Available from: www.bgis.sanbi.or/Projects/Detail/185

Lötter MC, Burrows JE (2012b) Gm 31 Long Tom Pass Montane Grassland. 2012 Vegetation Map—Biodiversity BGIS (sanbi.org). Available from: www.bgis.sanbi.or/Projects/Detail/185

Lötter MC, Burrows JE, Emery AJ (2002) Phytochoria: Centres and regions of endemism. In: Emery AJ, Lötter MC, Williamson SD (eds) Determining the conservation value of land in Mpumalanga. Mpumalanga Parks Board, Nelspruit, pp 48–59

Lötter M, Burrows J, Jones K et al (2023) Historical vegetation map and red list of ecosystems assessment for Mozambique—Version 2.0—Final report. USAID / SPEED+, AFD/FFEM, Maputo

Louw WJ (1951) An ecological account of the vegetation of the Potchefstroom area. Mem Bot Surv S Afr 24:1–105

Low AB, Rebelo A[T]G (eds) (1996) Vegetation of South Africa, Lesotho and Swaziland. A companion to the vegetation map of South Africa, Lesotho and Swaziland. Department of Environmental Affairs and Tourism, Pretoria

Lubke RA (1983) A survey of the coastal vegetation near Port Alfred, eastern Cape. Bothalia 14:725–738

Lubke RA, De Villiers G (1991) The vegetation of the Joan Muirhead Nature Reserve, Kenton-on-Sea. The Naturalist 35:21–30

Lütge BU, Hardy MB, Hatch GP (1996) Plant and sward response to patch grazing in the Highland Sourveld. Afr J Range Forage Sci 13:94–99

Malan PW (1993) Plantsosiologie van die Bloemfontein-Wes Distrik. Thesis, Department of Botany and Genetics, University of the Orange Free State, Bloemfontein

Malan PW (1998) Vegetation ecology of the southern Free State. Thesis, Department of Botany and Genetics, University of the Orange Free State, Bloemfontein

Malan PW, Venter HJT, du Preez PJ (1994) Phytosociology of the Bloemfontein West District: vegetation of the Ae land type. S Afr J Bot 60:245–250

Malan PW, Venter HJT, du Preez PJ (1995) Plant communities of the western part of the Bloemfontein district: the Ca land type. S Afr J Bot 61:306–311

Malan PW, Venter HJT, du Preez PJ (1998) Vegetation ecology of the southern Free State: shrubland communities of the rocky outcrops. S Afr J Bot 64:1–18

Malan PW, Venter HJT, du Preez PJ (1999) Vegetation ecology of the southern Free State: 1. Plant communities of the Zastron area. S Afr J Bot 65:260–269

Marloth R (1902) Notes on the occurrence of alpine types in the vegetation of the higher peaks of the south-western region of the Cape. Trans S Afr Phil Soc 11:161–168

Martens JC, Morris CD (1994) Classification of the grass layer of semiarid rangeland in the Smaldeel area of the Eastern Cape. Afr J Range Forage Sci 11:61–68

Martens JC, Danckwerts JE, Zacharias PJK (1996) Species responses to grazing in the Smaldeel area of the Eastern Cape. Afr J Range Forage Sci 13:29–36

Martin K (2017) Potential eco-physiological and phytosociological impacts of fracking on the vegetation of the Karoo, Eastern Cape, South Africa. Thesis, Faculty of Science, Nelson Mandela University, Port Elizabeth

Martin NL (1986) Classification and ordination of the plant communities of Lesotho. Thesis, University of Idaho, Moscow

Masubelele ML, Hoffman MT, Bond W et al (2013) Vegetation change (1988–2010) in Camdeboo National Park (South Africa), using fixed-point photo monitoring: The role of herbivory and climate.

Koedoe 55(1):Art.#1127. https://doi.org/10.4102/koedoe.v55i1.1127
Matthews WS (1991) Phytosociology of the North-eastern Mountain Sourveld. Thesis, Department of Botany, University of Pretoria, Pretoria
Matthews WS, Bredenkamp GJ, van Rooyen N (1991) The vegetation of the Black Reef Quartzite and associated large rocky outcrops in the north-eastern mountain sourveld of the Transvaal escarpment. S Afr J Bot 57:143–150
Matthews WS, Bredenkamp GJ, van Rooyen N (1992a) The phytosociology of the high altitude hygrophilous vegetation regions of the northeastern mountain sourveld in the Transvaal, South Africa. Phytocoenologia 20:559–574
Matthews WS, Bredenkamp GJ, van Rooyen N (1992b) The vegetation of the dry dolomitic regions of the north-eastern mountain sourveld of the Transvaal escarpment. Phytocoenologia 20:467–488
Matthews WS, van Wyk AE, Bredenkamp GJ (1993) Endemic flora of the north-eastern Transvaal Escarpment, South Africa. Biol Conserv 63:83–94
Matthews WS, Bredenkamp GJ, van Rooyen N (1994) The phytosociology and syntaxonomy of relatively low-altitude areas in the North-eastern Mountain Sourveld, in the eastern Transvaal Escarpment region. Koedoe 37:73–87
McDonald DJ, Oliver EGH, Linder HP et al (1993) Is there alpine vegetation on the mountains of the south-western Cape? Veld Flora 79(1):17–19
Midgley JJ, Cowling RM (1993) Regeneration patterns in Cape subtropical transitional thicket; where are all the seedlings? S Afr J Bot 59:496–499
Miller AJ, Cramer MD (2004) Root nitrogen acquisition and assimilation. Plant Soil 274:1–36
Mills AJ, Milewski AV, Sirami C et al (2012) Aerosol capture by small trees in savannas marginal to treeless grassland in South Africa. Geoderma 189–190:124–132
Mills AJ, Milewski AV, Rogers KH et al (2013) Boundary of treeless grassland in relation to nutrient content of soils on the Highveld of South Africa. Geoderma 200–201:165–171
Mochesane MV (2018) Plant diversity in grasslands of selected nature reserves and adjacent grazing areas within the Gauteng province. Thesis, University of South Africa, Pretoria, South Africa
Mokgakane TJ (2021) Grass species composition, distribution, biomass production potential and nutritional value in three selected communal rangeland areas under Msukaligwa Local Municipality, Mpumalanga Province. Thesis, North-West University, Potchefstroom
Mokgakane TJ, Mlambo V, Ravhuhali KE et al (2021) Contribution of soil type to quantity and nutritional value of grass species on the South African Highveld. Resources 10:106. https://doi.org/10.3390/resources10100106
Mokuku C (1991) Classification of the alpine plant communities of Mafika-Lisiu. Thesis, Department of Biological Sciences, University of Zimbabwe, Harare
Morris C (1990) Reconnaissance survey of the terrestrial vegetation of Sehlabathebe National Park, Lesotho. Department of Grassland Science, University of the Witwatersrand, Johannesburg
Morris CD (1994) The influence of environment and livestock grazing on the mountain vegetation of Lesotho. Thesis, University of Natal, Pietermaritzburg
Morris CD, Scott-Shaw R (2019) Potential grazing indicator forbs for two mesic grasslands in South Africa. Ecol Indic 107:105611. https://doi.org/10.1016/j.ecolind.2019.105611
Morris CD, Tainton NM (1996) Long-term effects of different rotational grazing schedules on the productivity and floristic composition of Tall Grassveld in KwaZulu-Natal. Afr J Range Forage Sci 13:24–28
Morris CD, Tainton NM, Boleme S (1993) Classification of the eastern alpine vegetation of Lesotho. Afr J Range Forage Sci 10:47–58
Morris JW (1973) Automatic classification and ecological profiles of southwestern Transvaal Highveld grassland. Thesis, Department of Botany, University of Natal, Pietermaritzburg
Morris JW (1976) Automatic classification of the highveld grassland of Lichtenburg, south-eastern Transvaal. Bothalia 12:267–292
Mostert JWC (1958) Studies of the vegetation of parts of the Bloemfontein and Brandfort districts. Mem Bot Surv S Afr 31:1–226
Mucina L (2018) Vegetation of Brazilian campos rupestres on siliceous substrates and their global analogues. Flora 238:11–23
Mucina L (2023) Biomes of the Southern Hemisphere. Springer Nature Switzerland, Cham
Mucina L, Rutherford MC (eds) (2006) The vegetation of South Africa, Lesotho and Swaziland. SANBI, Pretoria
Mucina L, Hoare DB, Lötter MC et al (2006a) Grassland Biome. In: Mucina L, Rutherford MC (eds) The vegetation of South Africa, Lesotho and Swaziland. SANBI, Pretoria, pp 348–437
Mucina L, Adams JB, Knevel IC et al (2006b) Coastal vegetation of South Africa. In: Mucina L, Rutherford MC (eds) The vegetation of South Africa, Lesotho and Swaziland. SANBI, Pretoria, pp 658–697
Mucina L, Rutherford MC, Palmer AR et al (2006c) Nama-Karoo Biome. In: Mucina L, Rutherford MC (eds) The vegetation of South Africa, Lesotho and Swaziland. SANBI, Pretoria, pp 324–347
Mucina L, Lötter MC, Rutherford MC et al (2022) Forest biomes of Southern Africa. New Zeal J Bot 60:377–428
Muir J (1929) The vegetation of the Riversdale area, Cape Province. Mem Bot Surv S Afr 13:1–82
Müller DB (1986) Plantekologie van die Willem Pretorius Wildtuin. Thesis, Department of Botany, University of the Orange Free State, Bloemfontein
Müller M (2002) Die fitososiologie van die sentraal Vrystaat. Thesis, Department of Botany and Genetics, University of the Free State, Bloemfontein

Munyai N, Ramoelo A, Adelabu S et al (2023) Effects of environmental factors on plant productivity in the mountain grassland of the Mountain Zebra National Park, Eastern Cape, South Africa. Ecologies 4:749–761. https://doi.org/10.3390/ecologies4040049

Myburgh WJ, Breytenbach PJJ, Theron GK et al (1992) Die fitososiologie van die Bankenveld in die Grootvlei-omgewing, Suid-Transvaal. Bothalia 22:245–254

Ndamane NG, Kraai M, Tsvuura Z et al (2023) Community perceptions on the transformation of a high-altitude grassland through invasion by *Leucosidea sericea* at Vuvu, South Africa. Human Ecol 51:439–454

Novellie P, Gaylard A (2013) Long-term stability of grazing lawns in a small protected area, the Mountain Zebra National Park. Koedoe 55(1):Art. #1108. https://doi.org/10.4102/koedoe.v55i1.1108

Nsibande TS, Nkosi SE (2023) Plant communities and veld condition of Glencoe farm in the North-Eastern Cape Grasslands Priority areas, South Africa. S Afr J Bot 161:161–170

Nuttall-Smith GD, Parker DM (2023) Limited impact of elephants on trees and shrubs in semi-arid thicket vegetation in southern Africa. Plant Ecol 224:931–941. https://doi.org/10.1007/s11258-023-01348-8

O'Connor TG, Martindale G, Morris CD et al (2011) Influence of grazing management on plant diversity of Highland Sourveld Grassland, KwaZulu-Natal, South Africa. Rangel Ecol Manage 64:196–207

Olivier MC (1983) An annotated systematic checklist of the Angiospermae of the Cape Receife Nature Reserve, Port Elizabeth. J S Afr Bot 49:161–174

Oosthuysen M, Strauss WM, Somers MJ (2023) The relationship between mammalian burrow abundance and bankrupt bush (*Seriphium plumosum*) encroachment. Bothalia 53(1):a11. https://doi.org/10.38201/btha.abc.v53.i1.11

Palmer AR (1981) A study of vegetation of the Andries Vosloo Kudu Reserve, Cape Province. Thesis, Department of Botany, Rhodes University, Grahamstown

Palmer AR (1988) Vegetation ecology of the Camdebo and Sneeuwberg regions of the Karoo Biome, South Africa. Thesis, Department of Botany, Rhodes University, Grahamstown

Palmer AR (1989) The vegetation of the Karoo Nature Reserve, Cape Province. I. A phytosociological reconnaissance. S Afr J Bot 55:215–230

Palmer AR (1991a) A syntaxonomic and synecological account of the vegetation of the Eastern Cape midlands. S Afr J Bot 57:76–94

Palmer AR (1991b) Vegetation/environment relationships in the central area of the Cape midlands, South Africa. Coenoses 6:29–38

Palmer AR, Crook BJS, Lubke RA (1988) Aspects of the vegetation and soil relationships in the Andries Vosloo Kudu Reserve, Cape Province. S Afr J Bot 54:309–314

Parker DM (2017) The composition and complexity of the woody and succulent components of Albany thicket with and without elephants. S Afr J Bot 112:19–28

Partridge TC (1997) Evolution of landscapes. In: Cowling RM, Richardson DM, Pierce SM (eds) Vegetation of southern Africa. Cambridge University Press, Cambridge, pp 1–20

Pedro JG, Barbosa LAG (1955) A vegetação. In: Esboço do reconhecimento ecológico-agrícola de Moçambique. vol 2, pp 67–224. Centro de Investigação Científica Algodeira Memorias e Trabalhos 23. Lourenço Marques [Maputo]

Pentzhorn BL, Olivier MC (1974) A systematic checklist of flowering plants of the Addo Elephant National Park. Koedoe 17:121–136

Pentzhorn BL, Robbertse PJ, Olivier MC (1974) The influence of the African elephant on the vegetation of the Addo Elephant National Park. Koedoe 17:137–158

Pierce SM, Cowling RM (1984) Phenology of fynbos, renosterveld and subtropical thicket in the southeastern Cape. S Afr J Bot 3:1–16

Potgieter JW (1982) 'n Plantekologiese studie van die Golden Gate Hoogland Nasionale Park, Clarens, Oranje Vrystaat. Thesis, Department of Botany, University of the Orange Free State, Bloemfontein

Potts G, Tidmarsh CE (1937) An ecological study of a piece of Karoo-like vegetation near Bloemfontein. J S Afr Bot 3:51–92

Pule HT, Tjelele JT, Tedder MJ (2018) The effects of abiotic factors in South African semi-arid grassland communities on *Seriphium plumosum* L density and canopy size. PLoS ONE 13(8):e0202809. https://doi.org/10.1371/journal.pone.0202809

Pule HT, Tjelele JT, Hattas D et al (2023a) Evaluating the potential use of *Seriphium plumosum* L. as forage in the semiarid grassland of Gauteng Province, South Africa. Rangel Ecol Manage 87:30–33

Pule HT, Tjelele JT, Tedder MJ (2023b) Post-fire soil nutrient dynamics in *Seriphium plumosum* L. encroached semi-arid grassland of Gauteng Province, South Africa. Agriculture 13:1971. https://doi.org/10.3390/agriculture13101971

Rahlao M, Seleteng-Kose L, Chatanga P et al (2023) Ecological impacts of *Rosa rubiginosa* on rangelands: case of Mphaki, in Quthing district, Lesotho. S Afr J Bot 163:376–388

R Core Team (2022) R: A language and environment for statistical computing. R Foundation for Statistical Computing, Wien. https://www.r-project.org

Rebelo AG, Cowling RM, Campbell BM et al (1991) Plant communities of the Riversdale Plain. S Afr J Bot 57:10–28

Rebelo AG, Boucher C, Helme N et al (2006) Fynbos Biome. In: Mucina L, Rutherford MC (eds) The vegetation of South Africa, Lesotho and Swaziland. SANBI, Pretoria, pp 52–219

Reddy RA, Balkwill K, McLellan T (2001) Is there a unique serpentine flora on the Witwatersrand? S Afr J Sci 97:485–495

Robesson RAJ (1998) Phytosociology of north-western KwaZulu-Natal. Thesis, Department of Botany, University of Pretoria, Pretoria

Roos PC, Allsopp N (1997) Soil nutrient ecology associated with *Acacia sieberiana* at different tree densities in a South African savanna. Afr J Range Forage Sci 14:39–44

Roussouw LF (1983) 'n Ekologiese studie van die boomgemeenskappe van Bloemfonteinomgewing, Oranje Vrystaat. Thesis, Department of Botany, University of the Orange Free State, Bloemfontein

Rubin F, Palmer AR (1996) The physical environment and major plant communities of the Karoo National Park, South Africa. Koedoe 39:25–52

Rutherford MC, Westfall RH (1986) Biomes of southern Africa—an objective categorization. Mem Bot Surv S Afr 54:1–98

Rutherford MC, Mucina L, Powrie LW (2006) Biomes and bioregions of Southern Africa. In: Mucina L, Rutherford MC (eds) The vegetation of South Africa, Lesotho and Swaziland. SANBI, Pretoria, pp 30–51

Rutherford MC, Powrie LW (2011) Can heavy grazing on communal land elevate plant species richness levels in the Grassland Biome of South Africa? Plant Ecol 212:1407–1418

Rutherford MV, Powrie LW, Husted LB (2012) Plant diversity consequences of a herbivore-driven biome shift from grassland to Nama-Karoo shrub steppe in South Africa. Appl Veg Sci 15:14–25

Rutherford MC, Powrie LW, Husted LB (2014) Herbivore-driven land degradation: consequences for plant diversity and soil in arid subtropical thicket in south-eastern Africa. Land Degrad Develop 25:541–553

Scholes RJ (1979) The vegetation of the Blouberg, north-western Transvaal. BSc(Hons) Project, Department of Botany, University of the Witwatersrand, Johannesburg

Scholes RJ (1997) Savanna. In: Cowling RM, Richardson DM, Pierce SM (eds) Vegetation of southern Africa. Cambridge University Press, Cambridge, pp 258–277

Scheepers JC (1975) The plant ecology of the Kroonstad and Bethlehem areas of the Highveld Agricultural Region. Thesis, Department of Botany, University of Pretoria, Pretoria

Scoggins PF, Theron GK (1990) An application of multivariate techniques to dry-weight-rank data from a Bankenveld nature reserve. S Afr J Bot 56:648–653

Scott-Shaw R (1999) Rare and threatened plants of KwaZulu-Natal and neighbouring regions. KwaZulu-Natal Nature Conservation Service, Pietermaritzburg

Scott-Shaw CR, Bourquin O, Porter RN (1996) The conservation status of Acocks' veld types in KwaZulu-Natal. Lammergeyer 44:50–63

Seaman M (1987) Pans of the Orange Free State. Afr Wildl 41:237–238

Seleteng-Kose L, Kobisi K, Pool-Stanvliet R et al (2021) A rapid biodiversity assessment of Lesotho's first proposed Biosphere Reserve: a case study of Bokong Nature Reserve and Tšehlanyane National Park. Bothalia 51(2):a6. https://doi.org/10.38201/btha.abc.v51.i2.6

Siebert SJ (2001) Vegetation on the ultramafic soils of the Sekhukhuneland Centre of Endemism. Thesis, Department of Botany, University of Pretoria, Pretoria

Siebert SJ (2011) Patterns of plant species richness of temperate and tropical grassland in South Africa. Plant Ecol Evol 144:249–254

Siebert F, Siebert SJ (2005) Dolomitic vegetation of the Sterkfontein Caves World Heritage Site and its importance in the conservation of Rocky Highveld Grassland. Koedoe 48:17–31

Siebert SJ, van Wyk AE, Bredenkamp GJ (2001) Endemism in the flora of ultramafic areas of Sekhukhuneland, South Africa. S Afr J Sci 97:529–532

Siebert SJ, van Wyk AE, Bredenkamp GJ (2002a) The physical environment and major vegetation types of Sekhukhuneland, South Africa. S Afr J Bot 68:127–142

Siebert SJ, van Wyk AE, Bredenkamp GJ et al (2002b) Grasslands and wetlands of the Sekhukhuneland Centre of Plant Endemism, South Africa. Bothalia 32:211–231

Siebert SJ, van Wyk AE, Bredenkamp GJ et al (2003) Vegetation of the rock habitats of the Sekhukhuneland Centre of Plant Endemism, South Africa. Bothalia 33:207–228

Siebert SJ, Victor JE, van Wyk AE (2002c) An assessment of threatened plants and conservation in Sekhukhuneland. PlantLife 26:7–18

Sigwela AM (1999) Goats and kudu in subtropical thicket dietary competition and seed dispersal efficiency. Thesis, Department of Zoology, University of Port Elizabeth, Port Elizabeth

Sigwela AM, Lechmere-Oertel RG, Kerley GIH et al (2004) Quantifying the costs of unsustainable domestic herbivory for biodiversity and ecosystem functioning in succulent thicket, Eastern Cape, South Africa. In: Allsopp N, Palmer AR, Milton SJ et al (eds) Proceedings of the VII International Rangelands Congress. Document Transformation Technologies, Durban, pp 1521–1526

Smit CM (1992) Phytosociology of the Newcastle-Memel-Chelmsford Dam area. Thesis, Department of Botany, University of Pretoria, Pretoria

Smit CM, Bredenkamp GJ, van Rooyen N (1992) Phytosociology of the B land type in the Newcastle-Memel-Chelmsford Dam area. S Afr J Bot 58:363–373

Smit CM, Bredenkamp GJ, van Rooyen N (1993a) Phytosociology of the Ac land type in the foothills of the Low Drakensberg in north-western Natal. S Afr J Bot 59:203–214

Smit CM, Bredenkamp GJ, van Rooyen N (1993b) Plant communities in the Ad land type in the Newcastle-Memel-Chelmsford Dam area. S Afr J Bot 59:116–122

Smit CM, Bredenkamp GJ, van Rooyen N (1995a) Grassland vegetation of the Low Drakensberg Escarpment in the north-western KwaZulu-Natal and north-eastern Orange Free State border area. S Afr J Bot 61:9–17

Smit CM, Bredenkamp GJ, van Rooyen N (1995b) The vegetation of the Ea land type in north-western KwaZulu-Natal. S Afr J Bot 61:18–28

Smit CM, Bredenkamp GJ, van Rooyen N (1995c) The vegetation of the upper Klip River Valley in the northeastern Orange Free State. Navors Nasion Mus Bloemfontein 11:37–58

Smit CM, Bredenkamp GJ, van Rooyen N et al (1997) Vegetation of the Witbank Nature Reserve and its importance in the conservation of threatened Rocky Highveld Grassland. Koedoe 40:85–104

Smits NAC (1997) The vegetation of old fields in Transkei. Thesis, Department of Botany, University of Pretoria, Pretoria

Smits NAC, Bredenkamp GJ, Mucina L et al (1999) The vegetation of old fields in Transkei. S Afr J Bot 65:414–420

Smith VR (1972) An ecological survey of the high altitude vegetation types of the southern section of the Blyde River Canyon National Park with vegetation map. BSc(Hons) Project, Department of Botany, University of the Witwatersrand, Johannesburg

Stalmans M (1990) Vegetation survey for the scientific management of the Lekgalameetse Nature Reserve. Thesis, School of Animal, Plant and Environmental Sciences, University of the Witwatersrand, Johannesburg

Stalmans M, Beilfuss R (2008) Parque Nacional da Gorongosa, Moçambique. Landscapes of the Gorongosa National Park. Report. Condensed Version. Gorongosa National Park

Stalmans M, De Klerk G (1991) A semi-quantitative survey of the vegetation of Bewaarkloof Nature Reserve. KaNgwane Parks Corporation, Nelspruit

Stalmans M, De Klerk G (1992) Protea landscapes in the Bewaarkloof Mountains, Transvaal. Protea Atlas Newsl 14:8–9

Stalmans M, Mentis MT (1993) Development of monitoring procedures for the herbaceous layer on the northeastern Transvaal escarpment. Afr J Range Forage Sci 10:129–134

Stalmans M, Balkwill K, Mentis MT (1997) An analysis of the flora of Lekgalameetse Nature Reserve, Northern Province. S Afr J Bot 63:305–329

Stalmans M, Robinson ER, Balkwill K (1999) Ordination and classification of vegetation of Songimvelo Game Reserve in the Barberton Mountainland, South Africa for the assessment of wildlife habitat distribution and quality. Bothalia 29:305–325

Stephen RJ, Chatanga P, Seleteng-Kose L et al (2023) Herbaceous vegetation changes along a bush encroachment intensity gradient in a montane area. Afr J Ecol 61:907–918. https://doi.org/10.1111/aje.13192

Story RA (1952) Botanical survey of the Keiskammahoek District. Mem Bot Surv S Afr 27:1–228

Strydom T, Grobler BA, Kraaij T et al (2021) Pre-and post-fire architectural guilds of subtropical dune thicket species in the southeastern Cape Floristic Region. J Veg Sci 32:e13079. https://doi.org/10.1111/jvs.13079

Stuart-Hill GC (1992) Effects of elephants and goats on the Kaffrarian succulent thicket of the eastern Cape, South Africa. J Appl Ecol 29:699–710

Stuart-Hill GC, Aucamp AJ (1993) Carrying capacity of the succulent Valley Bushveld of the Eastern Cape. Afr J Range Forage Sci 10:1–10

Talbot MMB, Bate GC (1991) The structure of vegetation in bushpockets of transgressive coastal dunefields. S Afr J Bot 57:156–160

Taylor HC (1970) IBP Checksheet for the Riversdale Nature Reserve. Botanical Research Institute, Pretoria

Taylor HC (1996) Cederberg vegetation and flora. National Botanical Institute, Pretoria

Taylor HC, Boucher C (1993) Dry coastal ecosystems of the South African South Coast. In: van der Maarel E (ed) Ecosystems of the world 2B. Dry coastal ecosystems. Africa, America and Oceania, Elsevier, Amsterdam, pp 89–107

Taylor HC, Morris JW (1981) A brief account of coast vegetation near Port Elizabeth. Bothalia 13:519–525

Timberlake JR, Darbyshire I, Wursten B et al (2016) Chimanimani Mountains: Botany and conservation. Royal Botanic Gardens, Kew

Todkill WB (2001) Towards the rehabilitation of degraded succulent thicket in the Addo Elephant National Park. Thesis, Department of Botany, University of Port Elizabeth, Port Elizabeth

Turner BJ (1989) A phytosociological study of the southeastern Transvaal Highveld grasslands. Thesis, Centre for Wildlife Management, University of Pretoria, Pretoria

van der Walt PT (1980) A phytosociological reconnaissance of the Mountain Zebra National Park. Koedoe 23:1–32

van der Walt M, Berner JM, Breed CA (2024) The vitality of native grassland plants in current urban climatic conditions in Gauteng. South Africa. Ecol Indic 158:111332. https://doi.org/10.1016/j.ecolind.2023.111332

van der Meulen F (1979a) A vegetation map of the western Transvaal Bushveld. Bothalia 12:731–735

van der Meulen F (1979b) Plant sociology of the western Transvaal Bushveld, South Africa: a syntaxonomic and synecological study. A. Gantner, Vaduz

van der Walt PT (1965) A plant ecological survey of the Noorsveld. Thesis, Department of Botany, University of Natal, Pietermaritzburg

van der Schijff HP (1963) A preliminary account of the vegetation of the Mariepskop Complex. Fauna & Flora 14:42–53

van Rooyen L (1981) 'n Fitososiologiese studie van die Rocherpan-natuurreservaat. BSc(Hons) Project, Department of Nature Conservation, University of Stellenbosch, Stellenbosch

van Staden PJ (2002) An ecological study of the plant communities of Marekele National Park. Thesis,

Centre for Wildlife Management, University of Pretoria, Pretoria
van Staden PJ, Bredenkamp GJ (2005) Major plant communities of the Marekele National Park. Koedoe 48:59–70
van Wyk AE, Smith GF (2001) Regions of floristic endemism in southern Africa. A review with emphasis on succulents. Umdaus Press, Hatfield
van Wyk S (1983) A plant ecological study of the Abe Bailey Nature Reserve. Thesis, Department of Botany, Potchefstroom University for C.H.E., Potchefstroom
van Wyk S, Bredenkamp GJ (1986) A Braun-Blanquet classification of the vegetation of the Abe Bailey Reserve. S Afr J Bot 52:321–331
Vareschi V (1980) Vegetationsökologie der Tropen. Verlag Eugen Ulmer, Stuttgart
Vlok JHJ, Euston-Brown DIW (2002) Subtropical thicket ecosystem planning project (STEP). Biological survey report (plants and birds). Terrestrial Ecology Research Unit, University of Port Elizabeth, Port Elizabeth
Vlok JHJ, Euston-Brown DIW, Cowling RM (2003) Acocks's Valley Bushveld 50 years on: new perspectives on the delimitation, characterisation and origin of subtropical thicket vegetation. S Afr J Bot 69:27–51
Vukeya LR, Mokotjomela TM, Malebo NJ et al (2003) The vegetation cover dynamics and potential drivers of habitat change over 30 years in the Free State National Botanical Garden. South Africa. Reg Environ Change 23:24. https://doi.org/10.1007/s10113-022-02024-9
Vukeya LR, Mokotjomela TM, Malebo NJ et al (2021) Interspecific competition in germination of bird-dispersed seeds in a habitat with sparse tree vegetation in South Africa. Bot Stud 62:10. https://doi.org/10.1186/s40529-021-00317-6
Vukeya LR, Mokotjomela TM, Malebo NJ et al (2022) Seed dispersal phenology of encroaching woody species in the Free State National Botanical Garden, South Africa. Afr J Ecol 60:723–735. https://doi.org/10.1111/aje.13013
Vukeya LR, Mokotjomela TM, Malebo NJ et al (2023) The vegetation cover dynamics and potential drivers of habitat change over 30 years in the Free State National Botanical Garden. South Africa. Reg Environ Change 23:24. https://doi.org/10.1007/s10113-022-02024-9
Wakeling JL, Cramer MD, Bond WJ (2010) Is the lack of leguminous savanna trees in grasslands of South Africa related to nutritional constraints? Plant Soil 336:173–182
Wakeling JL, Cramer MD, Bond WJ (2012) The savanna-grassland 'treeline': why don't savanna trees occur in upland grasslands? J Ecol 100:381–391
Weisser P, Cooper KH (1993) Dry coastal ecosystems of the South African east coast. In: van der Maarel E (ed) Ecosystems of the world 2B. Dry coastal ecosystems. Africa, America and Oceania. Elsevier, Amsterdam, pp 109–128
Werger MJA (1973a) An account of the plant communities of Tussen die Riviere Game Farm, Orange Free State. Bothalia 11:165–176
Werger MJA (1973b) Phytosociology of the Upper Orange River Valley, South Africa. Thesis, Catholic University of Nijmegen, Nijmegen
Werger MJA (1980) A phytosociological study of the Upper Orange River Valley. Mem Bot Surv S Afr 46:1–98
Werger MJA (1983) Vegetation geographical patterns as a key to the past, with emphasis on the dry vegetation types of South Africa. Bothalia 14:405–410
Westfall RH (1981) The plant ecology of the farm Groothoek, Thabazimbi District. Thesis, Department of Botany, University of Pretoria, Pretoria
Wild H, Barbosa LAG (1967) Vegetation map of the Flora Zambesiaca area. M.O. Collins, Salisbury [Harare]
Wieland RG (1982) Vegetation and ecosystem classification and range inventory of Molumong Prototype, Lesotho, southern Africa. Thesis, Washington State University, Pullman
Willis C, Smith G, Kose L (1999) Sehlabathebe National Park—Lesotho's mountain paradise. SABONET News 4:147–158
Young MM (1987) The Alexandria dunefield vegetation. Thesis, Department of Botany, University of Port Elizabeth, Port Elizabeth

Southern African Biomes: Caveats, Perspectives, Future Tasks

10

Ladislav Mucina and Michael C. Rutherford

Abstract

This chapter briefly summarizes the caveats of the presented system map and descriptions of the zonal biomes of Southern Africa. It also identifies the space of the future improvements of the biome system for the area. We recognise the suboptimal coverage and quality of vegetation mapping for many parts of Namibia, Botswana, and Zimbabwe as the major impediment to the precision of the biome definition in Southern Africa at present. These countries, unlike South Africa, Lesotho, eSwatini and Mozambique, have not engaged yet in comprehensive programmes focusing on mapping vegetation at fine scales. We shall welcome further scrutiny of our extensive bioclimate-based system.

Keywords

Biome · IUCN ecosystem hierarchy · Vegetation mapping · Vegetation unit

Our new biome system for Southern Africa relies on two primary sources of data, including (1) regional vegetation maps and (2) bioclimatic data. The former is well served for South Africa, Lesotho, and eSwatini (formerly Swaziland) through the VegMap2006 and its VegMap2018 version (Mucina and Rutherford 2006; Dayaram et al. 2019, respectively). A new vegetation map has been finalised and launched for Mozambique recently (Lötter et al. 2023). The basic vegetation mapping unit of the latter is, in principle, identical to the concept of 'vegetation unit' (VU) as used in VegMap2006 (Mucina and Rutherford 2006). There is one major deviation, however, and that concerns the grouping of the VUs in the Mozambique vegetation map. This does not follow either the logic of the hierarchy as applied in VegMap2016 (VUs grouping into Bioregions, bioregions grouping into Biomes) or the biome hierarchy as introduced by Mucina et al. (2022) and applied in the biome account as presented in this book. The vegetation map of Mozambique uses the IUCN hierarchy (for a critique of this hierarchy, see Mucina 2023). Nevertheless, for the sake

L. Mucina (✉)
Harry Butler Institute, Murdoch University, 90 South Street, Murdoch, 6150 Perth, Australia
e-mail: Ladislav.Mucina@murdoch.edu.au

Department of Geography and Environmental Studies, Stellenbosch University, Private Bag X1, Matieland, Stellenbosch 7602, South Africa

M. C. Rutherford
High Peak, Derbyshire, United Kingdom

Department of Botany and Zoology, Stellenbosch University, Private Bag X1, Matieland, Stellenbosch 7602, South Africa

L. Mucina and M. C. Rutherford (eds.), *Zonal Biomes of Southern Africa*, Biome Ecology 2, https://doi.org/10.1007/978-3-031-63731-5_10

of achieving commensurability of both major vegetation classification systems (VegMap2006 and VegMap2018 on the one hand and the VegMapMoz on the other), the VUs of the VegMapMoz have been translated into the hierarchical biome system involving RBs, CBs, GBs, ZBs, and associated ZEs.

We recognise that our map of the zonal biomes of Southern Africa is not free from several problems. The vegetation mapping data, which are an excellent source of spatial information serving delimitation of regional biomes, are of suboptimal coverage and quality for many parts of Namibia, Botswana, and Zimbabwe because these countries have not engaged so far in detailed national mapping programmes at fine-scale levels. However, despite the unavoidable use of some old sources featuring large-scale units here, we have also made full use of the many available well-founded regional or local studies in these countries, as well as providing original delimitations based on the best available environmental and surrogate information as well as, in some instances, on certain informal accounts and personal observation.

Undoubtedly, the countries mentioned above require comprehensive national mapping initiatives for the sake of the management of their precious vegetation resources. It is a complex task, yet achievable, provided that either governmental or international nature-conservation granting bodies would decide to invest wisely in such projects. Due to the progress of remote sensing and GIS technology, these projects are poised to be more feasible than would have been the case in the past when extensive and very expensive field-based data collection or using geographic proxy data (as in the case of the VegMap2006) were the dominating sources of vegetation mapping data and modelling.

Each system of biotic communities is only as good as the criteria of classification delimitation of the biome units and the data (and data analytical tools employed) fitting those criteria. We envisage that our system will (and should) undergo scrutiny. We suggest that our extensive bioclimate-based system is amenable to such tests, especially tests focused on functional aspects of the ecosystems in the light of their past, present, and future. Here, the tools of functional trait analysis and those looking into the phylogenetic assembly of the biomes are the most promising ones.

Some specific caveats of this study are mentioned below.

As stated earlier, the existing well-researched vegetation types in South Africa (Dayaram et al. 2019) were almost always used as the unmodified basic building blocks for many regional biomes. However, further re-evaluation of climatic ranges within certain vegetation types (for example, SVcb16 Western Sandy Bushveld; Dayaram et al. 2019) is likely to reveal a need for minor refinements of a few regional biome borders.

The focus of this book on zonal biomes has necessarily resulted in the dissolving of known azonal areas and replacing these with the nearest neighbouring zonal units on the map. This anomaly only becomes accentuated in the extensive azonal areas of, for example, the Cuvelai in northern Namibia. Consequently, this has led to a small overestimation of the zonal mopane woodland area.

In a few areas, the modelled precipitation values used in this book deviate significantly from those contained in Mucina and Rutherford (2006). This is primarily due to the current use of more recent climatic modelling sources, which are fully independent of the ones used in 2006.

A few important regional biome descriptions are incomplete. The use of the conventional northern boundary of the Southern African subcontinent, along the Cunene and Zambezi Rivers, unavoidably results in the incomplete descriptions of regional biomes that continue north of this convenient, working boundary. Here, regional biomes, such as the Northern Namib Desert, Kaokoland Mopane, Baikiaea Woodland, Zambezian Mopane, and Tropical East Coast Woodland, should be mentioned.

In savanna, the concept of 'mountain savanna' that was used for some regional biomes does very effectively delimit cohesive environmental complexes. However, we recognise that these are, at the same time, sometimes particularly diverse and could benefit from more detailed attention in future.

Clearly, geology (in particular rock type) has generally played a lesser role in delimiting regional biomes in the arid areas such as the deserts and semi-deserts than in some of the less arid areas of the MBSA. Importantly, these arid areas have a particularly high relative variability of annual rainfall (Nicholls and Wong 1990). In other words, they are subjected to highly unreliable rainfall with major droughts of uncertain length. In these stochastic environments, unpredictable moisture input poses a higher risk to obligate edaphic specialists than in more mesic regions. Consequently, a trend to greater generalisation across different substrates might be expected under these most variable climatic conditions (but see also Colles et al. 2009). However, considerably more research is required before one can properly evaluate the potential role of this complex geology (rock types) in governing populations of not only minor floristic components but also the full compositional range of vegetation in these extreme environments.

References

Colles A, Liow LH, Prinzing A (2009) Are specialists at risk under environmental change? Neoecological, paleoecological and phylogenetic approaches. Ecol Lett 12:849–863. https://doi.org/10.1111/j.1461-0248.2009.01336.x

Dayaram A, Harris LR, Grobler BA et al (2019) Vegetation Map of South Africa, Lesotho and Swaziland 2018: a description of changes since 2006. Bothalia 49(1):a2452. https://doi.org/10.4102/abc.v49i1.2452[VegMap2018]

Lötter M, Burrows J, Jones K et al (2023) Historical vegetation map and red list of ecosystems assessment for Mozambique. Version 2.0. Final report. USAID/SPEED+, AFD/FFEM, Maputo [VegMapMoz]

Mucina L (2023) Biomes of the Southern Hemisphere. Springer Nature Switzerland, Cham

Mucina L, Rutherford MC (eds) (2006) The vegetation of South Africa, Lesotho and Swaziland. SANBI, Pretoria [VegMap2006]

Nicholls N, Wong KK (1990) Dependence of rainfall variability on mean rainfall, latitude, and the Southern Oscillation. J Clim 3:163–170

The manufacturer's authorised representative in the EU is Springer
Nature Customer Service Centre GmbH, Europaplatz 3, 69115 Heidelberg,
Germany. If you have any concerns regarding our products, please
contact ProductSafety@springernature.com

Printed and bound by CPI Group (UK) Ltd, Croydon, CR0 4YY
01/10/2026
02225906-0019